TELECOMMUNICATIONS DEMAND MODELLING
An Integrated View

CONTRIBUTIONS TO ECONOMIC ANALYSIS

187

NORTH-HOLLAND
AMSTERDAM • NEW YORK • OXFORD • TOKYO

TELECOMMUNICATIONS DEMAND MODELLING
An Integrated View

Edited by:

Alain DE FONTENAY
Bell Communications Research
Livingston, New Jersey, U.S.A.

Mary H. SHUGARD
Bell Communications Research
Livingston, New Jersey, U.S.A.

David S. SIBLEY
Bell Communications Research
Morristown, New Jersey, U.S.A.

1990

NORTH-HOLLAND
AMSTERDAM • NEW YORK • OXFORD • TOKYO

ELSEVIER SCIENCE PUBLISHERS B.V.
Sara Burgerhartstraat 25
P.O. Box 211, 1000 AE Amsterdam, The Netherlands

Distributors for the United States and Canada:

ELSEVIER SCIENCE PUBLISHING COMPANY INC.
655 Avenue of the Americas
New York, N.Y. 10010, U.S.A.

Library of Congress Cataloging-in-Publication Data

Telecommunications demand modelling : an integrated view / edited by
Alain de Fontenay, Mary H. Shugard, David S. Sibley.
p. cm. -- (Contributions to economic analysis ; 187)
Selection from the papers given at a conference sponsored by Bell
Telecommunications Research held on Oct. 23-25, 1986.
Includes bibliographical references.
ISBN 0-444-88539-0 (U.S.)
1. Telecommunication--United States--Mathematical models--Congresses. 2. Consumer behavior--United States--Mathematical models--Congresses. 3. Demand functions (Economic theory)--Congresses. I. De Fontenay, Alain, 1940- . II. Shugard, Mary Helen, 1948- . III. Sibley, David S. (David Summer), 1947- . IV. Bell Communications Research, Inc. V. Series.
HE7775.T35 1990
384'.041--dc20 89-25663
CIP

ISBN: 0 444 88539 0

PRINTED IN THE NETHERLANDS

INTRODUCTION TO THE SERIES

This series consists of a number of hitherto unpublished studies, which are introduced by the editors in the belief that they represent fresh contributions to economic science.

The term "economic analysis" as used in the title of the series has been adopted because it covers both the activities of the theoretical economist and the research worker.

Although the analytical methods used by the various contributors are not the same, they are nevertheless conditioned by the common origin of their studies, namely theoretical problems encountered in practical research. Since for this reason, business cycle research and national accounting, research work on behalf of economic policy, and problems of planning are the main sources of the subjects dealt with, they necessarily determine the manner of approach adopted by the authors. Their methods tend to be "practical" in the sense of not being too far remote from application to actual economic conditions. In additon they are quantitative.

It is the hope of the editors that the publication of these studies will help to stimulate the exchange of scientific information and to reinforce international cooperation in the field of economics.

The Editors

PREFACE

This volume contains a selection from the papers given at a conference entitled "Telecommunications Demand Modeling: an Integrated View" sponsored by Bell Communications Research, which took place on October 23-25, 1986. The conference was attended by over 300 researchers, practitioners and regulators involved in telecommunications demand analysis and in closely related disciplines. The structure of the conference, and the selection of papers for this volume was guided by two principles. First, that demand analysis in telecommunications will benefit from the study of market structure, as well as from a focus on the details of customer choice. Second, a wide variety of techniques from various disciplines should be brought to bear.

With these criteria in mind, the papers in this volume make for a broad scope. The volume is divided into three sections. Section I covers end user demand and focuses on the individual choice problem. The techniques discussed cover both experimental psychology and the usual economic choice models, and range from theoretical models of the choice process to econometric and measurement issues. In an effort to bring a diversity of approaches to the attention of telecommunications specialists, papers which apply novel techniques to other industries have been included. Section II covers the staple area of telecommunications demand analysis, the econometrics of market behavior. Three outstanding papers in econometric theory have been included, as well as novel applications to the long distance market. In Section III the scope of analysis is broadened to include aspects of telecommunications demand as they affect market structure and public policy towards regulated telecommunications firms.

It is hoped that these papers will provide those interested in telecommunications demand in all its aspects a selection of novel papers which will provoke further work of even higher quality.

SECTION I: END USER DEMAND

EXPERIMENTAL PSYCHOLOGY

A major issue which has received little attention in the economic modeling of demand is the information available to the consumer in the decision process. This problem is particularly acute when market conditions are imperfect and society finds it necessary to bring about a change in consumption patterns. In their paper "Effective Consumer Information Interventions: Concepts, Design and Impacts Using Field Experiments" Winett, Kramer, Waller, Malone, and Lane consider this problem from experience surveyed in two fields: energy and health, with implications for telecommunications. They consider and synthesize a certain number of approaches to analyze the impact a controlled information flow can have on consumers. These approaches include Kahneman and Tversky's cognitive studies of choice behavior [1984], Bandura's social cognitive theory [1986], Wright and Huston's communication theory, social marketing, and behavior analysis [1983].

The experiments involved attempts to modify the consumer's consumption levels of energy, in one case, and health services, in the other case, in the presence of a low observed short run price elasticity. They develop the concept of a "hierarchy of effects" and present it as a disaggregation of the steps taken by the consumer in the decision process and emphasize the difference between changes in belief and attitudes on the one hand and changes in behavior on the other hand, where belief and attitudes and behavior are two distinct levels in the hierarchy. They show how it is the change of behavior which matters, if the goal is indeed a change in consumption, and that changes higher up in the hierarchy, such as changes in belief are neither completely necessary *ex ante* nor sufficient to bring about changes in behavior.

Winett *et al.*'s experimental work offers an empirical justification to the approach of Cross in "Dynamics of Adaptive Demand.' In fact Cross' model of experience-driven learning is based on the assumption that consumer behavior change comes through experience rather than through intellectual control — the essential proposition on which his theory of adaptive economic behavior rests is that consumers select their consumption actions from a set of available alternatives, but the selections are governed by probability distributions rather than conscious maximization. Cross goes to great length to link his proposed approach with the conventional neoclassical demand modeling and, in most situations, he is able to present it as an extension of the theory.

Cross considers in turn a number of factors which are inherent to an experience-driven learning model and typically neglected in a conventional approach. He notes that such a model can explain why consumers may be very slow to respond to a change in market conditions such as a new time-of-day pricing scheme even though past consumption patterns may be clearly suboptimal in light of the new pricing scheme. Another dimension which he addresses is the decoupling of the consumption act, such as a particular telephone call, from its economic consequence, a bill which covers all telephone calls over a period of a month and does not associate the subscriber's payment to any particular call. The implication of his analysis relative to conventional models is in the speed of adjustment to long run equilibrium, rather than in the nature of that equilibrium.

Imitation also takes new forms in his approach. Rather than being a theory of information management, it becomes a theory of behavior. While once more it does not typically affect the long run outcome of a market stimulus, it does profoundly affect the path to this new equilibrium. In considering imitation, he stresses a dimension which has wide implications for telephony, namely "visibility." A change in the telephone consumption pattern of a household has little visibility to the outside and offers little opportunity for imitation This is not the case of, say a telephone answering machine; in this latter case, visibility plays a large role. The dimension of risk is the dimension which may make the neoclassical model diverge from Cross' model, not only in the adjustment path but also in the eventual equilibrium. In the neoclassical approach, it is sufficient to consider the choice which maximizes the expected utility. By contrast, stochastic learning models may not converge at all, and even when they converge, they need not converge for the same choice as that given by an expected utility model. This result follows from the fact that all past events have equal weight in the neoclassical model while more recent events tend to be weighed more clearly in Cross' model. In fact one could even observe significant effects resulting from impossible events. This fundamental characteristic of consumer market stimulus response modeling as developed by Cross emphasizes that it is the environment at the time of the stimulus which matters rather than some long run "expected" environment.

In their paper, "A methodology for specification and Aggregation in Product Concept Testing," Gautschi and Rao analyze the role of prices in marketing experiment. Price may play two roles in these experiments: its usual one, as an allocative device, and a signalling role, in which it serves as a proxy for unobservable attributes, such as quality or prestige. They suggest experimental procedures to separate these two effects and illustrate with an experiment in which executives evaluate hypothetical profiles of portable microcomputers.

ECONOMIC APPROACHES TO CONSUMER CHOICE

In their paper, "The Effects of User Cost on the Demand for Telecommunication Service", Dunn and Oh examine the role of user cost in demand. An example of user cost could be the cost of time waiting in a queue to purchase a particular product. Consumers are assumed to make their consumption decisions based on the "full price", the sum of user cost plus money price. They derive the properties of demand functions with respect to the money price when user costs are assumed to be significant and when consumers have heterogeneous valuation of time. These demand functions be shown to have surprising properties. The reason for this is that as money price changes, equilibrium queue length may change, too. For example, if the price of a highly congested service rises, consumers with a low value of time (low user cost) may desert the service by so much that queue length falls by so much that the full price to users with high values of time falls, too. Thus, an increase in price could lead to larger equilibrium demand!

In "A Theory of the Consumption Decision: The Stochastic Commodity Model," Curien and deFontenay present a model of consumer behavior under uncertainty in which the consumer is assumed to be faced with the necessity to make certain binding consumption decisions at a point in time well before he (she) knows what his actual demand will be. To maximize expected utility, the consumer picks an optimal probability distribution function for future consumption, and once it is picked, day-to-day consumption decisions are simply random draws from that distribution. This approach leads to a class of consumer choice equations of which well known models such as logit and tobit are special cases.

ECONOMETRICS/DISCRETE CHOICE

Demand analysis for local telephone services is now increasingly playing a key role both in the policy arena, focusing on universal service, and in the business arena, with the introduction of new service options. At the same time, economists have also increasingly moved toward using more and more disaggregated data. The papers presented in this section all significantly contribute to the analysis and pioneer in that direction.

Taylor and Kridel's paper addresses "The Demand for Access to the Telecommunications Network," the same issue as Perl (1983) but they develop an original approach to the nondisclosure problem of census data. As a result of nondisclosure, the census provides place of residence only for areas of 100,000 population or more. Such areas will typically be characterized by multiple wire centers, and, as a consequence, by multiple local rates. The approach adopted here is to forego data on individual household for data aggregated at the census tract level, making it possible to achieve a better match between rates and place of residence. To minimize the information loss associated with aggregation, a formulation is developed which makes it possible to take into account differences in the distribution of income across census tract.

While Taylor and Kridel are primarily addressing service universality, Atherton *et al.*, in "Microsimulations of Local Residential Telephone Demand Under Alternative Service Options and Rate Structures" address the multiplicity of service options as an issue in the demand for local telephone services. Their model extends current models in three major areas. First of all, they internalize the household's choice decision, making it possible to analyze not only the impact of the service choice on calling patterns, but also the opposite impact of calling patterns on service choice. Second, they are able to model a wide variety of service options. Finally they are able to consider the household calling patterns at a disaggregated level, by average duration of calls and by time of day and destination. The latter disaggregation is essential to the analysis of the wider spectrum of service options since it is the only way to establish the cost to the individual households of the various options.

The level of disaggregation posed by this paper is such that the number of possible "portfolios" (a particular number and average duration of calls at each time of day to each distance zone) is extremely large. This problem is tackled through the use of a sample of calling portfolios drawn from the full set of alternatives.

Kling and Van Der Ploeg's paper "Estimation of Local Exchange Elasticities" specifies a service choice and usage model for local telephone service which endogenizes the stochastic dimension of the decision process. They hypothesize that for any one household a set of calls is associated with a purpose. Such a purpose is assumed to be generated by a Poisson distribution while the number of calls associated with each purpose is itself associated with a stochastic process, here a Poisson process, producing a Polya-Aeppli distribution of calls per day. The observation that most subscribers' demands for local telephone usage are satiated or nearly satiated is endogenized by assuming that, for a given household, the mean level of satiation usage is an unobservable random variable with a gamma distribution. Finally the probability that a household chooses measured rate service is modeled with a logit function of the systematic part of the difference of expected consumer surplus between measured rate and flat rate service. While their estimate of the local call price elasticity is consistent with a range of local elasticities published by various studies, they suggest that the price response of demand may be understated in an environment where local call prices significantly exceed their historical range.

MEASUREMENT

Economists have addressed the various inconsistencies in an economic agent's repeated choices and in choices made by economic agents belonging to the same type; namely that, given a choice set, it is generally the case that different options are selected by customers that appear the same in their observable features. This problem which links psychology and economics is addressed here by De Soete and Carroll, in their paper "Probabilistic Multidimensional Choice Models with Marketing Applications" who propose two novel approaches, when the options are restricted to two choice objects. Those approaches are the Wandering Vector and the Wandering Ideal Point.

In the former, an agent is associated with a multivariate normal distribution independent across situations and/or agents in an n-dimensional space. A choice situation consists of the random draw of a vector from the distribution and the selection of the option which is derived from the maximization of the orthogonal projection of each option on the vector randomly drawn. The latter addresses the issue of saturation, e.g., when one considers, for instance, the demand for coffee, it is clear that both too much and too little bitterness are undesirable properties, an observation which implies that there should be an optimal level of bitterness for coffee. Now the agent's random vector associated with a given choice situation can be conceived of as the agent's highest utility level (hence the name "ideal point," here the ideal level of coffee bitterness).

Transitivity in preferences is a cornerstone of most econometric and psychometric models. The Harshman and Lundy paper, "Multidimensional Analysis of Preference Structure" is able to overcome the problem by starting from a "data-analytic" and descriptive approach, DEDICOM. By assuming that the scalar which quantifies the preference of option a over option b is the negative of the scalar which quantifies the preference of option b over option a, and that any discrepancy one might observe are due to measurement errors, their method is designed around skew-symmetric matrices (i.e., where $x_{ij} = -x_{ji}$).

They are thus able to evaluate such hypotheses as linearity and additivity in preference scales and the presence of multiple hierarchies. Their method makes it possible to provide an independent and useful descriptive representation of the underlying preference process behind the available observations, a feature which can be used, for instance, to validate a theoretical model.

Harshman and Lundy consider also an interesting application to telecommunications, an application which is complementary to Larsen *et al.*'s paper on point to point demand. By applying their method to point-to-point traffic (here x_{ij} is the imbalance in the telephone traffic between i and j), they study the pattern of traffic between major areas, showing a relation which appears to order the regions (in terms of originating to terminating traffic) in terms of urbanization and population, where increased urbanization and larger population implies a higher ratio of originating traffic.

The latter problem is the object of Feger's contribution, "Ordinal Network Scaling." Just as Harshman and Lundy's telecommunication application, he starts from a communication matrix. However he makes no hypothesis as to possible imbalances and concerns himself solely with detecting patterns. He considers various nontelecommunications applications, such as, given communications withing a group segmented according to "status," whether intra-segment communications is monotonically related to status and whether inter-segment communication is monotonically related to how close any two status are. It would be of interest to see whether Feger's analysis would yield different results were it to be applied to the telecommunications data used by Harshman and Lundy.

SECTION II. ECONOMETRICS OF MARKET BEHAVIOR

THEORY

There are a number of technical issues which arise in the econometric analysis of demand and which have too often received only scant attention. This chapter addresses three such issues. First, Ghysels addresses the seasonality issue, in "A Study Towards a Dynamic Theory of Seasonality for Economic Time Series." Applied demand analysis routinely takes into account the seasonality inherent to telecommunications time series of data. However this is done in a very ad hoc manner and outside of the economic modeling itself. It is treated as an added problem which he can handled separately. Second, to select a model, the applied demand analyst, in a standard manner, will use a more or less extensive battery of tests to evaluate the proposed model. However, very little thought is given to the testing process. This is the problem which is addressed by Davidson and MacKinnon in "Testing the Specification of Econometric Models." Third, Ruud's paper, "Consistent Estimation of Limited Dependent Variable Models Despite Misspecification of Distribution" examines the important problem of parameter estimation when the dependent variable(s) of the model is imperfectly observed.

Seasonal analysis and econometrics have mostly developed independently of one another. Typically economists have failed to develop an economic theory to model seasonal fluctuations. This is the issue which Ghysels addresses in his paper in the context of a dynamic model. Given a model with seasonality in exogenous and endogenous variables, he shows that the endogenous variable may have power at all frequencies of the spectrum induced by the exogeneous variables' seasonality, so that the traditional decomposition in terms of mutually orthogonal components in which the seasonal component has its power only at the seasonal frequencies is inappropriate.

Davidson and MacKinnon address the interpretation of the tests used to decide whether to accept a model as valid. They focus on the first category of the following three broad categories of tests; tests in regression directions, tests in higher moment directions, and tests in mixed directions, which combine both regression and higher moment components. In this context, they consider the power of these tests, noting the dependence on three factors: the model being tested, the process that generated the data, and the direction in which the test is looking. This is done when both the null hypothesis and the process that generated the data are regression models and the direction is the regression data. In addition the setting is general with both the null hypothesis and the process by which the data are being generated described by likelihood functions while no restriction is imposed on the test's direction.

In telecommunications demand modeling, parameter estimates carry a lot of weight both in business decisions regarding such issues as pricing and in regulatory proceedings. Typically, some of the variables of interest have been observed imperfectly. Ruud develops a weighted M-estimator that provides consistent estimates (up to an unknown scaling factor) of the slope function of an imperfectly observed dependent variable. One of the interesting aspects of this estimator is its not having to make stringent demands for knowledge of the relationship between the observed dependent variable and its latent counterpart. In addition, the computation is a modification of popular estimator which can be readily computed. The consistency he obtains is robust to misspecification of the distribution of the latent dependent variable. One of the important applications to telecommunications demand analysis is the discrete choice modeling of option decisions where the customer's choices are not perfectly observed.

APPLICATION

Conventional applied telecommunications demand modeling leaves a considerable number of issues incompletely treated. This section consists of a set of papers which addresses a range of issues and extend applied demand analysis.

Aigner and Fiebig address the time of-day-tariff design problem in "Sample Design Considerations for Telephone Time-of-Use Pricing Experiments with an Application to OTC-Australia" in a context in which an improperly set time of day tariff is costly to everyone involved. It is costly to the end user in welfare losses and to the telephone company in revenue losses. A learning-by-doing approach to correct the initial miscalculation is itself costly in the uncertainty and disruption it creates for the consumer. In addition, it does not offer any guarantee that the learning process will be convergent. It is therefore desirable for a telephone company to evaluate carefully its approach. Available approaches may consist of any one of the following: sample surveys, test market, transferability of data from another region/country, and designed experiment. The latter is the one which is addressed here. The object is to design an experiment, i.e., a sample size, a segmentation, and a range of prices which will be useful in estimating the impact on demand of the introduction of time of day tariffs. The constraints of the experiment include the resources allocated to complete the testing the range of plausible econometric models to be considered, the impact of extreme values, and the confidence level associated with the design. The design which is presented in this paper draws heavily from the authors' experience gained in time-of-day pricing applied to electricity.

The issue of interdependence between caller and callee is treated by Larson *et al.* in their paper, "A Theory of Point-to-Point Long Distance Demand." They postulate a model in which the call, whether received or originated, produces information in which it is that information which is the argument in the consumer's utility function. The implication of their model is that given two locations, a and b, the demand for call terminating in b for customer residence in say a is function not only of price and income but also of the calls originating from b. They also show that the demand is very sensitive to the process by which information is produced (e.g., whether the production process is characterized by perfect substitution, i.e., an information content model or by fixed coefficient, i.e., a reciprocity model). Empirical evidence indicates that conventional models tend to underestimate elasticity by failing to account for the indirect impact of a price change and supports the reciprocity model over the information content model.

Conventional applied demand models have modeled messages or minutes, but not both. The interdependence between those is addressed by Veitch using empirical Bayesian techniques. The approach is based on a bivariate relation between total minutes and message frequency obtained from the conditional distribution of total minutes given message frequency. The analysis is based on the assumption that the expected holding time for an individual neither depends on the number of messages nor on the charges. The former hypothesis is suggested by the empirical observation that the overall average call length is essentially independent of the number of messages. The latter is in fact the object of Cameron and White's investigation in "Generalized Gamma Family Models for Long Distance Telephone Call Duration."

While the data set available to Cameron and White makes interpretation of the elasticity numbers an inadequate test of Veitch's hypothesis, the econometric methodology they develop is a useful development for existing telecommunications demand analysis. Their use of a generalized gamma distribution was sparked by essentially empirical research carried out over a number of years involving the distribution of both local and toll call duration.

SECTION III: USER MARKET STRUCTURE

In this section, authors discuss on a theoretical level particular distinct features of telecommunication demand and their impacts on market structure and regulatory policy. This discussion makes it clear that the traditional focus of telecommunications demand analysis on estimating price elasticities is too limited. Such broader phenomena as consumption extermobility, "lock-in" costs with suppliers, user costs and strategic interests by competitors all have profound effects on demand for local exchange carriers, the nature of market equilibrium and public policy towards regulation and deregulation. It is to be hoped that the econometric studies of telecommunications demand will take such factors into account.

ACCESS AND MARKET STRUCTURE

Joseph Farrell's paper, "Competition with Lock-in" explores the implications for suppliers, buyers and market equilibrium of the phenomenon of lock-in. Lock-in is said to occur when a buyer's technology is built around a product or service produced by one and only one particular seller. To switch to another seller involves extra cost or stranded investment. Farrell discusses the efficacy of long-term contracts and vertical integration as institutions to cope with the seller's incentives for *ex post* price gouging.

Hohenbelken and West's paper, "Algorithmic Analyses of Oligopolistic Competition in Space" emphasizes the importance of location in oligopolistic competition, with an interexchange carriers' selections of points of presence (POP) being an obvious example. They discuss algorithms by which the natural geographic market boundaries around groups of customers could be selected to illustrate with an application to supermarket location in Edmonton, Alberta.

In "Carrier Bypass and Competitive Strategy," Craig Buxton and Peter Cartwright apply the same approach, albeit heuristically, to analyze AT&T's choice of POPs in the San Francisco area. Buxton and Cartwright also analyze Pacific Bell's demand for carrier access by performing a competitive analysis of AT&T's strategic goals. In this case, AT&T refused to change its choice of POPs even after Pacific Bell offered it substantial discounts on transport charges. This leads Buxton and Cartwright to believe that AT&T was choosing POPs near large groups of customers so as to be physically in a better position to offer value-added services to end users. AT&T's ability to do this would be considerably reduced if end users were to be connected directly with Pacific Bell, rather than AT&T.

"Forecasting Technology Adoption with an Application to Telecommunications Bypass," by Richard Gilbert and Jeffrey Rohlfs, surveys the literature on demand-side and supply-side barriers to technology adoption and contains case studies on such innovations as the T-1 carrier, PBXs and centralized automatic message accounting. In each of these cases Gilbert and Rohlfs list the innovation barriers involved and the time it took for each innovation to become widespread. Then they perform a prospective analysis of user-owned bypass facilities and conclude that barriers to innovation are much lower than for the innovations cited above. They conclude that facilities bypass could become widespread within a decade or so. This paper points up the importance of considering dynamic factors in telecommunications demand.

PRICING & REGULATION

In "Consumption Externalities in Telecommunications Services," Benjamin Bental and Menachem Spiegel carry the analysis of implications of consumption externalities beyond the well-known work of Rohlfs [1974], and Littlechild [1975]. These authors focussed on the problem of optimal pricing for a regulated monopoly in the presence of network externalities. Bental and Spiegal repeat this analysis for a network which has only a single type of switch. They go on to consider equilibrium price and network size with (a) unregulated monopoly of local distribution and trunking, (b) unregulated local distribution and (c) unregulated interexchange carriers. The welfare comparisons across market structures involve both price and network size. An unregulated monopolist achieves the socially optimal network size because by so doing he/she can internalize the network externality; the monopoly price, of course, is higher than the socially optimal level of externality-adjusted marginal cost price. Various types of competitive market structures achieve lower equilibrium prices, but also network sizes smaller than the social optimum, since no single firm can fully capture the externality. This gives rise to the possibility that unregulated monopoly could be welfare-superior to certain forms of competition in the presence of the network externality. This suggests that econometric investigation of the magnitude of the externality is a topic of some relevance to public policy.

"A Theoretical Framework for Analyzing Access Charges," by Michael Einhorn, and "Performance Indicators for Regulated Industries" by Erwin Diewert, illustrate the use of demand information in regulated pricing. Einhorn considers the optimal second-best two part tariff for a local exchange carrier, given the presence of a bypass technology that appeals to high volume users. He shows that this two part tariff will generally involve a usage charge less than the marginal cost of the local exchange carrier. This is of some interest, especially considering the charges of cross subsidy or predatory pricing that would ensure if a local exchange carrier were to propose such a tariff. It is also noteworthy that this result does not come about because of network externalities of the sort analyzed by Bental and Spiegel. Diewert's paper, although concerned with regulated pricing, is much different in spirit from Einhorn's. Rather than having a regulator set prices which are second-best optimal, the regulator can be thought of as delegating price-setting to the firm, while ensuring economically efficient pricing by giving the firm a specially-chosen subsidy which is positive if it reduces prices, compared with some historical benchmark set of prices, and is negative (a tax) if the firm raises prices. Interestingly, Diewert's mechanism induces increases in total welfare without requiring the regulator to know more about demand than its equilibrium levels. Knowing the form of the demand function is not necessary.

The Editors

REFERENCES

Bandura, A., *Social Foundations of Thought and Action: A Social Cognitive Theory* (Prentice-Hall, New York, 1986).

Kahneman, D. and Tversky, A., Choices, values and frames. *American Psychologist, 29*, (1984), 341-350.

Littlechild, S. C., (1975). Two part tariffs and consumption externalities. *Bell Journal of Economics*, 6: 661-670.

Rohlfs, J. H., (1974). A theory of interdependent demands for a communications service. *Bell Journal of Economics and Management Science*, 5 (1): 16-37.

Wright, J. C. and Huston, A. C., A matter of form: Potentials of television for young viewers. *American Psychologist, 38*, (1983), 835-843.

TABLE OF CONTENTS

SECTION III: USER MARKET STRUCTURE

CONTRIBUTORS

Dennis J. Aigner
Terry Atherton
Moshe Ben-Akiva
Benjamin Bental
Craig P. Buxton
Trudy Ann Cameron
J. Douglas Carroll
Peter Cartwright
John G. Cross
Nicholas Curien
Russell Davidson
Alain de Fontenay
Geert De Soete
W.E. Diewert
Donald A. Dunn
Michael A. Einhorn
Joseph Farrell
Hubert Feger
Denzil G. Fiebig
David A. Gautschi
Eric Ghysels
Richard J. Gilbert
Richard A. Harshman
John P. Kling
Kathryn D. Kramer
Donald Kridel
M.K. Lane
Alexander C. Larson
Dale E. Lehman
Margaret E. Lundy
James G. MacKinnon
Steven W. Malone
Daniel McFadden
Hyung Sik Oh
Vithala R. Rao
Jeffrey H. Rohlfs
Paul A. Ruud
Menahem Spiegel
Lester D. Taylor
Kenneth Train
Stephen S. Van Der Ploeg
J.G. Veitch
Balder Von Hohenbalken
W. Bruce Walker
Dennis L. Weisman
Douglas S. West
Kenneth J. White
Richard A. Winett

SECTION I:
END USER DEMAND

I.1. Experimental Psychology

TELECOMMUNICATIONS DEMAND MODELLING
An Integrated View
A. de Fontenay, M.H. Shugard, D.S. Sibley (Editors)

Effective Consumer Information Interventions: Concepts, Design, and Impacts Using Field Experiments

Richard A. Winett, Kathryn D. Kramer,
W. Bruce Walker, Steven W. Malone, and M. K. Lane*
Psychology Department
Virginia Polytechnic Institute & State University
Blacksburg, VA 24061

*All research reported in this paper has been supported by grants to the first author.

Framework

Major tenets of economic theory and consumer policies revolve around the flow of information to affect consumer choice behaviors, i.e., consumer demand. However, apparently until recently, it often has not been appreciated by either theoreticians or policy-makers that some types of information can be markedly effective in modifying consumer behaviors, while other types often are ineffective (Winett & Kagel, [1]).

Several strains of research from different disciplines are converging to yield consistent concepts and procedures for effective information flow. These strains include:

Cognitive Studies of Choice Behaviors: These studies (e.g., Kahneman & Tversky, [2]) have shown how information framing and context influences choice behavior. While results of such studies are relatively consistent (e.g., present a choice as a positive gain and not a negative loss), the fact that most are laboratory studies merely involving verbal reports of potential choices raises the issue of external validity, i.e., generalization of principles and results to real settings and behaviors. Nevertheless, these studies cannot be ignored when designing information interventions.

Social Cognitive Theory: This approach integrates cognitive and behavioral perspectives by examining the interdependence of cognitive, behavioral, and environmental variables (Bandura, [3]). The use of modeling, feedback, and reinforcement principles, results in the design of information systems and messages with the objective of behavior change. This is different from other approaches where often the objectives entail knowledge, attitude, or belief change, with the expectation that such changes will ultimately affect behavior.

Communication Theory and Principles: Recent research has focused on the "formal features" (manipulable and distinct variables) of media that can be reliably used to increase attention to and comprehension of information (Wright & Huston, [4]). It is apparent that particular variables (e.g., speed of pacing of a TV spot) must be considered when designing information. Also under the rubric of communication research is an emerging body of studies of prosocial uses of TV and other media. When these studies are examined, consistent principles emerge (e.g., on simple message development; focus on discrete behaviors; Winett, [5]) that need to be incorporated into information efforts.

Social Marketing: Social marketing adapts the framework and variables of commercial marketing to the realm of causes, ideas, and social behaviors. Although the framework is not a perfect match for these new concerns, the emphasis on focusing on particular consumer segments, and tailoring messages to those segments is critical for effective information flows.

Behavior Analysis: Behavior analysis is a scientific paradigm with specific principles and methods. What appears most critical for designing information is the focus of this approach on environmental contingencies that foster and maintain (i.e., reinforce), or impede (punish), the enactment of particular behaviors. Such analyses can be undertaken at multiple levels and need to be done for effective information flows. For example, information about changing certain consumer behaviors must realistically depict barriers to change and strategies for overcoming barriers.

The first author (Winett, [5]6) has recently integrated these different concepts and strains of research to the study of information and behavior, through an overall framework called "behavioral systems". This integration appears timely not only because of the rapid development of new media and other technological changes (Rice, [6]), but because an interesting inconsistency exists within the communication and consumer behavior fields. On the one hand, there is the acknowledgement that many consumer information initiatives have neither informed consumers adequately, much less motivated consumer behavior change (Beales, Mazis, Salop, & Staelin, [7]1). There is also the pessimistic viewpoint of a number of communication scholars and researchers that "information does not affect behavior" (e.g., Robertson, Kelly, O'Neill, Wixom, Eiswirth, & Haddon, [8]). On the other hand, there is an emerging perspective, set of principles, and studies which attest that under certain circumstances, certain types of information, directed to specific consumer segments, can markedly affect specific behaviors.

The purpose of these next sections is to discuss select aspects of information and behavior. These sections will focus on a different emphasis in communication models, principles of effective media for behavior change, examples of studies using prescribed principles, and the development of an approach to information remedies using the entire behavioral systems framework.

Communication Model

Figure one shows the comprehensive communication model developed by McGuire [9]. There are several major points that will be made using this communication model including the nature of most, past communication research, the "hierarchy of effects" heuristic, and a redirection for communication research.

The pluses and minuses in Figure 1 follow from a review of much communication research (Winett, [5]). The pluses indicate where much research has been done, and the minuses indicate where there has been little research. Examination of the figure quickly reveals that few studies have actually focused on behavior either from the independent or dependent variables side. Instead, much of the research has consisted of studies of knowledge and attitude change, (e.g., as affected by source credibility) or verbal reports of possible decisions or intended actions. This has been true for a number of reasons: 1. many of these studies follow a laboratory paradigm as noted above; 2. a pessimistic

A BASIC COMMUNICATION/INFORMATION TECHNOLOGY MODEL

(From McGuire, 1981)

	INDEPENDENT COMMUNICATION VARIABLES				
Dependent Variables	Source	Message	Channel	Receiver	Destination
EXPOSURE	+	+	+	+	-
ATTENDING	+	+	+	+	-
LIKING	+	+	+	+	-
COMPREHENDING	+	+	+	+	-
SKILLS	-	-	-	-	-
YIELDING (ATTITUDE CHANGE)	+	+	+	+	-
MEMORY	+	+	+	+	-
INFORMATION SEARCH & RETRIEVAL	+	+	+	+	-
DECISION	+	+	+	+	-
INITIATE BEHAVIOR	-	-	-	-	-
REINFORCED BEHAVIOR	-	-	-	-	-
MAINTENANCE	-	-	-	-	-

+ Much research & consideration   - Minimal research & consideration perspective suggested that, at best, information flow may only affect knowledge and attitudes, which "later" may influence behaviors, because, 3. all the dependent variables noted in the figure are more or less linked in a sequence.

The last point, i.e., the "hierarchy of effects", is critical and deserves some discussion. The major supposition, as noted, is that these variables are more or less sequentially related. Much psychological research indicates that this is often not the case. For example, while it is obvious that information must be attended to and comprehended, is it the case that information must change beliefs and attitudes <u>before</u> behavior? Or quite specifically, in order to promote particular health behaviors (e.g., dental self-care such as flossing) must changes first occur in beliefs about susceptibility to dental illness and the viability of dental self-care? In the 1960's and 1970's, considerable investment was made in a "health beliefs model" (Becker, [10]) of this sort. However, it appears at this point that there is minimal evidence to support this model (Leventhal, Safer, & Panagis, [11]; Kegeles & Lund, [12]).

Beliefs, attitudes, and behaviors are often affected by different variables. For example, the first author does not believe he is susceptible to dental disease; has a highly favorable attitude toward preventive health efforts; religiously flosses every day, but does <u>not</u> regularly visit any dentist. Note that disbelief in susceptibility appears somewhat inconsistent with daily flossing, and favorable attitudes toward prevention seems markedly inconsistent with failure to visit a dentist.

We submit that rather than being an unusual and inconsistent pattern, that such patterns are typical <u>if</u> consistency is not expected and analyses of controlling variables are highly specific, idiosyncratic, and done at a more micro-level. Note in this example that the attitude measure (favorability to prevention health measures) is very general. The probability of a general attitude measure predicting a specific behavior is low (Bem & Allen, [13]), with very specific attitudinal measures yielding better predictions. However, note also in the example that even the two dental behaviors (flossing and appointments) are unrelated. Again, controlling conditions may be quite different. Flossing is simple, convenient, and inexpensive. Dental appointments can be fear provoking, inconvenient, and possibly expensive.

This example illustrates that while it may be important to examine all the dependent variables noted in Figure one in planning a communication-consumer behavior campaign, it is a mistake to only focus on variables early in the chain and assume that change in an early variable predicts the usual, ultimate objective, behavior. Attention must be directed, as will be discussed, to highly specific behaviors.

An alternative is to use an approach that explicitly focuses on highly specific behaviors of specific consumer segments. While data concerning segments' beliefs and attitudes toward preventive health and dental care are undoubtedly useful in framing information, principles and procedures are needed that can be used to affect changes in dental self-care behaviors. Note further that the example also indicates that information must be very clear about which behaviors are to be changed. Messages about flossing are unlikely to affect dental visits, nor does keeping dental appointments guarantee maintenance of flossing.

The perspective that has been developed in this paper also indicates that certain types of research paradigms that have been cornerstones of communication and consumer behavior and policy fields may be suspect. For example, studies that only focus on beliefs and verbal reports of decisions and intentions need to be questioned until it is demonstrated that a procedure which affected a reported intention actually affected real-life behavior.

In addition, in the social, health, and consumer behavior fields, the objectives are often more difficult. For example, today a nutritional campaign will focus on decreasing fat in the diet which entails diverse new selections from product classes and different meal preparations. This may be more complex than brand switching, a frequent objective of commercial campaigns. And, of course, consumer-oriented messages rarely have the depth or breadth of exposure of commercial ads, a point to be addressed at the end of this paper.

As also will be discussed and illustrated later in this paper, after considerable formative research, experimental field studies on communication and consumer behavior are required to change the minuses to pluses in Figure 1. Until that point, many information policies will not have a firm basis. However, such research endeavors need not start from scratch. There is a well-articulated set of principles to guide information-behavior change efforts.

Principles for Effective and Ineffective Information

It is possible to articulate principles that make information more effective for behavior change. Table 1, from Winett [5], is based on research done by Atkin [14], McGuire [9] and others, and summarizes effective and noneffective variables and elements of media/information campaigns. Following all these principles does not assure success because it is apparent that certain behaviors (e.g., drunk driving) will require an orchestration of legal, political, economic, and community forces to affect change. Sole reliance on information approaches is an obvious mistake. And, as has been emphasized, modest change is an appropriate goal. However, it is apparent that when one or more element in Table 1 is not carefully followed, noneffectiveness seems to be the rule (Mendelsohn, [15]).

Note also that the major points in Table 1 fit within the (integrated) behavioral systems framework described before. For example:

1. The message format and content require use of formal features ("communication") to promote attention ("cognitive") and specific types of content, e.g., modeling ("social cognitive theory") to affect specific behaviors.

2. The channel, source, and receiver variables require minute consideration and formative research with the intended audience to target information ("social marketing").

3. Proper conceptualization and the destination variable require extensive analyses of target behaviors including competing sources of information, personal barriers and environmental constraints to change, and reinforcers to maintain new behaviors ("behavior analysis").

When using an informational approach to behavior change, a critical point is to pick specific, relatively simple behaviors to change. These behaviors should be significant on an individual basis and extremely significant when relatively small effects with individuals are magnified by an information/media approach's ability to reach large numbers of people. For example, small changes in thermostat settings save meaningful amounts of energy on an individual basis and certainly on a collective bases (Winett, Leckliter, Chenn, Stahl, & Love, [16]). Likewise, a 10% shift in the carbohydrate-fat proportion of meals should enhance an individual's and the Nation's health (Matarazzo, [17]).

The next section will briefly review the first author's chain of studies in residential energy conservation that used price and information tactics. The focus will then shift to a recent study using TV programs that incorporated all the above points on effective media and experimental analysis to affect significant savings in residential energy use.

Experimental Field Studies on Information and Residential Energy Conservation

Complete reviews of the first author's studies on energy conservation are available elsewhere (e.g., Winett & Kagel, [1]). Therefore, this section will only note major points before reviewing the last study in this chain in detail.

All the studies from 1973 to 1983 were field experiments with random assignment of households to different conditions. Procedures were developed so that a large percentage (50%) of households in a designated area participated in each study, thus, avoiding problems associated with studies using highly selected volunteers. Studies lasted from a few months to over a year, and the average number of participants in each study was about 75.

The value of smaller-scale studies is in the series rather than in one particular study. This series followed the systematic replication approach wherein each study further developed concepts and procedures,

Elements and Variables Related to Ineffective and Effective Information Campaigns

	Ineffective	Effective
Message (format & content)	Poor quality decreases attention	High quality to enhance attention (e.g., comparable to commercial ad)
	Overemphasize quantity vs. quality	High quality may overcome some problems in limited exposure
	Vague or overly long messages: drab, e.g., "talking head."	Highly specific messages; vivid, e.g., behavioral modeling
Channel	Limited exposure (e.g., late night PSA's)	Targeted exposure
	Inappropriate media (e.g., detailed print for behavior change)	Appropriate media (e.g., TV for behavior change)
Source	Not well attended to	Trustworthy, expert or competent; may be also dynamic and attractive
Receiver	Little formative research to understand audience characteristics	Much formative research to target message to audience
Destination	Difficult or complex behaviors resistant to change	Simple behaviors, or behaviors in a sequence, changeable and where long-term change can be supported by the environment
Conceptualization	Inappropriate causal chain (e.g., early events as predictors of later ones)	More appropriate causal chain and emphasize behavior change
	Little analysis of competing information and environmental constraints	Analysis of competing information and environmental constraints used to design messages
Goals	Unrealistic, i.e., expect too much change	Realistic, specific, limited

but each study also replicated prior procedures. Often, a procedure that was replicated several times (e.g., feedback) served as a benchmark for evaluating new procedures (e.g., behavioral modeling).

The major measure in all studies was energy use recorded at least several times per week by research staff from outdoor household meters. Most often, homes selected were all-electric, and, thus, the major measure was electricity consumption.

Other measures focused on energy strategies currently used and those later adopted; attitudes and beliefs, and, of course, demographics and household characteristics. A series of four studies from 1979-1983 also focused on comfort so that temperature and humidity were continuously measured (with hygrothermographs) in the homes, and verbal reports on comfort ratings and clothing worn were available several times per week.

The most important findings from the comfort research were that:

1. The experimental field studies provided an important alternative to the usual short-term laboratory studies for developing comfort standards.

2. In particular, comfort was reported by participants at temperatures (65 -67 F) far below ASHRAE standards used world-wide (and with minimal clothing adjustments).

3. It was possible to develop a "substitution strategy" so that more expensive and high energy use practices (e.g., excessive use of air conditioning) were replaced by much less expensive and very low energy use practices (e.g., window fans, closing down house most of the day, natural ventilation). The comfort data showed that with these strategies, not only were substantial energy savings possible, but comfort could be preserved because temperature and humidity readings with low energy use were about the same as with more energy use.

This last point is critical for other types of consumer-information programs. The approach in the energy conservation studies was virtually perfected enough so that specific actions could be prescribed which not only saved money, but <u>were not aversive</u>, i.e., costly in other ways. Change plans in any realm must develop similar off-setting strategies, i.e., one set of behaviors is increased, another set is decreased. In addition, the typically low-cost, no-cost strategies were always presented in a positive way with positive language (e.g., "energy efficiency" and not "energy conservation"). In marketing parlance, an appealing and adaptable "product" was developed.

It is also important to note that the studies were not first developed within the information-behavioral systems framework described in this paper. Rather, operant psychology and economic principles were their first bases, later replaced by social learning theory. From the earlier studies it was found that:

1. Only large (and not cost-beneficial) incentives or price changes could yield energy savings of about 15%. In these studies, different monetary incentives and price changes were experimentally evaluated.

2. Feedback on energy use proved to be a powerful procedure. Written feedback on energy use, given almost everyday also resulted in average reductions of about 15%, but up to 30% on peak use days. The effectiveness of feedback was related to such parameters as feedback frequency, reduction goal selection, and the ability to fine-tune feedback information (e.g., a very accurate "weather correction" system was developed). Not surprisingly, feedback strategies were more effective when salience was high, i.e., when budget share for energy was higher. Participants could also be taught to give themselves feedback through self-monitoring their energy use, but with some loss of effectiveness. Although maintenance of effects was shown for several months after feedback or self-monitoring ceased, feedback in these forms was expensive and cumbersome.

3. Through three studies a procedure was developed using "behavioral modeling". In this procedure, 25-minute videotapes were developed where within a story-line the specific energy conservation strategies were demonstrated. Eventually, the production qualities of the videotapes resembled network fare. The three studies showed that behavioral modeling could approximate feedback in effectiveness even though the programs were only viewed once, and the studies suggested that modeling was the critical content format variable in the programs. Further, the tapes' effectiveness was not dependent on viewing in a group context or social interaction. The programs were effective when viewed alone, suggesting it was feasible to simply "deliver" them through cable television.

Cable Television Study

It is important to reiterate that conducting the prior series of studies allowed for the development of a highly acceptable and effective set of conservation strategies and sophisticated research methods. It is doubtful that state-of-the-art media can succeed with a poorly developed product or research methods which are not sensitive enough to detect small, but meaningful changes in dependent measures. It is also important to add that this knowledge-base does need to be combined with state-of-the-art media. At Virginia Tech, we have been fortunate to be able to work with its internationally known Learning Resource Center.

In this study (Winett, [16]), the program was specifically targeted to middle-class homeowners in southeastern cities. Participants were recruited from one central area (neighborhood) that best represented the audience characteristics. Using a door-to-door recruitment approach, about 50% of the households in the area were enlisted for the project.

During a baseline period of about a month, readings were taken three times per week on outdoor energy meters (requiring no contact with participants) and some other measures (measures of perceived comfort; continuous measurement of temperature and humidity using special devices; clothing worn; knowledge of energy conservation, strategies used) that required contact with participants and were pertinent to other aspects of the project (developing new comfort standards), were taken in 60% of ("contact") participant households. Persons hired to collect these measures (forms) on a weekly basis were blind to participants' treatment condition.

After the baseline period, and just prior to the first airing of the program, households were randomly assigned to one of five conditions:

1. No-contact control (n = 30): Households were simply informed that they were in a control condition for an energy conservation study. Only outdoor energy meters continued to be read, and these households were not involved in other measurements in the study. That is, contact ceased.

2. Contact Control (n = 30): This condition was the same as the prior condition except other measures which required contact were continued.

3. No-Contact Media (n = 30): This condition was the same as the no-contact control except that mail and phone prompts were used to inform participants about the time of the program.

4. Contact Media (n = 30): This condition was the same as the no-contact media except the other measures requiring contact continued to be taken.

5. Contact Media-Visit (n = 30): This condition was the same as the contact media condition except that a few days after the program a home visit lasting about 30 minutes was conducted. The purpose of the visit was to further explain procedures, adapt them, if necessary, to the participants' situation, and develop more interest and commitment.

The television program, "Summer Breeze," was about 20-minutes long and was shown on the public access station of the cable system at 7 p.m. on Monday, Tuesday, and Friday, and 9 p.m. on Thursday. This schedule provided excellent reach; 98% of the participant households viewed the program at least once. Viewing was verified by a special insert at the end of the program which required participants to mark a form with an "X" if they had watched the program. The procedure and mark was only noted in the television program. Only 2% of the control households watched the program, meaning that the integrity of the experimental design was maintained.

The most significant formal features, production qualities, and marketing aspects of the program were: (1) a theme song, "Summer Breeze," was used for audience recognition (it was a popular song), to tie program parts together and to represent the main point of the program--the use of natural ventilation and fans could replace much air conditioning; (2) in the beginning, a two-sided visual and auditory format was used to counter beliefs that energy conservation was no longer important, and only resulted in discomfort; (3) characters and locations closely fit the audience; (4) fairly rapid pacing was used to hold attention; (5) voiceovers and captions were used to emphasize main points and strategies were depicted (modeled) a minimal of three times, and summaries at different points were used.

The program had a story line based on modeling and diffusion principles, and provided some minimal drama and conflict. A young couple (early 30's) was dismayed at their high summer electric bill ($138) and was in a quandary about what to do. They recalled that their older neighbors who lived in a very similar home had very low summer and winter bills. After a number of scenes with social situations, the younger couple asked the older couple to help them to save energy and money. The rest of the program revolved around the older couple demonstrating the simple strategies. The couple tried each strategy and were pleasantly surprised how easy the strategies were, how much money they saved on their next bill, and how they were able to maintain comfort. At a final scene, the younger couple suggested that the other couple could help them in the winter, which served as a lead-in for the second part of the study conducted six months later.

The results of the study showed that households viewing the program (regardless of contact or home visit) reduced their overall electricity use by about 11% (compared to control conditions) for a period of two months (end of the summer) after the program. The study's design strongly indicated that simply seeing the program <u>once</u> was the crucial factor.

Six months later a similar winter program was delivered over the cable access station following the No-Contact Media condition for all experimental households. For this program, the results showed about a 10% reduction in energy use (electricity or natural gas). Finally, with <u>no</u> program shown, the following summer outdoor meters were again read. There was still some evidence for savings (about 5.5%) one year later for experimental households.

The results of this study were seen as quite positive. One program showing (i.e., very low "dose" and duration) was sufficient to produce change. The outcomes, 10% - 11% reductions, are meaningful for energy conservation (on an aggregate level), and also represented savings of <u>20% - 25%</u> on the major targets for the program, cooling or heating. Despite these changes, as in the first author's other studies, there were no reported changes in comfort level which remained quite acceptable. This is because the programs offered viewers a set of simple strategies which could offset reduced air conditioning or heating. Indeed, the hygrothermographs indicated that, for example, in the summer the electricity savings were achieved with a constant (baseline through intervention) temperature in the homes of about 77 degrees F.

In addition, the programs were shown at a time (1982-1983) when there was not the great interest in energy conservation, and in an area where energy prices were quite low. Thus, the programs seemed potent despite the lack of a strong supporting context (Winkler & Winett, [18]).

To balance these positive, optimistic results, it must be noted that participants only tended to adopt the simplest no-cost strategies. More effortful or costly procedures (e.g., insulate the hot water heater) tended not to be adopted. In addition, the simplicity of the procedures and lack of strong barriers to change (i.e., barriers existed to attempt change, but not to actual enactment of the behaviors) may not be general conditions for other problems (e.g., health).

Information and Consumer Policy: Nutritious and Economic Food Purchases

In the course of doing the series of studies on energy conservation, it was realized that the information strategies that were developed had applicability to the broader issues of information and consumer policy. The provision of information to consumers, is at the heart of benignly remedying market failures (e.g., the wide price discrepancy of the same food products in different stores) and attempting to noncoercively change other detrimental consumer practices (e.g., high-fat diets). Perhaps most information remedies fit within the antiregulatory spirit of the times (i.e., information disclosure on products as distinct from regulating products' content).

A detailed framework for regulatory policy based on Beales et al. [7] and Mazis, Staelin, Beales, & Salop [19] is available elsewhere: Suffice to say, that:

1. Information provision to consumers by neutral third-party sources or government is often needed.

2. Information must be effective, but it is unclear what are the most effective information variables and ways to deliver information.

Earlier in this paper, Table 1 showed variables and principles thought effective for behavior change media. Clearly, the previously described cable TV study followed all the points in Table 1 and the framework developed in the introduction of this paper. However, it appeared extremely critical to do the following:

1. Demonstrate that the same type of approach would work with seemingly more complex and challenging behaviors and practices.

2. More closely delineate key information content and context variables.

Thus, a project has been designed by the authors and sought to influence nutritional and economic food purchases. Video and feedback procedures similar to those used in energy conservation were tried in this new domain.

This information intervention is also quite different from other nutritional and shopping interventions, many of which have been point-of-purchase. Because of the wide array of food items in modern supermarkets, competing information, and the limited time available for most food shopping, a decision was made to try to arm the consumer with a few simple bits of information and rules prior to store entry. Another basic premise was that the strategies could be used in any food store at any time. However, the development of any effective information requires a considerable amount of various types of formative and pilot research to develop appropriate information for particular audience segments. What first follows are some results of this research and how this work fits into the design of information.

At the outset, consistent with points in Table 1, our goals in this project were relatively modest. We wanted participants to decrease fat consumption by 8% to 10%, increase complex carbohydrate consumption by the same amount, and have participants adopt three money-saving shopping tactics (use of complete shopping lists, buying store brands, and decreasing impulse buying) that could reduce costs by 15% to 20%. Dietary changes are based on National Cancer Institute recommendations and the shopping tactics have been found effective, but at best, only partially used by shoppers.

The project used these aspects of the behavioral systems framework:

1. Extensive study and use of effective communication variables and attempts to fit together cognitive and behavior change.

2. Use of social cognitive theory perspectives (<u>how</u> to fit cognitive and behavior change together) and principles, particularly modeling.

3. Development of specific programs for specific audience segments.

4. A consideration of incentives and barriers to behavior change including barriers created by cultural preferences and settings.

The following were the formative research steps we undertook, the results, and the directions each step has provided:

1. <u>Survey</u> - A one-hundred item survey was given to about 180 lower-middle class residents. Items concerned knowledge, beliefs, cultural preferences, and practices and focused on the nutritional and shopping goals of the project. Respondents were most often women. The results showed that what can be called "superficial knowledge" (e.g., "a diet high in fat has a lot of cholesterol and

is related to heart disease") was excellent. Definitive knowledge was fair to poor. For example, most respondents greatly overestimated the need for protein, and did not know that most complex carbohydrate foods (e.g., potatoes, rich, pasta, bread) are low in calories and are ideal foods. The belief that "starch is bad" appeared to prevail. Reported composition of dinners almost always followed high-fat content; i.e., 6 oz. or more of meat, small potato with fatty add-ons, a small salad drenched with dressing, a large dessert. Respondents did report moderate use of complete shopping lists, and store brands. Barriers to dietary and shopping change included perceptions of more time and effort and negative reactions of family and co-workers. Almost all the shopping and food preparation in the household was done by women.

2. Pilot videos and focus groups - The survey was used to develop two, brief (7 - minute) pilot videos which presented a rationale for change (health, costs) and used a voice-over technique to show a couple making simple changes in shopping and meal composition to reduce fat and costs. Home, supermarket, and meal scenes were used in the "visual" video. A companion "talking-head" tape only conveyed the identical information. One objective of the overall study was to ascertain the effect of information contnet ("talking head") alone compared to when such content was combined with appropriate psychological and communication formats and variables ("modeling").

Several important points emerged from these showings with small groups whose demographics were similar to people in the larger project: disbelief that carbohydrates ("starch") were good for you and low in calories; the personal and sensitive nature of nutrition; the need for many model meals; the resistance by men to meatless dinners; and the need to amplify current concerns (e.g., calories) rather than long-term health issues. These points were then included in the extended videotapes. It was also clear that the visual-modeling tape was preferred to the "talking-head" tape.

3. Nutritional/Shopping Analyses - Using a computer-based program, market basket analyses, and reports of typical shoppings and meals, several important and, indeed startling conclusions were reached. These included: 1. Many families continue to have meat at every dinner (and, even every lunch); 2. Meals could easily be designed by either just changing proportions (to small steak and large potato), substituting chicken or fish for meat, or using a complex carbohydrate as the main part of the meal (e.g., spaghetti); 3. These meals were obviously much lower in calories then the usual fare, and had more ideal fat-carbohydrate proportions; but, 4. The cost of shopping for and eating a high complex carbohydrate diet could be 20% to 30% less than the usual diet. These were potential savings before the use of any shopping tactic.

The display of more ideal meals, lower calorie eating, and substantial monetary savings were made main points in our extended video programs.

4. Food Shopping Feedback - A prototype weekly shopping/nutrition feedback procedure was developed by the third author. The procedure entails entry of all weekly food purchases into a specially designed computer-based nutritional program. The participant is given a breakdown of (percent) protein, simple and complex carbohydrates, saturated fat, and total fat of their purchases. These figures are then compared to their baseline and goals (usually the nutritional guidelines of 12% protein, 58% carbohydrates, 10% saturated fat and 20% other fat.) Part of the feedback shows suggestions for more closely meeting nutritional guidelines. Another part of the form entails price feedback by showing their baseline food expenditures, goal, and money savings. Suggestions for saving will also be given.

   We were also able to develop procedures to assure that project participants' reports of their weekly food purchases were accurate. For example, the use of shopping receipts and other means showed that data were acceptably reliable, obviously a critical finding.

5. Food Purchase Measure - Although this project used an array of measures to tap the interplays of cognitive and behavior change, it was clear that the "bottom-line" was food purchases. Unlike the energy studies, where there was an easily accessible and superbly reliable and valid measure of energy use from meter readings, this is not the case for food shopping. Eight different prototype forms were developed for people to record all their weekly food purchases. The final form only required checking items and filling in very minimal details, although the task was still perceived by some pilot and project participants as arduous.

Highly accurate recording of food purchases is essential for a project aiming to affect and detect relatively small shifts in behaviors. At a minimum, our pilot tests and various recording forms resulted in a final format that was acceptable to most participants and yielded reliable data.

Our various pilot efforts lead us to one general conclusion for the development of other consumer information programs. Formative research needs to be more systematized and much more time and money needs to be given to those efforts. It is difficult to conceive of information programs being successful without numerous and integrated pilot studies.

The present more limited pilot and formative research allowed us to develop and refine procedures. Further, in our pilots, we found that participants were consuming about 15% less carbohydrates, and 50% more protein and fat than recommended. In the pilot studies, our rudimentary procedures were effective in significantly changing food purchases and nutritional proportions (e.g., 4%-6% fat reduction)in healthful directions. Also, there was evidence for reduced expenditures for food shopping (15%).

The concerted effort of the formative and pilot research paid dividends in the larger study involving about 150 consumers (Winett & Kramer, [20]). In the study, the combination of the modeling videotape and weekly feedback proved to be more effective than modeling alone, and the talking head videotape with or without feedback. For example, in this condition, the fat content of food purchased was reduced about 7% with monetary savings of about 26%. The fat reduction amounted to more than a 50% reduction toward the National Cancer Institute's 1990 guideline.

Conclusions and Regulatory and Access Considerations

We believe that it is possible to develop effective information strategies that not only inform but activate consumer response. Two critical aspects of the development stage are careful attention to all elements in the behavioral systems framework and complimentary and extensive formative research. The preliminary results of the nutrition and food shopping project also support the generality of the overall approach.

We also believe that it is possible for communication, health, and consumer behavior professionals to develop information approaches that are superior to current commercials. This is necessary because, as noted, the task of many consumer information efforts can be more difficult than commercials. That is, the goal is often pronounced changes in product classes purchased, and not simply a shift in brand choice.

However, it will be futile to be able to develop potentially effective consumer information programs, but not have them viably compete against commercials. Current regulatory policy confuses technological change and types of media/information outlets with diversity of media content and access to the media (Le Duc, [21]). For example, exactly how and where could noncommercial spots on nutrition and economical food purchases be shown so they could compete with the onslaught of daily, TV food commercials, many of which undermine good nutritional practices?

Although we have not given up the notion of gaining access to TV and other media in some creative ways (Winett, [22]), we are also exploring other possibilities. For example, we are currently developing plans for brief, in-store, interactive modeling videos coupled with instantaneous nutrition/shopping feedback systems (Winett & Kramer, in press). Such systems will build on our current project and the behavioral systems framework and, perhaps, offer some competitive advantages to collaborative corporations.

References

1. Winett, R. A. & Kagel, J. H., The effects of information presentation format on demand for resources in field settings. Journal of Consumer Research, 14 (1984) 655-667.

2. Kahneman, D. & Tversky, A., Choices, values, and frames. American Psychologist, 29, (1984) 341-350.

3. Bandura, A., Social foundations of thought and action: A social cognitive theory (Prentice-Hall, New York, 1986).

4. Wright, J. C., & Huston, A. C., A matter of form: Potentials of television for young viewers. American Psychologist, 38 (1983) 835-843.

5. Winett, R. A., Information and behavior: Systems of influence (Erlbaum Associates, Inc., Hillsdale, NJ, 1986).

6. Rice, R. E. New media: Communication, research, and technology (Sage, Beverly Hills, 1984).

7. Beales, H., Mazis, M. B., Salop, S. C., & Staelin, R., Consumer search and public policy. Journal of Consumer Research, 8 (1981) 11-22.

8. Robertson, L. S., Kelley, A. B., O'Neill, B., Wixom, C. W., Eiswirth, R. S., & Haddon, W., A controlled study of the effect of television messages on safety belt use. American Journal of Public Health, 64 (1974) 1071-1080.

9. McGuire, W., Theoretical foundations of campaigns, in: Rice, R. E. and Paisley, W. J. (eds.), Public communication campaigns (Sage, Beverly Hills, 1981) pp. 67-83.

10. Becker, M. H., The heatlh belief model and personal health behavior, Health Educational Monograph, 2 (1974) 236-473.

11. Leventhal, H., Safer, M. A., & Panagis, D. M., The impact of communications on the self-regulation of health beliefs, decisions, and behavior. Health Education Quarterly, 10 (1983) 3-29.

12. Keyeles, S. S. & Lund, A. K., Adolescents' acceptance of caries-preventive procedures, in: Matarazzo, J. D., Weiss, S. W., Herd, J. A., Miller, N. E., and Weiss, S. M. (eds.), Behavioral health: A handbook of health enhancement and disease prevention (Wiley, New York, 1984) pp. 895-909.

13. Bem, D. J. & Allen, A., On predicting some of the people some of the time: The search for cross-situational consistencies in behavior. Psychological Review, 81 (1974) 506-520.

14. Atkin, C. K., Mass media information campaign effectiveness, in: Rice, R. E. & Paisley, W. J. (eds.), Public communication campaigns, (Sage, Beverly Hills, 1981) pp. 127-148.

15. Mendelsohn, H., Some reasons why information campaigns can succeed, Public Opinion Quarterly, 37 (1973) 50-60.

16. Winett, R. A., Leckliter, I. N., Chinn, D. E., Stahl, B. H., & Love, S. Q., The effects of television modeling on residential energy conservation, Journal of Applied Behavior Analysis, 18 (1985) 33-44.

17. Matarazzo, J. D., Behavioral health's challenge to academic, scientific, and professional psychology, American Psychologist, 37 (1982) 1-14.

18. Winkler, R. C., & Winett, R. A., Behavioral interventions in resource conservation: A systems approach based on behavioral economics, American Psychologist, 37 (1982) 421-435.

19. Mazis, M. B., Staelin, R., Beales, H., & Salop, S., A framework for evaluating consumer information regulation, Journal of Marketing, 45 (1981) 11-21.

20. Winett, R. A., & Kramer, K. D., A behavioral systems framework for information design and behavior change, in: Salvaggio, J. and Bryant, J., (eds.), Media use in the information age (Erlbaum Associates, Inc., Hillsdale, NJ, in press).

21. LeDuc, D. R., Deregulation and the dream of diversity, Journal of Communication, August (1982) 164-178.

22. Winett, R. A., Prosocial television for community problems: Framework, effective methods, and regulatory barriers, in: Jason, L. A., Felner, R. D., Hess, R., & Mortisugu, J. N. (eds.), Communities: Contributions from allied disciplines (Howarth Press, New York, in press).

TELECOMMUNICATIONS DEMAND MODELLING
An Integrated View
A. de Fontenay, M.H. Shugard, D.S. Sibley (Editors)

DYNAMICS OF ADAPTIVE DEMAND

John G. Cross*

The standard theory of consumer behavior, based upon utility maximization in a full information context, has proven over the years to be a valuable source of insights into the characteristics of market demand. It has provided the foundation for innumerable empirical studies of demand, and it is the centerpiece for theoretical analyses of market efficiency and welfare maximization. There are aspects of demand behavior, however, for which the standard theory is less well suited—or at least for which it needs supplementation. These include consumer responses to advertising, the determination of lag structures in dynamic adjustment models, and the phenomena of innovation and consumption of untried commodities. The purpose of this paper is to discuss the usefulness of an experience-driven learning model as one means for enlarging the theory so that it applies as well to these other areas.

As it is usually presented, the traditional model of consumer behavior presumes full information—consumers are assumed to be fully aware of price levels and of the levels of utility that will be conveyed by all possible consumption bundles. There are a number of models that weaken these assumptions, but they usually do so at the expense of an increase in the complexity of the decision problem. For example, the assumption of full information may be replaced by Bayesian decision rules. In such an analysis, prices or levels of satisfaction are unknown only to the extent that possible values are characterized by probability distributions. These probability distributions in turn are either known with certainty or are held as beliefs by consumers who then update these beliefs on the basis of experience. In either case, responses to changed market circumstances are immediate within this Bayesian context. Moreover, calculation of the appropriate modes of behavior requires the consumer to engage in often very sophisticated probability analysis, and this in turn may imply the use of computational techniques that extend well beyond the capacity of most "real" individuals.

The theory that we are to discuss here is presented in detail in J. Cross: *A Theory of Adaptive Economic Behavior*, (Cambridge, 1983). In this short space, we cannot develop the formal theory to any extent, and we will have to confine ourselves largely to verbal summaries. The essential proposition on which the theory rests is that consumers select their consumption actions from a set of available alternatives (defined just as in traditional theory), but the selections themselves are governed by probability distributions rather than conscious (or unconscious) maximization. These probability distributions are determined in turn from previous direct consumption experience. If we describe the set of available choices with the vector $A_i, \ldots, A_n$, then each possible action A_i will be selected during a period t with a probability $P_{i,t}$. If some A_i is chosen, and if the consumption experience is found to be "successful" in

* Department of Economics, University of Michigan, Ann Arbor, MI 48109

the sense that it conveys positive utility, then the likelihood that the action A_i will be repeated in the future is increased, where the extent of this increase is determined by the level of utility afforded by the consumption experience. Conversely, if the consumption experience is not satisfactory, or if it falls short of expectations, then the likelihood of repetition of the consumption decision is reduced. To the extent that the probability of A_i increases or decreases, the likelihoods of the alternatives, $P_k, k \neq i$, fall or rise to preserve the condition that the sum of all selection probabilities is equal to 1.

This view of demand behavior as a stochastic process is consistent with material found in the psychological learning literature. Experimental learning data are usually reported as frequency distributions, and indeed, the bulk of the experimental evidence indicates that one would be very hard pressed to produce conventional consumption situations in which choices could be predicted with certainty. This observation may of course be no more than a reflection of our own imperfect understanding of consumer behavior—if our science were more advanced, we might very well be able to come much closer to reliable prediction. At the present time, however, specification of choice behavior in stochastic terms appears to be the best that we can do.

The model of experience-driven learning that develops is not incompatible with conventional equilibrium theory. Intuitively, it is plausible that regardless of initial conditions, a mechanism that increases the selection likelihoods of consumption alternatives that have been found to be successful in the past, and that does so in approximate proportion to the utility levels that they produce, will eventually converge to a state in which utility-maximizing alternatives are the only ones with significant selection probabilities. Formal proofs of this proposition turn out to require relatively few assumptions, and one can go on to show that the equilibrium properties of simple models based upon learning behavior are identical to those obtained from maximization. That means that we may treat adaptive learning as an addendum to existing theory rather than as a wholesale replacement, and that the bulk of the (comparative static) propositions derived from the standard theory remain in force.

The identity between learning models and optimization theory does break down, however, if we introduce elements of risk into the environment, or if we address short run dynamic problems. These differences have a significant impact on our understanding of market demand behavior, particularly in the analysis of responses to altered market conditions. In this paper, we wish to describe the differences between short run dynamic behavior in a stochastic learning model and comparative static behavior in an equilibrium model, providing examples of how these differences influence our understanding of empirical demand behavior. There are a number of variables that are relevant to the learning theory models but that do not occur in comparative static theory, and of course it is the values taken by these variables that control the extent to which dynamic demand behavior will differ from that predicted by optimization. In the following sections, we describe five of the most important of these, together with examples of their relevance.

I. HETEROGENEITY OF BEHAVIOR

An experience driven learning model is inherently dynamic in nature. Adjustment to changed market circumstances is not instantaneous because it takes time for enough experience to be accumulated to bring about significant changes in behavior. The speed of this adjustment is dependent upon circumstances—the frequency with which the learning experience is reiterated, and the frequency with which suitable alternatives to current behavior are likely to be tested. The second of these two is the more subtle. Even if there are many opportunities to test alternative forms of behavior, if those opportunities are not taken, no learning can take place. Thus if the consumer has "learned" that the choice A_i is the one to take, and if he/she repeats that choice every time the occasion arises, then it will be impossible to discover if the choice A_k is actually better. This means in effect that consumers may be very slow in responding to changed circumstances because previous experience has led them to avoid other alternatives, even though some of these may now have become superior.

This same problem arises in statistical (Bayesian) models. A well-known example is provided by the so-called "two-armed bandit" problem: two slot machines, A and B, stand side by side. They appear to be identical, but in fact machine B offers a significantly higher payoff probability. Suppose that a statistician runs a series of trials on each machine in order to estimate its expected payoff value. No matter how long the series of trials, the random payoff property of the machines guarantees that there is a non-zero probability that the statistician will wrongly attribute the higher payoff expectation to machine A. It is then rational to go on to play A for the sake of its return. Even if one continues to update the estimate of the payoff value of A after every play, if the estimate of the value of B is low enough, one will never come to play machine B again.

The point is that even in a fully rational decision model, it is possible to become trapped permanently in an inferior mode of behavior. It is an interesting property of learning models that because choice probabilities are generally bounded away from zero, they do not permit inferior choices to persist indefinitely. Nevertheless, choice probabilities can become very small, and when they do, behavior very similar to that described by the two-armed bandit example can persist. Once someone has become familiar with a single "best" pattern of consumption, alternatives come to be tried only very rarely. A change in market circumstances that would turn one of these alternatives into a preferred choice may go untested for quite a long time, and as a consequence, adaptation to the change may be very slow.

The general principle is straightforward: If a consumer's choices are heterogeneous, ranging over a wide variety of alternatives, adaptation to change will be relatively rapid. If the consumer's choices are repetitive and confined to a narrow range of alternatives, adaptation will be much slower. The popular belief that young people are more responsive to change than are older persons who have become more "set in their ways" is entirely consistent with this description (whether or not the conventional perception is valid). Similarly, the market for a relatively new product will respond much more quickly to price and quality changes than will the market for a product that has already acquired an established place in everyone's consumption habits.

An interesting case in which these principles may be applied is found in the number of experiments that have been performed recently on "time-of-day" electricity pricing. The object of these tests has been to ascertain whether changes in rate structure will alter the pattern of electricity use over the day. Suppose, for example, that the relative price of electricity during the afternoon hours is made much higher. The view implicit in optimization theory is that once the utility has announced the new rates, households will use that information to adapt their consumption. Thus if the new optimal consumption pattern requires less use of air conditioning during the afternoon, that change in behavior will come about relatively quickly. It is because of this implication of the theory that most of the studies of time-of-day pricing have concentrated on estimates of the elasticity of demand as though it were a simple comparative-static problem.

From the perspective of experience-driven learning, the problem of time-of-day electricity pricing is much more difficult. In most households, electricity consumption has become a matter of unconscious routine, and as a consequence, adaptive learning can only take place if the household happens to choose a different pattern of consumption and then notices the resulting change in the electricity bill. Awareness of the change in rates may help (we will discuss this matter later), but it is fundamental to our theory that behavior change comes through experience rather than through intellectual control. This is particularly true in a case such as this: the computation of optimal electricity consumption in the presence of a time-of-day rate structure can be quite complex, and it is likely to be expensive compared to the relatively small dollar benefits that are available. Even "rational" households might refuse to take the trouble, and this leaves experience as the only guide to behavior. Since adaptation in this case may require modification of long-standing habits and experimentation with consumption patterns that are not yet a part of experience, we have every reason to expect the response to the introduction of time-of-day pricing to be extremely sluggish.

The implication of all of this is that studies of the response to time-of-day pricing that only cover a year or two are likely to underestimate by a substantial margin the long run effectiveness of this device in rationing electricity consumption. For the same reason, the short run benefits to be obtained from the institution of such a program are likely to be quite small. If we wish to smooth the peaks from electricity consumption a decade hence, this device may be quite useful. If we wish it to do so now, we will be disappointed.

II. FREQUENCY OF EXPERIENCE

Speed of adjustment in a learning model depends upon the frequency with which a particular choice situation arises. If a particular consumption opportunity only arises once a year, adaptation to a change in market conditions will be much slower in real time than it will if the consumption opportunity arises weekly or even daily. Infrequency of experience is always an impediment to learning.

If we are to measure the frequency with which learning opportunities arise, we must specify with some care just what we mean by a consumption "experience." Formally, we have specified that selection of the choice A_i leads to the payoff R_i, and thus we define the "experience" as the pair (A_i, R_i). There are many cases, however, in which an action A_i may be repeated several times before any payoff is received, or in which the payoff is divided into two more parts, so that a portion of the return is not received until sometime in the future. The entire experience, then, takes place over time, and the entire time interval must become the unit in which the frequency of experience is measured.

Suppose we return to our example of time-of-day electricity pricing. The experience of being comfortably cool during a hot afternoon may be a daily occurrence, but the electric bill may only come due once a month. The time interval that is to be used as the base for determining a rate of adjustment must then be one month rather than one day. Each daily experience encourages the use of air conditioning, but this is only part of the payoff, and the remainder arrives much less frequently. Thus adaptation to a change in rate structure can be expected to proceed very slowly in real time. If the electricity bill were to arrive once a week, or if it came every day, the speed of adjustment might be very much greater.

This speed of adjustment effect is offset, at least in part, by the fact that payoff magnitudes are necessarily made larger by longer accounting intervals. If the householder spends \$100 per month on electricity, then one monthly bill will be four times as large as a weekly bill would be, and one expects that the impact on behavior of a \$100 charge would be considerably greater than that of one bill for \$25. Nevertheless, it is possible that four bills for \$25 will have greater cumulative effect than will a single \$100 charge.

Although the theory does postulate a positive relation between magnitude of payoff and speed of adjustment, this relation cannot be strictly linear. Suppose that the alternative A_i has a selection probability of $P_{i,t}$ during period t. If A_i is chosen, and encounters a payoff value of $R_i > 0$, then the transition from $P_{i,t}$ to a new higher value $P_{i,t+1}$ can be described by means of some function of the form:

$$P_{i,t+1} = P_{i,t} + L(P_{i,t}, R_i), \tag{1}$$

where $L(\cdot)$ incorporates the effect of learning.

Because $P_{i,t+1}$ is a probability variable, its value must be bounded between 0 and 1. Thus although the function $L(\cdot)$ is increasing in R_i, a strictly proportional relation is impossible—otherwise very large values of R_i would drive $P_{i,t+1}$ above its upper bound of 1. If we accept the principle that $L(\cdot)$ must be monotonically increasing and non-cyclic in R_i, then we must conclude that $L(\cdot)$ is concave in R_i—if $L(\cdot)$ is continuous in R, its second derivative with respect to R must be negative.

Intuitively, then, one might expect that a series of four \$25 charges would have a greater cumulative impact than a single \$100 charge. As it happens, however, this is not necessarily true. Suppose the function $L(\cdot)$ has the simple (linear operator) form:

$$L(P_{i,t}, R_i) = g(R_i)(1 - P_{i,t}), \tag{2}$$

where the function $g(\cdot)$ is strictly concave in R.

Now consider two cases: a) the choice A_i is repeated n times, receiving a payoff value R on each occasion, and b) the choice A_i is repeated once, receiving a payoff value nR on that occasion. In the case of electricity billing, for example, n might be equal to four, while R included the benefits (and \$25 costs) associated with a weekly billing period. Then nR would describe the consequences of going to a monthly billing. If P' describes the probability that the choice A_i will be made at the end of the entire period in case (a), and P'' describes the same probability in case (b), then it is easy to show using (1) and (2):

$$P' = P_o + (1 - P_o)[1 - (1 - g(R)]^n \tag{3}$$

$$P'' = P_o + (1 - P_o)g(nR) \tag{4}$$

Case (a) will lead to a greater increase in the probability of selection of A_i whenever $P' > P''$, and this requires:

$$[1 - g(R)]^n < 1 - g(nR) \tag{5}$$

It is easy to show that if $g(R)$ is quadratic in form, then (5) is satisfied, but if $g(R)$ is logarithmic (which it could be over only a finite range) then (5) is not satisfied. Thus whether or not a partition of a single experience into a series of smaller units will increase rates of adjustment depends upon the form of the adaptation function. If condition (5) is satisfied, then we can divide a single learning experience into a number of smaller cases such that the effect on behavior will be greater, even if in terms of payoff, the aggregate effect of the smaller cases is equivalent to the original one. If we wish, for example, to accelerate the effects of time-of-day pricing on household behavior, we would alter the billing cycle at the same time, moving from a monthly to a weekly or even a daily cycle. In the long run, response to the changed pricing schedule is always the same (because the experience-driven model does converge ultimately to the same outcome that would be predicted by the optimization model), but the speed of adjustment to the new equilibrium could be quite different.

III. ASSOCIATION OF PAYOFF WITH ACTION

Naturally, it would be impossible for one to learn what are the most appropriate courses of action if there were no reliable way to associate payoffs with the actions that bring them about. Usually, this is not a problem when we purchase food at a supermarket or acquire a new automobile, because we are aware, more or less simultaneously, of both the costs and the benefits of our purchases. There are exceptions to this rule. When we make credit card purchases, we experience the benefits of the commodities separately, but their costs are aggregated at the end of the month into a single lump. This naturally dilutes the association of any specific benefit with its own cost. This is not to say that the use of credit cards leads to uncontrolled spending—the large monthly bill may be enough to prevent that. Indeed, our fundamental theorem that the adaptation process converges eventually to a comparative static optimum still stands. Thus even though the expenses for a variety of unrelated consumption items may be aggregated into one single bill, the final equilibrium consumption bundle is identical to that which would have been chosen had everything been billed separately. Nevertheless, the association of specific costs with specific benefits does become looser, and that slows the learning adaptation process.

It is worth demonstrating that these conclusions hold even if all the payoffs from a bundle of commodities are aggregated into one return. Suppose, for example, that there are two types of action, A and B, with choice alternatives $A_1, \ldots, A_n$; $B_1, \ldots, B_m$. These actions have payoffs $R_1, \ldots, R_n$ and $S_1, \ldots, S_m$ respectively, but these payoffs are always aggregated into a single lump $(R_i + S_j)$ rather than being delivered separately. We use $P_1, \ldots, P_n$ and $Q_1, \ldots, Q_m$ to describe the selection probabilities of elements of A and B respectively. For the sake of brevity in this paper, we will use the simple linear operator form of the adaptation function that is given in (2). Using this version, if the pair (A_i, B_j) is chosen during period t then:

$$\begin{aligned} P_{i,t+1} &= P_{i,t} + g(R_i + S_j)(1 - P_{i,t}) \\ Q_{j,t+1} &= Q_{j,t} + g(R_i + S_j)(1 - Q_{j,t}) \\ P_{k,t+1} &= P_{k,t} + [1 - g(R_i + S_j)] \quad \text{for} \quad k \neq i \\ Q_{\ell,t+1} &= Q_{\ell,t}[1 - g(R_i + S_j)] \quad \text{for} \quad \ell \neq j \end{aligned} \tag{6}$$

These values of $P_{i,t+1}$ and $Q_{j,t+1}$ are conditioned upon the selection of A_i and B_j. If we apply the probabilities with which each element of A and B are chosen, multiply by expressions (6), and sum, we may calculate an overall expected value for $P_{i,t+1}$:

$$E[P_{i,t+1}] = P_{i,t} + P_{i,t} D_i \tag{7}$$

where

$$D_i = \sum_{\ell \neq 1}^{m} Q_{\ell,t} g(R_i + S_\ell) - \sum_{k=1}^{n} P_{k,t} \sum_{\ell=1}^{m} Q_{\ell,t} g(R_k + S_\ell)$$

One way to describe the effect of bundled payoffs on adaptive behavior is to consider the effect on $P_{i,t+1}$ of introducing additional payoffs into the sum $R_i + S_j$. We also would like to know the effect on $P_{i,t+1}$ of increases in values of S_j. That is, what happens to the adjustment rates for elements of A when their payoffs become small relative to the payoffs to choices in B? Both these questions may be addressed by looking at the derivative of equation (7) with respect to some S_j. This requires only that we consider the derivative of the expression D with respect to S_j:

$$\frac{dD}{dS_j} = Q_{j,t}\left[g'(R_i + S_j) - \sum_{k=1}^{n} P_{k,t} g'(R_k + S_j)\right] \tag{8}$$

Suppose A_i is the optimal alternative in A because R_i is maximal in the vector $R_1, \ldots, R_n$. Because $g(\cdot)$ is monotonically increasing in its argument, it follows that:

$$\sum_{\ell=1}^{n} Q_{\ell,t} g(R_i + S_\ell) > \sum_{\ell=1}^{n} Q_{\ell,t} g(R_k + S_\ell) \quad \text{for all} \quad k \neq \ell \tag{9}$$

It follows from (9) that $D_i > 0$ unless a) $P_{i,t} = 1$, or b) there are several alternatives whose payoffs are maximal and equal to one another, and the sum of their selection probabilities is equal to 1. Thus the presence of bundled payoffs does not interfere with the conclusion that the adjustment mechanism converges to a comparative static optimum.

We have also specified that the function $g(\cdot)$ must be concave. It follows that $g'(R_i + S_j) < g'(R_k + S_j)$ for any A_k that is not optimal. It is clear from (8), then, that $dD_i/dS_j < 0$, and this means that convergence to the comparative static optimum is slowed by the introduction or increase of payoffs to unrelated choices.

In summary, we do not need to be concerned that forms of billing or the bundling of commodities might prevent convergence to the traditional comparative static optimum. We do need to be concerned, however, that these factors affect the speed of convergence. Utility bills, for example, which typically aggregate the charges for a variety of different services, can not be effective in inducing rapid adjustment to altered conditions. Not only are all expenses lumped together in monthly bills, but the consumer does not even have the experience of signing separate chits for components of the total purchase (as he does have in the case of credit card purchases). The relevant information is certainly available on the statements themselves, but it has to be sought out and evaluated at an intellectual level rather than an experiential level. From the perspective of optimization theory, this is of no consequence—utility-maximizing consumers are willing and able to assimilate information and employ it immediately in the establishment of efficient consumption plans. From the perspective of learning theory, however, it makes an enormous difference because abstract knowledge is no substitute for direct experience: the monthly aggregated bill muddies the waters, obscuring the relationship between specific acts of consumption and their costs, and although this may not distort the final equilibrium, it certainly retards convergence to it.

IV. OPPORTUNITIES FOR IMITATION

Suppose that a consumer is not fully informed regarding the properties of some commodity. Many of the goods that we buy—such as houses, automobiles, and insurance policies—are only fully understood by experts of one sort or another, and we cannot expect "ordinary" consumers always to be able to make optimal choices on their own. Instead, most of us come to rely upon the experts, and upon the reports of satisfaction or dissatisfaction that we receive from friends and acquaintances. This behavior is not in any way inconsistent with conventional theory. The cost of information and the determination of optimal courses of action is often high enough so that the best procedure is to imitate the decisions of others rather than to undergo the total cost of calculation oneself. (See, for example, Conlisk, 1980).

This interpretation of imitatative behavior is adequate for cases in which we ask a friend for advice on insurance policies, or seek a recommendation for a good house painter, but it does not apply so easily to the apparent ability of nationally known athletes to influence our beer drinking habits or to alter our selection of rental car agencies. Within the traditional theory, imitation is a reasonable substitute for conscious optimization only when information is expensive to acquire, or when the computational costs of full maximization exceed any possible benefits, and the evaluation of beer, cigarettes, or rental car agencies does not fall into any such category. To deal with cases of this sort, I have postulated that individuals treat the experiences of others as surrogates for experiences of their own. In effect, the experiences of others act upon our own behavior in a fashion parallel to the process that has already been described. If we observe what appears to be a successful consumption decision on the part of someone else, then the likelihood that we will make a similar choice ourselves is increased. The effect on our behavior might not be as large as it would be if we had undergone the experience first hand, but it does move us in the same direction. Of course, if everyone had identical tastes, then imitation of this sort would be fully "rational," but we do not restrict ourselves to such an assumption, and it is possible, therefore, that imitation will guide us to inferior forms of behavior.

Within the context of optimization theory, imitation is a theory of information management. In our adaptive model, imitation is a theory of behavior. Nevertheless, there is an important informational element to be found in the adaptive model as well. Imitation that leads to consumption of some commodity necessarily generates first-hand experience with the commodity itself, and this first hand experience will also influence future behavior. Imitation can never be wholly passive over the long run because it forces consumption experience upon us, and our future behavior is necessarily conditioned by the information that we have received from engaging in that behavior ourselves.

Obviously, imitation can lead to faddish and conventional behavior. If consumers are more inclined to purchase those goods that they see being consumed by large numbers of their acquaintances, then they may join in, and thereby even expand the number of examples to be followed by still others. It is very easy to construct models along these lines, generating cases of explosive growth in consumption of highly visible goods, converging to a state in which everyone that is susceptible to the mechanism of imitation is engaged in similar consumption behavior. The effect is damped, however, by the information feedback. In order to participate in a fad, one must engage in the relevant consumption behavior, and necessarily one "learns" about the intrinsic qualities of the good. Fads can be successfully modelled, therefore, only in the cases of goods that are relatively unimportant in ones expenditure plans, or that are very nearly optimal anyway—otherwise the information uncovered by the faddish consumption would lead to experience-determined rejection of the good.

The phenomenon of imitation, which can lead to faddish behavior and resistance to change, can also accelerate innovation in consumption. Suppose a new type of commodity comes on the market—one with which consumers have no previous experience. Naturally, if one has no occasion to try out a particular commodity, then there is no opportunity for an experience-driven mechanism to add that commodity to consumption bundles. Even if the new commodity is superior to others that are already available, consumers may remain trapped in inferior modes of behavior through lack of experience. Imitation offers a surrogate for this experience, encouraging the acquisition of first hand experience with the product. If the new alternative is indeed superior, this mechanism provides the foundation for revised consumption behavior in equilibrium.

The same principles apply to cases of changed market conditions. We have noted that a consumer in isolation who has well established consumption habits will only infrequently select consumption alternatives that could reveal important changes in the environment. From the perspective of such a consumer, experimentation with a revised consumption pattern is something of an innovation. Even if one does not try a revised consumption bundle oneself, one can observe others who do, and if the apparent payoff is large, one will be induced to make a first-hand "experiment" with the revision. As usual, if the experiment is not a success, this will serve to offset the original stimulus. The consumer may return entirely to the old consumption pattern, although if the new pattern is widely adhered to, and one is still inclined toward imitation, there may always be a tendency to repeat it occasionally, even if it is not optimal.

An important qualification to all this is that imitation cannot take place without visibility. One cannot imitate what one cannot see. It is easy to find examples of the relevance of this factor. A new clothing fashion may spread rapidly throughout a society while reactions to a change in utility pricing are negligible. A new type of jeans may be obvious to any observer, and it is easy for others to recognize when their use has been chosen and to try them out for themselves. This is even true for new commodities on supermarket shelves, because shoppers may well notice the purchasing choices of one another. On the other hand, if a household successfully adapts its consumption of electricity to a new rate structure, or if it changes its use of long distance telephone service, these new patterns are not obvious to any outsiders, and the imitation mechanism cannot work. Thus factors that ordinarily speed adjustment to changed circumstances are not available in these cases, and our general conclusions that adaptive behavior will lead to long response times hold.

"Visibility" cannot be provided simply by informative advertising. The view incorporated into our adaptive model is that behavior is modified by experience and not by calculation. Communication of the "facts" does not provide any behavioral norm for consumers to imitate. This is a possible explanation why successful advertising so often seems to emphasize experience rather than data. The apparent tendency for certain product advertisements to stress the pleasures of use rather than the physical attributes of a product suggests that experiential imitation is at least as important a determinant of behavior as is conscious optimization. Thus it does no good to tell a consumer (over and over) that an electric rate structure has changed and that appropriate patterns of electricity consumption have thereby been altered. Instead, one would have to provide a model of what the new behavior might be, together with graphic descriptions of its benefits.

V. FREEDOM FROM RISK

The expected utility hypothesis provides the cornerstone for the theory of optimal decisions under risk. Given a known (or subjectively believed) probability distribution over outcomes, the single best choice is the one that maximizes expected payoff. Thus even though the environment is random, the optimal decision is determinate. The operation of adaptive learning models under risk is entirely different. Because they are already stochastic in nature, even in the absence of risk in the environment, they forecast a variety of different choices in the short run, for it is only after convergence to a long run (comparative static) equilibrium that such a model will settle upon a single choice. Under conditions of risk, stochastic learning models may not converge at all. This is not to say that anything can happen—choices that are inferior under any circumstances will come to be rejected just as they would be under a certainty model. Moreover, certain linear formulations of the feedback process (such as that described in equation (2))can be shown to converge to a single choice that is similar in character to the optimum under expected utility maximization.

Even when adaptive behavior under risk does converge to a single choice, however, that choice is not exactly the same as the one that would have been selected under conscious optimization. As a general rule, adaptive behavior does not deal efficiently with conditions of risk. Suppose, for example, that we face a set of m distinct states of nature $S_1, \ldots, S_m$, and that these occur subject to the stationary probability distribution $Q_1, \ldots, Q_m$. Any choice A_i receives a payoff R_{ij} whose magnitude depends on which state j actually occurs. The optimal choice i is that which maximizes the sum $V_i = Q_1 R_{i1} + \cdots + Q_m R_{im}$. This expression is a weighted average calculation using frequency measures for the states as weights. Adaptive models do not employ such simple averages because they always put the greatest weight upon the most recent experience: the importance of any particular event in determining the current selection probability vector $P_1, \ldots, P_n$ always declines the further that experience falls into the past. Due to the risk in the environment, the choice A_i will sometimes look good, and sometimes look bad, and rather than treating A_i as though it had an overall payoff V_i, the adaptive consumer will vacillate, sometimes behaving as though A_i had a high payoff (following a series of events which did yield relatively high payoffs to that choice), and sometimes modifying behavior so as to avoid it.

It is clear that behavior of this sort can lead to what appear to be wholly irrational activities from the perspective of optimization theory. For example, it is possible to show that even risk-averse consumers (that is, those with concave utility functions) will engage in the purchase of some lottery tickets. Similarly, we can show the existence of a stock market equilibrium in which "speculators" continue to buy and sell despite continuing losses. The reason for this is that the occasional successes that these activities encounter prevent the behavior from being rejected entirely despite overall and persistent losses.

These observations have relevance to our understanding of markets that are not ordinarily thought of as containing "risky" returns. The payoff to almost any consumption decision entails some risk. The dishwasher may develop mechanical trouble, the new air conditioner may be purchased at the beginning of an unusually cool summer, or the new automobile may be arrive just in time to facilitate a change of residence. An ideal optimization model would take account of such possibilities as these, but if their likelihoods are low, the determination of the best choice would probably be unaffected. In an adaptive model, even improbable events can have a significant effect on long run behavior.

A few years ago, one of the electric utilities in Michigan announced that a special low rate was available for central air conditioning. To qualify, a homeowner must permit the installation of a small radio receiver that enables the company to turn off the air conditioning (for no more than 20 minutes per hour) should overall electricity demand threaten to place a strain on the system. Moreover, the homeowner must have a second service meter installed that will keep separate account of electricity used for the central air conditioning, and this involves an expense of 2 to 3 hundred dollars. The optimizing homeowner would estimate his average annual consumption of electricity for air conditioning and compare the expected saving from the lower rate over a period of several years to the initial cost of the additional equipment. If the saving appeared to be significant (in a relatively cool state such as Michigan, many homeowners would find that it was not), then the value of this saving would still have to be compared to the expected inconvenience associated with the occasional loss of air conditioning for 20 minutes per hour. Only after passing both these tests would be final decision be positive.

The adaptive/imitating homeowner, on the other hand, would seek out some neighbor or acquaintance who had installed the system and ask about his experience. The response would inevitably depend upon conditions during the most recent summer. If the summer was cool, the information would be that the expense was not justified. If it was hot, the investment would be described as a success—unless the discretionary shut off was used so much as to be a serious inconvenience. Thus the number of households joining the program would depend upon what the weather happened to be the preceding summer, and whether or not the electric system was placed under any strain. Over the long run, this behavior might lead to a number of households participating in the program that was similar to the number that would find the new system desirable on the basis of fully informed calculation. Nevertheless, the two behavioral scenarios are entirely different, and surely the descriptive plausibility of the adaptive version is the greater of the two.

SUMMARY

What we have described here is only a summary of the dynamic properties of an adaptive model of behavior, and we have been able to do no more than bring out the flavor of the theory. It is interesting to consider the effectiveness of alternative strategies for introducing new products or new pricing systems into markets in the light of these results. One overriding conclusion that emerges is that simple information dissemination is not an effective way to bring about desired changes in consumer behavior. If all we do is provide data, then we are implicitly expecting the consumer to do all the calculation necessary to hit upon effective choices. He/she must evaluate the consequences of untried alternatives, disentangle the consequences of simultaneous decisions, and take full account of risk inherent in the circumstances. Even in the case of major purchases, these calculations may be impossible for the average consumer, and in the cases of minor purchases, they are certainly not worth the trouble. If we wish to alter behavior quickly, then we must present a graphic version of the allegedly superior course of action, and we must do so in a way that isolates new choices and their effects from other elements of consumption that are simultaneous but unrelated.

A second set of issues relates to our interpretation of various studies of demand—particularly those that do not take care to specify the time frame in which the demand function under investigation is supposed to function. In the short run, demand is subject to a number of influences that do not play any role in comparative static equilibrium, and if these influences are not properly taken into account, we may be seriously misled concerning the elasticity properties of long run demand. If imitation is an important behavioral factor, for example, short run elasticities will be significantly overstated because imitation has the effect of magnifying changes in consumption behavior. If imitation is not important, and if consumers are engaging in long-established habitual behavior, then measured elasticities will understate elasticities and take an unduly pessimistic view of the effect of price changes over demand.

REFERENCES

Conlisk, J.: "Costly Optimizers Versus Cheap Imitators" *Journal of Economic Behavior and Organization*, 1980, pp 275-93

Cross, J.: *A Theory of Adaptive Economic Behavior*, Cambridge University Press, 1983

TELECOMMUNICATIONS DEMAND MODELLING
An Integrated View
A. de Fontenay, M.H. Shugard, D.S. Sibley (Editors)

A METHODOLOGY FOR SPECIFICATION AND AGGREGATION IN PRODUCT CONCEPT TESTING

David A. Gautschi
Associate Professor of Marketing
European Institute of Business Administration (INSEAD)
Fontainebleau, F-77305
FRANCE

and

Vithala R. Rao
Professor of Marketing and Quantitative Methods
Johnson Graduate School of Management
Cornell University
Ithaca, New York 14853, U.S.A.

While consumer researchers have employed various methods for eliciting individuals' preferences toward new product concepts, it is not clear that the methods permit easy generalization to consumer behavior. In this paper we propose a methodology to deal with the common problem of limited information in conjoint analysis methods used to measure preferences. We also present a pre-estimation stage aggregation method based on the notion of the representative consumer. We identify appropriate methods of preference measurement, aggregation, and specification of utility functions of multi-attributed choice alternatives and describe an empirical study to illustrate the methods.

1. INTRODUCTION

A common activity of marketing researchers involves the difficult task of studying consumers' preferences for a product concept or product modification before market introduction. A crude estimate of the demand for the concept is derived from individuals' preferences under one or more scenarios of competitive behavior, market conditions, and assumed rules for consumer choice. Typically, marketing researchers engage in this activity as part of a firm's new product development program where it is important to guide research and development toward potential market opportunities that promise a reasonable chance of commercial success.

---

*The authors are grateful for a grant from the INSEAD Research and Development Committee for support of the data collection.

In economics, Lancaster's reformulation of consumer theory (Lancaster [15]), wherein objective product characteristics rather than commodity quantities are taken as arguments of the utility function, permits the estimation of the demand for new combinations of a set of characteristics describing existing alternatives in the market. The Lancasterian approach to demand estimation for new products is based on the revealed preferences of consumers.

By definition, transactions data are not available in product concept testing and may be limited in the case of new products. Even when transactions data are available - from a market test, for example - they may not reveal true preferences. It is widely documented in marketing (see Kotler [14], for example) that consumers often elicit a series of responses to the marketing of a (new) product. Indeed, in most cases, a consumer does not transact (i.e., choose an alternative on the market) until he is sufficiently aware that it exists, has acquired a sufficient understanding of it, has developed a sufficient liking for it, resulting in a sufficiently strong intention to buy it. If these responses can be seen collectively as a response chain, then a low market share for a new product could be the result of insufficient consumer awareness of insufficient comprehension. Of course one must also assume that consumers intending to buy the product can find it!

In practice, these are usually not minor issues. For example, the French PTT at present is attempting to encourage the adoption of Minitel by residential and commercial subscribers. Presumably, the low penetration (less than 10% at the time of this writing) is a result of collection of responses among PTT subscribers in the aggregate. Some subscribers are not aware of Minitel, some do not understand what it is or what it can do. If some of these subscribers were made aware and knowledgeable, they might prefer Minitel over other alternatives on the market. Some subscribers intend to obtain a terminal, but they cannot even place an order for one until they receive a voucher indicating availability of a terminal. And some subscribers knowing about Minitel, preferring it over alternatives, and intending to obtain it, have successfully transacted.

In the case of product unavailability, there is insufficient production or distribution, so that supply does not match demand and the market is in disequilibrium. In the case of insufficient awareness or comprehension, a good number of consumers do not have a complete preference ordering over a choice set that includes the new product. In either case, the indirect utility function derived from the estimation of a choice model on transactions data could be misleading, suggesting to the producer that something may be "wrong" with the product when, in fact, the product may be "right" assuming the consumer knows about it and can find it.

Eventually the market may settle into equilibrium, and consumer's preference orderings may become complete. Of course, long before such eventualities, the product may cease to qualify as "new." Hence, limitations of transactions data in new product and product concept contexts present incentives for collecting survey or experimental data. In marketing one often applies conjoint analysis to measure directly consumers' preferences for alternative combinations of product characteristics and then to infer choices (Green and Wind [9]). Conjoint-like experiments have also been used recently to attempt to estimate a choice model directly and to infer an (indirect) utility function (e.g., Beggs et al. [2]).

In spite of the increasing rigor of these analyses applied to the measurement of product concept preferences, there is no escaping the fact that the exercise in which the respondent in such studies participates is basically hypothetical. Indeed, there is a strong possibility that serious measurement error in the hypothetical setting would prevent accurate generalization to a market setting. Hence, in designing the experiment, the burden upon the researcher is to strike a balance between replicating a real environment and presenting the respondent with a manageable task.

With respect to the design of realistic experiments, two issues must be addressed. First, before a product is presented to the market it is often difficult for the marketing researcher to know precisely how a representative buyer defines the set of alternatives from which he intends to choose. In fact, consumers often use "evoked sets" of a limited number of alternatives in a well-defined product category and will not choose among all possible close substitutes (cf. Silk and Urban [27]). Moreover, the composition of a consumer's evoked set from one product category could influence his preference ordering on alternatives in his evoked set from another product category. This possibility is rarely acknowledged in product concept studies.

In many circumstances, the second issue regarding the design of the task follows from the first issue. If one represents choice alternatives as bundles of characteristics or attributes, then unavoidably the researcher must omit some attributes of alternatives in the experimental setting, if only to make the task for the respondent possible. Under the low information conditions of a conjoint analysis exercise price could be included in the preference function to proxy for unobservable qualities of alternatives. However, one must recognize that price also performs its conventional function as an allocative mechanism, given the individual's limited resources. Hence, the two influences of price--as a proxy and as an allocative tool--are likely to be confounded in such laboratory measurements of preferences. In a real market setting, the influence of price as a signal for hidden qualities of a product is likely to be significantly diminished, if not eliminated entirely, because information in a market setting is likely to be more complete than in a laboratory setting.

As it is generally important to determine the influence of price on demand (in the market), the two ways that price may influence preferences in the experimental setting must be distinguished from each other. In other words, it is not sufficient just to achieve satisfactory predictive validity in the experiment. In order for an experiment to be useful in guiding the eventual commercialization of a product concept, the models calibrated in the experimental setting must have true explanatory value as well. We propose a simple, multi-stage procedure to accomplish this. The procedure requires the collection of two preference orderings from the individual respondent, namely, an ordering in the presence of a budget constraint and another ordering in the absence of any constraint.

The structure of the paper is as follows. In the next section we specify the model of a representative individual's preferences and choices and address two potential specification problems. In the third section we discuss the procedures to identify representative individuals prior to the estimation of their utility functions. The estimation procedures suggested ensure separation of the informational and allocative effects of price in the utility function. An empirical illustration is presented next, and managerial implications are discussed in the last section of the paper.

2. SPECIFICATION OF THE REPRESENTATIVE CONSUMER'S PROBLEM

2.1. Modeling of Preferences

The representative consumer derives utility from consuming goods from a wide variety of product groups and categories. We may represent this utility, U, by a general utility function, G,

(1) $U = G(\chi_1, \chi_2, \ldots, \chi_I)$

where χ_i is a vector of alternatives (e.g., brands) in product category i. In marketing one imposes (often implicitly) a condition of weak separability on $G(\cdot)$ such that U may be written as:

(2) $U = G[G_1(\chi_1), G_2(\chi_2), \ldots, G_I(\chi_I)]$

and the subutility function for some group of alternatives, $G_i(\chi_i)$, is modeled without explicit reference to $G(\cdot)$. Furthermore, it is assumed that a mapping exists between a set of characteristics, Z_i, and the group of alternatives χ_i such that every alternative, $\chi_t^i \in \chi_i$, may be described as a combination of the elements in Z_i. Likewise, it is assumed that the individual's <u>utility</u> for any alternative in χ_i may be modeled as some combination of the individual's utilities for each of the elements in Z_i. So the utility function for alternatives in χ_i may be expressed generally as

(3) $U_i = G(\chi_i) = h_i(Z_i)$.

We shall express $h(\cdot)$ in the form of a random utility function

(4) $U_i = h_i(Z_i) = V_i(Z_i) + e(Z_i)$

where V is a utility component corresponding to the representative consumer and e represents the idiosyncratic deviation of the individual's utility from V in modeling U. In the context of a product concept test, one obtains evaluations on a set of alternatives X that represent some χ; and each alternative in X is described in terms of a vector of characteristics Z^o, where $Z^o \subset Z$. Accordingly, the researcher is likely to encounter some conventional specification problems in modeling individual level preferences using X and Z^o. We now address the two potential sources of misspecification of the preference model.

2.2. Misspecification (1): Incorrect X

If $\chi \cup X - \chi \cap X$ is non-empty, then the researcher is required to make some assumptions about the individual's preferences for alternatives in χ that are not elements of X. Given the separability assumption on G, an expedient means of accounting for preferences for excluded alternatives is to introduce a budget constraint explicitly into the analysis. For convenience, one can assume that the individual uses a two-stage budget procedure (cf. Deaton and Muellbauer [5]) such that in the first stage he apportions a global budget B according to intended allocations within each $\chi_1, \chi_2, \ldots, \chi_I$. We denote the budget portions as $\mathbf{b}_1, \mathbf{b}_2, \ldots, \mathbf{b}_I$. If we obtain the individual's preference ordering for the alternatives in some $X_i \in \chi_i$ in the presence of $\mathbf{b}_i$, then the individual could register his preference for each alternative in X_i and for $\mathbf{b}_i$. The individual's preference for $\mathbf{b}_i$ could be interpreted as his preference for alternatives outside of the set X_i.

2.3. Misspecification (2): Incorrect Z

When representing choice alternatives parsimoniously as bundles of characteristics, it is inevitable that the researcher will omit some elements of Z in an experimental setting. Indeed, to make the respondent's task of evaluating alternatives manageable, the researcher may choose to manipulate only a small number of characteristics (3 or 4) over delimited ranges even though additional characteristics or the same set of characteristics defined over wider ranges would better describe the alternatives. We partition $Z = (Z_1, Z_2, \ldots, Z_K)$ as $Z = [Z^o, Z^u]$ where $Z^o = [z_1^o, \ldots, z_k^o]$ are observed (included) characteristics in the product concept test and Z^u is a vector of unobserved (excluded) characteristics. Using this simple partitioning of Z, we re-express (4) as

$$(5) \quad U = V(Z^o, Z^u) + \epsilon(Z^o, Z^u).$$

If one attempts to estimate U with no knowledge of Z^u, then the estimators of V are likely to be biased. An obvious approach to controlling for Z^u is to find some kind of proxy variable. Srinivasan [29], for example, has argued that price can be used as a proxy variable for Z^u. Indeed, in the marketing research literature, a substantial set of empirical studies exploring the influence of price on consumers' evaluations of products has established that individuals tend to use price as a cue for unobservable qualities of products when available information about products is limited and where the salience of price varies inversely with the amount of information available to the consumer (cf. McConnell [17]; Tull, Boring, and Gonsior [31]; Stafford and Enis [30], and in economics, Gabor and Granger [6], for example).

The problem of using price as a proxy for Z^u is that price also performs a conventional function of allocating the individual's resources. It is this latter function that is most appropriate in attempting to extend the analysis in the product concept test to a market setting. Thus, the confounding of the Z^u-effect and the allocative effect of price must be reduced when preference measures are obtained by means of a conjoint analysis exercise. The procedure that we propose to reduce the confounding of the price effects requires that two preference orderings on X be obtained from each individual. These are labeled unconstrained and constrained preferences, respectively; see Rao [22] for an earlier use of these constructs.

2.4. Preference Ordering 1 (Unconstrained Preferences)

Denote by $U(\mathbf{b}^*)$ a preference ordering on X obtained under no budget constraint. The alternatives in X may be thought of as possible prizes in a lottery, and the individual is merely asked to express his preference for each alternative under the assumption that he wins the lottery. Under this scenario, price cannot perform an allocative function, and if it has any influence on the individual's preferences, then it must be as a signal for unobservable qualities of the alternatives. Price as proxy for Z^u is denoted as $\hat{Z}^u$.

If we adopt an additive form for $U(\mathbf{b}^*)$, then we have

$$(6)\quad U(\mathbf{b}^*) = V^*(Z^o, P) + \epsilon\mathbf{b}^*$$

where $\frac{\partial V^*}{\partial Z^o}$ and $\frac{\partial V^*}{\partial P}$ are the marginal utilities of Z^o and $\hat{Z}^u$, respectively, P denotes price, and $\epsilon\mathbf{b}^*$ is the idiosyncratic deviation of the individual's utility from $V^*(Z^o,P)$. If, for example, $U(\mathbf{b}^*)$ is linear and additive in Z^o and P, then $u(\mathbf{b}^*) = \alpha_0 + \alpha_1 Z^o + \alpha_2 P + \epsilon\mathbf{b}^*$. If $\alpha_2 \neq 0$, then we conclude that Z^u is non-empty.

2.5. Preference Ordering 2 (Constrained Preference)

Denote by $U(\mathbf{b})$ a preference ordering on X obtained under the budget constraint $\mathbf{b}$. This ordering is conditioned on the event that the individual would decide to choose from X. One could ask, for example, "which of these alternatives would you most prefer to buy, if you were to buy an alternative from the set" under the scenario of a prespecified budget constraint and a set of prices for the alternatives? If we adopt an additive form for $U(\mathbf{b})$, then we have

$$(7)\quad U(\mathbf{b}) = V(Z^o,P) + \epsilon\mathbf{b}$$

where $\frac{\partial V}{\partial Z^o}$ is the marginal utility for Z^o and $\epsilon\mathbf{b}$ is the idiosyncratic deviation from $V(Z^o,P)$. In this case, $\frac{\partial V}{\partial P}$ accounts for the confounded effects of price, i.e., the Z^u-effect <u>and</u> the allocative effect of price. One can attempt to isolate the latter effect by expressing the difference between (7) and (6). Thus,

$$(8)\quad U(\mathbf{b}) - U(\mathbf{b}^*) = V(Z^o,P) - V^*(Z^o,P) + \epsilon\mathbf{b} - \epsilon\mathbf{b}^*.$$

Assuming an additive representation in Z^o and P for V and V^*, we may write equation (8) as:

$$(9)\quad U(\mathbf{b}) - U(\mathbf{b}^*) = [V_1(Z^o) - V_1^*(Z^o)] + [V_2(P) - V_2^*(P)] + \epsilon\mathbf{b} - \epsilon\mathbf{b}^*.$$

Here we interpret the term $[V_2(P) - V_2^*(P)]$ as the allocative effect of price. To illustrate the procedure, consider a situation with one product feature, Z_1, and price, P.[2] A possible functional form for (6) and (7) would be

$$(6')\quad U(\mathbf{b}^*) = \alpha_0 + \alpha_1 Z_1 + \alpha_2 P + \epsilon\mathbf{b}^*$$

$$(7')\quad U(\mathbf{b}) = \beta_0 + \beta_1 Z_1 + \beta_2 P + \epsilon\mathbf{b}.$$

The difference equation, (8), becomes

$$(8')\quad U(\mathbf{b}) - U(\mathbf{b}^*) = (\beta_0-\alpha_0) + (\beta_1-\alpha_1)Z_1 + (\beta_2-\alpha_2)P + (\epsilon\mathbf{b}-\epsilon\mathbf{b}^*).$$

In this case, one need only estimate equation (6') and equation (8') constraining $(\beta_1-\alpha_1)$ to zero. The main allocative effect of price is then revealed by the estimate of $(\beta_2-\alpha_2)$. The signaling effect is reflected in the estimate of α_2. In general, we expect that the signaling effect to be positive with high values associated with low levels of information available on product concepts. Similarly, we expect that the allocative effect to be negative for normal range budgets.

3. IDENTIFYING REPRESENTATIVE CONSUMERS

In conjoint analysis (cf. Moore [20]) one typically aggregates the responses of individuals in conjunction with the estimation of the preference or choice model. The most compelling reason for aggregating is to improve the efficiency of estimation. Hagerty [10] has shown that in most cases the gain in efficiency outweighs the concomitant increase in bias. One assumes that there exists some vector, Y, that controls for individual differences. Adopting the expression in (4) of individual level utility this approach suggests that the inclusion of Y in the empirical expression for U controls for elements of Z^u that are correlated with elements of Z^o. If one assumes that V is linear in its arguments (Z^o and Y) and Y includes socio-economic variables, then generalization to the level of a market segment is accomplished by expressing

$$(9) \quad \hat{U} = V(Z^o, \bar{Y})$$

where $\bar{Y}$ corresponds to the average values of the socio-economic characteristics of a (pre-defined) market segment. Aggregation to the total market is then accomplished by taking the weighted average of different $\hat{U}$'s, where the weights reflect the proportional representation of different segments (socio-economic groups) in the relevant population. Because the vector Y is introduced to minimize the bias in the estimates of the parameters of Z^o, the implicit assumption in conjoint analysis is that V is common to all individuals in the sample (and, ultimately, in the population).

The procedure that we propose as an alternative to the conventional aggregation methods is based on the concept of the <u>representative consumer</u>. Briefly, our procedure aims to group individuals in terms of their stated preferences before proceeding to the estimation stage. We recommend the pre-estimation aggregation on the grounds that the preference orderings should allow us to aggregate individuals into different <u>preference</u> groups. In terms of the general individual level utility model in (4), this means that each group should have a unique V. Hence, a unique representative consumer corresponds to each group. A set of background characteristics, Y, may be associated with each group after it has been formed.

3.1. Tests for Homogeneity of Preferences

Our procedure calls for ascertaining the degree of homogeneity of preferences of the individuals in the sample. We propose two diagnostic tests for consistency and transitivity for this purpose. (The same tests can be applied for any subgroup of individuals in the sample.)

The basic data for these tests are the preference measures (constrained or unconstrained) for the set of alternatives obtained from N individuals in the sample. First, for each individual, n, construct a dominance matrix, D_n, such that the i,j-th cell is defined as follows

$$(10)\quad d_{ij}^{n} = \begin{cases} 1, & \text{if } x_i \text{ is preferred to } x_j \\ 0, & \text{otherwise} \end{cases}$$

using the measures of preference.

3.2. Consistency Test

Construct the summary dominance matrix, SD, of the total sample of N individuals by combining additively the individual dominance matrices, D_n, such that the i,j-th cell is defined as follows:

$$(11)\quad sd_{ij} = \begin{cases} \sum_n d_{ij}^{n}, & \text{if } \sum_n d_{ij} > N/2, \\ N/2, & \text{if } \sum_n d_{ij} = N/2 \text{ and } i > j \text{ and} \\ 0, & \text{otherwise.} \end{cases}$$

Cells in which the frequency of individuals preferring x_i to x_j equals the frequency of individuals preferring x_j to x_i are non-zero only in one triangle of the SD matrix. The middle condition of (11) places (arbitrarily) the non-zero equal frequency entries in the upper triangle.

We define the consistency score for the total sample of individuals as

$$(12)\quad \overline{CS} = \sum_{\substack{i,j \\ i \neq j}} sd_{ij} / \binom{A}{2}$$

where A is the number of choice alternatives. Note that this sum effectively excludes the zero cells of the SD-matrix and that only A(A-1)/2 cells are added. The maximum possible score is N and the minimum possible score is N/2. The score can be viewed as an index of the "goodness of grouping", where a score of N indicates that the total sample is composed of individuals with perfectly consistent preferences and a score of N/2 indicates that the total sample is composed of individuals with minimally consistent preferences.

One test (cf. Corstjens and Gautschi [4]) of the consistency of the preferences among the individuals in the total sample would entail measuring deviations of $\bar{CS}$ from the mean of a chance distribution of the choice of the dominant alternative in each pair of alternatives. With N individuals and a 50/50 chance of choosing either alternative in any pair, the chance distribution is binomial folded over N, N-1, N-2, ..., (N+1)/2 [if N is odd] or N/2 [if N is even]. The mean of this distribution is

$$(13)\quad \mu_{Bin} = \sum_{k}^{N} k\, P_{Bin}(r=k;\ N;\ p=0.5)/(1- \sum_{0}^{k-1} P_{Bin}(r=k;\ N;\ p=0.5)$$

where k=N/2 for N even and (N+1)/2 for N odd. The variance is

$$(14)\quad \sigma^2_{Bin} = [\sum_{k}^{N} k^2 P_{Bin}(r=K;N;P=0.5)/(1- \sum_{0}^{k-1} P_{Bin}(r=k;N;p=0.5))] - \mu^2_{Bin}$$

The relevant null hypothesis is H_0: $\bar{CS} = \mu_{Bin}$ versus H_1: $\bar{CS} > \mu_{Bin}$. Rejection of H_0 gives a crude indication that the preferences of the individuals in the total sample are sufficiently consistent to obviate a disaggregated analysis of individuals in the sample. We call this test a crude test because of its low power, i.e., it is difficult <u>not</u> to reject H_0, when N is large.

The consistency score in (12) depends upon N, the number of individuals in the sample. One way of comparing this measure across samples is to use $\bar{CS}/N$. An alternative way is to convert it to the 0-1 scale by the appropriate normalization.

3.3. Transitivity Tests

The "average" preference ordering of the total sample might be viewed as the preferences of the "benchmark" representative consumer.

Because all individuals in the total sample will not likely have identical preference orderings, it would be useful to determine how stable the preferences of the benchmark representative consumer are. Under the worst possible case the representative consumer's preferences would appear to be randomly generated. The more random the preferences appear to be, the less confidence one should put in the predictions on the total sample. We propose an index of the stability of the benchmark representative consumer's preferences based on the incidence of intransitivity among triples of alternatives.

For each triple of alternatives, there are eight possible sets of pairwise orderings, of which two are explicitly intransitive at the level of the representative consumer. That is, from any D_n the two possible intransitive orderings for any three alternatives, x_i, x_j, x_k are defined as:

(15) $d_{ij} > d_{ji}$ and $d_{jk} > d_{kj}$ and $d_{ki} > d_{ik}$

and

$d_{ji} > d_{ij}$ and $d_{kj} > d_{jk}$ and $d_{ik} > d_{ki}$.

At any aggregate level, the detection of the two possible intransitive orderings must be sensitive to the stochastic nature of the data. Coombs [3] suggests three tests--strong, moderate and weak--to detect intransitivity from a stochastic dominance matrix, such as SD. Using $\succ$ to denote strict preference, for any three alternatives x_i, x_j, x_k, if

$$\text{Prob}(x_i \succ x_j) \geq \text{Prob}(x_j \succ x_i)$$

and

$$\text{Prob}(x_j \succ x_k) \geq \text{Prob}(x_k \succ x_j)$$

then the ordering on x_i, x_j, x_k is intransitive if

(16) $\text{Prob}(x_i \succ x_k) < \max[\text{Prob}(x_i \succ x_j), \text{Prob}(x_j \succ x_k)]$, or if

(17) $\text{Prob}(x_i \succ x_k) < \min[\text{Prob}(x_i \succ x_j), \text{Prob}(x_j \succ x_k)]$, or if

(18) $\text{Prob}(x_i \succ x_k) < 0.5$.

The condition in (16) is referred to as the strong test, the condition in (17) is referred to as the moderate test, and the condition in (18) is referred to as the weak test for intransitivity (cf. Coombs [3]. In reference to the matrix SD, the relative frequency sd_{ij}/N is the maximum likelihood estimate of $\text{Prob}(x_i \succ x_j)$.

Under the worst possible case of a random preference ordering the distribution of the number of intransitive triples follows a binomial ($T; \pi=0.25$). The mean of the distribution is

$$\mu_{Int} = T\pi = \binom{A}{3} \times 0.25$$

where A is the number of alternatives in the choice set X, and $T = \binom{A}{3}$ is the number of triples. Denoting the number of observed intransitivities (using either the strong, moderate or weak test) by I, we can state the relevant hypothesis to test as:

$$(19)\quad K_0:\ I = \mu_{Int} \text{ versus } K_1:\ I > \mu_{Int}$$

The Z-scores and the significance probability associated with that level of I at which K_0 cannot be rejected become useful indicators of the stability of the benchmark representative consumer's preferences.

3.4. Procedure

Our methodology involves four steps labeled A through D as follows.

Step A. Assess the preference homogeneity of the total sample.

Step B. Two-stage grouping for identifying representative individuals.

Step C. Assess the preference homogeneity of the subgroups.

Step D. Estimate the two price effects.

We will elaborate on each of these.

Step A. This step involves conducting the two tests discussed in the previous section using the constrained measures of preference for the sample as a whole. Usually, these tests will indicate that the sample is not homogeneous in which case Steps B and C are necessary.

Step B. Construct groups $C_1^*, C_2^*, \ldots, C_K^*$ of individuals according to the strength of the correlation of their unconstrained preferences, $U(\mathbf{b}^*)$, for the alternatives in X. (A variety of clustering algorithms are available to accomplish this. See, for example, Romesburg [25],

Anderberg [1] and Hartigan [11]. Recently, Hagerty [10] has demonstrated that a form of Q-factor analysis may be more accurate than clustering procedures). For each group, C_k^*, construct sub-groups $C_{k1}, C_{k2}, \ldots, C_{kM_k}$ of individuals according to the strength of the correlation of their <u>constrained</u> preferences, $U(\mathbf{b})$, for the alternatives in X. Let the size of a typical group, km be N_{km}.

<u>Step C</u>. For each sub-group, C_{km}, construct the dominance matrix, D_{km}, from the dominance matrices of all individuals in the sub-group using the definition in (11). For each sub-group C_{km} compute the consistency score $\overline{CS}_{km}$, using the definition in (12). The consistency score for any given C_{km} can be compared with that of the benchmark representative consumer (i.e., the total sample). The difference in the proportion of consistent individuals in the subgroup and the total sample becomes a suitable index of the "goodness-of-grouping" for the sub-group. Indeed, if this proportion for any group significantly exceeds that of the total sample, then, the resulting sub-group qualifies as a <u>specific</u> representative consumer for purposes of estimating the preference functions. Moreover, for each resulting sub-group, one can search for background characteristics, Y, that distinguish individuals in any one group from individuals in other groups.

It will be of interest to test the hypothesis that the subgrouping procedure has improved the total consistency taking all the subgroups together compared with that of the ungrouped case of the total sample. For this purpose, we can use the measure, $R = \sum_k \sum_m \overline{CS}_{km}$ which approximately measures the "degree of consistency" in the subgroups taken together and set up the null hypothesis:

(20) L_0: $R = \overline{CS}$ against the alternative, L_1: $R > \overline{CS}$.

Noting that the statistic, R, can be built up from several binomial variables corresponding to each subgroup, we can compute the variance of R as

$$\mathrm{Var}(R) = \sum_k \sum_m N_{km} P_{km} (1 - P_{km})$$

where $P_{km} = \frac{2}{N_{km}} (\overline{CS}_{km} - \frac{N_{km}}{2})$ if N_{km} is even; and

$$\frac{2}{(N_{km}-2)} (\overline{CS}_{km} - \frac{N_{km}}{2} - 1) \text{ if } N_{km} \text{ is odd.}$$

A one-sided Z-test can be performed with the statistic, $Z = (R-\overline{CS})/\sqrt{Var(R)}$.

To assess the stability of the preferences of each sub-group one can conduct the intransitivity tests of (19). The larger the significance probability for the rejection of K_0, the less confident should one be in generalizing the empirical utility functions, calibrated in the laboratory, to the ultimate market. In some sense, a high incidence of intransitivity may indicate a propensity for individuals in the corresponding sub-group to switch among brands or alternatives in the actual market.

The advantages of the subgrouping and the concept of the representative consumer may be illustrated with an example. Assume there are six individuals who (each) have evaluated six alternatives in the following manner (preference ranks in body of table).

Alternatives/Individuals	A	B	C	D	E	F
Q	1	2	1	2	3	3
R	2	4	2	4	6	5
S	3	6	3	6	5	6
T	4	1	4	1	2	4
U	5	3	5	3	4	2
V	6	5	6	5	1	1

The dominance matrix SD for these data for the total sample is shown below with zero entries omitted.

	Q	R	S	T	U	V
Q	-	6	6		5	4
R		-	5			4
S			-			
T	4	4	4	-	5	4
U		4	4		-	4
V			4			-

Then, the average consistency score for the total sample = 67/15 = 4.467; equivalently, the $\bar{CS}$/SIZE = 77.4%.

The following subgroupings would yield maximum consistency:

Subgroup	Individuals	$\overline{CS}$	$\overline{CS}$/SIZE
1	A, C	2.00	100%
2	B, D	2.00	100%
3	E, F	1.73	86.5%

In this example, the overall consistency has improved by the process of subgrouping. Maximum consistency is achieved for Subgroups 1 and 2 so that representative consumers clearly correspond to each of the subgroups for the purpose of estimation. The consistency score for Subgroup 3, though not maximum, exceeds that of the total sample so that one could treat the individuals in Subgroup 3 as a single unit as well.

Step D. Estimate the two price effects using the equations (6') and (8') described earlier.

3.5. Summary

The essential aspects of our methodology consist of changes in data collection, aggregation and estimation. These are schematically shown in Figure 1, which is quite self-explanatory.

4. EMPIRICAL ILLUSTRATION

4.1. Overview

To enable an empirical examination of the issues raised by our methodology, we have conducted a small empirical study, patterned after traditional conjoint analysis. In this study, executives evaluated hypothetical profiles of portable microcomputers; it provided an opportunity for the subjects to rely more on price for inferring the qualities of the product. The data collection and other procedures described in the paper are followed in this illustration.

4.2. Study Design

Forty-five executives attending an executive development program at INSEAD, France provided evaluations of twelve hypothetical portable microcomputers in this study. Each microcomputer was described on three attributes in addition to price. The attributes and levels were as follows:

Manufacturer:	IBM; IBM-compatible
Expandability:	Yes; No
Country of Manufacture:	Japan; France
Price (in 000's Francs):	15; 22.5; 30

The unconstrained preferences were obtained under the scenario of a lottery as before and the constrained preferences were obtained using budget amounts estabished idiosyncratically by each respondent. Both the measures were ranks.

4.3. Analysis

The first step in our analysis was to form representative individual consumers and to test for the consistency and transitivity of responses as discussed above. Next, the effects of price--informational and allocative--were estimated using the linear specification of the utility functions.

4.4. Results

We have identified eight representative individuals in these data. The grouping results are displayed in Table 1 show that we have accounted for about one-half of the variation in these data by the disaggregation procedure.

The consistency and transitivity tests shown in Table 2 clearly indicate that we have identified the subgroups who are highly homogeneous within and different from one another. In every case, the subgroups pass both the consistency and transitivity tests. In this illustration, the $\overline{CS}$/SIZE for each subgroup is larger than that of the total sample indicating the clear advantages of disaggregation. The Z-statistic for the hypothesis (20) is 5.71, which is very highly significant showing that the subgrouping process has generated subgroups which are more consistent than the total sample.

The estimates of informational and allocative effects of price for these subgroups are shown in Table 3. The fits of the model of unconstrained preference are excellent in almost every case.

The informational (signalling) effects of price are very strong with appropriate signs (i.e., positive) for all but two subgroups. One of these subgroups (#3), is too small to be of any consequence. The signaling effect of price for subgroup #4 is negative indicating that these individuals are generally suspicious of the quality of higher priced concepts.

The fits of this model for estimating the allocative effect of price are generally acceptable except one case suggesting that the individuals in that subgroup are price insensitive within the price range presented. The allocative effect is significant for every subgroup, and its sign is negative (as expected) for all Subgroup #4. We have examined closely the preference data for Subgroup (#4) and have found that these individuals most prefer a collection of low priced and higher priced alternatives under the constrained situation. This suggests that the preference functions are probably U-shaped, thus not conforming to the linear functions we have used in the estimation.

5. SUMMARY

This empirical illustration shows the viability of our methodology in dealing with the specification problems raised with regard to conjoint analysis. We have also shown how to segregate a sample of individuals into subgroups, each corresponding to a specific representative consumer, that is, each subgroup has a unique V. Furthermore, we have demonstrated that the informational and allocative effects are <u>not</u> necessarily the same for these subgroups. The illustration also shows that the price effects are very strong when only limited information on the choice alternatives is provided to respondents in the conjoint experiment.

6. DISCUSSION

This paper presented a methodology to deal with the two specification issues relevant to applications of conjoint analysis for testing product concepts. The issues of specification of the set of competing items and incomplete information in the profile description can be handled by collecting two sets of preference data under no constraint and under a budget constraint. While price can be used a proxy for the information not included in the profile, the two effects of price--allocative and informational--are usually confounded. Our methodology shows how these two effects can be separately estimated.

The empirical illustration on portable computers shows how the methodology can be applied to practical problems. The subgrouping procedure and corresponding tests worked very well. The study also showed positive results on the effects of price. The effects of price--informational and allocative--are quite pronounced for all subgroups. One anomaly detected is possibly due to a non-linear preference function for one of the eight subgroups.

Our methodology offers a defensible way of identifying representative individuals (subgroups) since it is based on the complete vectors of preferences for the concepts. This approach is highly consistent with the marketing concept. We have proposed and implemented various tests for preference homogeneity which provide confidence in the stage of estimation.

Several directions for future research may be identified. First, the relationship between the tests on consistency and transitivity and the potential for switching brands should be explored. Once this relationship is established, our test procedure will provide a powerful way of identifying target markets for a new product concept. The effectiveness of the segmentation scheme should be compared to more standard schemes using background characteristics.

Another research direction is to devise additional statistical tests on the preference homogeneity with higher power. While we have utilized only the linear function in estimating price effects, the implications of nonlinear functional forms should be investigated further.

The estimates of allocative and informational effects of price for the representative individuals can be directly employed in identifying market targets. For example, the groups with a negative <u>informational</u> effect may be skeptical consumers prone to feeling "ripped off" in the marketplace while the groups with a positive allocative effect may be gullible consumers. The latter group may be influenced by snob appeal or may place high confidence in products with high prices due to uncertainty of perceptions of concepts.

The investigation of the effects of varying budgets on the constrained preferences for concepts will be another worthwhile pursuit in the future. These studies will show how the allocative effect of price varies with changes in budget arising possibly from borrowing.

Figure 1

ESSENCE OF OUR METHODOLOGY

Issues		Solution of the Methodology	
A. SPECIFICATION OF UTILITY			
1.	Set of Competing Alternatives is Not Well Defined.	1.	Obtain Two Preference Orderings (Unconstrained and under a budget constraint).
2.	Information on Alternatives is Not Complete.	2.	Use price as a proxy variable.
B. IDENTIFYING REPRESENTATIVE INDIVIDUALS			
3.	Methods of Aggregation Using Background Characteristics Are Not Necessarily Consistent With the Precepts of Economic Theory.	1.	Two-Stage Grouping Procedure (Pre-Estimation) Using Unconstrained Preferences First, Followed by Constrained Preferences.
		2.	Perform Various Diagnostic Tests on the Goodness of Grouping and Identification of Representative Individuals.
C. ESTIMATION OF THE TWO PRICE EFFECTS			
4.	The Two Price Effects Are Usually Confounded in the Traditional Methods of Estimation; Separation of Informational and Allocative Effects of Price is Not Apparent.	1.	Estimate the Informational Effect of Price from Unconstrained Preference and the Allocative Effect of Price from Difference of Constrained and Unconstrained Preference for the Representative Consumer.

Table 1

SOME STATISTICS ON GROUPING FOR COMPUTER DATA

Stage 1. Unconstrained Preferences

Number of Groups	% Trace	Size
1	100	45
2	75.6	33 12
3	58.1	18 18 9
*4	47.4	18 8 12 7

*Used in subsequent analysis.

Stage 2. Constrained Preferences

Group	Number of Subgroups					
	2		3		4	
	% Trace	Sizes	% Trace	Sizes	% Trace	Sizes
1	47.8	9*,9*	40.7	9,6,3	29.4	8,5,3,2
2	59.8	2*,6*	31.9	5,2,1	27.0	5,1,1,1
3	54.7	4*,8*	43.6	8,3,1	33.3	7,3,1,1
4	47.7	3*,4*	20.0	4,2,1	15.3	3,2,1,1

*Used as representative individuals.

Table 2

CONSISTENCY AND TRANSITIVITY TESTS FOR SUBGROUPS FOR COMPUTER DATA

		CONSISTENCY TEST			TRANSITIVITY TEST		
Subgroup	SIZE	$\overline{CS}$	Z	$\overline{CS}$/SIZE	STRONG	MODERATE	WEAK
1	9	8.33	3.03	92.5%	-6.70	-8.41	-8.56
2	9	7.99	2.64	88.8	-6.07	-8.25	-8.56
3	2	2.00	1.41	100.0	-8.56	-8.56	-8.56
4	6	5.22	1.89	87.0	-0.31	-8.10	-8.10
5	4	3.67	1.72	91.8	-2.02	-7.63	-7.78
6	8	7.47	2.86	93.4	-3.58	-8.56	-8.56
7	3	2.34	0.21	78.0	2.34	-7.01	-7.16
8	4	3.65	1.69	91.3	-4.52	-8.10	-8.10
All	45	29.86	2.33	66.3	7.94	-4.05	-6.38

Table 3

ESTIMATES OF INFORMATIONAL AND ALLOCATIVE EFFECTS OF PRICE BY SUBGROUP FOR COMPUTER DATA

		Informational Effect		Allocative Effect	
Subgroup	Size	Estimate (t-value)	R^2 (F;P-value)	Estimate (t-value)	R^2 (F;P-value)
1	9	.05 (2.86)	0.91 (255.4;0.0001)	-0.06 (-2.40)	0.05 (5.77;0.018)
2	9	.10 (4.23)	0.82 (114.3;0.0001)	-0.51 (-11.62)	0.56 (134.96;0.0001)
3	2	-0.13 (-∞)	1.0 (∞;0.0001)	-0.35 (-3.90)	0.41 (15.18;0.0008)
4	6	-0.25 (-7.28)	0.75 (51.09;0.0001)	0.20 (3.63)	0.16 (13.15;0.0005)
5	4	0.20 (5.94)	0.84 (56.28;0.0001)	-0.64 (-10.40)	0.70 (108.15;0.0001)
6	8	0.13 (5.71)	0.85 (127.49;0.0001)	-0.15 (-4.57)	0.18 (20.92;0.0001)
7	3	0.28 (2.85)	0.23 (2.26;0.0418)	-0.16 (-2.12)	0.12 (4.47;0.0001)
8	4	0.47 (12.56)	0.81 (45.77;0.0001)	-0.96 (-17.99)	0.88 (323.76;0.0001)
All	45	0.09 (5.560)	0.55 (165.11;0.0001)	-0.28 (-12.91)	0.24 (166.56;0.0001)

ACKNOWLEDGEMENTS

Thanks are due to Eric Wruck, a doctoral student in marketing at Cornell University, for his assistance in analysis. The authors are also grateful to the constructive criticisms of Daniel McFadden on an earlier version of this paper.

(FOOT)NOTES

1. Further, it is assumed that the consumer has no opportunity for resale of the item.
2. While we have used linear functional forms here, the general argument will extend to other forms as well. For example, in a model with interactions between Z_1 and P, there will be two effects of price--main allocative effect, $\beta_2 - \alpha_2$ and its effect on the interaction, $\beta_3 - \alpha_3$ where α_3 and β_3 are the coefficients for the product term, Z_1P.
3. We briefly examined the role of budget in our study on portable computers. Each respondent was asked to indicate the amount of budget they would allocate to buying a portable computer (conditional upon buying one); the choice set consisted of five items: stereo, television, winter vacation, summer vacation and portable computer. Then each individual later indicated whether (s)he would choose the most preferred profile of portable computer or the budget specified earlier. In all, sixteen individuals chose the product in preference to the cash budget. Further, there was considerable variation in this preference among the eight subgroups identified earlier. This result points to the importance of considering budget in choice models designed to predict market share of a new product.

REFERENCES

[1] Anderberg, M.R., Cluster Analysis for Applications (Academic Press, New York, 1973).

[2] Beggs, S., Cardell, S., and Hausman, J., Journal of Econometrics 16 (September, 1981) 1.

[3] Coombs, C.H., A Theory of Data (John Wiley & Sons, New York, 1964).

[4] Corstjens, M. and Gautschi, D.A., Management Science 29 (December, 1983) 1393.

[5] Deaton, A. and Muellbauer, J., Economics and Consumer Behavior (Cambridge University Press, Cambridge, 1980).

[6] Gabor, A. and Granger, C.W.J., Econometrica 33 (1966) 43.
[7] Green, P.E. and Rao, V.R., Journal of Marketing Research 8 (1971) 355.
[8] Green, P.E. and Srinivasan, V., Journal of Consumer Research 5 (1978), 103.
[9] Green, P.E. and Wind, Y., Harvard Business Review 53 (1975), 107.
[10] Hagerty, M.R., Journal of Marketing Research 22 (May, 1985), 168.
[11] Hartigan, J.A., Clustering Algorithms (John Wiley & Sons, New York, 1975).
[12] Hauser, J. and Simmie, P., Management Science 27 1 (1981) 33.
[13] Hauser, J. and Urban, G., Prelaunch Forecasting of New Consumer Durables: Ideas on a Consumer Value-Priority Model, in: Srivastave, R. and Shocker, A., (eds.), Analytic Approaches to Product and Marketing Planning (Marketing Science Institute, Cambridge, Massachusetts, 1982) pp. 276-296.
[14] Kotler, P., Marketing Management: Analysis, Planning and Control (Prentice-Hall, Inc., Englewood Cliffs, New Jersey, 1984).
[15] Lancaster, K., Consumer Demand: A New Approach (Columbia University Press, New York, 1971).
[16] Levitt, T., Harvard Business Review 38 (1960) 24.
[17] McConnell, J., Journal of Marketing Research 5 (1968) 300.
[18] McFadden, D., Conditional Logit Analysis of Qualitative Choice Behavior, in: Zarembka, P., (ed.), Frontiers in Econometrics (Academic Press, New York, 1974).
[19] McFadden, D., Annals of Economic and social Measurement 5 (1976) 363.
[20] Moore, W.L., Journal of Marketing Research 17 (November, 1980) 516.
[21] Rao, V.R., Conjoint Measurement in Marketing Analysis, in: Sheth, J., (ed.), Multivariate Methods for Market and Survey Research (American Marketing Association, Chicago, 1977) pp. 257-286.
[22] Rao. V.R., A Model for Brand Choice Under Price-Quality Hypothesis (Combined Proceedings of the American Marketing Association, Chicago, Illinois, 1972) pp. 366-371.
[23] Rao, V.R. and Gautschi, D.A., The Role of Price in Individual Utility Judgments, in: McAlister, L., (ed.), Choice Model for Buyer Behavior, Research in Marketing, Supplement 1 (JAI Press, 1982) pp. 57-80.
[24] Ratchford, B., Journal of Marketing Research 17 (February, 1980) 14.

[25] Romesburg, H. Charles, Cluster Analysis for Researchers (Lifetime Learning Publications, Belmont, California, 1984).

[26] Shepard and Srinivasan, (1969)

[27] Silk, A. and Urban, G., Journal of Marketing Research 15 (May, 1978) 171.

[28] Spence, M., Journal of Economic Theory 7 (1974) 296.

[29] Srinivasan, V., Comments on the Role of Price in Individual Utility Judgments, in: McAlister, L., (ed.), Choice Models for Buyer Behavior, Research in Marketing, Supplement 1 (JAI Press, 1982) pp. 81-90.

[30] Stafford, J.E. and Enis, B.M., Journal of Marketing Research 6 (1969) 456.

[31] Tull, D., Boring, R., and Gonsior, M., Journal of Business 37 (1964) 186.

SECTION I:
END USER DEMAND

I.2. Economic Approaches to Consumer Choice

TELECOMMUNICATIONS DEMAND MODELLING
An Integrated View
A. de Fontenay, M.H. Shugard, D.S. Sibley (Editors)

The Effects of User Cost on the Demand for Telecommunication Service*

Donald A. Dunn
Professor
Engineering-Economic Systems Department
Stanford University

Hyung Sik Oh
Assistant Professor
Industrial Engineering Department
Seoul National University, Korea

ABSTRACT

Users of a telecommunication service incur costs in using the service in addition to the price. The user's own time costs involved in learning to use the service, waiting for the service, and making use of the service are typically greater than the price of such telecommunication services as telephone service or electronic message service. Other user costs can also be important but are not treated explicitly in this paper.

The question addressed in this paper is, can thinking about user time cost be helpful in understanding consumer responses to price changes? Cases of special interest are large changes in the prices of existing products that go beyond the range of existing elasticity data, setting the price for new products for which no demand data exists at all, and pricing in congested markets.

Consumer behavior is analyzed in three different situations: (1) when user time cost is large in comparison with price; (2) when user time cost is a function of the total amount of service provided, in a congested market in which all users have the same value of time; and (3) same as (2), except that there are two user groups with different values of time instead of a single user group.

*This research was supported by the National Science Foundation under Grant MCS 78-02272 to Stanford University.

1. Introduction

Users of a telecommunication service incur costs in using the service in addition to the price. The user's own time costs involved in learning to use the service, waiting for the service, and making use of the service are typically greater than the price of such telecommunication services as telephone service or electronic message service. Other user costs can also be important but are not treated explicitly in this paper.

Total user cost per unit of service, u, can be expressed as the sum of price, p, and time cost, c, which is the product of user time value, ω, in dollars per hour, and the amount of time required to consume one unit of service, τ, in hours. Thus,

$$u = p + c = p + \omega\tau \qquad (1)$$

Users view u as the decision variable in choices with respect to the quantity of service purchased, rather than p. Service providers, on the other hand, can only control price. Providers, therefore, seek information about consumer behavior in the form of conventional demand curves which plot quantity purchased as a function of price.

The question addressed in this paper is, can thinking about user time cost be helpful in understanding consumer responses to price changes? Cases of special interest are large changes in the prices of existing products that go beyond the range of existing elasticity data, setting the price for new products for which no demand data exists at all, and pricing in congested markets. In the usual theory of consumer behavior, the demand curve is derived from a theory of consumer choice in which utility is maximized subject to a budget constraint. Cairncross [2] and Becker [1] have suggested that a theory of consumer behavior that includes the effects of user time could be derived by including both a budget and a time constraint in the theory. One result of such an analysis is the form of the user decision variable given in Eq. (1) as the sum of price and time cost.

2. Consumer behavior when time costs are high relative to price

Consumer behavior in a number of different situations can be explained by considering the relative values of p and c in Eq. (1).

If user time cost, c, is large in comparison with p, the following kinds of consumer behavior would be expected.

- an upper limit to individual usage without regard to price
- low price elasticity of demand
- user inertia in switching from old to new products
- limited effects of peak load pricing on demand

A characteristic demand curve shape for products that require that the user incur substantial time costs in their consumption is shown in Fig. 1. This curve shape can be justified in terms of both qualitative arguments about user cost and estimated elasticity data indicated in Fig. 1. As the

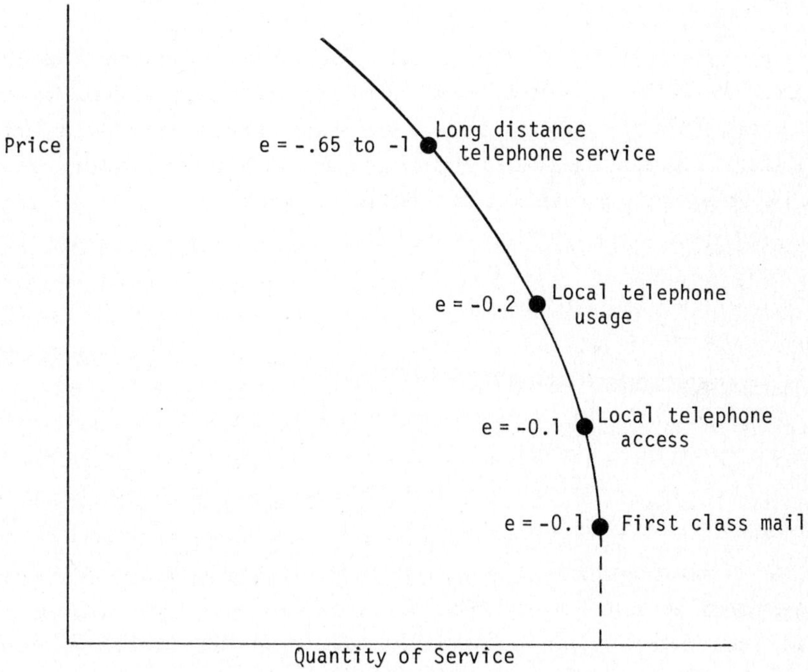

Fig. 1. "Universal" demand curve for message services. Estimated elasticities and operating points for several services are indicated.

price of the service in comparison with user time cost falls, the elasticity also falls, as suggested by the elasticities indicated.

Data on long-term trends in the demand for telephone and postal service are also consistent with this curve shape. For the last 80 years the usage of the postal service, in terms of pieces of mail per capita, has been increasing. During this same period, revenues per piece of mail have also been increasing in constant dollars. While a number of phenomena are involved in this situation, the data suggests a very low price elasticity of demand and is consistent with what would be expected for a product with c much greater than p. Indeed, the time cost of composing, typing, and filing a first class letter is typically $10 to $20, while the price is $0.22. The substantial demand for express mail at a price of about $10 is also consistent with the time cost, as well as the high user value, associated with personal correspondence and other business mail.

The situation for telephone calls is similar, but more complex, because the price of a toll call has been decreasing as traffic has increased, while the price of a local call has been bundled with the price of access to the system, which has increased slowly. Nonetheless, usage has increased rapidly over the last 20 years. Total revenue for all calls has risen steadily as traffic has increased. Again, there are a number of phenomena involved in this data, and only in the last 10 years or so has the price of a toll call fallen below the time cost of an average business user. Local calls have had a marginal price of zero, which further complicates the argument. However, in the future, as prices are brought more in line with costs in the telephone industry, the price of toll calls should drop further. As the price of a toll call falls further below the user's time cost, both the elasticity and the rate of growth of demand would be expected to drop.

A related phenomenon is the high user willingness to pay for small improvements in telephone terminal equipment that has been demonstrated since about 1976. User purchases of modern PBX's and associated multifunction telephone sets has demonstrated a willingness to pay for improvements that reduce total user costs by increasing the price of service and reducing the time required to consume one unit of service. User willingness to pay for new services that increase user satisfaction, without clearly reducing user time costs, is more difficult to estimate, but is related to the value users place on their time [5]. Some of the purchases of new terminal equipment may be based more on a high valuation of user benefits, such as the ability to leave a message when the called party is not available, than on user time cost savings.

3. Consumer behavior in congested systems

A congested information service typically involves time costs associated with waiting for service that are comparable to or greater than the time cost associated with actually using the service. The theory of a market in which congested services are offered can be developed by considering user cost in the form:

$$u_j = p_j + c_j = p_j + \omega\tau_j(x_j) \tag{2}$$

where $\tau_j(x_j)$ is the time required to consume one unit of the service, provided by the jth provider, expressed as a function of the total amount of service delivered by the jth provider to all users, x_j. When the total output of the jth provider, x_j, becomes comparable to the available capacity of the jth provider's system, waiting time for that system rises rapidly. In the usual case, the individual user consumes a quantity small in comparison with the total output, x_j, and is unaware of his contribution to waiting time. It will be assumed that $\tau_j(x_j)$ is an increasing function of x_j, so that $\tau_j'(x_j) > 0$, and that the marginal cost of congestion increases with the amount of service provided, because $\tau_j''(x_j) > 0$. Two cases will be considered: (1) a case in which all users have the same time value, ω; and (2) a case in which there are two classes of users, each with a different value of time. In both cases, at least two different sources of services are assumed to be available, with potentially different prices but no other differences in service quality, other than those that users create through their use of the services and the resulting differences in $\tau_j(x_j)$ that are produced.

In the first case, Eric, Lecaros, and Dunn [3] have shown that, for any given set of prices, the sources that are used will all operate at the same user cost at equilibrium. The equilibrium in this situation is a demand equilibrium, defined as a condition in which, for any given set of prices, no user has an incentive to change either the quantity of service purchased or (implicitly) the distribution of quantities purchased from individual firms. A demand equilibrium is arrived at by each user minimizing user cost, according to

$$u^* = \min_j u_j = \min_j [p_j + c_j(x_j)] \qquad (3)$$

where u^* is the user cost at equilibrium. Eq. (3) states that, at a demand equilibrium, the cost of services for all users is the lowest cost available in the market. Any firm that supplies services at equilibrium will do so at u^*.

$$u^* = p_j + c_j(x_j) \quad (x_j > 0) \qquad (4)$$

Since the congestion cost functions $c_j(x_j)$ are increasing functions of x_j, the inverse function, $c_j^{-1}(u^*)$ exists, and firm j supplies a quantity x_j at equilibrium given by

$$x_j = c_j^{-1}(u^* - p_j) \qquad (5)$$

To clear the market, supply equals demand, so the equilibrium condition is

$$X^D(u^*) = \sum_j c_j^{-1}(u^* - p_j) \qquad (6)$$

where $X^D(u)$ is the aggregate demand, aggregated over all users and all firms. Eq. (6) can be solved for u^*, and from u^*, the amount of service supplied by each firm can be determined from Eq. (5). Fig. 2 shows supply and demand curves at demand equilibrium for the case of two firms supplying the market, with the values x_1 and x_2, as well as X^D, indicated.

The demand equilibrium condition is seen to be a result of user optimization, and can occur for any set of prices set by firms offering services in a market for a good that is homogeneous except for the effect of congestion. Congestion cost acts as an equilibrating mechanism in this situation. Firms may post different prices, but users will adjust their demands among the firms in a way that brings the sum of price and congestion cost to the same value for all firms that supply any service. This

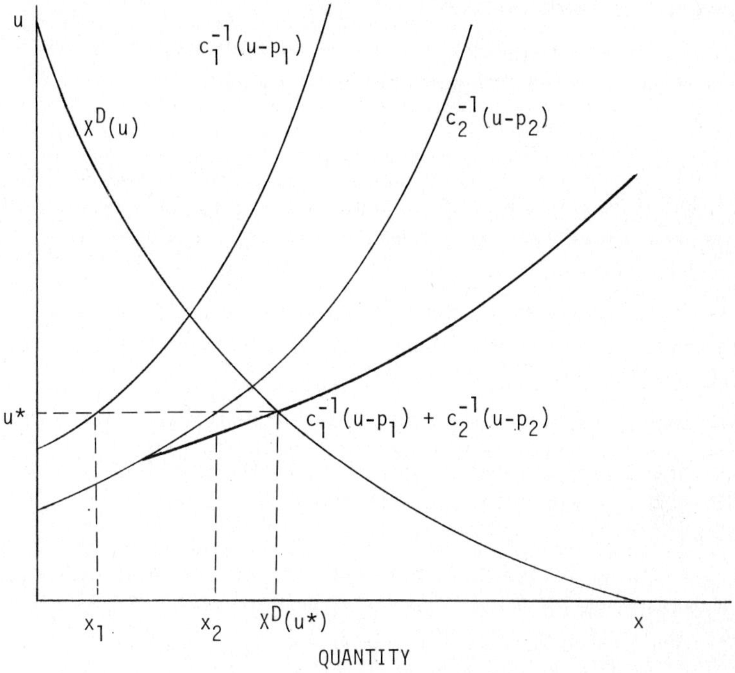

Fig. 2. User cost vs. quantity. Total demand, $X^D(u)$, and supply curves for the two firms individually and together are shown for the case of two firms supplying the market. Demand equilibrium occurs at u*.

model is appropriate for services markets where inventories are not carried over. The model is limited to cases in which users are homogeneous with respect to their value of time and their time to receive a unit of service. An example of such a situation might be a group of timeshared computer service providers offering service to a user population in which all of the users have similar abilities to make use of the service and are paid about the same salary.

Price dispersion can exist in this case at equilibrium, without any difference in the products offered other than price, in the presence of perfect information. Price provides a mode of product differentiation, but users adjust their consumption of each service in such a way that all services consumed are available at only one user cost, the minimum user cost, so no user has an incentive to change sources.

4. Consumer behavior in congested markets with multiple user time values

In the case discussed above, where all users had the same value of time, it was shown that demand equilibrium was reached when all users incurred the same total user cost for the service. In the case discussed here, two user groups exist, each of which has a different value of time. In this case, the users with higher time values incur a higher user cost than the users with a low time value. In this analysis as in the analysis of the preceding case, it is assumed that the individual user ignores his own effect on congestion when making consumption decisions.

Let u_j^k be the unit cost of service j to user k. User k has time value ω^k per unit time and takes time τ_j to consume one unit of service, where τ_j is taken to be a function of the total amount of service provided by provider j but independent of the user k. Then

$$u_j^k = p_j + c_j^k = p_j + \omega^k \tau_j(x_j) \qquad (7)$$

One service is said to be more congested than another, if the unit congestion costs, τ_j, of the two services have the same functional form and the more congested service has a higher unit congestion cost.

The user's decision problem will be stated under the assumption that user k has a benefit function B^k which is his monetary willingness to pay for the total quantity of service q^k that he consumes. Given a set of n services, user k maximizes net benefit, the difference between B^k and user cost. His decision problem, assuming a concave benefit function, is then

$$\max_{q_1^k,\ldots,q_n^k} \quad B^k(q^k) - \sum_j u_j^k q_j^k \qquad (8)$$

subject to

$$q^k = \sum_j q_j^k \, , \qquad j = 1,\ldots,n \qquad (9)$$

$$u_j^k = p_j + \omega^k \tau_j \tag{10}$$

$$q_j^k \geq 0 \;, \qquad j = 1,\ldots,n \tag{11}$$

The optimization leads to the relationships

$$q_j^k \geq 0 \qquad \text{if} \qquad \frac{\partial B}{\partial q_j^k} = u_j^k \tag{12}$$

$$q_j^k = 0 \qquad \text{if} \qquad \frac{\partial B^k}{\partial q_j^k} < u_j^k \tag{13}$$

If u^k is the equilibrium (unit) user cost, then u^k is the minimum user cost

$$u^k = \min_j u_j^k \tag{14}$$

At such a demand equilibrium, no user has an incentive to modify his choice of services or the quantities of service purchased, and supply equals demand.

$$x_j = \sum_k q_j^k \tag{15}$$

$$q^k = \sum_j q_j^k \tag{16}$$

At such an equilibrium, the demand of every user group is allocated among the services in a way that yields the minimum user cost.

Oh [4] derives several properties of the demand equilibrium that will be stated here without proof.

1. At equilibrium, if a user chooses to consume at all, the user's user cost must be equal to the minimum user cost for users with the same value of time.

2. In congested markets, a high priced service is always less congested than a low priced service.
3. Users with a high value of time incur higher user costs than users with a low value of time.
4. Users with a high value of time consume a higher priced (less congested) service than users with a low value of time.

Consider now a simple case in which only two services are offered, each of which is characterized by a price, p_j, and a unit congestion cost, τ_j, which is a function of the amount of service provided, x_j. Congestion costs are further assumed to be linear functions of the amount of service provided

$$\tau_j = \alpha_j x_j , \qquad j = 1,2 \tag{17}$$

where the coefficient α is positive.

In order to explore some of the effects of price on demand, it will temporarily be assumed, further, that both services have the same coefficient α. This condition will be relaxed, in making a sample computation below.

The quantity of service demanded by user group k, q^k, is assumed to be linear in user cost as follows.

$$q^k = a_k - b_k u^k , \qquad k = 1,2 \tag{18}$$

where a^k and b^k are positive constants. Thus, the users are inhomogeneous both with respect to the value of user time and their demand functions.

Four different cases can exist at demand equilibrium, depending on the prices and user time values.

<u>Case 1</u>: Service 1 (the lower priced service) captures all the demands for services by both user groups.

In general, a reduction in price may not capture all the demand in a congested market. But in this case, one service has captured the entire market demand. Since users choose services by user cost, under this case both user groups incur a least user cost for service 1. The demand for service 2 is thus zero and the demand for service 1 is the sum of the demands by both user groups. The corresponding demand equilibrium conditions are

$$u^1 = p_1 + \omega^1 \alpha x_1 \leq p_2 + \omega^1 \alpha x_2 \tag{19}$$

$$u^2 = p_1 + \omega^2 \alpha x_1 \leq p_2 + \omega^2 \alpha x_2 \tag{20}$$

$$x_1 = \sum_{k=1}^{2} (a_k - b_k u^k) \tag{21}$$

$$x_2 = 0 \ . \tag{22}$$

Substituting u^1 and u^2 from Eqs. (19) and (20) into Eq. (21) and solving for x_1 yields

$$x_1 = (1 + b_1\omega^1\alpha + b_2\omega^2\alpha)^{-1} \left[(a_1+a_2) - (b_1+b_2)p_1\right] \tag{23}$$

which is equal to the total market demand for services. The user costs, u^1 and u^2 can be obtained as a function of p_1 by substituting x_1 from Eq. (23) into Eqs. (19) and (20), respectively:

$$u^1 = (1 + b_1\omega^1\alpha + b_2\omega^2\alpha)^{-1} \left[(a_1+a_2)\omega^1\alpha + (1 + b_2\alpha(\omega^2-\omega^1))p_1\right] \tag{24}$$

$$u^2 = (1 + b_1\omega^1\alpha + b_2\omega^2\alpha)^{-1} \left[(a_1+a_2)\omega^2\alpha + (1 + b_1\alpha(\omega^1-\omega^2))p_1\right] \ . \tag{25}$$

The user demands, q^1 and q^2 can be obtained as a function of p_1 by substituting u^1 and u^2 from Eqs. (24) and (25) into Eq. (18):

$$q^1 = (1 + b_1\omega^1\alpha + b_2\omega^2\alpha)^{-1}$$

$$\times \left[a_1 + a_1b_2\omega^2\alpha - a_2b_1\omega^1\alpha - b_1(1 + b_2\alpha(\omega^2-\omega^1))p_1\right] \tag{26}$$

$$q^2 = (1 + b_1\omega^1\alpha + b_2\omega^2\alpha)^{-1}$$

$$\times\ [a_2 + a_2 b_1\omega^1\alpha - a_1 b_2\omega^2\alpha - b_2(1 + b_1\alpha(\omega^1 - \omega^2))p_1] \ . \qquad (27)$$

Define R_1 as the set of prices for which this case can occur, then we can find the set of prices R_1 which satisfies constraints (19), (20), and the nonnegativity of the user demands, q^1 and q^2 defined in Eqs. (26) and (27), respectively.

Case 2: User group 1 consumes both services, and user group 2 consumes service 2 only.

This case can happen if user group 1 incurs the same user cost for both services and user group 2 incurs a least user cost for service 2. The corresponding demand equilibrium conditions are

$$u^1 = p_1 + \omega^1\alpha x_1 = p_2 + \omega^1\alpha x_2 \qquad (28)$$

$$u^2 = p_2 + \omega^2\alpha x_2 \leq p_1 + \omega^2\alpha x_1 \qquad (29)$$

$$x_1 + x_2 = \sum_{k=1}^{2} (a_k - b_k u^k) \ . \qquad (30)$$

Following a procedure similar to that used in Case 1, the market demand functions, x_1 and x_2, and the user demands, q^1 and q^2, can be obtained. The total market demand x can be obtained by summing x_1 and x_2 or q^1 and q^2:

$$x = (2 + b_1\omega^1\alpha + b_2\omega^2\alpha)^{-1}$$

$$\times\ [2(a_1+a_2) - \frac{b_1\omega^1\alpha + b_2\omega^2\alpha}{\omega^1\alpha} p_1 + \frac{b_2\alpha(\omega^2 - 2\omega^1) - b_1\omega^1\alpha}{\omega^1\alpha} p_2] \ . \qquad (31)$$

Define R_2 as the set of prices for which this case can occur, then we can find the set of prices R_2 which satisfies constraint (29), and the nonnegativity of x_1, x_2, q^1, q^2, and q_2^1 defined above.

The market demand for service 1 (service 2) decreases as the price of service 2 (service 1) decreases. Since both services are same, except for the degree of congestion, the two services compete with each other as if they were substitutes in this case.

Next, it is of interest to examine the signs of the cross derivatives of the user demand functions. The demand of user group 1 increases as either service 1's price or service 2's price decreases. The demand of user group 2 increases as service 1's price decreases, although user group 2 consumes only service 2. This is because, as a result of a decrease in service 1's price, some users who purchase service 2 switch to service 1, resulting in a decrease in service 2's congestion level. Hence user group 2 incurs a lower user cost than before, even with a fixed price for service 2. The effect of a price change of service 2 on the demand of user group 2 is indeterminate. A price increase of service 2 can lead to an increase in user group 2's demand under certain conditions. This is likely to occur when the demand elasticity of user group 1 and the congestion coefficient are high. The economic reasoning is as follows. As service 2's price increases, some of the users in user group 1 who consume service 2 turn to service 1. This results in a decrease in the congestion level of service 2. If a decrease in the congestion level of service 2 compensates for the price increase, user group 2 finds it has a lower user cost for service 2 as a result of a price increase of service 2, thus increasing user group 2's demand for service 2.

Finally the effects of price changes on total market demand are examined. As service 1's price increases, total demand decreases. But an increase in service 2's price can result in an increase in total demand if $b_2\ \alpha(\omega^2-2\omega^1) - b_1\omega^1\alpha > 0$. This property may lead to the case where a total surplus maximizer sets service 1's price below marginal cost. Therefore cross subsidization may exist under total surplus maximization.

Case 3: User group 1 consumes service 1, and user group 2 consumes service 2.

The demand conditions are

$$u^1 = p_1 + \omega^1 \alpha x_1 \leq p_2 + \omega^1 \alpha x_2 \tag{32}$$

$$u^2 = p_2 + \omega^2 \alpha x_2 \leq p_1 + \omega^2 \alpha x_1 \tag{33}$$

$$x_k = a_k - b_k u^k , \qquad k = 1, 2 . \tag{34}$$

The market demand functions, x_1 and x_2, can be expressed in terms of prices by substituting u^1 and u^2 from Eqs. (32) and (33) into Eq. (34) and solving for x_1 and x_2, respectively:

$$x_1 = q^1 = (1 + b_1 \omega^1 \alpha)^{-1} (a_1 - b_1 p_1) \tag{35}$$

$$x_2 = q^2 = (1 + b_2 \omega^2 \alpha)^{-1} (a_2 - b_2 p_2) . \tag{36}$$

Define R_3 as the set of prices for which this case can occur, then we can find the set R_3 which satisfies the constraints (32), (33) and the nonnegativity of x_1 and x_2.

Case 4: User group 2 consumes both service 1 and service 2, and user group 1 consumes only service 1.

User group 2 incurs the same user cost for both services, and user group 1 incurs a least user cost for service 1. The corresponding demand equilibrium conditions are

$$u^1 = p_1 + \omega^1 \alpha x_1 \leq p_2 + \omega^1 \alpha x_2 \tag{37}$$

$$u^2 = p_2 + \omega^2 \alpha x_2 = p_1 + \omega^2 \alpha x_1 \tag{38}$$

$$x_1 + x_2 = \sum_{k=1}^{2} (a_k - b_k u^k) . \tag{39}$$

Following the same procedure as before, x_1, x_2, q^1, and q^2 can be obtained. The total market demand x can be obtained by summing x_1 and x_2 or, equivalently q^1 and q^2:

$$x = (2+b_1\omega^1\alpha+b_2\omega^2\alpha)^{-1}\ [2(a_1+a_2) + \frac{b_1^{\ 1} - 2b_1^{\ 2} - b_2^{\ 2}}{\omega^2\alpha}\, p_1$$

$$- \frac{b_1\omega^1\alpha + b_2\omega^2\alpha}{\omega^2\alpha}\, p_2] \qquad (40)$$

Define R_4 as the set of prices for which this case can occur, then we can find the set R_4 which satisfies constraint (37), and the nonnegativity of the user demands and market demands.

A number of interesting observations can be made from examining the structure of market demand functions and user demand functions.

The effect of price changes on the market demand for each service is considered first (see Table 1). The market demand for each service decreases as its price increases. The market demand for the high-priced (less-congested) service may increase as a result of the decrease in the price of the more-congested service. In terms of cross-elasticity, this corresponds to the case where the services are complements. However, a more-congested service acts always as a substitute for a less-congested

Table 1. The Effects of Price Changes on Market Demands

	x_1		x_2		$x\ (= x_1 + x_2)$	
	$\partial x_1/\partial p_1$	$\partial x_1/\partial p_2$	$\partial x_2/\partial p_1$	$\partial x_2/\partial p_2$	$\partial x/\partial p_1$	$\partial x/\partial p_2$
Case 1	-	0	0	0	-	0
Case 2	-	+	+	-	-	?
Case 3	-	0	0	-	-	-
Case 4	-	+	?	-	-	-

service; the demand for a more-congested service increases as the price of a less-congested service decreases.

The effect of price changes on the demand by each user group is considered next (see Table 2). The quantity demanded by users with a high time value may increase as the price of the more-congested service increases. This is because as the price of the more-congested service increases, the demand by users with a low value of time decreases. The decrease of the demand by these users reduces the equilibrium level of congestion. The user cost incurred by users with a high value of time may therefore decrease, and consequently their demand can increase in spite of the increase in the price of the more-congested service.

Table 2. The Effects of Price Changes on User Demands

	q^1		q^2	
	$\partial q^1/\partial p_1$	$\partial q^1/\partial p_2$	$\partial q^2/\partial p_1$	$\partial q^2/\partial p_2$
Case 1	-	0	-	0
Case 2	-	-	-	?
Case 3	-	0	0	-
Case 4	-	-	?	-

In summary, the market demand function for each service has been derived as a function of prices, user time values, and other user demand characteristics. Since the market demand for each service is a result of user self-selection process, for any given set of services, prices set by providers, user time values, and user demands, the corresponding demand faced by each provider can be uniquely obtained.

An example with two user groups having different values of time

In this example, a provider of service, such as a university, is providing a service, such as timeshared computing service, initially to a user population at zero price. It is assumed that the user population is divided into two user groups (possibly users with an immediate deadline and users with a more remote deadline for completing their work), each of which has different time values. In this case, let $\omega^1 = 1$ and $\omega^2 = 2$. The user demand and congestion functions are assumed to be linear as follows:

$$q^1 = 10 - u^1$$

$$q^2 = 20 - u^2$$

$$\tau_1(x) = \alpha_1 x_1$$

where α_1 denotes the congestion coefficient of an existing facility, and is assumed to be 1. The cost per unit time of providing these services is assumed to depend on the value of the congestion coefficient of the system. This cost is assumed to be $(10/\alpha_1)$. Since the services are provided at zero price, the university incurs a net cost of 10 in providing the services in this case. Given the user demands of the two user groups and the existing system, the equilibrium congestion level and user demands are (see case 1 above)

$$\tau_1(x_1) = 7.5$$

$$q^1 = 2.5 \qquad u^1 = 7.5$$

$$q^2 = 5 \qquad u^2 = 15$$

The university feels that the level of congestion is too high and decides to increase the budget from 10 to 15. With the increased budget, two alternatives are considered: the existing zero price policy and a nonzero price policy. Under the zero price policy, a new facility with a congestion coefficient 2 will be made available, in addition to the existing facility, and the net cost involved in providing the services will be 15. Under this zero price policy, the equilibrium congestion level for each facility and user demands are (see the case 4 above)

$$x_1 = 6.68$$

$$x_2 = 3.33$$

$$\tau_1(x_1) = x_1 = 6.67$$
$$\tau_2(x_2) = 2x_2 = 6.67$$
$$q^1 = 3.33 \qquad u^1 = 6.67$$
$$q^2 = 6.67 \qquad u^2 = 13.33$$

where x_1 denotes the quantity of services provided from the existing facility, x_2 denotes the quantity of services provided from the new facility, τ_1 denotes the equilibrium congestion level for the existing facility, and τ_2 denotes the equilibrium congestion level for the new system. The congestion coefficient for each service is different. However, the equilibrium congestion level (unit cost) for each service is the same. The user cost has been reduced for both user groups and the quantity of service provided to each group has been increased, as a result of the increased budget allocation.

We now consider a nonzero price policy. Under this policy, the university will charge for services from the new facility and use the revenues collected from the new facility to cover some of the cost involved in providing the services from the new facility. The university decides to set the congestion coefficient for the new facility and the price for services from the new facility to the values that maximize the user surplus subject to the budget allocated:

$$\max_{p_2,\alpha_2} \int_{u^1}^{\infty} q^1(v)dv + \int_{u^2}^{\infty} q^2(v)dv$$

subject to

$$p_2 x_2 = (10/\alpha_2) - 5$$
$$u^1 = p_1 + \omega^1 \alpha_1 x_1$$
$$u^2 = p_2 + \omega^2 \alpha_2 x_2$$

where x_1 is the demand for services from the existing facility, x_2 is the demand for services from the new facility, and p_1, α_1, ω^1, and ω^2 are set to 0,1,1, and 2 respectively.

Solving this optimization problem for p_2 and α_2, we obtain the congestion coefficient for the new facility, the price for services from the new facility, equilibrium congestion level for each facility, user demand of each user group:

$$p_2 = 4.15$$

$$\alpha_2 = 0.19$$

$$\tau_1(x_1) = 5$$

$$\tau_2(x_2) = 2.18$$

$$q^1 = 5 \qquad u^1 = 5$$

$$q^2 = 11.47 \qquad u^2 = 8.53$$

Again, both user groups have reduced user costs and increased consumption levels in comparison with previous cases. But the budget allocation is still the same as in the preceding case.

The cost involved in providing services for the new facility will be $52.63 = (10/\alpha_2)$, which is the sum of the revenues collected from the facility, 47.63, and the increased budget, 5. The use of the revenues collected from the high time value users has permitted the university to acquire a facility with a substantially higher capacity than was possible before. The comparisons between the two alternatives are summarized in Table 3. From Table 3, both user groups are better off under the nonzero price policy. This example shows that total welfare is increased simply by charging for services from the new facility and allowing the users to self-select which facility to use. Total welfare is defined here as the difference between user surplus and the university's net cost.

Since the university incurs the same net cost under the two new system alternatives, the comparison of these two alternatives is appropriate. The users with low time values still obtain service at zero price, and their time costs are reduced, because the nonzero price facility draws off high time value users. The users with high time values must pay a nonzero price, but their total user costs are reduced.

In this case, the higher priced product attracts the user group with the higher value of time, and a form of discriminatory pricing is made possible in which users self-select themselves. Benefits to both the user group with the low user time value and the user group with the high user time value are greater for positive pricing. It appears that there may be opportunities for increasing both user benefits and provider profits in congested systems serving user groups with different time values, simply by offering the same service through multiple sources at different prices.

Table 3

The Comparison of a Zero Price Policy and an Nonzero Price Policy

	Existing System	New System	
	Zero Price Policy	Zero Price Policy	Nonzero Price Policy
p_1	0	0	0
p_2	-	0	4.15
α_1	1	1	1
α_2	-	2	0.19
x_1	7.5	6.67	5
x_2	-	3.33	11.47
$\tau_1(x_1)$	7.5	6.67	5
$\tau_2(x_2)$	-	6.67	2.18
u^1	7.5	6.67	5
u^2	15	13.33	8.53
University Net Cost	10	15	15
Welfare	21.25	40.58	141.98

REFERENCES

[1] Becker, G. S. (1965), "A theory of the allocation of time," Economic J., v. 75, pp. 493-517, Sept.

[2] Cairncross, A. K. (1958), "Economic schizophrenia," Scottish J. Political Economy, Feb.

[3] Eric, M. J., F. Lecaros, and D. A. Dunn (1979), "Equilibrium conditions in congested markets," Report No. 18, Program in Information Policy, Engineering-Economic Systems Dept., Stanford University, Stanford, CA, Feb.

[4] Oh, H. S. (1983), "Product differentiation in markets with congestion," Report No. 46, Program in Information Policy, Engineering-Economic Systems Dept., Stanford University, Stanford, CA, May.

[5] Sharp, Clifford (1981), The Economics of Time, Martin Robertson, Oxford.

TELECOMMUNICATIONS DEMAND MODELLING
An Integrated View
A. de Fontenay, M.H. Shugard, D.S. Sibley (Editors)

A THEORY OF CONSUMPTION DECISION: THE STOCHASTIC COMMODITY MODEL

Nicolas Curien

Direction Generale des Telecommunications, France*

Alain de Fontenay

Bell Communications Research, 435 South Street, Morristown, N.J. 07960, U.S.A.

INTRODUCTION

In the present paper we propose a new framework, namely the stochastic commodity model, which generalizes the neoclassical theory of demand. The main rationale for this new framework is to provide a relevant level of analysis in the description of individual demand, intermediate between the traditional aggregate level where only the global amount of consumption over a given period is taken into account in the context of a deterministic decision process, and the most disaggregate one where the demand for every individual and elementary consumption decision for each instant and possible state of nature over that period are being considered.

At least for a number of commodities, the individual consumer can reasonably be assumed to allow for uncertainty and to consider more information than his/her global consumption when optimizing *ex-ante* his/her economic behavior. On the other hand, it seems as reasonable that he/she does not plan either his/her future consumption on a consumption event-by-consumption event basis. In the stochastic commodity model we introduce, it is fundamentally assumed that the individual's behavior is best described as *if* he/she selected *ex ante* an optimal probability distribution function, and that, during the consumption period, his/her day-to-day purchases can be associated with random draw from that distribution. The loss of information which is incurred when considering the probability distribution function rather than the itemized consumption provides the desired pattern of intermediate aggregation.

The paper is divided into three main sections. In section 1., we explicit the passage from the neoclassical theory of demand to the stochastic commodity model. The neoclassical approach is first recalled (section 1.1) and ti is shown that while this approach can deal with several levels of aggregation of an individual's consumption over time, it however turns out to be inappropriate when the relevant aggregation acts on some measured attribute of the elementary purchases, rather than on the time processing of those purchases. Then, the precise formulation of the stochastic commodity model is developed (section 1.2), introducing a utility functional the argument of which is the probability distribution of consumption with respect to the demanded commodity: In the stochastic commodity model, the preferences no longer operate on consumption itself but on its distribution. It is shown that the stochastic commodity model generalizes the neoclassical analysis. It is also shown that in the particular case where the utility functional is linear, i.e. when it can be written as the expectation of a sub-utility acting directly on the consumption space the stochastic commodity reduces to the standard neoclassical model. When the utility functional is non-linear, the solution of the utility maximization program is explicited, in the general case and in the particular case when the form of the consumption probability distribution is pre-imposed. Applications of the stochastic commodity model in telecommunications economics are eventually reviewed.

---

* This work was completed while N. Curien was a resident visitor with Bell Communications Research, July-September 1985.

In section 2, we discuss the epistemological grounds of the stochastic commodity model. First, the link with the theory of choice under uncertainty is mentioned (section 2.1). In the later, as in the stochastic commodity model, one considers preferences orderings among random variables as represented by their probability distribution functions. As in the standard theory of expected utility preferences are linear, we investigate conditions under which the non-linearity of preferences can be assumed to hold, as required for the stochastic commodity model to actually generalize the neoclassical analysis. Then, analogies between the stochastic commodity model and similar models in other fields of science, chiefly physics, are being developed (section (2.2). The place of the stochastic commodity model with respect to the neoclassical model is compared to the one of wave mechanics with respect to newtonian mechanics, for in both cases the idea is introduced that *objective* functions (such as utility, or energy) do no longer depend directly on deterministic variables (such as consumption, or spatial position and speed), but depend on probability distributions defined over those variables. The bi-hierarchical structure of the stochastic commodity model is further emphasized: in a first stage an optimal probability distribution is determined by the individual so as to maximize a utility functional while can be assumed to be deterministic, whereas in a second stage the real time elementary consumptions are associated with random draws according to that distribution. An analogy is suggested with the modelization of a gas in statistical thermodynamics, where one introduces two levels of description, respectively macroscopic and microscopic, and where the distribution of energy among molecules plays between these two levels a role of information interface analog to the role of the distribution of demanded quantities in the stochastic commodity model. More fundamentally, it is argued that the conceptual framework of the stochastic commodity model is relevant in economics as in any other scientific discipline, whenever a boundary has to be drawn between an aggregate level where the analyst intends to formulate casual and explicit laws, and a disaggregate level which is by definition beyond the scope of analysis and which appears to behave at random under the overall control of a probability distribution optimized at the aggregate level. The later consideration leads to a new conception in the treatment of error in economic models: whereas in standard econometric practice the error term is not incorporated in the model itself but is rather added *ex-post* on statistical grounds, in the stochastic commodity model it is on the contrary part of the very model and *internalized* in the economic analysis when defining the stochastic disaggregate level and its relationship to the deterministic aggregate level.

In section 3., we present an application of the stochastic commodity model to the modelization of discrete choice. The standard stochastic utility approach to that issue is first recalled and discussed (section 3.1). Particular emphasis is given to the behavioral interpretation of stochastic utility, according to which the random component of utility does not only stand as an extrinseque representation of errors or omitted variables, but also as the intrinseque reflection of whims in the consumption decisions. Moving in the same direction, the stochastic commodity model is then proposed as an alternative to stochastic utility in discrete choice. Whereas in the latter approach differences between choices made by identical individuals were explained by differences in realizations when drawing utility *prior* to optimization, in the stochastic commodity model variations in choices are due to differences in drafts *posterior* to optimization: similar individuals would first choose the same discrete probability distribution defined over the set of the available options and determined so as to maximize utility, and would then draw different choices according to that distribution. The observed ex post difference results from the individuals making independent draws from the same distribution function. It is shown that if the choice frequencies of the discrete options are invariant when the prices of all options are raised by the same amount, then the stochastic commodity model and the stochastic utility model are two equivalent and dual approaches to discrete choice. However, in the general case where the stability of choices under a price translation does not hold, the two types of modelization can be differentiated and the stochastic commodity model is shown to yield an easier scheme of validation from empirical data.

In brief, the stochastic commodity model can be conceived of as a generalization of the standard neoclassical theory of demand which both provides a relevant intermediate pattern of aggregation when describing individual consumption, and allows the analyst to deal intrinsequely (rather than extrinsequely through an error term) with situations where, as in discrete choice, presumably identical individuals make different demand decisions.

1. FROM THE NEOCLASSICAL THEORY OF DEMAND TO THE STOCHASTIC COMMODITY MODEL

1.1 The Deterministic Neoclassical Theory of Demand

1.1.1 The Conventional Model

In the neoclassical microeconomic modelization of demand, it is typically assumed that the consumer faces a n-dimensional commodity space X. The consumer is assumed to be a utility maximizer faced by a budget constraint y. The consumer maximizes utility in two stages. In one stage he/she chooses a subspace X_n among all possible subspaces contained in X to select in the other stage the quantity demanded for each commodity in X_n. The consumer is characterized by his/her preferences which is described by a utility function $u(.)$. The utility function maps the commodity space onto the reals. The arguments of $u(.)$, x_i, $i = 1, ..., N$, where N denotes the dimension of X, are quantities indexed by the decision period t. The later, by definition, is some future period relative to the decision point, i.e., the date of the utility maximizing decision. In other words, at the decision point, the consumer is assumed to select quantities to be consumed in some future period t in order to maximize utility for that period. Commodities can be obtained on the X_n market by paying an entry fee $A_n \geq 0$ and by paying a linear price p, per unit of commodity i. Given the entry fee A_n, commodities from the complementary commodity subspace XX_n cannot be purchased by the consumer. The price p_i may be conditional on the entry fee A_n, thus it could be that, given two commodity subspaces X_l and X_k such that X_l is a proper subspace of X_k, $A_l => A_k$ and $p_{il} \leq p_{ik}$. However, in this paper it is assumed that the entry fee for a proper subset is never higher than the entry fee for the set itself and that the commodity price is independent of the commodity subspace selected. Finally, the consumer is assumed to be a price taker, i.e., he/she takes all entry fees A_n and all prices p_i, $i = 1, ..., n^*$ as given. The the consumer's utility maximization program, constrained to the commodity subspace X_n, can be represented as follows:

$$v\{\mathbf{p}, y - A_n\} = \max_{\mathbf{x}}[u(\mathbf{x}) | \mathbf{p}^T.\mathbf{x} \leq y - A_n, \mathbf{x} \in X_n] \tag{1}$$

where:

$\mathbf{p}$ is the price vector with element p_i, $i = 1, ..., n$.

The unconstrained utility maximization problem is given by:

$$v\{\mathbf{p}, y\} = \max_{\{X_n\}}[v[\mathbf{p}, y - A_n\} | \mathbf{X}_n \in \mathbf{X}] \tag{2}$$

Under proper regularity conditions[1] the above program can be solved for $\mathbf{x}$ and the resulting vector of utility-maximizing quantity demanded $\mathbf{x}^*$ can be derived through Roy's identity as a function of the income y, the entry fee A_n, and the price vector $\mathbf{p}$:

$$\mathbf{x}^* = -\frac{v_p\{\mathbf{p}, y - A_n\}}{v_y(\mathbf{p}, y - A_n\}} \tag{3}$$

where:

v_z is the partial derivative of the indirect utility function with respect z.

This program provides a description of the individual's economic behavior. The decision variables in this program can be viewed as aggregates in the sense that for any commodity i the quantity demanded is considered a the total quantity x_i^* purchased and consumed during the decision period. The program is deterministic in the sense that the quantity vector demanded $\mathbf{x}^*$ which results from the utility maximization program is exactly the quantity vector which will be available to the consumer to be consumed over the decision period. In fact, most commodities are consumed by the individual throughout the period.[2]

1.1.2 A Disaggregated Neoclassical Model The previous model is highly aggregated in that it does not take into account the intra-decision period consumption path selected by the consumer. It is reasonable to suspect that consumers are not indifferent between various consumption paths and that the consumption path itself is the argument of the consumer's utility maximization program. This is particularly relevant where the unit price, other things equal, varies throughout the period such as say with restaurant rates for meals which vary depending on the time of day. The same situation arises in telecommunications where much of the tariffs are time-of-day specific. The price may also be assumed to vary with the consumption flow. Then the commodity space becomes a space of functions defined on the time period $t, X(t)$, with as elements consumption flow vectors $\dot{\mathbf{x}}(t) = (x_i, i = 1, ..., N)$ defined as a function of time, the integral of which over the reference period is the vector $\mathbf{x}$ of aggregate consumption:

$$\mathbf{x} = \int_0^t \dot{\mathbf{x}}(\tau)\, dtau \tag{4}$$

in this formulation the utility function $U\{\dot{\mathbf{x}}(.)\}$ is a functional of the vectorial function $\dot{\mathbf{x}}(t)$, and the utility maximization consists in the following constrained and unconstrained programs:

$$V\{\mathbf{p}(.), y - A_n\} = \max_{\dot{\mathbf{x}}(.)}[U\{\dot{\mathbf{x}}(.)\} | \int_0^t \mathbf{p}(\dot{\mathbf{x}}(\tau), \tau).\, \dot{\mathbf{x}}(\tau)\, d\tau \leq y - A_n] \tag{5}$$

$$V\{\mathbf{p}(.), y\} = \max_{\{X_n\}}[V\{\mathbf{p}(.), y - A_n\}] \tag{6}$$

where

$\mathbf{p}(\dot{\mathbf{x}}(t), t)$ *is the vectorial price function with* $p_i(\dot{\mathbf{x}}(t), t)$ *as its i th element*

This program is to be solved for the demand function $\dot{\mathbf{x}}^*(t)$, which can be shown to derive from the following generalized Roy's identity, analogous to (2):

$$\dot{\mathbf{x}}^*(t) = -\frac{\delta_p V\{\mathbf{p}(.), y\}}{V_y(\mathbf{p}(.), y\}} \tag{7}$$

where

$\delta_z f(.)$ *denotes the variation of* $t(.)$ *with respect* z.

Thus, a comprehensive description of demand over time is now provided by the model.

1.1.3 An Intermediate Model The object of this section is to formulate the utility maximization problem at a level of aggregation which is intermediary between the conventional neoclassical model and the disaggregated model presented in the preceding section. The intuition behind the model developed here builds on the assumption that the conventional model does not contain adequate information to describe the consumer's decision process, while the disaggregated model is rather information-intensive and would appear to impose an undue burden on the consumer's decision process.

While the disaggregated model is required when the price function $\mathbf{p}(\dot{\mathbf{x}}(t), t)$ is not constant, at least over some interval of time, most time-dependent prices are step functions and constant over all intervals but for their points of discontinuity at least for given consumption flows $\dot{\mathbf{x}}$. In this case, the customer needs not specify the consumption path. Nevertheless, as long as the price is a function of the consumption flow, i.e., $\mathbf{p}(.) = \mathbf{p}\{\dot{\mathbf{x}}(.)\}$, one cannot reduce the disaggregated problem to the conventional one. Further justification for this intermediary level of analysis will be developed in the following sections of this paper.

The intermediate solution which is proposed, while economizing on the information required, provides enough detail to be reasonably used by the consumer in a wide class of utility maximization problem. Provided the time period is not too long, it is assumed that the argument the consumer optimizes with respect to is the frequency of these consumption levels with respect to the quantity consumed. For instance, telephone calls can be assumed not be taken into account on a call by call basis by the consumer in the utility maximization process but rather the consumer can be assumed solely to care about an overall allocation of his/her calls with respect to duration, i.e., the number of calls of each duration. Thus, for instance, we assume that, typically, two calls of five minutes are not equivalent to ten calls of one minute, or even one one-minute call and one nine-minutes call.

Denoting the relative frequency of the consumption flow $\dot{x}_i$; $i = 1, ..., n$ by a function $f_i(\dot{x}_i)$, one can define the vectorial function $\mathbf{f}(\dot{\mathbf{x}})$ with $f_i(x_i)$ as its i th element. Then the intermediate program is given as follows:

$$V\{\mathbf{p}(.), y - A_n\} = \frac{\max}{\mathbf{f}_n}(.)[U\{\mathbf{f}_n(.)\}| \int_0^x \mathbf{r}(\dot{\mathbf{x}}).\mathbf{f}_n(\dot{\mathbf{x}})d\dot{\mathbf{x}} \leq y] \tag{9}$$

where

$\mathbf{r}(.)$ *is a vectorial outlay flow function with* $r_i(\dot{x}_i)$ *as its i th element,* $i = i, ..., n$, *and* $r_i(\dot{x}_i) = p_i, \dot{x}_i$

The conventional program appears as a special case of this disaggregated program once it is assumed that:

1. $\mathbf{p}(\dot{\mathbf{x}}(t), t) = \mathbf{p}$, i.e., when the price does not vary with the consumption flow nor through time;
2. $U\{\dot{\mathbf{x}}(.)\} = \int_0^t u\{\dot{\mathbf{x}}(\tau)\}\, d\tau$, i.e., the utility functional can be written as the integral of a "utility" function $u\{.\}$ which depends solely upon the quantity consumed at the instant τ;
3. Preferences are strictly convex.

The constant consumption path $\dot{\mathbf{x}}^*$ is optimal and, from equation (3), one has $\dot{\mathbf{x}} = \frac{\mathbf{X}}{t}$ and the disaggregated program can be written as:

$$V\{\mathbf{p}(.), y - A_n\} = \max_{\dot{\mathbf{x}}}[U(\dot{\mathbf{x}})| \mathbf{p}^{\tau}.\dot{\mathbf{x}}) \leq \frac{y}{t}] \tag{8}$$

which formally coincides with the first program it t is conventionally equated to unity.

The new modelization framework that we shall introduce in section 1.2, namely the stochastic commodity model, precisely provides a way of aggregating into a distribution pattern the elementary consumption acts that are preformed by an individual over time.

1.2 The General Formulation of the Stochastic Commodity Model

As in the disaggregated neoclassical model, a consumption intensity function $\dot{\mathbf{x}}(t)$ is considered in the stochastic utility model. However, in this latter model contrary to the former, the observed consumption intensity is not assumed to be a demand function deriving from program (6) of utility maximization, but is rather assumed to be the realization of some stochastic process. For the sake of simplicity, this consumption process will be supposed to be uncorrelated over time and stationary. Then, at every instant, the individual draws his/her consumption intensity out of the consumption space according to some joint probability density function $f(\dot{\mathbf{x}})$, which is defined over that space and invariant over time.

In the stochastic commodity model, as the series of elementary consumption acts is not seen as being a deterministic sequence due to be optimized *ex ante*, but as being instead a probabilistic sequence generated from a stochastic process, one has now to specify the level at which the optimization behavior takes place. It is postulated that the information which is optimized by the individual is the one embedded in the probability distribution function $f(\dot{\mathbf{x}})$. Hence, in the stochastic commodity model, the objective function is no longer set as a utility functional $u(\dot{\mathbf{x}}(.))$ of the consumption intensity, but it is made a functional $U(f(.))$ of the density function according to which the actual consumptions are drawn. For the sake of convenience, and though the perspective is now quite different than the one in the standard neoclassical approach, we shall still refer to $U(f(.))$ as to a utility functional. It should nevertheless be borne in mind that the preferences ordering represented by $U(f(.))$ does not operate directly on the consumption space, but indirectly on the space of the probability densities over that space. Such utility functionals acting on probability distributions have already been considered in economics, namely in the literature on choice under uncertainty. In section 2.1, we shall recall this literature and further discuss the meaning of utility as the representation of a preferences ordering among distributions. However, an intuitive understanding of the expression $U(f(.))$ is sufficient to the present development of the stochastic commodity model.

The shift in the analysis from the consumption space toward the probability distribution space, which characterizes the stochastic commodity model, is associated with a loss of information and thus yields the aggregation pattern mentioned in section 1.1. More precisely, the model states that the individual does not consider *ex ante* each of his/her purchases over the reference period separately, but that he/she aggregates them into classes, every purchase in a given class corresponding to a given level of consumption: two purchases are not distinguished *ex ante* if they lead to buy the same quantity. The *ex-ante* optimization procedure then consists in determining the frequency at which each level of consumption will be selected, when making purchases over the course of time. Provided that the process which generates purchases can be assumed to be stochastic, uncorrelated over time, and stationary, this amounts to predetermining an optimal probability distribution function $f^*(\dot{\mathbf{x}})$.

The optimization is subject to the constraint that the integral over time of the outlay intensity expectation ought not exceed the disposable income. Under the non-correlation and stationarity assumptions, this results in:

$$\int_0^T dt \int r(\dot{\mathbf{x}}(t))\, f(\dot{\mathbf{x}}(t))\, d\dot{\mu} \leq y \tag{10}$$

or, in an equivalent way, in:

$$\int r(\dot{\mathbf{x}})\, f(\dot{\mathbf{x}})\, d\dot{\mu} \leq \frac{y}{T} \tag{11}$$

where

$\dot{\mathbf{x}}$ *is the vector of consumption intensity*
$f(.)$ *is the joint density function defined on the consumption space*
$r(.)$ *is the outlay intensity function*
$d\mu = d\dot{\mathbf{x}}_1 ... d\dot{\mathbf{x}}_n$ *is the Lebesgue measure of integration in the* $\dot{\mathbf{x}}$*-space*
$\mathbf{y}$ *is the disposable income*

Eventually, the optimization program in the stochastic commodity model can be written:

$$V(r(.), y) = \underset{f(.)}{Max}\, [U(f(.)) \mid \int r(\dot{\mathbf{x}})\, f(\dot{\mathbf{x}})\, d\dot{\mu} \leq \frac{y}{T}] \tag{12}$$

where $V(r(.), \mathbf{y})$ is the indirect utility functional associated with U.

The stochastic commodity model, as summarized by equation (12), can be shown to be a generalization of the standard neoclassical model, as summarized by equation (9). More precisely, it appears that (12) degenerates into (9) when the utility functional $U(f(.))$ in the stochastic commodity model can be written as the expectation, with respect to the probability distribution $f(.)$, of the utility function $u(.)$ in the neoclassical model, i.e. when:

$$U(f(.)) = \int u(\dot{\mathbf{x}})\, f(\dot{\mathbf{x}})\, d\dot{\mu} \tag{13}$$

Technically, the result is due to the fact that when (13) holds, then program (12) consists in optimizing with respect to $f(.)$ a linear functional subject to a linear constraint. The solution of such a linear program is:

$$f^*(\dot{\mathbf{x}}) = \delta(\dot{\mathbf{x}} - \dot{\mathbf{x}}^*) \tag{14}$$

where $\delta(.)$ is the Dirac distribution and $\dot{\mathbf{x}}^*$ is the solution of program (9).

Thus, in the case of a linear utility functional, the stochastic utility model yields a sharp peaked distribution $\delta(.)$ for the consumption intensity $\dot{\mathbf{x}}$. As a result, the consumption process is uniform over time, i.e. the consumption intensity $\dot{\mathbf{x}}^*$ is selected at every instant with probability one. In brief, in this linear case, the stochastic commodity model conveys no more information than the aggregate form (9) of the neoclassical model: both models yield a constant consumption intensity $\dot{\mathbf{x}}^*$ over the period (O, T) or, in an equivalent way, yield the global consumption $\mathbf{x}^* = T\dot{\mathbf{x}}^*$ over that period.

Then degenerating of the stochastic commodity model into the neoclassical model requires further comments. In fact, according to equation (13), everything takes place as though the utility functional yielded by distribution $f(.)$ were the expectation of a sub-utility function $u(.)$, which is directly defined on the space of consumption intensities $\dot{\mathbf{x}}$, and which does not depend on the form of the distribution $f(.)$. It is the very existence of such a sub-utility intresequely acting on the consumption space which causes the model to degenerate. Convesely, this means that in order to be more informative than the standard neoclassical analysis, the stochastic commodity model must involve a utility functional $U(f(.))$ which depends on$f(.)$ in a non-linear way, i.e. which cannot be written as the expectation of some sub-utility. In section 2.1, we shall discuss in more details the type of assumptions under which such a non-linearity is achieved. Non-linearity, which thus appears as an essential feature of the stochastic commodity model, amounts to postulate that the items to be evaluated by the individual when optimizing *ex-ante*, are not directly the consumptions (i.e. there exists in general no sub-utility function $u(\dot{\mathbf{x}})$), but they are instead the probability distributions according to which consumptions will be drawn during the forthcoming period. In brief, the individual optimizes the stochastic procedure which rules his/her consumption, rather than this consumption itself.

When $U(f(.))$ is a non linear functional, then program (12) can be solved for a non-trivial probability distribution $f^*(\dot{\mathbf{x}})$, according to which the consumption intensity $\dot{\mathbf{x}}(t)$, no longer uniform over time, is drawn at every instant. More precisely and through a variational calculus, the optimal distribution can be shown to satisfy the following generalized Roy's identity, to be compared with (7):

$$\int \delta r(\dot{\mathbf{x}})\, f(\dot{\mathbf{x}})\, d\dot{\mu} = -\frac{\delta_t V}{V_y} \tag{15}$$

where:

$\delta r(.)$ *is the variational increment of the outlay function*
$\delta_t V$ *is the variation of the indirect utility functional* V *with respect to function* $r(.)$
V_y *is the marginal utility of income*

When the density function $f(.)$ is restricted to belong to some given class of probability distributions depending on a vector of parameters λ, then the stochastic commodity model can be formulated as a usual neoclassical model where λ takes the place of the consumption vector. More precisely, let:

$$f(.) \equiv \phi(\boldsymbol{\lambda}, .) \tag{16}$$

Let:

$$\hat{U}(\boldsymbol{\lambda}) \equiv U(\phi(\boldsymbol{\lambda}, .)) \tag{17}$$

and let:

$$\hat{r}(\boldsymbol{\lambda}) \equiv \frac{1}{T} \int r(\dot{\mathbf{x}})\, \phi(\dot{\mathbf{x}}, \boldsymbol{\lambda})\, d\dot{\mu} \tag{18}$$

Then, program (12) can be written:

$$\hat{V}(\boldsymbol{\lambda}, y) \equiv \underset{\boldsymbol{\lambda}}{Max}[\hat{U}(\boldsymbol{\lambda}) |\, \hat{r}(\lambda) \leq y] \tag{19}$$

which is similar to the standard utility maximization program (1) and can be solved for an optimal $\boldsymbol{\lambda}^*$. According to specification (19), the goods that are demanded *ex-ante* by the individual appear to be the parameters which define the shape of the distribution of his/her future purchases. Those parameters being related to the moments of the distribution, it can be said as well that the individual buys *ex-ante* the statistics of his/her consumption: the mean, the variance..

Though not developed extensively, the stochastic commodity model has already been introduced in the literature on telecommunication demand, when attempting to modelize the distribution of telephone calls with respect to duration. Curien[3] specified a "continuous Cobb Douglas" utility functional of the following form:

$$U(f(.)) = \exp[\int_0^x \mu(\dot{x})\, Ln\, f(\dot{x})\, d\dot{x}] \tag{20}$$

where the scalar consumption intensity $\dot{x}$ represents the duration of a call and where the calls' distribution under a duration non-sensitive pricing, i.e. $u(\dot{x})$, was set to be exponential. In de Fontenay et al[4] a more general form was considered, namely:

$$U(f(.)) = \int_0^x u(f(\dot{x}), F(\dot{x}), \dot{x})\, d\dot{x} \tag{21}$$

where the cumulative distribution function $F(.)$ associated with the density $f(.)$ was assumed to be Weybull.

In our presentation of the stochastic commodity model, we have stated so far that the consumption intensity $\dot{\mathbf{x}}$ is the random variable which is of interest to the individual ex-ante. The model can also be formulated in an equivalent integral form, where the random variable relevant for the ex-ante economic decision is assumed to be the global consumption $\mathbf{x}$ over the reference period, rather than the consumption flow $\dot{\mathbf{x}}$. Denoting by $G(.)$ the probability law of $\mathbf{x}$, and by $R(.)$ a global outlay function, the optimization program can then be written:

$$V(R(.), y) = \underset{G(.)}{Max}[U(G(.)) | \int R(\mathbf{x})\, G(\mathbf{x})\, d\mu \leq y] \tag{22}$$

where $d\mu$ is the Lebesgue measure of integration $dx_1 \cdots dx_n$ in the $\mathbf{x}$-space. Such an aggregate specification of the model has been used by de Fontenay and White[5] in order to build a model of demand for telecommunication services.

2. THE EPISTEMOLOGICAL GROUNDS OF THE STOCHASTIC COMMODITY MODEL

2.1 The Link With Choice Under Uncertainty

The idea of introducing a preferences ordering among probability distributions, rather than directly among consumptions, is not really new in the economic literature and already came in with the theory of choice under uncertainty. In the latter theory (see Friedman and Savage 1948),[6] one considers that the individual receives a payment through a lottery, i.e. as the realization of some positive random variable s depending on the occurrence of states of nature and defined by a probability density $f(s)$. A proper set of axioms is then postulated, the chief one being the independence axiom or the sure thing principle. According to the latter, if two lotteries systematically lead to the same payments on some subset of the states of nature, then whatever are these common payments, the lotteries will be compared by the individual on the only basis of their outcomes on the complementary sub-set of states, where they do not coincide. The axioms being supposed to hold, it is further proven that a preferences ordering can be defined among the lotteries, which is represented by the following linear utility functional acting on the probability distribution $f(.)$:

$$U(f(.)) = \int_0^x u(s)\, f(s)\, ds \tag{23}$$

The interpretation of equation (23) is that everything occurs as though the consumer were given over the set of payments a cardinal system of preferences represented by function $u(.)$, and as though his/her resulting utility, when facing the lottery, were the expectation of the cardinal subutility $u(s)$ with respect to the distribution function $f(s)$.

Turning back to the stochastic utility model and considering that the random variable at stake is the consumption intensity $\dot{\mathbf{x}}$ instead of the payment s, then equation (23) is strictly analog to equation (13). As we have shown in section 1.2 that whenever the linear equation (13) holds, then the stochastic commodity model degenerates into the standard neoclassical model, the former model generalizes the latter only on the condition that a satisfactory meaning can be given to non linear functionals such as $U(f(.))$.

In this direction, an attempt has been recently made by Yaari[7] to modify the set of axioms in the standard theory of choice under uncertainty, in order to obtain a more general and non-linear form for the utility functional $U(f(.))$. To that purpose, the sure thing principle is replaced by the so called axiom of monotonic cancellation. The latter states that, whenever a given lottery is preferred to some other lottery, then adding the outcome of a third lottery to both outcomes of the first two lotteries does not alter the order of preference of the latter, provided that the added lottery is not a hedge against either the others (co-monotonicity). Under this axiom, and others borrowed to the standard theory of choice under uncertainty, it is proven that there exists a preferences ordering among lotteries which is represented as follows by a non-linear functional of the distribution function $f(.)$:

$$U(f(.)) = \int_0^x \hat{u}[\int_s^x f(\sigma)\, d\sigma]\, ds \tag{24}$$

where $\hat{u}(.)$ is a continuous non decreasing real function defined on [0, 1], such that $\hat{u}(0) = 0$ and $\hat{u}(1) = 1$. Yaari further shows that, as a consequence of the monotonic cancellation axiom, the marginal utility of wealth is constant in his model so that the non-linear form (24) makes it possible to account for situations where the individual is risk-averse, i.e. function $\hat{u}(.)$ is concave, although his/her marginal utility of wealth is not decreasing. Such a scheme could not be obtained within the standard expected utility framework, in which decreasing marginal of wealth and risk-aversion are bound together and both implied if function $u(.)$ featuring in equation (23) is concave. Removing the assumption of constant marginal utility of wealth, similar work to Yaari's wold presumably yield non linear functionals more general than (24).

The latter incursion into the literature on choice under uncertainty helps to understand the kind of assumptions which are needed in order that the utility can be put as a non-linear functional in the stochastic commodity model. Beyond that point however, the stochastic commodity model fundamentally departs from the framework of choice under uncertainty: whereas in the latter case the individual is assumed to have no control over the lottery which generates the states of nature, in the stochastic commodity model it is postulated on the contrary that the individual optimizes $U(f(.))$ with respect to the probability distribution $f(.)$. In the stochastic commodity model, the purchased commodities are assumed to appear stochastic to the eyes of the consumer when he has to optimize his economic behavior for the next period: *ex-ante*, the individual has not direct control over the realization of his/her consumptions, for these depend on the occurrence of unforeseeable states of nature, but he/she can at least perform some overall behavioral control by selecting the shape of the distribution function according to which his/her elementary consumption acts will be drawn during the reference period. Though these elementary acts are not intresequely of a probabilistic nature and will in time presumably give rise to some deterministic micro-decision procedures in reaction to the perception of states of nature, the stochastic commodity model states that these real time procedures are irrelevant to the individual, when making *ex-ante* an overall economic decision: all the information useful to such a decision is assumed to be summed up in the probability distribution of consumption, i.e. what is at stake in this decision is not the consumption itself but the procedure which generates that consumption.

2.2 Analogies With Models In Non-Economic Scientific Fields

The principles which found the stochastic commodity model in the context of economic theory are very similar to the ones which characterize wave mechanics within theoretic physics. Somehow, the stochastic commodity model can be seen as being to the neoclassical model of demand what wave mechanics is to newtonian mechanics. In newtonian mechanics, the hamiltonian function which describes the move of a particle depends on this particle's position in the state space (i.e. the cartesian product of the geometric space and of the speeds' space), whereas in wave mechanics the hamiltonian is an operator, no longer defined over the state space but over the space of probability distributions defined over that space. Similarly, whereas in the neoclassical microeconomic model an individual's utility depends directly on the position of this individual in the space of consumption, in the stochastic commodity model the utility depends on a probability distribution function according to which the position in the consumption space is then drawn. In wave mechanics,[8] the Schroëdinger's equation yields the distribution function of the particle's presence but cannot be solved for the precise trajectory of the particle in the state space. In the same way, the maximization of the utility functional in the stochastic commodity model yields the probability distribution of consumption, but not the consumption intensity function itself. Just as various trajectories may be observed for a given particle with given initial conditions, various sequences of purchases may be observed for an individual with a given stochastic demand function, depending on the realizations that occur when drawing according to the optimized probability distribution.

A metaphore borrowed to another field of theoretic physics may help to understand how, in the stochastic commodity model, the simulation of the consumption decisions involves two stages of analysis, namely: an aggregate level where the probability distribution is optimized, and a disaggregate level where the actual decisions are drawn according to that distribution. In fact, such a two-stage structure of modelization is introduced in statistical thermodynamics[9] where the physicist deals with two levels in describing a molecular gas: the microscopic level, at which each molecule could theoretically be identified individually, and the macroscopic level, at which the molecules can only be labeled by their energy level (two molecules with the same energy cannot be distinguished at that level). Any macroscopic state is defined by its associated distribution of energy levels (how many molecules belong to each level), whereas any microscopic state is specified by the precise allocation of all individual molecules into the set of energy levels. The *entropy* of a given macroscopic state is then defined as the number of microscopic complexions which are consistent with the realization of that state. Through a combinatorial calculus, the entropy appears as being a functional of the distribution function of energy, and can be seen as a measure of the likelihood of observing its associated macroscopic state: the higher is the entropy, the more numerous are the microscopic complexions leading to that state. Due to the very large number of molecules in any observable sample of gas, it can be shown that the most likely macroscopic state is characterized by so much a higher level of entropy than any other state that this state occurs with probability almost equal to unity. In brief, everything looks to the observer as though, at some aggregate level, the molecular system had maximized entropy *ex-ante* so as to produce an optimal distribution function of energy, while at a lower level, the individual molecules would draw their speeds and their spatial positions according to that probability distribution. Formulated in this way, the situation is very similar to the one incurred in the stochastic commodity approach to consumption decisions: in the latter case everything looks to the analyst as though, at an aggregate level, the individual had maximized utility *ex-ante* so as to produce an optimal distribution of consumption, while at a disaggregate leve, he/she were selecting the sequence of his/her elementary purchases under the overall control of the pre-optimized distribution. In the analogy, utility corresponds to entropy, elementary purchases to individual molecules, and consumption distribution to energy distribution.

Further metaphores or analogies could be developed between the stochastic commodity model in economics and models within other scientific fields, as for instance information theory or biology. In fact, the conceptual framework which underlies the stochastic commodity odel, and other similar models, is relevant whenever a scientist has to address the issue of characterizing hierarchically the system which is at study, i.e. whenever a boundary has to be drawn between two levels of description. On the one hand, an aggregate level where the behavioral laws (of economics, physics, biology,...) are applicable to the system, and on the other hand, a disaggregate level which is considered as being by definition beyond the scope of casual analysis, but which cannot be ignored, for it interacts with the aggregate level through an exchange of information. More precisely, the system's behavior may be seen in terms of an interactive communication between the two levels of the bi-hierarchical structure: at first, the disaggregate level provides the aggregate level with some global informations (labels) which are relevant to that latter level, i.e. variables such as the level of energy or the purchased quantity, in our previous examples; then, the aggregate level determines an optimal probability distribution function defined over those variables and transmits the resulting information to the disaggregate level; eventually, the disaggregate level behaves in a way which appears to the aggregate level as being stochastic and ruled by the optimization distribution.

In the latter general perspective, it is important to note that the probabilistic character of the disaggregate level of description is not necessarily meant as the reflection of an intresequely stochastic reality. This probabilistic character is rather a way for the analyst to specify the limit beyond which the behavior of the system (the individual, in our economic concern) will no longer be explicitly described as deriving from casual laws. This certainly does not imply that no such laws exist at the disaggregate level, which could well become the aggregate level of description in some more in depth analysis. It only means that identifying laws at the disaggregate level is out of the scope of the study and that the consequences of these laws can be efficiently described as being the realizations of some stochastic process. Then, this stochastic process somehow stands for the unexplicited laws of the disaggregate level. More precisely, when the process can be assumed to be uncorrelated and stationary over time, the generating probability distribution acts as an interface between the explicit and the implicit levels of the analysis: this distribution is both supposed to summarize properly the information embedded in the underlying laws of the disaggregate level, and to be relevant input to the optimization program which is performed at the aggregate level.

The latter considerations lead to perceive in a new manner the treatment of error in economic models. The standard way of introducing an error term is to assume the model to hold exactly (or at least up to some pre-specified approximation as with the flexible form approach), and to attribute the discrepancy between observed and fitted values to various sources of noise, such as errors of measurement, errors of specification and omitted variables. Traditionally, a random variable is used to take account of the observed discrepancy. This type of approach in modeling the error structure is fundamentally extrinseque: a deterministic model is first postulated on microeconomic grounds while some class of random variables is then postulated on statistical grounds, and one searches to minimize the variance over all random variables within that class so as to produce the best statistical fit, or to maximize the likelihood of the observed data conditionally on the model's parameters. In the stochastic commodity model, the perspective is different in the sense that the treatment of error is so to speak *internalized*, i.e. incorporated into the very structure of the model: the presence of a random term is not added *ex-post* as standing for the analyst's ignorance, but is part of the economic model itself and is essential to its formulation. As we already mentioned, the random character of the disaggregate level in the stochastic commodity model can be interpreted as the apparent manifestation of some hidden behavioral laws, which rule the disaggregate level but are not explicitly perceived at the aggregate level of the studied system, i.e. at the level of the analyst.

3. STOCHASTIC COMMODITY AS AN ALTERNATIVE TO STOCHASTIC UTILITY IN DISCRETE CHOICE

3.1 The Stochastic Utility Approach To Discrete Choice

Let us consider the economic situation of discrete choice where individuals have to chose their consumption within a set of n exclusive options $k = 1 ,..., n$. For a given consumer, say i, this choice may be irreversible, i.e. once he/she has selected an option the individual is bound to that option over the whole reference period, or it may be reversible, i.e. the individual may alternately select the available options over time. Assuming that the option selection process is uncorrelated and stationary over time and/or across the population, individual i's choice can be described by a discrete probability distribution $P_{i1} ,..., P_{in}$. When considering irreversible choices choice probabilities are revealed by the shares taken by the various options across the population at a given date, whereas in the case of reversible choices those probabilities are also observable at the individual's level by measuring the selection frequencies of the options. As just described, the discrete choice issue seems liable to be analyzed through the stochastic commodity concepts, as we discussed them in section 2.2. However, a different type of approach, namely the stochastic utility model, has been developed in the literature in order to handle that issue. In this section, we shall recall the principles of the stochastic utility model, while in the next section, we shall recall the principles of the stochastic utility model, while in the next section, we shall introduce the stochastic commodity model as an alternative approach to discrete choice and shall analayze the links between the two types of modelization.

In the stochastic utility model, the individual choices are explained both by the socio-economic characteristics of the consumers i, and by the attributes describing the eligible options k. If the preferences ordering was assumed to be deterministic over the consumption space, then, in accordance with the standard neoclassical framework, an individual of a given type would at first maximize his/her utility with respect to the non-optional consumptions, subject to the income constraint, and conditionally on the realization of each particular option k. This would result in the conditional indirect utilities:

$$v_{ik} = \psi_{ik}(y_i - p_k, \mathbf{q}) \tag{25}$$

where p_k is the price of option k (assumed to be a fixed access fee), y_i is the income of individual i, and $\mathbf{q}$ is the price vector of non-optional commodities. Now optimizing across options, all individuals of a given type i would select the same option, i.e. the one k_i^* which yields to them the highest indirect conditional utility, namely:

$$v_{ik^*} = \psi_i(y_i - p_1, \ldots, y_i - p_n, \mathbf{q}) = \underset{k}{Max}\, \psi_{ik}(y_i - p_k, \mathbf{q}) \tag{26}$$

where $\psi_i(\;)$ is the resulting indirect utility function of individual i. As one generally observes that for given measured characteristics choices may be different from individual to individual, or from date to date for a given individual, in order to account for that variability one is compelled to move out from the standard deterministic framework of neoclassical analysis. One way of doing so could be to adopt the stochastic commodity framework, as we shall show in section 3.2., while another way is to make stochastic the preferences ordering and its representative utility. In this later case, a random component ε_{ik} is added to conditional utilities, resulting in the following stochastic indirect utility function:

$$\tilde{\psi}_i(y_i - p_1, \ldots, y_i - p_n, \mathbf{q}) = \underset{k}{Max}[\psi_{ik}(y_i - p_k, \mathbf{q}) + \varepsilon_{ik}] \tag{27}$$

The randomness of utility generates inter-individual or inter-temporal differences in choices across a given socio-economic type i, and unables the analyst to compute the choice probability of a given option k as being the probability P_{ik}^* that this particular option happens to yield the highest utility to consumer i through the utility maximization process, i.e.:

$$P_{ik}^* = \Pi_{ik}(y_i - p_1, \ldots, y_i - p_n, \mathbf{q}) = Pr[\tilde{\psi}_i(y_i - p_1, \ldots, y_i - p_n, \mathbf{q}) = \tilde{\psi}_{ik}(y_i - p_k, \mathbf{q})] = E[\mathbf{1}_{ik}] \tag{28}$$

where $E[.]$ is the mathematical expectation operator and $\mathbf{1}_{ik}$ is the random variable which takes the value 1 when individual i demands option k, and the value 0 otherwise.

The usual rationale for the randomness of utility is as follows: some characteristics of the individuals, and/or some attributes of the options, that are relevant to the choice process have not been measured, so that the individual preferences are partly unknown to the analyst and in fact vary across a population of apparently identical individuals, or over time for a given individual. These discrepancies imply choice differences that are not explicitly explained by the model and that are subsumed in the random term ε_{ik} of utility. According to this view, the role of randomness in the stochastic utility model is very similar to the role of noise in the statistical model of regression, i.e. it stands as the representation of an error or an ignorance. Moving in the direction indicated in section 2.2, i.e. towards a more intrinseque treatment of error, some authors[10] have attempted to internalize the random component in the stochastic utility model by interpreting it not only as an extrinseque error term, but rather as an intrinseque character selecting the apparent randomness of the consumer's behavior. To that purpose, the individual is assumed to make impulsive and capricious choices in the sense that, prior to optimization, he/she would draw a particular utility function out of a continuum according to some probability measure. Once the utility function is drawn, the choice is supposed to be deterministic and rational in the standard neoclassical sense. The differences between choices made by similar individuals, or by a given individual over time, are then explained by differences in realizations when drawing utility.

In order to be comprehensive and to be tested statistically, the latter interpretation of the stochastic utility model should be made more explicit by specifying the nature of the stochastic process which generates the selection of utility. The stochastic commodity approach to discrete choice, that we shall develop in the next section, precisely allows the analyst to overcome this difficulty by changing the place of the random term from outside to inside the utility function: so doing, and according to the bi-hierarchical structure of analysis discussed in section 2.2, the stochastic character of a modelization is removed from the casual level of the analysis, i.e. the level of utility ex-ante maximization, and is places at the level where explicit rationality is no longer considered, i.e. the level of actually making purchases.

3.2 The Stochastic Commodity Approach To Discrete Choice

In the same direction as the one indicated by McFadden and Heckman[11] within the stochastic utility approach, the stochastic commodity approach intends to give a behavioral interpretation of the observed discrepancies in choices, which are considered as being an error term in the standard discrete choice analysis. However, as we already mentioned, the stochastic commodity approach fundamentally differs from the stochastic utility one: whereas in the latter the introduction of a utility noise ε_{ik} denies the deterministic character of preferences which is usually postulated in neoclassical analysis, the former approach essentially preserves the deterministic character of utility, while in return commodities become stochastic.

The basic principle of the stochastic commodity approach to discrete choice consists in assuming that the individual process of selecting an option can be broken down into two stages, the first stage being deterministic and the second one being stochastic: at first the consumer determines a discrete probability distribution $P_{i1},..., P_{in}$, i.e. he/she defines a lottery, according to which he/she then draws the option to be actually selected. The draft in the second stage of the consumption process accounts for the observed differences in choices between identical individuals, whereas in the first stage similar consumers (or the same consumer over the course of time) demand the same lottery. This optimal lottery $P_{i1}^*,..., P_{in}^*$ derives from the maximization of a deterministic direct utility function, defined over the space of choice probability distributions and over the space of non-optional consumptions, i.e.:

$$U_i = \Phi_i(P_{i1},..., P_{in}, \mathbf{x}_i) \tag{29}$$

where $\mathbf{x}_i$ is the consumption vector of individual i in non-optional commodities. The maximization of that utility is performed subject to the constraint that the outlay expectation does not exceed the disposable income, according to the following program:

$$\psi_i(y_i - p_1,..., y_i - p_n, \mathbf{q}) = \underset{P_{i1},..., P_{in},\mathbf{x}}{Mzx} [\Phi(P_{i1},..., P_{in}, \mathbf{x}_i) | \sum_{k=1}^{n} P_{ik}(y_i - p_k) - \mathbf{q}: \mathbf{x}_i \geq 0] \tag{30}$$

The optimal choice frequencies then derive from program (30) through Roy's identity, i.e.:

$$P_{ik}^* = \Pi(y_i - p_1,..., y_i - p_n, \mathbf{q}) = -\frac{\partial \psi_i}{\partial p_k} / \frac{\partial \psi_i}{\partial y_i} = \frac{\psi_{i(k)}}{\sum_{m=1}^{n} \psi_{i(m)}} \tag{31}$$

where $\psi_{(i(m)}$ denotes the partial derivative of function ψ_i with respect to its mth argument.

Although the perspectives are different, a close link exists between the two approaches to discrete choice, respectively the Stochastic Utility Model (SUM) and the Stochastic Commodity Model (SCM). More precisely, under some restrictions imposed to the form of the individual utility function, a SUM can be shown to be associated with an equivalent SCM. To that purpose, let us consider a SUM in which the stochastic indirect utility is additively separable with respect to income, i.e. a SUM in which equation (27) can be written in the following way:

$$\tilde{\psi}_i(y_i - p_i,..., y_i - p_n, \mathbf{q}) = \underset{k}{Max} \frac{y_i - p_k - \alpha_{ik}(\mathbf{q}) + \varepsilon_{ik}}{\beta_i(\mathbf{q})} \tag{32}$$

Then, by averaging this indirect utility across the sub-population of individuals of type i, i.e. by taking the mathematical expectation, one derives:

$$\bar{\psi}_i(y_i - p_1,..., y_i - p_n, \mathbf{q}) = E[\tilde{\psi}_i(y_i - p_1,..., y_i - y_n, \mathbf{q})] = \frac{y_i + G_i(p_1,..., p_n, \mathbf{q})}{\beta_i(\mathbf{q})} \tag{33}$$

in which:

$$G_i(p_1,..., p_n, \mathbf{q}) = -E[\underset{k}{Max}(p_k + \alpha_{ik}(\mathbf{q}) - \varepsilon_{ik})] \tag{34}$$

is a social surplus function for a sub-population i, associated with the social utility $\bar{\psi}_i$. It follows from equations (34) and (28) that:

$$\frac{\partial G_i}{\partial p_k} = -E[\mathbf{1}_{ik}] = P_{ik}^* \tag{35}$$

and then, from equation (33):

$$P_{ik}^{*} = -\frac{\partial \overline{\psi}_i}{\partial p_k} / \frac{\partial \overline{\psi}_i}{\partial y_i} \quad (36)$$

Comparing equations (31) and (36), it eventually appears that the social utility $\overline{\psi}_i(\ )$ can be identified with the indirect utility function of a stochastic commodity model equivalent to the initial stochastic utility model. Therefore, everything is as though every individual within the segment of population i were given the social utility function $\overline{\psi}_i(\ )$, or the social surplus function $G_i(\ )$, and were deriving choice frequencies from these functions according to (35) or (36). In brief, the initial SISUM (Separable Income Stochastic Utility Model) has been exactly re-formulated in the terms of an equivalent SFSCM (Surplus Function Stochastic Commodity Model). Conversely, an equivalent SISUM can be associated with any SFSCM. Although interpreted here is an original way, this result is not new and is known in the literature on discrete choice as the Williams-Daly-Zachary theorem.[12]

It is also proven in the Williams-Daly-Zachary theorem that a SISUM and its equivalent SFSCM yield a Translation Invariant Probability Choice System (TPCS), i.e. choice probabilities remain unchanged when the prices of all options are raised by the same amount. First noting from (34) that the surplus function is additive, namely:

$$G_i(p_1 + \theta, ..., p_n + \theta, \mathbf{q}) = G_i(p_1, ..., p_n, \mathbf{q}) = \theta \quad (37)$$

Then the stability of choice frequencies under a price translation immediately follows from (35). Conversely, whenever a Probability Choice System (PCS) satisfies the TPCS condition, then it derives from a SISUM or from its equivalent SFSCM. It finally results that, when empirically observed choice frequencies can be checked to be TPCS, then they indistinctively lead whether to a SISUM or to a SFSCM. In this particular case, the stochastic utility and the stochastic commodity backgrounds turn out to be two dual and equivalent theorizations of reality.

However, in the general case where choice frequencies are not TPCS, the SUM and SCM depart from one another: a SUM cannot necessarily be transformed into a SCM and conversely, unless these models respectively belong to the restricted sub-classes of SISUMs and SFSCMs. As a consequence, one should be able to differentiate the types of models, namely SUM and SCM, when testing them from empirical data on choice frequencies. One knows[13] that any PCS, unless it is a TPCS, is not systematically consistent with stochastic utility maximization. However, no analytical integrability conditions similar to the ones existing in the standard neoclassical analysis have been so far produced, that could be used to prove an observed PCS to be consistent with a SUM. Though a necessary and sufficient condition for consistency is given by the theorem of revealed stochastic preferences,[14] the practical implementation of this condition is uneasy. This is the reason why most of the SUMs which are developed in the econometric literature (Logit, Probit, Preferences tree model..) belong the sub-class of SISUMs, where the income separability of utility and the stability of choices under a translation of prices are assumed to hold. Less difficulties in the empirical validation should be met in the stochastic commodity approach: as in the case the utility maximization is deterministic according to program (30), integrability conditions similar to the ones in the neoclassical model can be derived, from which one could check whether or not a PCS, as a function of prices and income, is consistent with a SCM.

In quite a different domain of consumption, the individual would probably not plan in advance with accuracy when exactly he will go to the restaurant and what will be the expense each time; he/she would rather plan the global number of times he/she will go for the next period and care as well for an approximate range in which the bill should fall each time, i.e. he/she would consider the moments of the distribution of his/her elementary purchases.

FOOTNOTES

1. See for instance Varian, Hal R. (1984)[1].
2. In addition it will also be assumed that commodities are not stored by the consumer during part of the period, i.e. that they are consumed as they are purchased, and that there is no intra-period saving.
3. Curien, Nicolas (1981), ''Modelisation de l'Effet des Tarifs sur la Consommation et le Trafic, et Application a Etude de la Taxation des Communications Locales a la Duree'', Working Paper,

Direction Generale des Telecommunications, paris.

4. Fontenay, Alain de, Debra Gorham, J. T. Marshall Lee, and Goerge Manning (1982), "Stochastic Demand for a Continuum of Goods and Services: the Demand for Long Distance Telephone Services", Paper presented at the meeting of the American Economic Association, New York.

5. de Fontenay, A. and R. G. White (1985), "A Model of Access and Usage Demand (MAUD)", paper presented at this conference.

6. Friedman, M. and L. J. Savage (1948), "The Utility Analysis of Choices Involving Risk", Journal of Political Economy, Vol. 56 pp. 279-304.

7. Yaari, M. E. (1985), "Risk Aversion Without Diminishing Marginal Utility, or the Dual Theory of Choice under Risk", Research Memorandum $N^0$65, Center of Research in Mathematical Economics and Game Theory, The Hebrew University, Jerusalem.

8. See for instance: Pauling, L. and E. Bright Wilson, J. R. (1935), Introduction to Quantum Mechanics with Applications to Chemistry, Reprent, Dover Publication Inc., 1985, New York.

9. See for instance: Rocart, Yves.....

10. See McFadden, D. (1981), "Economectric Models of Probabilistic Choice" and Heckman, J.J. (1981) "Statistical Models for Discrete Choice Data" in *Structural Analysis of Discrete Data with Econometric Applications*, Ed. by Manski, C. F. and D. McFadden, The MIT Press, 1981.

11. See: Manski, C. F. and D. McFadden, op. cit.

12. See: Manski, C. F. and D. McFadden (1981), op. cit.

13. See: Manski, C. F. and D. McFadden (1981), op. cit.

14. Se: McFadden, D. and M. Richter (1979), "Revealed Stochastic Preferences", Mimeographed, Department of Economics, Massachussets Institute of Technology.

REFERENCES

[1] Varian, Hal R. (1984), *Microeconomic Analysis*, 2nd edition, Norton, New York.

[2] Deaton, A. and J. Muelbauer, *Economics and Consumer Behavior*, Cambridge University Press, 1980.

SECTION I:
END USER DEMAND

I.3. Econometrics/Discrete Choice

TELECOMMUNICATIONS DEMAND MODELLING
An Integrated View
A. de Fontenay, M.H. Shugard, D.S. Sibley (Editors)

RESIDENTIAL DEMAND FOR ACCESS TO THE TELEPHONE NETWORK

Lester D. Taylor & Donald Kridel*

1. INTRODUCTION

The protection of a loosely defined concept -- universal service -- is a center of debate in the telecommunications industry. Most regulators and telephone companies define universal service in terms of the 1934 Communications Act:

> "... to make available, so far as possible, to all people of the United States a rapid efficient nationwide and worldwide wire and radio communictions service with adequate facilities at reasonable charges".

In practice, however, universal service seems to mean the level of service that we have today, and anything less is no longer universal service. Bypass, competition, divestiture, and access charges are viewed with alarm, for they threaten the subsidy from the toll market that has historically allowed local service, especially for residential customers, to be priced below cost. Most regulators have steadfastly refused to allow local rates to move toward cost, because they feel that universal service will be destroyed.

That higher local rates will cause some telephones to be disconnected is an accepted conclusion. Existing empirical evidence shows the price elasticity of network access demand to be small, yet different from zero. But this is only part of the story. The evidence also shows that the elasticity varies inversely with income, and regulators and consumer groups are concerned that the burden of higher local rates will fall disproportionately on certain socio-demographic groups with low incomes.

At present, most discussions of the quantitative effects of high local rates make use of the residential access demand model developed by Lewis Perl [1]. This model, which is an update of an earlier effort of Perl [2], was purposely designed with a long list of socio-economic-demographic groups in mind. The Perl model is a probit/logit model estimated with data for over 80,000 individual households from the Public Use Sample of the 1980 Census.

Ordinarily, a sample of this size is the stuff of dreams, but in this case there is a problem with nondisclosure. To forestall the identifying of any individual household, the Census disclose place of residence only

---

*The authors are at the University of Arizona and Southwestern Bell Telephone Company, respectively. We are grateful to D.R. Dolk for programming and use of his econometric program PERM and to Mary Flannery for word processing.

for areas of at least 100,000 population. This creates major problems in matching rates with households, as in virtually all cases areas of 100,000 or more population contain several wire centers. Since the rate data refer to wire centers, this means that it is impossible with the Public Use Sample to get an accurate matching of households and rates. The consequences of this may be unimportant when the focus is national, but problems clearly emerge where attention shifts to specific socio-demographic groups and areas.

An alternative to Perl's procedure is to use data aggregated to the level of a census tract. The census tract is the smallest geographical area that can be analyzed with census data, and provides a better match between rates and place of residence than is possible using data for individual households. Although aggregation necessarily leads to a loss of information, census tracts are sufficiently diverse in income levels and socio-demographic characteristics that the parameters of most interest can continue to be isolated.

The present paper, accordingly, reports on a model of residential access demand that uses census tracts as the unit of observation. The model has been developed at Southwestern Bell for use in estimating the impacts of higher local-service rates on its residential customers. The model that is estimated is a probit model similar to Perl's, but with an extension that takes into account differences in the distribution of income across census tracts. The model is estimated from a data set consisting of 8423 census tracts in the five states (Arkansas, Kansas, Missouri, Oklahoma and Texas) served by Southwestern Bell. The results corroborate existing views regarding the price elasticity for residential access demand. Perl's findings with regard to the importance of socio-demographic factors are also confirmed, but the present model provides a better vehicle for quantifying specific effects.

2. A PROBIT MODEL OF RESIDENTIAL ACCESS DEMAND

The point of departure for modeling telephone demand is the distinction between access and use. Use of the telephone network obviously requires access, but whether or not access is demanded depends upon the net benefits from use. Benefits in this context are most conveniently measured by consumer's surplus. Following Perl [1], we assume that the demand for use is given by

$$(1)\qquad q = Ae^{-\alpha p}y^{\beta}e^{u},$$

where q denotes use of the telephone network, p denotes the price of use, y denotes income, and u denotes a random error term.[1] We ignore at this point complications of network and call externalities, and we also ignore for now socio-demographic factors.

The consumer's surplus from q units of use will be given by

[1]Use for present purposes refers to local calling. The demand function in (1) accordingly allows for both flat-rate (p=0) and measured (p≠0) service.

$$(2)\qquad CS = \int_{p}^{\infty} Ae^{-\alpha z} y^{\beta} e^{u} dz$$

$$= \frac{Ae^{-\alpha p} y^{\beta} e^{u}}{\alpha},$$

or in logarithms,

$$(3)\qquad \ln CS = a - \alpha p + \beta \ln y + u,$$

where $a = \ln A/\alpha$. Let π denote the price of access. Access will be demanded whenever

$$(4)\qquad \ln CS > \ln \pi,$$

i.e., when

$$(5)\qquad a - \alpha p + \beta \ln y + u > \ln \pi,$$

or alternatively when

$$(6)\qquad u > \ln \pi - a + \alpha p - \beta \ln y.$$

If u is $N(0,\sigma^2)$, the probability of access will accordingly be given by

$$(7)\qquad P(\text{access}) = 1 - \Phi\left(\frac{\ln \pi - a + \alpha p - \beta \ln y}{\sigma}\right),$$

where Φ denotes the distribution function for the unit normal distribution. This is essentially the model that is applied by Perl to individual households.

However, with census tracts as the unit of observation, equation (6) must be aggregated over the households in a tract. Our procedure will be to aggregate equation (6) over the joint distribution of u and y, where we assume that y is lognormally distributed, independent of u, with mean μ and variance σ_y^2. Let $v = u + \beta \ln y$. Under our assumptions, it follows that v will be $N(\beta\mu, \sigma^2 + \beta^2\sigma_y^2)$. Let P_j denote the proportion of households in census tract j that have a telephone. This proportion will be given by

$$(8)\qquad P_j = P(v_j > \ln \pi_j - a + \alpha p_j),$$

or equivalently

$$(9)\qquad P_j = P(w_j > w_j^*)$$

$$= 1 - \Phi(w_j^*),$$

where:

$$(10)\qquad w_j = \frac{V_j - \beta\mu_j}{(\sigma^2 + \beta^2\sigma_{yj}^2)^{1/2}}$$

$$(11)\qquad w_j^* = \frac{\ln \pi_j - a + \alpha p_j - \beta\mu_j}{(\sigma^2 + \beta^2\sigma_{yj}^2)^{1/2}}.$$

Let F_j denote the probit for census tract j calculated from equation (9), and write:

(12) $$F_j = \frac{\ln \pi_j - a + \alpha p_j - \beta\mu_j}{(\sigma^2 + \beta^2\sigma_{yj}^2)^{1/2}} + \varepsilon_j .$$

Clearing the fraction, we have

(13) $$F_j(\sigma^2 + \beta^2\sigma_{yj}^2)^{1/2} = \ln \pi_j - a + \alpha p_j + \varepsilon_j^* ,$$

where $\varepsilon_j^* = (\sigma^2 + \beta^2\sigma_{yj}^2)^{1/2} \varepsilon_j$.

3. ESTIMATION

Equation (13) offers a number of challenges for estimation, but before discussing these the fact that customers in some areas can choose between either flat-rate or measured service must be taken into account. Three possibilities must be considered:

(1). Only flat-rate service is available;

(2). Only measured service is available;

(3). Both flat-rate and measured service are available.

In flat-rate only areas, access will be demanded if [from equation (6), since p = 0]

(14) $$u + a + \beta \ln y > \ln \pi_f ,$$

while in measured-rate only areas, access will be demanded if

(15) $$u + a + \beta \ln y > \ln \pi_m + \alpha p ,$$

where π_f and π_m denote the monthly fixed charges of flat-rate and measured services, respectively. In areas where both flat-rate and measured service are options, access will be demanded if *either* (14) *or* (15) hold. From these two inequalities, we see that access will be demanded if

(16) $$u + a + \beta \ln y > \min(\ln \pi_f, \ln \pi_m + \alpha p) ,$$

and that, given access is demanded, flat-rate will be selected if

(17) $$\ln \pi_f < \ln \pi_m + \alpha p ,$$

while measured-rate will be chosen if

(18) $$\ln \pi_m + \alpha p < \ln \pi_f .$$

While choice is deterministic (given the estimate of α), bill minimization is *not* assumed. Thus, flat-rate expenditures may exceed measured-rate expenditures, but equation (17) could still hold. To incorporate these choices, we let:

$$\delta_1 = \begin{cases} 1 & \text{if only flat-rate service is available} \\ 0 & \text{otherwise} \end{cases}$$

$$\delta_2 = \begin{cases} 1 \text{ if only measured service is available} \\ 0 \text{ otherwise} \end{cases}$$

$$\delta_3 = \begin{cases} 1 \text{ if both flat-rate and measured services are available} \\ 0 \text{ otherwise.} \end{cases}$$

With these dummy variables, we can rewrite equation (13) as

$$(19) \qquad F_j(\sigma^2 + \beta^2\sigma_{yj}^2)^{1/2} = \delta_1 \ln \pi_{fj} + \delta_2(\ln \pi_{mj} + \alpha p_j) + \delta_3 \min (\ln \pi_{fj}, \ln \pi_{mj} + \alpha p_j) - a - \beta\mu_j + \varepsilon_j^* .$$

This is the equation to be estimated.

Several things are to be noted about equation (19):

(a). The equation is nonlinear in β.

(b). σ^2, the variance of u, is a parameter to be estimated.

(c). Min $(\ln \pi_{fj}, \ln \pi_{mj} + \alpha p_j)$ must be calculated for the areas in which both flat-rate and measured service are available. To do this, however, requires knowledge of α.

We begin by rewriting the model as follows:

$$(20) \qquad z_j = - a - \beta\mu_j + \sum_i \gamma_i X_{ij} + \varepsilon_j^* ,$$

where:

$$(21) \qquad z_j = F_j(\sigma^2 + \beta^2\sigma_{yj}^2)^{1/2} - \delta_1 \ln \pi_{fj} - \delta_3 \min(\ln \pi_{fj}, \ln \pi_{mj} + \alpha p_j) ,$$

and where X_{ij} denotes the socio-demographic factors thought to be relevant.[2] If σ^2, α, and β were known, z_j could be calculated (using observed values for F_j, σ_{yj}^2, π_{fj}, π_{mj}, and p_j) and equation (20) could be estimated as a linear regression. However, these parameters are not known, and so an iterative scheme has been used.

The scheme consists of a sequential search procedure defined as follows[3]:

---

[2] As none of Southwestern Bells service territory has mandatory measured service, δ_2 is dropped from equation (19).

[3] A detailed description of the estimation logarithm, together with results of a Monte Carlo study evaluating it, is given in an appendix available from the authors. The results briefly are: (1) estimator is slightly biased; (2) bias falls slightly with sample size; and (3) variance of error term has little effect on accuracy of coefficients.

(i). The initial search is over values of σ^2 for fixed values of α and β. The value of σ^2 which is finally selected is the one which maximizes the correlation between the actual and predicted values of F_j.

(ii). The second step is to fix σ^2 at the value obtained in (i), keep β fixed as in (i), and search over α. Again the value of α selected is the one which maximizes the correlation between the actual and predicted values of F_j.

(iii). The final step involves iteration on β. An iteration consists of the estimation of equation (20), with z_j [equation (21)] calculated using the values of σ^2 and α obtained from (i) and (ii) and β from the preceding iteration. Iterations continue until a stable estimate of β is obtained.

4. EMPIRICAL RESULTS

The model summarized in equations (20) and (21) has been estimated, using the procedure just described, from a data set consisting of 8423 census tracts from the 1980 Census in the area served by Southwestern Bell in Arkansas, Kansas, Missouri, Oklahoma and Texas. The model actually estimated is as follows:

$$(22) \quad F_j(\sigma^2 + \beta^2\sigma_{yj}^2)^{1/2} - \delta_1 \ln \pi_{fj} - \delta_3 \min(\ln \pi_{fj}, \ln \pi_{mj} + \alpha p_j)$$

$$= -a - \beta\mu_j - \gamma_1 \text{ RENTER}_j - \gamma_2 \text{ RURAL}_j - \gamma_3 \text{ BLACK}_j$$

$$- \gamma_4 \text{ SPANISH}_j - \gamma_5 \text{ AMINDIAN}_j - \gamma_6 \text{ IMMOB}_j$$

$$- \gamma_7 \text{ AVGAGE}_j - \gamma_8 \text{ MILAGE}_j - \gamma_9 \text{ EMP}_j$$

$$- \gamma_{10} \text{ LINES} - \gamma_{11} \text{ AVESZHH}_j + \varepsilon_j \ ,$$

where:

F = probits corresponding to the proportion of households in a census tract that have a telephone

σ^2 = variance of error term for individual households (a nuisance parameter that must be estimated)

β = coefficient for income

σ_y^2 = within-census tract variance of income

δ_1, δ_3 = dummy variables as defined earlier

$\ln \pi_f$ = logarithm of flat-rate monthly service charge

$\ln \pi_m$ = logarithm of measured-rate monthly service charge

α = coefficient for price of a local call

p = price/minute for a local call

a = a constant

μ = geometric mean of income

RENTER = % of census tract that rents home

RURAL = % of census tract that is rural

BLACK = % of census tract that is black

SPANISH = % of census tract of hispanic origin

AMINDIAN = % of census tract that is American Indian

IMMOB = % of census tract that has not moved since 1975

AVGAGE = average age of census tract

MILAGE = % of census tract that pays a local-loop mileage charge

EMP = % of census tract that suffered no unemployment in 1979

LINES = number of lines that can be reached in "local-calling area"

AVESZHH = average number of people per household

ε = error term.

Before presenting the empirical results, some comments on other models that have been estimated are in order. At the beginning of the empirical analysis, the three standard discrete choice models -- linear probability, probit, and logit -- were estimated, but without taking into account differences in the variance of income across census tracts. The results were surprising, especially with regard to the price of access. The access price was significant in the linear probability model, but not in the probit or logit models. For reasons relating to the strength of the poverty in the earlier analysis, we singled out our simplistic treatment of income as the cause of this disturbing result. Considering the impact of the poverty point on the income distribution, we settled on the present model in which the full distribution of income is taken into account.

The estimated coefficients for equations (22) are given in Table 1. Except for the constant, the t-ratios are seen to be well in excess of two. Mileage has the smallest t-ratio (3.8), while the proportion of households that are black has the largest (24.9). The t-ratio for income is 19. T-ratios for the price coefficient and the "nuisance" variance (σ^2) cannot be obtained directly because of the way they are estimated.[4] As expected,

[4]Efforts to obtain standard errors for these coefficients by a jackknife technique have been attempted, but the results are too preliminary to report.

Table 1

Coefficient Estimates

Variable	Coefficient	t-statistic
Constant	.43	1.2
INCOME	.98	19.0
PRICE OF LOCAL MINUTE	-6.87	--
PRICE OF ACCESS	-1.00	--
RENTER	-1.56	14.3
RURAL	-.82	19.0
BLACK	-1.18	12.9
SPANISH	-2.78	24.9
AMINDIAN	-7.38	14.2
IMMOB	.51	4.7
AVGAGE	.05	9.9
MILAGE	-.45	3.8
EMP	2.88	14.3
LINES	.37	4.8
AVESZHH	.55	10.3
Nuisance Variance (σ^2)	5.6	--

$R^2 = 0.420$*

*The R^2 is calculated as the square of the correlation between the actual and predicted values for F_j.

the sign for income is positive, while the signs for the price terms are negative. Mileage charges also have a negative effect. Regarding the socio-demographic factors, development (i.e., the proportion of households having a telephone) is lower for black, hispanic, or American Indian households, lower for renters than for homeowners, and lower in rural areas than in urban. Unemployment also has a negative effect, while address longevity (IMMOB) has a positive effect, as do average age and the number of lines in the local-calling area.

Care is required in interpreting the coefficients. The coefficient for income, 0.98, represents the elasticity for *usage* with respect to income. Likewise, 0.51 times the mean (0.52) of IMMOB is the usage elasticity (0.27) with respect to the mobility variable, and similarly for the other variables. The *access* elasticities, on the other hand, cannot be observed directly, but must be induced by changing a variable and then calculating the resulting change in development. For example, consider a 10% increase in income. This yields a 0.42% increase in development, and hence an access elasticity of 0.042 with respect to income.[5]

Of special interest is the elasticity of access with respect to the number of lines in local-calling area, for this can be interpreted as a measure of the subscriber (or network) externality. The externality appears to by tiny at present levels of development for an increase of 10% in the number of lines implies an increase of 0.027 in development. Taken at face value, this small subscriber externality offers virtually no support for large access subsidies.

We turn now to the main purpose of the study, which is to quantify the impact of higher access charges on development. Tables 2-4 show the predicted effects of some of the most widely discussed pricing scenarios. Table 2 shows the predicted effects on development of a doubling of flat-rate and measured-rate access charges in each of the five states served by Southwestern Bell. The predicted impacts across several socio-economic-demographic groups in Texas are shown in Table 3.[6] Finally, Table 4 shows predicted impacts on development of the FCC-mandated end-user subscriber-line charge (EUCL) of $1.

Tables 2-4 make several points. In the first place, it is clear that access repression, while small, is not zero, and it is in the best interest of the telephone companies to acknowledge this up front in any regulatory proceeding. The threat to universal service has received tremendous attention from consumer groups, Congress, and state regulators,

---

[5]Two points should be kept in mind in making this calculation. First, since the variance of income is also an argument in the model, only small changes in income should be considered. Second, the implied income elasticity may seem small at first glance. However, since 93% of households (on the average) already have access, an increase in income can have no effect on their access decision.

[6]Similar tables could be prepared for the other states. Texas was chosen since it had the highest measured service availability. Hence, the effect of having lower priced alternative could be more readily seen.

Table 2

Development and Repression for 100% Access Price Increases*

	Actual Development	100% increase with lower priced alternative**	100% increase all prices***
AR	89.1	83.6 (-5.5)	83.2 (-5.9)
KS	95.3	--	93.0 (-2.3)
MO	95.4	92.3 (-3.1)	92.3 (-3.1)
OK	92.7	--	89.3 (-3.4)
TX	91.4	89.3 (-2.1)	87.7 (-3.7)
SWBT	92.5	89.6 (-2.9)	88.8 (-3.7)

Numbers in parenthesis are repression estimates in basis points.

*No measured service was available in Kansas or Oklahoma. Measured availability for the other three states was as follows: AR: 10.4%; MO: 2.2% and TX: 58.4%.

**Measured service, where available, is the lower priced alternative and its price is unchanged.

***Both flat rate and measured rate, where available, are doubled.

Table 3

Development and Repression for 100% Access Price Increases for Texas

	Actual Development	100% Increase for lower priced alternative*	100% Increase all prices*
POOR, RURAL**	80.4	73.9 (-6.5) [-6,400]	73.3 (-7.1) [-6,900]
POOR, URBAN	83.1	79.6 (-3.5) [-24,000]	75.4 (-7.7) [-53,300]
POOR, NONWHITE RURAL	75.5	67.5 (-8.0) [-3,800]	66.6 (-8.9) [-4,200]
NOT POOR, NOT NONWHITE, URBAN	94.6	93.7 (-0.9) [-20,300]	92.0 (-2.6) [-60,200]

Numbers in parenthesis are repression estimates in basis points. Numbers in brackets are estimates of number of housedholds affected.

*Same definitions as in Table 2

**"Poor" refers to the poorest 25% of the tracts. "Not poor" refers to other 75%, while "rich" would refer to the richest 25% of the tracts and all rural tracts. The other groups are defined similarly.

Table 4

Impact of "Current" FCC Access Charge Plan
Southwestern Bell

	Actual Predicted	$1 EUCL	$1 EUCL Poor exempted
Total # subscribers	6,608,500	6,569,300	6,579,900
ALL			
development	92.5	92.0	92.1
(change)		(-0.5)	(-0.4)
POOR, RURAL			
development	83.7	82.6	82.9
(change)		(-1.1)	(-0.8)
POOR, URBAN			
development	85.0	84.0	84.3
(change)		(-1.0)	(-0.7)
POOR, NONWHITE, RURAL			
development	78.5	77.0	77.5
(change)		(-1.5)	(-1.0)
NOT POOR, NOT NONWHITE, URBAN			
development	95.3	95.0	95.1
(change)		(-0.3)	(-0.2)

and any attempt by the telephone companies to increase local rates will surely bring an emotional response from at least one of these groups. If a filing is not *accompanied* with the telephone company's estimates of repression, together with a plan to minimize it, any subsequent response will appear hurried and politically motivated.

There is no denying (see rows 2-4 in Table 3) that certain socio-economic-demographic groups are more adversely affected by higher access prices than others. However, these groups do not appear to be as large as the emotional and political debates suggest. A lifeline offering, or other targeted subsidy, could substantially reduce the impacts on these groups. (Compare columns 2 and 3 in Table 3 and columns 2 and 3 in Table 4.) Any lower-priced alternative, as seen in Table 2 for Texas and Oklahoma, would help reduce repression.[7]

5. CONCLUSION

In current circumstances, regulators will likely continue to be wary of granting large local-rate increases to telephone companies that are already performing well-above expectations. But this performance is misleading, for its basis is a continuation of the pre-divestiture separation of toll revenues that has historically provided the subsidy to residential local service from toll revenues. In one form or another, this flow of revenue is going to end, and the loss of revenue will almost certainly be made up, again in one form or another, from end-users. The present study offers further evidence, in line with Perl [1], Bell Communications Research [3], and Brock [4], that a well-conceived end-user subscriber-line charge will not only lead to lower toll rates but preserve universal service as well.

REFERENCES

[1] Perl, L., Residential Demand for Telephone Service (1983).
[2] Perl, L.J., Economic and Demographic Determinants of Residential Demand for Basic Telephone Service, National Economic Research Associates, Inc. (1978).
[3] Bell Communications Research, The Impact of Access Charges on Bypass and Universal Telephone Service (1984).
[4] Brock, G.W., Bypass of the Local Exchange: A Quantitative Assessment, FCC, OPP Working Paper Series (1984).

[7] Compare columns 3 and 4 for Texas with column 4 for Oklahoma in Table 2. The difference in these columns for Texas is due to the availability of a lower-priced alternative.

TELECOMMUNICATIONS DEMAND MODELLING
An Integrated View
A. de Fontenay, M.H. Shugard, D.S. Sibley (Editors)

ESTIMATING LOCAL CALL ELASTICITIES WITH A MODEL OF STOCHASTIC CLASS OF SERVICE AND USAGE CHOICE

John P. KLING
Stephen S. VAN DER PLOEG

Michigan Bell Telephone Company
Detroit, Michigan
U.S.A.

We estimate local telephone usage elasticities from a model of class of service choice and usage demand which utilizes a panel data base. The class of service choice occurs in an environment of optional measured service choice. The estimated elasticities are implicit in the simultaneous class of service and usage choice. The model deals with stochastic variation in usage patterns over time by each household, in usage patterns across households, and in the class of service choice. The model is estimated with disaggregated data at the household level. Elasticities are projected over a range of potential local call prices.

1. INTRODUCTION

The divestiture of regional telephone operating companies from AT&T in 1984 was an institutional response to increasing competition in telecommunications markets that has been developing for several decades. This competition has begun a process of substantial price restructuring of the services sold by these companies. The pressures of revenue requirements in this process are creating renewed interest by the firms in the demand response of local telephone usage. Public utility commissions are similarly concerned about the demand response of local telephone usage, since they understand that price restructuring by the operating companies is likely to increase local service charges as prices of other services are reduced in response to differential competitive pressures. In this paper, we present new estimates of price elasticities for local telephone usage demand from panel data. These estimates are derived from a class of service and usage choice model applied to individual households. Since there is very little price variation in our panel data, the estimated elasticities are implicit in the class of service and usage choices made by the households.

This paper describes the Michigan Bell disaggregated residential telephone demand model and presents parameter estimates of this model. These parameter estimates are used to generate usage demand elasticities for local telephone service. These elasticities are compared to the existing literature on local telephone demand elasticities and conclusions are drawn.

Our research work at Michigan Bell makes extensive use of the SLUS (Subcriber Line Usage Sample) data base to investigate disaggregated models of household demand for telephone service. This approach to modeling is in part a response to the paucity of time series data which can currently be utilized to model telephone demand in the post-divestiture environment. There are, however, other reasons which favor a disaggregated approach. Individuals, households, and firms make decisions to purchase telephone services, rather than the aggregate units defined by aggregated data. Secondly, pricing decisions of telecommunications firms increasingly require knowledge of the responses of disparate groups of consumers in a competitively structured environment. Third, reductions in computational costs have made a disaggregated effort more attractive. Research in the telephone industry by Bell Laboratories, Bell Communications Research, General Telephone, the Rand Corporation, and several Bell Operating Companies has utilized disaggregated economic data.[1] Elsewhere in the profession, economists are also increasing their usage of disaggregated data to answer questions previously approached with aggregated data.

Our model is estimated from a SLUS data base of more than 1456 households on 1 party flat rate and 1 party measured rate service. The panel consists of 8 months of daily call volumes for each household in the sample during 1984 and 1985. These call records were matched to 860 records from a survey of households which provide demographic information on each household. The parameters of the model are well determined and indicate a price elasticity for local calls of about -0.17 near the current measured rate price of $.062 per call. This estimate of the local call price elasticity is near the center of a range of small local call elasticities that have been previously estimated [1,6,15,16,17,18]. Our model, however, suggests that these low measures of local call price elasticities may understate the price response of demand in an environment where local call prices significantly exceed their historical range.

Our model is developed for the household's choice of service and demand for local usage. The model of class of service (COS) choice and demand for local usage begins with the observation that most subscribers' demands for local telephone usage are satiated or nearly satiated. We adopt a semi-log demand function, consistent with this observation. The expected consumer surplus is derived for each household under both flat and measured rate service. The household is presumed to choose the class of service which yields the greater expected surplus. The household's demand for local usage is determined by its' demographic characteristics, by its' expected marginal calling price, and by the other chosen characteristics of its' class of service.

Stochastic elements enter the model in three ways: (i) as variation in the mean satiation level among households, (ii) in the variation of calling frequency over time by a household, and (iii) in variation of the probability that a household will choose a given class of telephone service, given the systematic part of the difference between expected surplus from the available service options. We assume that the household's mean level of satiation usage of local calls is an unobservable random variable with a gamma distribution. In maximum-likelihood estimation, this variable is "integrated out" of the likelihood function. We assume that the number of calls per day has a Polya-Aeppli distribution. The number of calls originating from a household in a day is usually zero or another small integer. The distribution of calls per day is marked by a frequent occurrence of zeros, and exhibits a contagious pattern. If we assume that the number of purposes for which calls are made has a Poisson distribution, and that the number of calls made for a particular purpose has a geometric distribution, then the number of calls made per day has a Poisson-geometric (Polya-Aeppli) distribution, a distribution used in epidemiology. This distribution has several features which make it attractive and tractable for disaggregated modeling. The variance is proportional to the mean, and the sum of i.i.d. Polya-Aeppli variables also has a Polya-Aeppli distribution. The final stochastic element in the model is the probability that the household chooses measured rate service as its' expected surplus maximizing choice. This is modeled with a simple logit function of the systematic part of the difference of expected consumer surplus between measured rate and flat rate service.

We now turn to a consideration of the theoretical ideas underlying the modeling of SLUS data. This is followed by explicit statistical modeling of the stochastic elements of the study and the reporting of the empirical results.

2. ECONOMICS OF TELEPHONE USAGE CHOICE

The economic theory of consumer behavior assumes that the decision making agent (perhaps an individual, household, or firm) acts as if he/she/it consistently maximizes a utility function. This utility may be thought of as a representation of satisfaction or well-being from consumption activity. For divisible services (e.g. telephone calls), the assumption of utility maximization implies that the decision maker purchases an additional unit of the service if the utility embodied in the purchase is greater than the utility of other goods represented by the expenditure. In a discrete choice such as the class of local telephone service, the principle of utility maximization is the same, although the application is to discrete rather than continuous choice variables.

With telephone usage, this choice principle is complicated by the stochastic nature of the consumption decision. The stochastic commodity model has been advanced as an appropriate application of the utility maximizing principle in the consumption of telephone calls [4]. In this view, the decision maker maximizes stochastic utility by choosing an optimal probability distribution of call frequencies over time, from which the observed calling distribution is a random draw. We conceive of the class of service choice as a selection of some of the parameters that describe the optimal probability distribution of calls. These parameters constrain the set of potential outcomes of the observed frequencies of calls per unit of time. This framework can be regarded as descriptive of the process of maximizing expected utility in the context of simultaneous class of service and usage choice.

If decision makers maximize utility stochastically, they choose a quantity of call consumption such that the expected utility of the marginal call is equal to the expected call price. We operationalize expected utility maximization via the concept of optimal expected consumer surplus. The optimal expected consumer surplus implies that the expected consumer surplus of the marginal call is zero, which implies maximizing of consumer surplus on the intramarginal units of consumption.

In our estimation procedures, we calculate the expected consumer surplus associated with alternative class of service choices and usage intensities per unit of time. The estimation of choice of class of service and associated usage intensity allows explicit estimation of the implicit usage price elasticity.

A peculiar feature of telephone call consumption is the frequency with which satiation occurs at finite call quantities. For flat rate service and often for measured rate service with call allowances, maximizing expected consumer surplus implies call satiation at an expected marginal call price of zero. To capture this feature of telephone call demand, we employ a semi-logarithmic specification of the (non-stochastic) demand for telephone calls:

$$X_i = \theta_i \exp(\beta_p P)$$

where X_i: messages demanded by the ith household;

- θ_i : unrepressed (satiation) usage for the ith household;
- β_p : the price response parameter, and
- P : the marginal price of a message.

θ will vary across households according to specific household characteristics. An expanded specification would include income and other household demographic characteristics in the semi-log demand function:[2]

$$X_i = \theta_i \exp(\sum_j \beta_j N_{ij} + \ldots + Y_i^{\beta y} \beta_p P)$$

where the additional variables are:

- N_{ij} : value of jth demographic variable for household i,
- Y_i : income of the ith household,
- β_j : the response coefficient of variable j, and
- β_y : the response coefficient of household income.

We implement the stochastic dimension of telephone usage by taking the expected value of demand, $E(X_i)$.

3. ELASTICITIES AND CONSUMER SURPLUS CALCULATIONS OF CLASS OF SERVICE CHOICES

The semi-logarithmic model allows direct calculations of elasticities from the price parameter β_p.[3] The estimates of this parameter, however, depend indirectly on the class of service (COS) choice since the observed usage depends on the chosen COS. Conversely, the COS choice also depends on expected usage. We expand on this simultaneity aspect of the model below. Since the model estimates the price parameter in the context of class of service choice, it is useful to express the relationship between COS consumer surplus calculations and the elasticity of demand.

The calculation of expected consumer surplus is complicated by the presence of call allowances in measured service offerings. For example, the Michigan Bell call allowance with measured rate service allows $3.10 in local calls a month (ie. 50 calls a month). However, the semi-log form of the demand function expedites a general expression of consumer surplus regardless of the call allowance.[4] Since the class of service choice depends on the difference in expected consumer surplus between two service offerings, we focus on the difference in consumer surplus between flat rate and any measured rate offering with an arbitrary call allowance. The following expression is based on the nonstochastic form of the semi-logarithmic specification without demographics because of its relative simplicity:

$$S_i^m - S_i^f = (P_f^f - P_m^m) + \frac{P}{\eta_p}(\Theta_i - X_i) + P'(A)$$
$$= \text{(i)} \qquad + \text{(ii)} \qquad + \text{(iii)}$$

where X_i = consumption of calls with measured service;
P' = the marginal call price to the household;
A = number of calls in the call allowance;
P = the usage price of a call charged by the firm;
η_p = the call price elasticity of demand;
Θ_i = the (satiation) volume of flat rate usage, and
P_f^f, P_m^f = the monthly access outlays for flat rate service and measured rate service, respectively.

The expression has three terms. The first term (i) is the dollar difference in the fixed monthly outlays corresponding to the two services. The second term (ii) summarizes the difference in the usage volume between the two services. Here the relationship between the class of service choice and the call price elasticity of demand is explicitly formulated. The third term (iii) is the measure of the contribution of the call allowance to consumer surplus with measured rate service, where this is evaluated at the ex post marginal call price to the consumer (P').

There are three distinct cases which arise from the relationships which may occur between satiated flat rate usage (Θ_i), repressed usage (X_i), and the call allowance, (A). FIGURE 1 shows a linear demand curve which helps to visualize these relationships. The shaded area shows the difference in surplus created when repressed consumption with measured rate service differs from flat rate satiated consumption. The size of this shaded area is influenced by the price parameter β_p, or alternatively, the price elasticity. A single call allowance (A) (corresponding to the third case) is shown.

The first case arises from the possibility that the call allowance exceeds the satiation level of usage ($A > \Theta_i$). In this case, the marginal call price (P') and the price of usage charged by the firm (P) are equal at $P = P' = 0$, and the difference in surplus consists only in the difference in the fixed outlay for the two services. The shaded area will not exist as a difference in surplus, being part of the surplus received from either service. In the second case, the call allowance is smaller than both satiated and repressed usage ($A < X_i < \Theta_i$). With consumption at X_i in the diagram, the marginal call price is equal to the usage price charged by the firm ($P' = P > 0$). The contribution of the call allowance to the surplus with measured rate service may be evaluated at

FIGURE 1

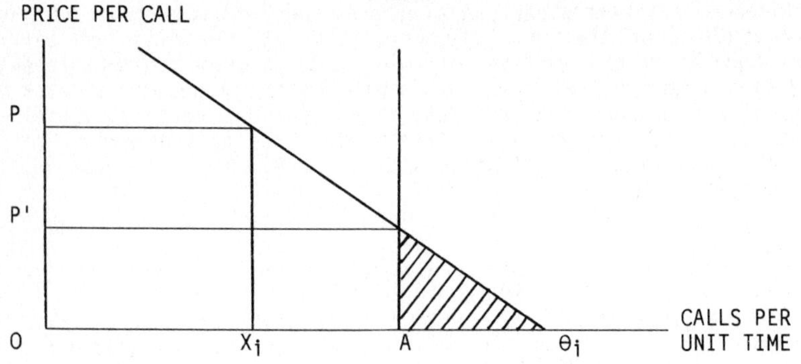

the usage price. In the third case (shown in FIGURE 1), the call allowance exceeds repressed usage but is smaller than the satiation level of usage ($X_i < A < \theta_i$). In this case, the usage price charged by the firm (P) exceeds the marginal call price P'.[5] The ex post contribution of the call allowance with measured rate service is evaluated at the latter price. All three terms in the above formulation contribute to the difference in surplus. Only in the first case will the elasticity of demand have no bearing on the class of service choice.

In summary, the expression for the difference in consumer surplus when measured rate service is packaged with a call allowance consists of three terms, and depends on four factors. The household is more likely to choose measured rate service over flat rate service (i) the greater the usage insensitive monthly outlay for flat rate service relative to measured rate service, (ii) the greater the ex post usage volume with measured rate service relative to flat rate service, (iii) the greater the ex post value of the call allowance attached to measured rate service, and (iv) the greater the absolute value of the price elasticity of demand, η_p.

We modify the general expression for the difference in consumer surplus to allow for the stochastic nature of surplus and call demand by taking expected values of both sides of the equation.

4. STOCHASTIC ASSUMPTIONS

We now turn our attention to the statistical modeling of the data that is required to calculate expected surplus for estimating the COS and usage choice parameters. Our data base consists of a subscriber line usage sample panel, currently including eight months of calling data. Several problems characterize this data base. The data have large fluctuations in the volume of usage from month to month. Due to the limited number of monthly observations, we have chosen to analyze the daily data in order to measure call variation through time. The daily call data also exhibit substantial variation in usage. There appears to be a contagious element to call frequencies, with one call leading to another call. A distribution which describes data with these characteristics is the Polya-Aeppli distribution. Sousa has recently used this distribution in predicting demand for items in which the quantity demanded in many periods is zero [21]. We emply the Polya-Aeppli distribution as our statistical model of the daily call behavior of the individual household [14]:

$$f = \begin{bmatrix} \Pr(X=0) = \exp(-\lambda) \\ \Pr(X=k) = \exp(-\lambda)\, \Gamma^k \sum_{j=1}^{k} \binom{k-1}{j-1} \frac{1}{j!} \left[\frac{\lambda(1-\Gamma)}{\Gamma}\right]^j \\ k = 1,2,\ldots,\infty \end{bmatrix}$$

This distribution is an appropriate model of the distribution of calls per day for an individual household when (1) the number of different purposes for communications has a Poisson distribution with parameter λ and (2) the number of calls required for any purpose has a geometric distribution with parameter Γ [7]. These parameters λ and Γ are in turn assumed to be distributed as gamma and beta variables, respectively, across households. The Polya-Aeppli distribution has mean $\lambda/(1 - \Gamma)$ and variance $\lambda(1+\Gamma)/(1-\Gamma)^2$. An attractive property is that the sum of n independent identically distributed Polya-Aeppli variables has mean $n\lambda/(1-\Gamma)$ and variance $\lambda(1+\Gamma)/(1-\Gamma)^2$. A convenient recursive formula allows efficient computation of Polya-Aeppli probabilities.[6]

We next consider the distribution of calls across households. The most prominent characteristic of this distribution is a great variation in average usage. As noted, we assume that the Polya-Aeppli parameter λ has a gamma distribution across households and the parameter Γ is assumed to have a beta distribution across households. The average monthly unrepressed usage $E(\Theta_i)$ is proportional to the daily household mean of unrepressed usage:

$$E(\Theta i) = n(\lambda_i/(1-\Gamma_i))$$

where n is the number of days in the month.

In our view of household calling behavior, the variation among households of the unrepressed usage is a function of the household's taste for telephone service, an unpredictable random variable. Although some of this variation can be explained by economic and demographic variables, much cannot be explained by observable variables. This random element of household usage variation is modeled by the gamma distribution of λ and the beta distribution of Γ, the Polya-Aeppli parameters. The non-random component of variation of usage across households is assumed to be a function of the expected marginal call price and the household demographic variables. The average monthly household usage $E(\Theta_i)$, where the effects of call price and household demographics have been taken into account may be written:

$$\tilde{E}(\Theta_i) = n\lambda_i^*/(1-\Gamma_i)$$

where $\lambda_i^* = \lambda_i \exp(\sum_j \beta_j N_{ij} + \ldots + Y_i^{\beta_y} \beta_p P')$ and where λ_i^* : is the modified Polya-Aeppli parameter for the ith household,

λ_i : is a random element (gamma variable) from the distribution across households,

Γ_i : is a random element (beta variable) from the distribution across households,

N_{ij} : is the value of the jth demographic usage variable for household i,

Y_i : is household income,

β_j : coefficient of the jth demographic usage variable, and

P' : is the expected marginal call price.

We now turn to the final stochastic element which concerns the class of service choice. The semi-log demand function makes the calculation of expected consumer surplus straightforward.[7] Since many individual household characteristics enter into the calculation of expected surplus, the choice of service equation is itself necessarily a stochastic relationship. An additional random variable is introduced through the logit specification which transforms the difference between expected surpluses into choice of service probabilities:

$$\Pr(\text{measured service}) = \pi \left[\frac{\exp(\beta_{logit} * E_i)}{1 + \exp(\beta_{logit} * E_i)} \right]$$

where $\tilde{E}_i = \text{Bias} + E_i + \sum_j \beta_j N_{ij}$,

$E_i = E(S_i^m - S_i^f)$: expected surplus difference,

β_{logit} : the logit parameter,

Bias : constant introduced to capture household preference for flat rate service relative to measured service not related to usage or pricing,

β_j : the coefficient of the jth demographic COS variable,

N_{ij} : the value of the jth demographic COS variable, and

π : the probability that the household knows that the measured rate choice exists.

The probability that household i makes x1 calls on day 1 (given household demographics and the choice of service) is calculated from the Polya-Aeppli distribution. The assumption of statistical independence of λ and Γ allows straight-forward calculation of the probability of the observed distribution of calls for a particular household. All the calculations are conditional upon a particular value of the unobservable variables λ and Γ. Unconditional probabilities are found by integrating out λ and Γ to obtain an expression for the probability of the joint occurrence of the choice of service and the distribution of calls actually observed. This probability is the likelihood function (L) of the sample. The likelihood function is:

$$L = \prod_{i=1}^{H} \int g(\lambda|\alpha_1, \alpha_2) \int h(\Gamma|v_1, v_2) \left[\sum_j \Pr(COS_j|\lambda_i^*) \Pr\left[(X_1^i, \ldots, X_n^i) | \lambda_i^* COS_j) \delta_{ij} \right] \right]$$

where δ_{ij} = 1 if household i has actual class of service j and the value 0 otherwise, and where
$g(\lambda|\alpha_1,\alpha_2)$: gamma density function given gamma parameters α_1 and α_2, adjusted for the expected calling price and household demographics,
$h(\Gamma|v_1,v_2)$: beta density function given beta parameters v_1,v_2,
H : the number of households in the sample,
COS_j : class of service j,
i : index of the household,
j : index of class of service choice, j=1,2,
n : the number of days, and
X_k^i : the number of local exchange calls on day k by household i, k=1,...,n.

We have evaluated the likelihood function by weighting each household in our sample (random within each class of service group) by the ratio of the number of households with COS_j in the household population to the number of households with COS_j in our sample.

5. THE SIMULTANEITY PROBLEM

One of our objectives in modeling local service is explanation of the variation in the volume of messages in response to price changes and to changes in demographics. A second objective is the explanation of and prediction of the choice of local service. The choice of service depends upon expected usage. Observed usage is a function of the expected marginal price which in turn depends on the class of service. It is necessary to explicitly model this interdependency between choice of service and actual usage in the estimation of these relationships.

A visual overview of how we have modeled the reciprocal dependence of service choice and the demand for usage above is presented in FIGURE 2. The random variables λ and Γ are randomly drawn from gamma and beta disibutions across households, respectively. For a household, the value of λ, Γ, and the value of household demographic variables determines the daily (Polya-Aeppli) distribution of calls. This distribution, together with the class of service and usage price structure, determines the difference between expected surplus for flat rate and measured service.

6. ESTIMATED PARAMETERS AND ELASTICITIES

The choice of service/local usage model has been estimated with data on 1456 1MR and 1FR households for the eight months of June, September, October, November, and December of 1984, and January, February, and March of 1985. We matched these records with responses on demographic variables from a survey of households. The sample of matched records from the two sources consists of 427 households. The dependent variable for class of service choice is binary, with values of 1 if a household has chosen measured service with a call allowance of 50 and 0 if the household has chosen flat rate service. Usage is measured by the mean of calls per day over the eight months of days. We specified the class of service and usage equations of the model by estimating a two stage least squares linear probability model as a preliminary step in estimating the

FIGURE 2
"CAUSAL" FLOW

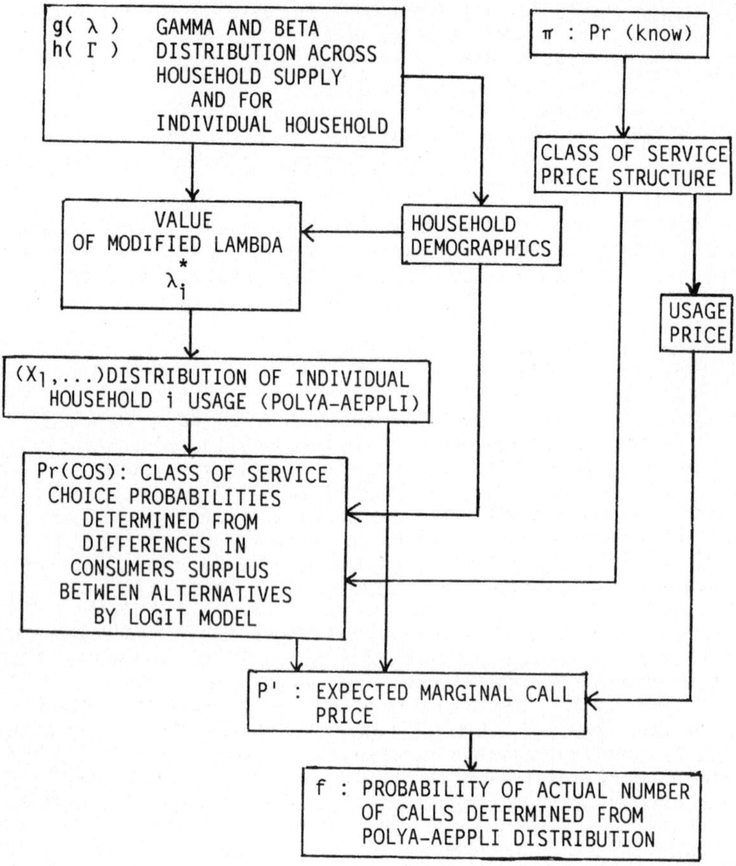

parameters of the full model. Substantial computer resources were utilized in estimation of the model. A quadratic hill climbing algorithm by Goldfeld, Quandt and Trotter was used to maximize the likelihood function [9].

The parameter estimates and t ratios are presented in Table 1. All of the parameters are statistically significant at the 99 per cent level except β_{logit} and β_o. The β_{logit} parameter is significant at the 95 per cent level. Only the constant β_o is not statistically different from zero.

Table 1

Coefficient	Estimate	t ratio
Gamma α_1	1.87	16.2
Gamma α_2	0.78	14.0
Beta v_1	4.10	20.0
Beta v_2	5.37	20.6
βp	-2.79	-21.4
β_{logit}	0.23	2.2
(Coefficients on attributes and Demographic Variables)		
Class of Service Choice:		
Ignorance	0.44	4.1
Constant (β_o)	0.09	0.6
COSprice	-0.66	-2.7
Income	-0.03	-3.0
Usage:		
Race	0.18	17.9
Adults	0.21	48.2
Seniors	-0.35	-46.2
Teens	0.20	46.3
Children	0.04	12.2
Distance	-0.08	-7.8
Busphone	0.41	27.9

The source of the data on demographic and attribute variables is the demographics survey.[8] All of the signs on these coefficients are reasonable in terms of intuitive theoretical expectations. Looking first at the usage parameters, black households make more telephone calls than non-black households. Additional adults in the household results in a greater demand for telephone calls. A qualitatively similar household size effect is observed for teenagers and children. For seniors (household members over 65 years old), however, fewer telephone calls are made, an age-specific effect. If most of the household's acquaintances reside in the local calling area, the number of telephone calls is smaller, perhaps reflecting a substitution of face to face communication for telephone communication. If someone in the household uses the telephone for business purposes, more telephone calls originate from the household.

A parameters with significant interpretive value in the class of service choice segment of the model is the ignorance parameter π. This coefficient in the expression for expected surplus restricts the opportunity to choose measured rate service to an estimated 44 per cent of the sample households. It captures a bias toward flat rate service attributed to ignorance of the measured rate service, and/or habit inertia.

Examining the remaining class of service choice coefficients, the negative sign on COSprice indicates that if a household has not explicitly examined the cost difference in the two classes of service, then it is more likely to choose flat rate service. This variable reflects a household taste variable, its' sensitivity to the household's cost of local telephone service. Households which are not sensitive to their local telephone service costs, as measured by this response variable, are more likely to choose flat rate service. Households with higher income also have a greater probability, ceteris paribus, of choosing flat rate service. Finally, the constant of 0.09 is statistically insignificant and indicates that most of the explicit preference for flat rate service has been captured in our specification of the model by the ignorance parameter.

The parameters of greatest interest are the β_{logit} parameter and the call price coefficient. The estimated β_{logit} parameter of 0.23 implies that the maximum probability of switching from flat rate to measured rate service in response to a $1.0 change in the difference in fixed outlays between the two services is approximately 0.06. This maximum probability is calculated for a probability of choice of measured service by the household of 0.5, assuming knowledge of both services. The effect of the ignorance parameter π becomes apparent in this context at the level of the sample. The magnitude of this probability response to an increase in the surplus differential of $1 is reduced to about 0.025 for the sample of households when the ignorance parameter is considered.

We now turn our attention to the principal empirical interest of this paper, the elasticity of demand for local telephone calls. Previous empirical research has found generally low price elasticities for local telephone calls, rarely exceeding 0.35. Estimated toll call price elasticities, on the other hand, are considerable higher. They tend to exceed 0.5 and may exceed 1.0. Time series estimates of toll call elasticities with Michigan Bell data suggest an elasticity of these calls at an average price of about $0.80 of approximately 0.7.

Similar work by AT&T reveals that elasticity estimates are lower in discount periods than peak periods. The most obvious economic difference between discount period calls and peak period calls on one hand, and local exchange calls and toll calls on the other, from the point of view of the consumer, is the level of the call price. A local telephone call is often priced below 7 cents per call. A typical toll call may be priced at about 80 cents. We expect the elasticity of demand for a telephone call to vary with the price of the call.

Other researchers have examined the relationship between the price of a call and the absolute value of the elasticity of demand. de Fontenay and Lee analyzed the variation in price response across toll mileage bands by estimating a translog function with Canadian data [5]. Their estimates confirm the hypothesis that price elasticities increase as the price of telephone calls rises.

These considerations prompted us to choose a semi-log functional form to estimate the demand for telephone calls. This form incorporates the hypothesis of a direct (proportional) relationship between the price elasticity and the call price.[3] Although this is a strong hypothesis, it is a practical formulation which lends itself to estimation in the context of our larger class of service and usage choice model.

Our estimated call price coefficient of -2.79 yields an estimated call

TABLE 2

Call Price Elasticities of Demand

base price	new price	Elasticity of of Demand
	.01	-0.03
	.02	-0.06
	.03	-0.08
	.04	-0.11
	.05	-0.14
.00	.05	-0.07*
	.06	-0.17
	.07	-0.20
	.08	-0.22
	.09	-0.25
	.10	-0.28
.05	.10	-0.21*

* arc elasticities; other elasticities are point elasticities.

price elasticity of -0.17 at the current Michigan Bell measured rate call price of $0.062 per call. This is in the middle of a range of local call elasticities previously estimated [1,6,15,16,17,18]. Over a range of low call prices, the estimated elasticity of demand varies from -0.03 to -0.28.

Our estimates of selected price elasticities are presented in table 2. The direct relationship between these elasticities and the call price is imposed by the semilog specification. The arc elasticities are somewhat lower than the point elasticities, since these measure an average elasticity over a price range of 5 cents below the reference price. Because our price data consists only of $0.0 and $0.062 per call, we have restricted our elasticity estimates to a range which does not depart far from our observed call prices.

7. CONCLUSION

We have presented a methodology for estimating local exchange elasticities from disaggregated (household level) telephone company usage data with severely limited variation of call prices. The methodology may be applied with or without demographic data. The projected usage price elasticities are proportional to the level of the call price. These elasticities range in absolute value from below 0.10 to above 0.2 for calls priced from 1 to 10 cents. These elasticities are derived from our estimated class of service choice and usage model with demographic variables included in the specification of the model.

Our price elasticity estimate (-0.17) at the current measured rate price of messages near 6 cents per call is in the center of the range of estimates that have been published elsewhere. Estimated elasticities of less than 0.1 have been obtained by Park, Wetzel, and Mitchell [18], Beauvais [1], and Morra and Bowman [17]. Estimates between 0.1 and 0.2 are credited to Kristan, Gaustad and Kosberg [15], and Jensik [13]. Estimates greater than 0.2 have been obtained by Doherty [6] and Mitchell [16]. The modest magnitudes of our price elasticity estimates reinforce, in the context of the work done by others, the conclusion that local call price elasticities are low at the prevailing low levels of measured rate prices. They also emphasize the hypothesis that they are not uniformly low, and are likely to be higher if local telephone calls are priced higher than in the past. These results have some implications for pricing of local telephone calls. For example, it is probable that changes in measured rate call prices in the region of current low call prices will not have large effects on social welfare. Welfare gains from repricing measured service at (low) marginal costs, while not negligible, are likely to be modest. On the other hand, our estimation in the context of a semi-log demand function recognizes the view that local call price elasticities are not generically low. Our projected elasticities suggest the likelihood of significantly greater elasticities than previously estimated as higher local call prices are experienced by consumers.

From an entirely different perspective, the prevailing view of low local call price elasticities which has been strengthened by our estimates, suggests that measured rate pricing (above marginal cost) could be used as an effective practical pricing method to generate revenues to finance the usage insensitive costs of providing telephone service. In this capacity, it would be an alternative to financing these costs with higher access prices. The evidence of low local telephone call elasticities, in this context as well, argues that the welfare cost of such repricing will be modest. However, these welfare costs will rise with the level of the call price not only because these prices are increasingly diverging from marginal cost, but because the elasticity of demand is rising with the call price. It is necessary, of course, to determine if these welfare costs are indeed smaller than those generated by access pricing above the marginal cost of access. This hypothesis needs to be evaluated through additional research.

In conclusion, the call price elasticities estimated in this paper confirm the prevailing impression that usage demand elasticities are low at the current low usage prices. The range of local usage elasticities estimated also suggests that usage elasticities would be higher than observed if the call prices were higher. Our elasticity estimates suggest a local call price elasticity as high as 0.28 at a call price of 10 cents.

FOOTNOTES

1. Our work has particularly benefited from previous research of Crew and Dansby [3], Brown and Sibley [2], and by Hausman, Hall and Griliches

[10]. Victoria Lazear's cross-sectional modeling effort at Illinois Bell has influenced our modeling strategy. The research of Bell Laboratories has been fundamental [11,12,19,20]. We are also thankful for valuable comments made by Richard Pfau, Socrates Tountas, and Nancy Brucken of Michigan Bell, and by Richard Spady, Roger Klein, and especially Alain de Fontenay of Bell Communications Research.

2. The treatment of income in the specification is asymmetrical. A requirement for correctly calculating consumer surplus is that the marginal utility of income must be constant for each household. If the income constraint is not effective, income is not a determinant of the quantity of calls demanded. Customers on flat rate service and those on measured service for whom the call allowance constraint is not binding fall within this group. While our assumption of constant marginal utility of income over the range of call demand experienced by a household is reasonable, this parameter is likely to change with movement across income classes. A more general model would also include a time constraint. In practice, many two (high) income households have less time available to make telephone calls.

3. The elasticity is proportional to the level of price:

$$\eta_p = \beta_p P$$

4. This expression is derived by integrating the demand for calls under flat rate service and subtracting the monthly outlay. This is given by:

$$S_i^f = \int_0^\infty \theta_i \exp(\beta_p P)dp - P_f^f$$

$$= -\frac{\theta i}{\beta_p} - P_f^f, \text{ where } P_f^f \text{ is the fixed outlay.}$$

The corresponding consumer surplus for measured rate service is:

$$S_i^m = \int_{P_m}^\infty \theta i \exp(\beta_p P)dp - P_m^f$$

$$= -\frac{\theta i}{\beta_p} \exp(\beta_p P_m) - P_m^f,$$

where P_m^f is the fixed outlay, and P_m is the call price.

With a call allowance, we simply add the value of the call allowance, P'(A) to this equation, where P' is the expected marginal call price.

5. P' is also the call price corresponding to the call allowance volume of calls. We can solve for P' from the demand function. The solution is:

$$P' = \ln(A/\theta_i)/\beta_p.$$

6. The recursive formula is:

$$Pr(X = k) = c1^{*}Pr(X = k-1) - c2^{*}Pr(X = k-2)$$

$$\text{where } c1 = \frac{2(k-1)\Gamma}{k} + \frac{\lambda(1-\Gamma)}{k}$$

$$c2 = \frac{-\Gamma^{2}(k-2)}{k}$$

7. In the general case of a call allowance (A) which applies to local exchange calls, the difference in expected surplus is:

$$E(S_i^m - S_i^f) = \text{Prob(case1)}E(S_f^f - S_m^f \text{ case1})$$
$$+ \text{Prob(case2)}E(S_f^f - S_m^f \text{ case2})$$
$$+ \text{Prob(case3)}E(S_f^f - S_m^f \text{ case3})$$

$$E(S_i^m - S_i^f) = P_f^f - P_m^f$$
$$+ \text{Prob(case2)}E\left[\frac{1}{\beta_p}(\theta_i - X_i)\right] + P_m A$$
$$+ \text{Prob(case3)}E\left[\frac{1}{\beta_p}(\theta_i - A) + P'A\right]$$

where Prob(case2) and Prob(case3) refer to the probabilities that the call allowance is less than or greater than repressed usage, respectively, and Prob(case1) is the probability that the call allowance exceeds the satiation volume of usage, θ_i. The expected marginal call price P' is made operational as shown in footnote 5. Evaluation of the surplus requires estimating these probabilities and conditional expectations. Of course, the distribution of θ_i will vary across households and will depend upon demographic characteristics.

8. COSprice is a dummy variable which has a value of 1 if respondents answered no to the question: Have you ever tried to determine if flat rate or measured rate service would be more economical for your household? Income is a categorical variable where each household was placed in one of 15 income categories based on its' response on the survey questionnaire. Stations is the total number of business and residential lines in the household's telephone exchange. Race is a dummy variable with value 1 for black, 0 otherwise. Adults is the number of adults in the household in addition to the respondent and/or spouse. Seniors is the number of persons 65 years or older. Teens is the number of persons aged 13-20. Children is the number of persons aged 0-12 years residing with the household. Distance is a dummy variable with value 1 if respondents indicated that most of their relatives and/or friends lived in the local calling area and 0 otherwise. Busphone is a dummy variable with value 1 if respondent indicated that someone in the household uses the telephone for business.

REFERENCES

[1] Beauvais, E., "The Demand for Residential Telephone Service under non-metered Tariffs: Implications for Alternative Pricing Policies." A paper presented at the Western Economic Association, Anaheim California, June, 1977.

[2] Brown, B., and D. Sibley, The Theory of Public Utility Pricing, Lexington Mass. Lexington Press, 1986.

[3] Crew, M., and R. Dansby, "Cost-Benefit Analysis of Local Measured Service." In M. Crew, (ed.), Regulatory Reform and Public Utilities, Lexington Mass: Lexington Books, 1982.

[4] Curien, Nicolas and Alain de Fontenay, "A Theory of Consumption Decision: The Stochastic Commodity Model." Paper presented at the Telecommunications Demand Modeling Conference, October 23-25, 1985, New Orleans, Louisiana.

[5] de Fontenay, A., and J. Lee, "BC/Alberta in Long Distance Calling" in L. Courville, A. de Fontenay, and A. Dobell (eds.), Economic Analysis of Telecommunications, North Holland, Amsterdam, 1983, pp. 199-229.

[6] Doherty, N., "Econometric Estimation of Local Telephone Price Elasticities," in H.M. Trebing, (ed.), Assessing New Pricing Concepts in Public Utilities , East Lansing, Mi: Institute of Public Utilities, Michigan State University, 1978.

[7] Evans, D., "Experimental Evidence Concerning Contagious Distributions in Ecology", Biometrika, vol. 40, 1953, pp. 186- 211.

[8] Fuss, M., and L. Waverman, "The Regulation of Telecommunications in Canada," Technical Report No. 7, Ottawa: Economic Council of Canada, March, 1982, p. 66.

[9] Goldfeld, S., R. Quandt and H. Trotter, "Maximization by Quadratic Hill Climbing", Econometrica, vol. 34 (1966), pp. 541- 551.

[10] Hausman, J., B. Hall and Z. Griliches, "Econometric Models for Count Data with an Application to the Patents-R&D Relationship," NBER Technical Paper no. 17, August, 1981.

[11] Hoffberg, M., and M. Shugard, "A Utility Theory Approach to Class of Service Choice and Usage Repression," BTL Technical Memorandum TM 81-59541-5,81-59543-7, October, 1981.

[12] Infosino, W., "Relationships Between the Demand for Local Telephone Calls and Household Characteristics," Bell System Technical Journal(59), July-August 1980, pp. 931-53.

[13] Jensik, J.M., "Dynamics of Consumer Usage," in Baude, J. and others (eds.), Perspectives on Local measured Service, Kansas City, Mo., Telecommunications Industry Workshop, 1979.

[14] Kendall, M. and A. Stuart, The Advanced Theory of Statistics, vol 1, London: Griffen, 4th ed., 1977.

[15] Kristan, B., O. Gaustad, and J. Kosberg, "Some Traffic Characteristics of Subscriber Catagories and the Influence from Tariff Changes." A paper presented to the Eighth International Teletraffic Congress, Australia, November, 1976.

[16] Mitchell, B., "Optimal Pricing of Local Telephone Services." American Economic Review, vol. 68 (September,1978), pp. 517-537.

[17] Morra, W., and G. Bowman, "The Demand for Local Telephone Use: Modeling the effects of Price Changes", in Award Papers in Public Utility Economics and Regulation, East Lansing, Mi.: Institute of Public Utilities Michigan State University, 1982.

[18] Park, R., B.M. Wetzel, and B. Mitchell, 1983, "Price Elasticities for Local Telephone Calls," Econometrica, Vol. 51, No. 6, pp. 1699-1730.

[19] Shugard, M., "Demand for Local Telephone Service: The Effects of Price and Demographics," BTL Memorandum for file MF-79-9541-42, Oct. 8, 1979.

[20] Shugard, W., "Choice Dependent on Usage: Residence Class of Service Choice", BTL Technical Memorandum TM:82-59543-14, Aug. 30, 1982.

[21] Souza, R., "Slow Moving Items Demand Forecasting", paper presented at The Fifth International Symposium on Forecasting, Montreal, Canada, June, 1985.

TELECOMMUNICATIONS DEMAND MODELLING
An Integrated View
A. de Fontenay, M.H. Shugard, D.S. Sibley (Editors)

MICRO-SIMULATION OF LOCAL RESIDENTIAL TELEPHONE DEMAND UNDER ALTERNATIVE SERVICE OPTIONS AND RATE STRUCTURES[1]

Terry Atherton, Cambridge Systematics
Moshe Ben-Akiva, MIT
Daniel McFadden, MIT
Kenneth Train, Cambridge Systematics and the
University of California, Berkley

We present a model of local telephone demand, emphasizing the use of the model in forecasting. We draw from a previous paper (Train, et al, 1987) in describing the specification and estimation of the model. The current paper extends the previous one by describing a method for forecasting with the model and applying the method to examine changes in service offerings and tariffs.

1. INTRODUCTION

Ongoing changes in the telecommunication industry have led to dramatic changes in the type and range of services offered. This paper is concerned with the options for local service that telephone operating companies offer residential customers. There are two general categories of service: flat service, under which a household can, for a fixed monthly charge, make an unlimited number of calls within a specified geographical area, and measured service, for which the household pays a lower fixed monthly fee but can make only a specified number (or dollar value) of free calls, after which charges are incurred for additional calls. Various flat services differ in the size of the geographical area in which calling is free, with higher monthly fees for larger areas. Measured services differ with respect to the threshold number (or dollar value) of calls beyond which the customer is charged.

The service option that a household chooses depends in general on the household's calling pattern, i.e., on the number and duration of calls the household makes by time of day and distance. Households who make numerous calls, or make relatively expensive calls (e.g., at high cost times of day or to relatively distant locations), tend to choose flat services, while households who make few, inexpensive calls tend to choose measured services. Causation also runs in the other direction. Once a household has obtained a particular service, the household's calling pattern is conditional upon that choice, since the marginal price that the household faces for calls is then given.

Previous analysis has examined the impact on economic welfare and households' calling patterns of a shift from flat to measured service. Theory suggests that economic efficiency is enhanced under measured service and that the extent of the welfare gain depends on the price elasticity of demand (Alleman, 1977; Mitchell, 1978). Empirical studies have indicated that the effect of the shift on demand is fairly small and depends on the demographics of the household (Pavarini, 1979; Wilkinson, 1983; Park, Wetzel, and Mitchell, 1983; Park, Mitchell, et al., 1983). We extend the previous empirical work by allowing for (1) voluntary choice of service by the household (thereby capturing the opposite direction of causality, from calling patterns to service choice), (2) a wider variety of service options, and (3) a more detailed delineation of households' calling patterns that includes the number and average duration of calls by time of day and destination zone. This third extension is particularly important given the first two since the detailed calling pattern of a household determines which of the various service options is least costly.

Modeling service option choice depends critically on being able to model accurately the calling patterns of households. One method for modeling individual households' calling patterns is to divide calls into several categories based on the time and distance of the call and to estimate regression equations for the number and average duration of calls within each category. The difficulties encountered with this approach are well-known and numerous. (1) For any reasonable number of distinct times of day and distance bands, the number of regression equations to be estimated is quite large. Allowing for a full set of cross-elasticities entails a generally unmanageable number of parameters. (2) The dependent variables in these equations are truncated at zero, and zero calls are usually observed for many of the time/zone categories for any particular household. Consequently, estimation of the regression equations by OLS leads to classic truncation bias (Amemiya, 1974; Heckman, 1976; Lee, 1981). Correction for this bias is complex, particularly given the number of equations and the interrelations among the equations. (3) The concept of price in these models is problematical. Under measured services the household is charged for calls beyond a threshold, but not for those below. As a result, the marginal price for a call in any category depends on the number of calls, in that and other categories, that the household has previously made during the billing period. Marginal price for any category of calls is thus endogenous to the individual household, not just with the outcome of the regression for that category but with the combined outcome for all call categories.

We present a model that describes households'interrelated choices of local service option and monthly calling pattern and does so in a way that avoids the difficulties described above. Each household is characterized as choosing a particular service option and a particular calling portfolio, where a portfolio of calls is defined as a particular number and average duration of calls at each time of day to each distance zone. The model is specified as nested logit, which is a type of probabilistic choice model especially designed for handling interrelated discrete choice situations (McFadden, 1978; Ben-Akiva and Lerman, 1985). With this specification, the probability of choosing a particular service option depends on the household's choice of portfolio (reflecting, for example, the tendency of households that place many calls to choose flat services), and the portfolio that the customer chooses in a given month depends on the chosen service option (reflecting the fact that households' calling patterns depend on the cost per call under their chosen service option).[2]

The set of portfolios among which the household chooses is immense. Estimation and simulation are therefore performed on a sample of portfolios randomly selected in accordance with a probability distribution that is similar to the distribution of observed portfolios in a sample of households. A correction factor is included in estimation to preserve consistency in the face of this sampling. A weighting scheme is employed in the simulation to obtain consistent predictions.

Sampling of alternatives for estimation and simulation of nested logit models has previously been utilized in several situations for which the choice set is very large. Examples include households' choices of make and model, or class, of automobile (Manski and Sherman, 1980; Berkovec and Rust, 1985; Mannering and Winston, 1985; and Train, 1986); dwelling location and unit (Friedman, 1975; Weisbrod, et al., 1980); and travelers' choice of destination (Silman, 1980; Daly, 1982). These studies have utilized sampling procedures that allow the correction factor to reduce to either a constant, such that it does not affect estimation, or a very simple function. We build upon this earlier work by utilizing a more flexible sampling procedure, that Ben-Akiva and Lerman (1985) call "importance sampling." This form of sampling has previously been used in empirical work by Watanatada and Ben-Akiva (R79), Ben-Akiva and Watanatada (1981) and Cambridge Systematics (1984).

The specification and estimation results of the models are described in the next sections, followed by a presentation of the micro-simulation procedure and the results obtained on the effects of several prototypical rate scenarios for a sample of residential customers of a local telephone operating company.

2. MODEL SPECIFICATION

Let the times of day be divided into distinct categories labeled $t = 1,\ldots,T$. The geographic areas, or zones, to which a person can call under local service are labeled $z = 1,\ldots,Z$. The number of calls that a household makes to zone z at time t is labeled N_{tz}, and the average duration of these calls is D_{tz}. A "portfolio" of calls is defined as a particular number and average duration of calls to each zone during each time of day. More precisely, a portfolio is a particular value of the vector whose elements are $(N_{11},\ldots,N_{TZ},D_{11},\ldots,D_{TZ})$.[3] Label the set of all possible portfolios as A and a particular portfolio as $i \in A$. Finally, let the available service options be indexed $s=1,\ldots,S$.

In our application, three service options are available to all households and two additional services are available to some households. These options are described in Table 1. A portfolio is defined on the basis of the twenty-one time/zone categories given in Table 2.

We observe a household choosing service option s and making portfolio of calls i during a time period.[4] We assume the probability of our observing a particular (s,i) combination given the available options and possible portfolios to be nested logit. Under this assumption, the appropriate nesting of alternatives (where, in this case, an alternative is an (s,i) combination) depends on the correlations across alternatives of unobserved factors. In particular, alternatives that are similar in unobserved factors are to be grouped together.

Most of the specific factors that relate to portfolio choice (such as where the household's friends and relatives live, the time and location of activities that require telephone use, and so on) are not observed; these unobserved factors are similar over all service options for any portfolio. Consequently, all alternatives with the same portfolio but different service option are nested together. On the other hand, the primary factor affecting service options choice is cost, which is observed. Alternatives with the same service option but different portfolios can be similar with respect to observed factors, but not, at least relatively, with respect to unobserved factors. Consequently, these alternatives are not nested together.

Under this nesting pattern, the nested logit probability of observing option s and portfolio i is

$$P_{is} = \frac{e^{Y_{is}}(\sum_{s'} e^{Y_{is'}})^{\lambda-1}}{\sum_j(\sum_{s'} e^{Y_{js'}})^{\lambda-1}} ,$$

where Y_{is} is is a parametric function of observed factors relating to service option s and portfolio i. This expression for Pis is convenient in that it can be rewritten as the product of two logit probabilities. Without loss of generality, we can decompose Y_{is} into two parts, one that varies over both i and s, and another that varies only over i:

$$Y_{is} = Wis + Vi/\lambda .$$

P_{is} can be written as the product of the marginal probability of portfolio i and the conditional probability of service option s given portfolio i:

$$P_{is} = P_i \cdot P_{s|i} , \qquad (1)$$

where

$$P_{s|i} = \exp(W_{is}) / \sum_{s'=1}^{S} \exp(W_{ix'}) , \qquad (2)$$

$$P_i = \exp(V_i + \lambda I_i) / \sum_{j\epsilon A} \exp(V_j + \lambda I_j) , \qquad (3)$$

where

$$I_i \equiv \ln(\sum_{s'=1}^{S} \exp(Wis')) .$$

The term Ii is called the "inclusive value" of portfolio i.

Note that the direction of conditionality in this specification is from portfolio to service option. This does not imply, however, that the household makes choices sequentially in this manner.[5] As in all nested logit models, the direction of conditionality reflects correlations among unobserved factors across alternatives; as such it arises from patterns in the researcher's lack of information rather than from the households' decision process.[6]

The coefficient of inclusive value, λ, measures substitutability across alternatives. If substitution is greater within than among nests, then $0<\lambda<1$; whereas, if among-nest substitution exceeds within nest substitution, then $1<\lambda$. Given our nesting pattern, the parameter is less than one if households shift to different service options more readily than they shift to different portfolios but greater than one if households shift to different portfolios more readily than they shift to different service options.[7]

This specification has several advantages. (1) Since P_s and $P_{s|i}$ are both logit, the model is relatively inexpensive to estimate and easy to interpret. (2) The cost of any portfolio under any service option is simply the bill that the household would obtain if it made that portfolio of calls and chose that service option.[8] Threshold values for free calling, based on either the number or dollar value of calls, enter the calculation of the cost of a portfolio under a service option in the same way as in the telephone company's calculation of bills. Consequently, the impact of changes in tariffs and thresholds can be readily and consistently examined. (3) Interrelations between calling patterns and service option choice are incorporated. The probability of choosing any particular portfolio changes as the tariffs and/or thresholds associated with any service option change. And the probability of choosing a particular service option depends on the portfolio of calls that the household makes.

3. MODEL ESTIMATION

3.1 Sampling of Alternatives

Estimation of the parameters entering P_i is complicated by the fact that the number of possible portfolios (elements of A) is immense. For example, even if number of calls and minutes of conversation are collapsed into four categories for each, with 13 calling band/time period combinations the number of alternative calling portfolios facing the household would be 16^{13}, or 6.7×10^{17}. As a result, both model estimation and model application are feasible only if carried out using a sample of calling portfolios drawn from the full alternative set. (See the discussion in Train et al., 1987, and the detailed presentation of these sampling methods in Ben-Akiva and Lerman, 1985.)

The sample of portfolios for each household is constructed by drawing from the set of all portfolios in accordance with a pre-specified probability distribution and (in estimation) adding the household's chosen alternative. With this procedure, unchosen portfolios were constructed by sampling the number of calls and average duration (independently) for each possible combination of calling band and time period available to a particular household. An importance sampling approach was employed in which selection probabilities were assigned to portfolios on the basis of aggregate usage distributions. Within each of four types of exchanges (e.g., metro, nonmetro, etc.), separate distributions were developed for each calling band/time period combination and for each of six market segments defined on the basis of the number of telephone users in the household. The number of distributions to be developed for number of calls and for average duration is given in Table 3. As shown, the number of distributions ranges from 18 for non-metro exchanges (3 time periods, 1 calling band, and 6 market segments) to 78 for metro exchanges (3 time periods for the local calling area plus 2 time periods for each of up to 5 additional calling bands, and 6 market segments). Overall, a total of 204 distributions were developed for both the number of calls and average duration.

Based on plots of the frequency distributions for the number of calls and average duration of observed portfolios, the exponential distribution was selected for sampling number of calls, and an Erlang distribution was selected for sampling average duration. To estimate the parameters of these distributions, it was necessary to calculate the mean number of calls and the mean and variance of average duration for each of the 204 calling band/time period/market segment combinations.

A total of nine unchosen portfolios were sampled for each household to accompany the one chosen portfolio for model estimation. Factors for use in model estimation were calculated to correct for the biases due to the non-uniform sampling procedure used to generate unchosen portfolios.

Denote by B the sample of portfolios constructed for a particular household. Denote by $\pi(B|i)$ the conditional probability of constructing the subset B given that the chosen portfolio is i.[9]

The logit form that is used to estimate P_i is

$$\frac{\exp(V_i + \lambda I_i + \ln \pi(B|i))}{\sum_{j \in B} \exp(V_j + \lambda I_j + \ln \pi(B|j))} . \qquad (4)$$

McFadden (1978) has shown that this procedure yields, under normal regularity conditions, consistent estimates of the unknown parameters. Note that the logit model (4) is the same as P_i except (i) the summation in the denominator is over all portfolios in the constructed set B rather than in the entire set A, and (ii) the exponentiated terms include an additive alternative-specific correction for the bias introduced by the sampling of alternatives; the coefficient of this variable is constrained to one.

3.2 The Data

The number and average duration of local calls made in November of 1984 in each time/zone categories were observed for a sample of residential customers from a local telephone operating company on the east coast. The sample is stratified random on the basis of households' locations. However, the sample is not representative for two reasons. (1) Some geographical areas are not represented, because local calling records are not available for households in these areas. (2) Socioeconomic characteristics of the sampled households were obtained through a survey, for which the non-response rate was fairly high. We rely in the analysis on the fact that estimation of logit models on non-representative samples is consistent if the sample is drawn on the basis of exogenous factors (Manski and McFadden, 1981.)

3.3 Estimation Results

Service Option Choice Conditional on Portfolio

Table 4 gives estimated parameters for each of three specifications of the function of cost that enters $P_{s|i}$. With each specification, cost has a significantly negative impact in that the probability of choosing a service option decreases as the cost of that option increases (holding the cost of other options constant). The best fit is obtained with ln (C_{is}), which obtains a coefficient of about two. This implies that the ratio of probabilities for any two service options is inversely proportional to the square of the ratio of their costs:

$$\frac{Ps\ i}{Ps'\ i} = k.\ \left(\frac{Cs'i}{Csi}\right)^{2.08},$$

such that the relative probabilities change at an increasing rate as the relative costs change. This result gives credence to the popular notion that households are relatively insensitive in their choice of service option to small cost differences and become increasingly sensitive as the differences become larger.

The option-specific constants are highly significant. Mechanically, the estimated values for these constants are those that result in the average probability (i.e., predicted share) for each option being equal to the actual (i.e., observed) share in the sample. Intuitively, these constants capture the average effect of all unincluded variables. Perhaps the most important unincluded variable relating to each service option is the insurance quality of the option. Under flat services, for an additional fixed charge the household is provided on upper limit on charges for calling within a certain geographic area. The wider the area of free calling, the more valuable is the insurance provided by the option.

The estimated constants are consistent with this concept. Budget measured service provides no insurance, since there is a charge for each call; its constant (which is zero by normalization) is lower than those estimated for all the other services. Standard measured service insures against cost variation within a range of calls (i.e., below the threshold), while local flat service provides complete insurance for all calls in a local area; the constant for local flat exceeds that for standard measured. Finally, metro flat provides insurance for a wider area that local flat and local extend flat, and its constant is consequently greater.[10]

Portfolio Choice

Table 5 presents the estimated parameters of the model with inclusive value based on service choice model 1 in which cost enters in log form.[11] All of the parameters enter with the expected signs and reasonable relative magnitudes.

There are two aspects of the parameter estimates that warrant discussion. (1) The coefficient of inclusive value exceeds one. This implies that substitution among nests (i.e., from one portfolio to another) occurs more readily than substitution within nests (i.e., from one service option to another.) Stated more directly, households respond to price changes by adjusting their calling patterns more than by shifting to different service options.

(2) The function Vi, and hence the probability of choosing portfolio i, is estimated to decrease in number of calls. This result is expected for two reasons. First recall that there are multiple time-of-day and destination zone categories. As the total number of calls increases, the number of portfolios that are possible with that number of calls increases (e.g., there is only one portfolio of no calls, but twenty-one portfolios, ignoring duration, associated with making a total of one call--a portfolio for each of the time/zone categories in which that one call could be made). Therefore, if the probability of making a certain total number of calls increases with number of calls, but increases less rapidly than the number of portfolios that are possible with that number of calls, then the probability of each portfolio must decrease in number of calls. This is what is occurring in the estimated model.

Second, it is reasonable to expect the probability of making particular number of calls to reach a maximum at some finite number of calls. The number of portfolios that are possible with a certain number of calls increases with the number of calls, but does so at a continuously decreasing rate. Consequently, if Vi decreases linearly with the number of calls (as the estimates in Table 5 indicate), then the probability of making a certain number of calls first increases with the number of calls (with the expansion of the number of possible portfolios dominating the decrease in Vi minus the cost of each portfolio) but eventually decreases (when the decrease in Vi, which is linear in the number of calls, starts to dominate the diminishing expansion in the number of portfolios).

4. FORECASTING

We employ a micro-simulation procedure based on the "sample enumeration" approach. The joint distribution of independent variables is represented by an appropriate sample of households (e.g., a sub-sample taken from a survey). The choice probabilities are forecast for each sampled household and expanded to represent the entire population. Thus, no arbitrary distribution need be assumed for the independent variables; the sample itself represents any distribution. Because the models are applied at the household level with this approach, it is free from aggregation bias. However, this approach, like any other stochastic simulation, introduces sampling errors.

The models are applied to predict local service choice probabilities and usage patterns for each sampled household. These individual household choice probabilities and usage patterns are then weighted by an appropriate factor and summed across all households in the sample. Alternative rate structure or service option scenarios can be evaluated by changing the appropriate explanatory variables to reflect the specific scenario of interest, and re-applying the choice models, resulting in a revised set of choice probabilities representing the demand response to the scenario under consideration.

With respect to the particular application being presented here, two areas warrant further discussion:

- Development of the factors needed to scale up the sample of households to population totals; and
- Development of the factors needed to estimate shares of services options and expected revenues from a sampling of alternative calling portfolios.

Household expansion factors - In our application households were sampled only from electronically switched exchanges and therefore would not necessarily be representative of all residential telephone customers served by the local operating company.

As a result, it was necessary to develop different expansion factors for different types of households such that the distributions of certain key variables in the sample matched the cor-responding distributions for the total population. These key variables included the following:

- Household income
- Household size
- Class of service

Separate distributions for each of these variables were developed for each of eight geographic areas. Household income and size were obtained from Census data updated to represent 1984 population characteristics; class of service shares were obtained from billing data.

Using the joint distributions of these variables from the sample of households together with the marginal distributions for the total population (from Census data), an iterative proportional fitting procedure was applied to estimate the joint distribution of these variables for the total population.

Sampling alternative calling portfolios - As in model estimation it is necessary to draw a sample of calling portfolios for use in the simulation model. These portfolios were constructed for each household using the same procedure as in sampling non-chosen portfolios for model estimation.

Household-level predictions from a sample of calling portfolios - Let X_{is} be an attribute of the combination of portfolio i and service option s such as the monthly cost of service. Then the expected value of X_{is} over all possible combinations of portfolios and service options is given by the following expression:

$$X = \sum_{s} \sum_{i \epsilon A} P_i P_{s|i} X_{is} \qquad (5)$$

The calculated value is a measure of interest such as the expected monthly telephone bill. If we replace X_{is} in this expression with Kronecker delta function $\delta_{ss'}$ which is equal to one for s = s' and is zero otherwise, we obtain:

$$X = \sum_{s} \sum_{i \in A} P_i P_{s|i} \cdot \delta_{ss'} \tag{6}$$

$$X = \sum_{i \in A} P_i P_{s'|i}$$

$$X = P_{s'}$$

Thus, by an appropriate definition of the attribute of interest X_{is} we also obtain from equation (5) the expected marginal probability of a service option.

Because A is a very large set we cannot practically evaluate the expression in (5). Let B be the subset of portfolios drawn from A according to a pre-specified probability distribution. Obtain an estimate of X, denoted by $\hat{X}$, by an expansion from the sample of calling portfolios $B \subseteq A$, as follows:

$$\hat{X} = \sum_{s} \sum_{i \in B} w_i \hat{P}_i P_{s|i} X_{is} \tag{7}$$

where w_i is the expansion factor for portfolio $i \in B$ and $\hat{P}_i$ denotes the estimate of P_i. P_i must also be estimated from the sample of portfolios as follows:

$$\hat{P}_i = \frac{\exp(V_i + bI_i)}{\sum_{j \in B} w_j \exp(V_j + bI_j)} \tag{8}$$

The expansion factors w_i, $i \in B$, are given by:

$$w_i = \frac{1}{f_i}, \quad i \in B$$

where f_i is the pre-specified selection probability of portfolio i that was derived previously. Note that these selection probabilities need be specified only up to a scalar multiple because they appear both in the numerator and the denominator of (7). Also note that

$$\sum_{j \in B} w_j \exp(V_j + bI_j)$$

is an unbiased estimator of the logit model denominator

$$\sum_{j \in B} \exp(V_j + bI_j),$$

P_i is a consistent estimator of P_i, and $\hat{X}$ is a consistent estimator of X.

The properties of this estimator were analyzed by Watanatada and Ben-Akiva (1979) who showed that the expected bias in $\hat{X}$ is inversely proportional to the sample size of portfolios, i.e., the number of elements in the set B. The standard error of $\hat{X}$ is approximately inversely proportional to the square root of the sample size. Thus, as the sample size increases the bias diminishes at a much faster rate than the random error. This analysis suggested that it may be possible with a relatively small sample to have a bias which is substantially smaller than the random error. However, it does not indicate what an appropriate sample size would be.

Since the actual relationship between the number of alternatives and the accuracy of these output measures is unknown a priori, an empirical approach was used to determine how many alternatives are needed to represent the full set of calling portfolios such that key output measures (e.g., class of service shares, revenues, etc.) are predicted with sufficient accuracy. The simulation procedure was applied to three different data sets: one with 9 alternative calling portfolios for each household, a second with 29 alternatives, and the third with 49 portfolios.

The predicted class of service shares for these three data sets are presented in Table 6. As shown, there is very little difference among class of service shares across these three data sets. Between the 29 and 49 alternative data sets, differences are present only at the second decimal place for percentage shares. While the differences observed between the 9 and 49 alternative data sets are somewhat larger, the largest difference is only 0.31 percent. As a result, using nine alternatives would seem quite reasonable for most purposes.

5. SCENARIO RESULTS

This section presents the results of applying the simulation procedure to predict the demand response to the following hypothetical rate scenarios:

- Increase local usage rates to match intra-LATA toll usage rates for given distance classes;
- Equalize the price of local area flat service across all rate groups; and
- Increase the threshold allowance of standard measured service.

The results of these scenarios are discussed below. Base case conditions for the simulation procedure are summarized in Table 7.

The output variables reported here are limited to class of service shares, message minutes, and revenues. It should be noted, however, that it is possible to accumulate virtually any variable used within the model. Further, in addition to total population values, these model outputs can be broken down on the basis of any variable in the model. In Table 7, for example, model outputs are disaggregated on a geographic basis (i.e., city versus suburban versus non-metro customers). Disaggregation on the basis of customer socioeconomic attributes could be done as well.

Increase local usage charges - In this scenario, the usage charges associated with calls to metropolitan calling bands 2 through 6 are increased to match the corresponding intra-LATA toll usage rates for a given distance. Residential customers living in metropolitan areas enjoy less expensive usage rates on a per-mile basis relative to customers living in non-metropolitan areas where calling band 2 through 6 are not defined and intra-LATA toll rates apply. The purpose of a scenario such as this would be to equalize usage rates across the state. The current and revised usage rates associated with this scenario are presented below:

		Current Rates		Revised Rates	
Calling Band	Approximate Distance (miles)	Initial Period	Additional Period	Initial Period	Additional Period
2	1-10	8	2	16	8
3	11-16	11	3	20	10
4	17-22	14	5	23	12
5	23-30	17	6	27	16
6	31-40	20	7	31	18

The changes in class of service shares, message minutes, and revenues associated with this scenario are presented in Table 8. As shown, the shares of both measured services decreases for suburban and city customers. (Non-metro customers are unaffected by the change.) The decreases for standard measured are somewhat less than those for budget measured. This could reflect households with calling portfolios that are well within the $4.00 threshold associated with standard measured service. For these households, increased usage rates would not increase their monthly telephone bill for the standard measured service option until this $4.00 threshold was reached. As a result, these increased usage rates would result in a greater increase in price (and therefore a greater reduction in share) for budget measured service.

The share of local area flat rate service decreases, by 9 percent among suburban customers but increases slightly (by 2 percent) among city customers. This could reflect a higher frequency of calls to calling bands 2 through 6 among suburban customers. The share of metropolitan area flat rate service increases by 62 percent and 64 percent for suburban and city customers respectively. Among all residential customers, total revenues are increased by 14 percent under this scenario.

Equalized local area flat - In this scenario the monthly charge associated with local area flat rate service is equalized across all rate groups. The current and revised rates associated with this scenario are presented below:

	Local Area Flat Rate	
Rate Group	Current	Revised
City	$9.00	$9.00
Suburban	6.90	9.00
Non-Metro 1	6.90	9.00
Non-Metro 2	5.20	9.00
Non-Metro 3	3.75	9.00

As shown, price increases are greatest among non-metro customers, while the rate for city customers remains unchanged.

The simulation results associated with this scenario are summarized in Table 9. As shown, among non-metro customers, the share of local flat service decreases by 16 percent, while the shares of both measured services increase by 71 percent. Increased shares are also predicted for extended area and metro area flat services. Similar patterns are also predicted for suburban customers. However, because the difference between current and revised rates is less than that for non-metro customers, the magnitude of these changes is somewhat smaller. Among all residential customers, total revenues are increased by 12 percent under this scenario.

Increase threshold allowance for standard measured service - Under this scenario the threshold allowance is increased from $4.00 to $6.00. In other words, for a fixed monthly charge of $2.50, residential customers would now be allowed $6.00 worth of free calling versus $4.00. The results of this scenario are presented in Table 10. As shown, among all residential customers, the share of standard measured service increases by 16 percent, while the shares of budget measured, local flat and metro flat decrease by 4, 3 and 4 percent respectively. Overall, revenues would be reduced by 3 percent under this scenario.

FOOTNOTES

[1]This work was performed through Cambridge Systematics, Inc. We gratefully acknowledge the support and comments of Chris Swann of Bell of Pennsylvania. The views expressed herin are our own and do not represent positions of Cambridge Systematics or Bell of Pennsylvania.

[2]Nested logit models have been used in numerous other contexts to capture the interrelations among discrete choices. In particular, empirical models have been estimated for the number and make/model, or class, of vehicles that a household chooses (Hensher and Le Plastrier, 1983; Hocherman, et at., 1983; Mannering and Winston, 1985; and Train, 1986); the number of vehicles to own and the mode of travel to work, such as auto or bus (Lerman and Ben-Akiva, 1976; and Train, 1980); mode of travel and destination (Ben-Akiva, 1973); related appliance choices, such as gas, electric, or oil space heating and central or room air conditioning (Goett, 1984; and Goett and McFadden, 1984); and housing location and type (Lerman, 1977). The specification of our model is within the tradition of these studies.

[3]Note that if data are available on the duration of each call, then duration categories $d = 1,\ldots,D$ can be defined and a portfolio defined as a particular value of the vector $(N_{111},\ldots,N_{TZD})$.

[4]A household can also choose not to acquire access to the phone system. That is, one option is "no service" with the only portfolio available under this option being no calls. The specification in the text allows for this possibility. However, it might be more reasonable to specify a three-level nested logit model with the "highest" level being whether to acquire any service and portfolio. We do not investigate this issue since our estimation sample consists only of households that have acquired service. The estimation is consistent under either method of handling the choice of access. (that is, estimation is consistent (1) on a subset of alternatives, or (2) as the "lower" two levels of a three-level nested logit.) However, the estimated model does not incorporate the factors and parameters that relate to the household's decision of whether to access the system. For simulation of situations in which access is relevant, the model can conceivably be combined with results from an empirical model of access choice with appropriate renormalization of parameters.

[5]If anything, the household probably does the opposite, choosing service option first and portfolio conditional upon service option.

[6]The specification can be generalized by describing the service option choice and/or portfolio choice as itself nested logit such that the complete model is multi-level nested. A similar approach has been proposed by Taylor (1979) for choice of service option using an elimination-by-aspects model (Tversky, 1972).

[7]A value of exceeding one can, depending on the range of data used in estimation, be inconsistent with a particular description of consumer behavior called the random utility model. There are, however, dynamic aspects of the choice of service option and portfolio that the random utility model does not represent (i.e., the observed portfolio is "built up" over a period of time as households choose to make additional calls, and the service options is chosen before the portfolio is revealed.) From a purely statistical perspective, the value of indicates relative substitutability within and among nests, and neither possibility can be ruled out **a priori**.

[8]Since cost varies over i and s, it enters as an element of W_{is} and its impact on choice of portfolio and service option is estimated in $P_{s|i}$.

[9]Since B necessarily includes the chosen portfolio, $\pi(B|j) = 0$ for $j \notin B$.

[10]Extended local flat cannot be compared with local flat since the former is available to only some households, so that the unincluded variables are averaged over different populations. The alternative specific constants in the service choice model were adjusted to reflect actual shares in the telephone company's service area. See Section 2.6 of Train (1986) for a discussion of the consistency of such adjustment.

[11]The estimated parameters of the portfolio choice model are essentially the same under all three specifications of the service choice model given in Table 2.

REFERENCES

Alleman, J., 1977, "The Pricing of Local Telephone Service," U.S. Department of Commerce, Office of Telecommunications Special Report 77-14

Amemiya, T., 1974, "Multivariate Regression and Simultaneous Equation Models when the Dependent Variables are Truncated Normal," Econometrica, 42, pp. 999-1012

Ben-Akiva, M. 1973, Structure of Passenger Travel Demand Models, Ph.D. Dissertation, Department of Civil Engineering, MIT, Cambridge, MA.

Ben-Akiva, M. and S. Lerman, 1985, Discrete Choice Analysis, Cambridge: MIT Press

Ben-Akiva, M. and T. Watanatada, 1981, "Application of a Continuous Choice Logit Model," in Manski, C. and D. McFadden (eds.)., Structural Anlysis of Discrete Data with Econometric Applications, MIT Press, Cambridge, MA.

Berkovec, J. and J. Rust, 1985, "A Nested Logit Model of Automobile Holdings for One Vehicle Households," Transportation Research, Vol. 19B, No. 4, pp. 275-285

Brandon, B. (ed.), 1981, The Effect of the Demographics of Individual Households on Their Telephone Usage, Ballinger Publishing Co., Cambridge, MA

Cambridge Systematics, Inc., 1984, Estimation and Application of Disaggregate Models of Mode and Destination Choice, report prepared for Direction des Etudes Generales, Regie Autonome des Transport Parisien, Paris

Daly, A., 1982, "Estimating Choice Models Containing Attraction Variables," Transportation Research, Vol. 16B, No. 1, pp. 5-15.

Friedman, J., 1975, "Housing Location and the Supply of Local Public Services," Ph.D. dissertation, Department of Economics, University of California, Berkeley

Goett, A., 1984, Household Appliance Choice: Revision of REEPS Behavioral Models, Electric Power Research Institute, Report EA-3409

Goett, A. and D. McFadden, 1984, "The Residential End-Use Energy Planning System: Simulation Model Structure and Empirical Analysis," Advances in the Economics of Energy and Resources, Vol. 5, pp. 153-210

Heckman, J., 1976, "The Common Structure of Statistical Models of Truncation, Sample Selection and Limited Dependent Variables and a Simple Estimator for Such Models," Annals of Economic and Social Measurement, 5, pp. 475-492

Hensher, D. and V. Le Plastrier, 1983, "A Dynamic Discrete Choice Model of Household Automobile Fleet Size and Composition," Working Paper No. 6, Dimensions of Automobile Demand Project, Macquarie University, North Ryde, Australia

Hocherman, I., J. Prashker, and M. Ben-Akiva, 1983, "Estimation and Use of Dynamic Transaction Models of Automobile Ownership," Transportation Research Record, No. 944, pp. 134-141

Infosino, W., 1980, "Relationship Between the Demand for Local Telephone Calls and Household Characteristics," Bell System Technical Journal, Vol. 59, pp. 931-953

Lee, L., 1981, "Simultaneous Equations Models with Discrete and Censored Dependent Variables," in C. Manski and D. McFadden (eds.), Structural Analysis of Discrete Data with Econometric Applications, Cambridge, MA, MIT Press

Lerman, S., 1977, "Location, Housing, Automobile Ownership, and Mode to Work: A Joint Choice Model," Transportation Research Record, No. 610

Lerman, S. and M. Ben-Akiva, 1976, "A Behavioral Analysis of Automobile Ownership and Modes of Travel," Report No. DOT-05-3005603, prepared by Cambridge Systematics, Inc., for the U.S. Department of Transportation, Office of the Secretary

McFadden, D., 1978, "Modelling the Choice of Residential Location," in A. Karquist, et al. (eds.), Spatial Interaction Theory and Planning Models, Amsterdam: North-Holland Press

McFadden, D., 1984, "Econometric Analysis of Qualitative Response Models", in Z. Griliches and M. Intriligator, eds., Handbook of Econometrics, Vol.II., New York: North-Holland.

Mannering, F. and C. Winston, 1985, "A Dynamic Empirical Anlysis of Ownership and Utilization," Rand Journal of Economics, Vol. 16, No, 2, pp. 215-236.

Manski, C. and D. McFadden, 1981, "Alternative Estimators and Sample Design for Discrete Choice Analysis," in Manski and McFadden (eds.), Structural Analysis of Discrete Data with Econometric Applications, Cambridge, MIT Press.

Manksi, C. and L. Sherman, 1980, "An Empirical Analysis of Household Choice Among Motor Vehicles," Transportation Research, Vol. 14A, No. 5-6, pp. 349-366

Mitchell, B.M., 1978, "Optimal Pricing of Local Telephone Service," American Economic Review, Vol. 68, pp. 517-537

Park, R., B. Wetzel, and B. Mitchell, 1983, "Price Elasticities for Local Telephone Calls," Econometrica, Vol. 51, No. 6, pp. 1699-1730

Park, R., B. Mitchell, B. Wetzel, and J. Alleman, 1983, "Charging for Local Telephone Calls: How Household Characteristics Affect the Distribution of Calls in the GTE Illinois Experiment," Journal of Econometrics, Vol. 22, No. 3, pp. 339-364

Pavarini, C., 1979, "The Effect of Flat-to-Measured Rate Conversions on Local Telephone Usage," in J. Wenders (ed.), Pricing in Regulated Industries: Theory and Application, Mountain States Telephone and Telegraph Co., Denver

Silman, L., 1980, "Disaggregate Travel Demand Models for Short-Term Forecasting," Information Paper 81, Israel Institute of Transportation Planning and Research, Tel-Aviv

Taylor, L., 1980, Telecommunications Demand: A Survey and Critique, Ballinger Publishing Co., Cambridge, MA

Taylor, L., 1979, "Modeling the Choice of Class of Service by Residential Telephone Customers," Working Paper, Department of Economics, University of Arizona, Tucson

Train, K. 1986, Qualitative Choice Analysis, Cambridge: MIT Press

Train, K., 1980, "A Structured Logit Model of Auto Ownership and Mode Choice," Review of Economic Studies, 47, pp. 357-370

Train, K.E., D.L. McFadden and M. Ben-Akiva, 1987, "The Demand for Local Telephone Service: A Fully Discrete Model of Residential Calling Patterns and Service Choices," Rand Journal, Spring 1987.

Tversky, A., 1972, "Elimination by Aspects: A Theory of Choice," Psychological Review, Vol. 79, No. 4

Watanatada, T. and M. Ben-Akiva, 1979, "Forecasting Urban Travel Demand for Quick Policy Analysis with Disaggregate Choice Models: A Monte Carlo Simulation Approach," Transportation Research, Vol. 13A, pp. 261-268.

Weisbrod, G., S. Lerman, and M. Ben-Akiva, 1980, "Tradeoffs in Residential Location Decisions: Transportation versus Other Factors," Transport Policy and Decisionmaking, Vol. 1, No. 1, pp. 13-26

Wilkinson, G., 1983, "The Estimation of Usage Repression Under Local Measured Service: Empirical Evidence from the GTE Experiment," in L. Courville, A. de Fontenary, and R. Dobell (eds.), Economic Analysis of Telecommunications: Theory and Applications, Amsterdam: North-Holland Press

Table 1

SERVICE OPTIONS

Name		Availability	Charges for Calls to Nearest Zone	Charges for Calls to Other Zones
1	Budget measured	Available to all customers	Each call charged seven cents	Each call charged at rate that varies with time, distance, and duration of call
2	Standard measured	Available to all customers	$4.00 worth of calling is free, then each call is charged as under budget measured (option 1)	
3	Local flat	Available to all customers	Free	Each call charged at rate that varies with time, distance, and duration of call
4	Extended local flat	Available to customers in only some exchanges	Free	Free for calls to some charged as under local calls to other exchanges
5	Metro flat	Available to non-rural customers	Free	Free

Table 2

TIME/ZONE CATEGORIES

ZONES		TIMES	
Number	Description	Times of Week During Which Different Rates are Chared for That Zone	Rates During That Time
1	Zone immediately surrounding the household's residence	• 9AM-9PM, Monday-Sunday	Full tariffs
		• 7AM-9AM and 9PM-Midnight, Monday-Sunday	50% off
		• Midnight-7AM, Monday-Sunday	86% off
2-6	Geographic bands successively more distant from household's residence	• 9AM-9PM, Monday-Friday	Full tariffs
		• All other times	50% off
7	Specific exchanges outside of zones 1-6, applicable only to households in certain exchanges within a metropolitan area	• 9AM-9PM, Monday-Friday	12% off, on average
		• 7AM-9AM, 9PM-Midnight, Monday-Sunday	48% off, on average
		• Midnight-7AM, Monday-Sunday	60% off, on average
		• 9AM-9PM, Saturday-Sunday	56% off, on average
8	Remainder metropolitan area in which household resides, applicable only to households in certain exchanges within a metropolitan area	• 9AM-9PM, Monday-Friday	12% off, on average
		• 7AM-9AM, 9PM-Midnight, Monday-Sunday	48% off, on average
		• Midnight-7AM, Monday-Sunday	60% off, on average
		• 9AM-9PM, Saturday-Sunday	56% off, on average

TABLE 3

DISTRIBUTIONS FOR SAMPLING OF ALTERNATIVE PORTFOLIOS

Area/Exchange Type	Calling Band/ Time Combinations?	Household Types	Number of Distributions
Non-Metro (ex. Perimeter)	3	6	18
Perimeter (w/ extended)	11	6	66
Perimeter (w/o extended)	7	6	42
Metro	13	6	78
TOTAL			204

1 Calling band/time period combinations are defined as follows:

- non-metro: 3 discount periods for local calling area.
- perimeter: 3 discount periods for local calling area plus 4 discount periods for calling bands corresponding to areas served by extended and metro area flat rate service.
- metro: 3 discount periods for local calling area plus 2 discount periods for each of up to 5 additional calling bands.

Table 4

LOGIT MODEL OF SERVICE CHOICE GIVEN PORTFOLIO

Service Options	Number of Customers Who Alternatives Available	Number of Customers Who Chose the Alternative
1 Budget measured	2963	579
2 Standard measured	2963	855
3 Local flat	2963	1120
4 Extended local flat	84	20
5 Metro flat	1873	389

	Estimated Parameters (t-statistics in parentheses)		
	Model 1: ln (cost)	Model 2: cost	Model 3: cost divided by income of household in thousands of dollars
Cost of portfolio under designated service option (includes monthly fixed fee and charges for calls, in 1984 dollars; specified differently in each model)	-2.081 (23.87)	-0.09111 (17.42)	-0.4538 (14.51)
Option-specific constants:			
Standard measured	1.228 (17.83)	0.6135 (11.04)	0.5079 (9.282)
Local flat	2.635 (24.74)	1.576 (20.94)	1.081 (18.07)
Extended local flat	2.254 (7.880)	1.123 (3.922)	0.8614 (3.279)
Metro flat	3.757 (21.82)	3.474 (17.23)	1.317 (13.60)
Number of households	2963	2963	2963
Initial log likelihood	-3812.4	-3812.4	-3812.4
Log likelihood at convergence	-3356.0	-3487.8	-3562.3

Table 5

LOGIT MODEL OF PORTFOLIO CHOICE MARGINAL OVER SERVICE OPTIONS

Alternative Set: The househld's chosen portfolio plus nine portfolios selected randomly from the set of all available portfolios.

Explanatory Variable	Coefficient	t-statistic
Benefits of information (θ)		
N log D for calls to zone 1	0.0239	11.8
N log D for calls to zones 2-6	0.0410	13.10
N log D for calls to zones 7-8	0.0474	11.11
$\sum_{\text{zones}} \left(\begin{array}{c}\text{Population}\\ \text{of zone in}\\ \text{millions}\end{array}\right)$ N log D for zone	0.00752	14.58
$\left(\begin{array}{c}\text{Income of household}\\ \text{in thousands of \$}\end{array}\right)\left(\begin{array}{c}\text{N log D for}\\ \text{all zones}\end{array}\right)$	0.213×10^{-3}	3.13
$\left(\begin{array}{c}\text{Number of telephone}\\ \text{users in household}\end{array}\right)\left(\begin{array}{c}\text{N log D for}\\ \text{all zones}\end{array}\right)$	-0.755×10^{-3}	4.035
Rate of information transfer ($-\gamma$)		
Total number of calls N to all zones	-0.890	8.525
Opportunity cost of conversation minutes ($-\alpha$)		
Total duration at 9 a.m.-9 p.m. to zone 1	-0.00472	10.11
Total duration 7 a.m.-9 a.m. and 9 p.m.-midnight to zone 1	-0.00438	8.475
Total duration midnight-7 a.m. to zone 1	-0.00504	5.158
Total duration for 9 a.m.-9 p.m. to zones 2-8	-0.00141	1.866
Total duration for 9 p.m.-9 a.m. to zones 2-8	-0.0111	12.56

Table 5 continued

Explanatory Variable	Coefficient	t-statistic
(Household income in thousands of $) (Total duration all zones and times)	-0.0606x10-3	3.22
Other variables		
Inclusive value of service option choice (utilizing model 1 of Table 2)	4.178	13.68
Sampling correction factor (coefficient is constrained to 1.0)	1.0	-.-
Number of households	3038	
Log likelihood at zero	-7125.9	
Log likelihood at convergence	-6242.8	

TABLE 6

CLASS OF SERVICE SHARES VERSUS
NUMBER OF ALTERNATIVE PORTFOLIOS SAMPLED

	Portfolio Sample Size		
Service Option	9	29	49
Budget measured	5.43%	5.49%	5.50%
Standard measured	14.85	14.72	14.70
Local area flat	73.23	73.54	73.54
Extended area flat	1.35	1.36	1.34
Metro area flat	5.13	4.89	4.92

TABLE 7

BASE CASE SCENARIO FOR SIMULATION PROCEDURE

	Non-Metro Customers	Suburban Customers	City Customers	All Residential Customers
CLASS OF SERVICE SHARES				
Budget Measured	6.3	4.8	3.8	5.5
Standard Measured	9.4	11.3	24.4	14.8
Local Area Flat	77.5	68.6	67.9	73.3
Extended Area Flat	4.7	NA	NA	1.3
Metro Area Flat	2.1	15.4	3.9	5.1
MESSAGE MINUTES (monthly average)	360	324	387	354
REVENUES (monthly average, $ per customer)	$ 9.50	$15.77	$12.55	$11.72

TABLE 8

PREDICTED DEMAND RESPONSE FOR INCREASED LOCAL USAGE RATES

	New Value/(Percent change from base)			
	Non-Metro Customers	Suburban Customers	City Customers	All Residential Customers
CLASS OF SERVICE SHARES				
Budget Measured	6.3 (0%)	3.6 (-25%)	3.1 (-18%)	5.1 (-7%)
Standard Measured	9.4 (0%)	8.8 (-22%)	21.0 (-14%)	13.3 (-10%)
Local Area Flat	77.5 (0%)	62.6 (-9%)	69.4 (+2%)	72.4 (-1%)
Extended Area Flat	4.7 (0%)	NA	NA	1.3 (0%)
Metro Area Flat	2.1 (0%)	25.0 (+62%)	6.4 (+64%)	7.9 (+55%)
MESSAGE MINUTES Monthly Average	360 (0%)	319 (-2%)	368 (-5%)	347 (-2%)
REVENUES Monthly Average ($ x customer)	$ 9.50 (0%)	$20.34 (+29%)	$14.78 (+18%)	$13.34 (+14%)

TABLE 9

PREDICTED DEMAND RESPONSE FOR EQUALIZED LOCAL AREA FLAT RATES

	New Value/(Percent change from base)			
	Non-Metro Customers	Suburban Customers	City Customers	All Residential Customers
CLASS OF SERVICE SHARES				
Budget Measured	10.8 (+71%)	5.8 (+21%)	3.8 (0%)	7.9 (+44%)
Standard Measured	16.1 (+71%)	13.6 (+20%)	24.4 (0%)	18.5 (+25%)
Local Area Flat	65.1 (-16%)	61.8 (-10%)	67.9 (0%)	66.2 (-10%)
Extended Area Flat	5.4 (+15%)	NA	NA	1.5 (+15%)
Metro Area Flat	2.6 (+24%)	18.9 (+23%)	3.9 (0%)	6.0 (+18%)
MESSAGE MINUTES Monthly Average	316 (-12%)	324 (0%)	387 (0%)	333 (-6%)
REVENUES Monthly Average ($ per customer)	$11.52 (+21%)	$17.31 (+11%)	$12.55 (0%)	$13.12 (+12%)

TABLE 10

PREDICTED DEMAND RESPONSE FOR INCREASED THRESHOLD LEVEL FOR STANDARD MEASURED SERVICE

	New Value/(Percent change from base)			
	Non-Metro Customers	Suburban Customers	City Customers	All Residential Customers
CLASS OF SERVICE SHARES				
Budget Measured	6.2 (-2%)	4.6 (-4%)	3.5 (-8%)	5.3 (-4%)
Standard Measured	10.2 (+8%)	14.1 (+25%)	29.3 (+20%)	17.2 (+16%)
Local Area Flat	76.9 (-1%)	66.4 (-3%)	63.5 (-5%)	71.3 (-3%)
Extended Area Flat	4.7 (0%)	NA	NA	1.3 (0%)
Metro Area Flat	2.1 (0%)	14.9 (-3%)	3.6 (-8%)	4.9 (-4%)
MESSAGE MINUTES Monthly Average	362 (-1%)	321 (-1%)	383 (-1%)	350 (-1%)
REVENUES Monthly Average ($ per customer)	$ 9.45 (-1%)	$15.36 (-3%)	$11.98 (-5%)	$11.45 (-3%)

SECTION I: END USER DEMAND

I.4. Measurement

TELECOMMUNICATIONS DEMAND MODELLING
An Integrated View
A. de Fontenay, M.H. Shugard, D.S. Sibley (Editors)

PROBABILISTIC MULTIDIMENSIONAL CHOICE MODELS FOR MARKETING RESEARCH

Geert De Soete

Department of Psychology, University of Ghent
Henri Dunantlaan 2, B-9000 Ghent, Belgium

J. Douglas Carroll

AT&T Bell Laboratories
600 Mountain Avenue, Murray Hill, NJ 07974

Two recently developed probabilistic multidimensional models for analyzing pairwise choice data are discussed. The first one, the wandering vector model, has originally been suggested by Carroll and was further elaborated by De Soete and Carroll. The second model, called the wandering ideal point model, is a recently proposed unfolding analogue of the wandering vector model.

A maximum likelihood estimation method for fitting the models is illustrated, as well as a statistical test for assessing the goodness-of-fit. Finally, some potential applications to marketing research, particularly involving telecommunications demand modeling are discussed.

1. INTRODUCTION

The purpose of this paper is to present some recently developed psychological choice models that might be useful in marketing research. The methods suggested in this paper are mainly theory-based as opposed to a more descriptive, data-analytical approach as elaborated for instance in Harshman and Lundy (1987). In the first section, we discuss some empirical and practical considerations that should be taken into account by a choice model in order to be (a) realistic from a psychological point of view and (b) applicable in marketing research. Next, two models which satisfy these requirements are discussed in considerable detail. Finally, an illustrative application of the models is presented and some issues regarding parameter estimation and model validation are discussed.

2. SOME EMPIRICAL AND PRACTICAL CONSIDERATIONS

2.1 Choice behavior is often inconsistent

It is well-known that if we repeatedly present the same set of choice objects to a subject in an identical situation, he or she will not always respond in a consistent manner. Also,

---

Geert De Soete is supported as "Bevoegdverklaard Navorser" of the Belgian "Nationaal Fonds voor Wetenschappelijk Onderzoek" at the University of Ghent, Belgium.

as Tversky (1972) remarks, a subject often experiences considerable uncertainty when making preference judgments. One way to model this inconsistency is to conceive each response as an event in a probability space, leading to the development of stochastic choice models. These models are intended to predict the probabilities of selecting each object out of an offered set (called a feasible set) as the most preferred alternative.

Although it is in principle possible to construct models accounting for choices on feasible sets consisting of more than two alternatives, we restrict ourselves to probabilistic choice models for pairwise choices, mainly for two reasons. Calculating the choice probabilities predicted by a model for feasible sets larger than two objects is usually far from easy because it involves numerical evaluations of multiple integrals. Furthermore, enormous amounts of data are required for validating such models. This makes fitting these models highly impractical in many marketing research situations. Therefore, we resort to the time-honored method of paired comparisons as a data gathering procedure. In this paradigm, a subject is presented two choice objects at a time and is required to choose one. Models fit to such paired comparisons data can then be used, however, to predict choice probabilities from larger feasible sets (e.g., for predicting probabilities of first choices in an actual purchase situation in which all alternatives are presented as options).

Thus, because of the inconsistency of choice behavior, probabilistic models are needed. Given our concern for models that are routinely applicable in a variety of applied settings, we restrict ourselves to probabilistic models for paired comparisons data.

2.2 Similarity effects

The two most popular psychological models for representing paired comparisons data are Thurstone's (1927) Law of Comparative Judgment (LCJ) Case V and the Bradley-Terry-Luce (BTL) (Bradley & Terry, 1952; Luce, 1959) model. Denoting the probability of preferring object a to object b by $p(a, b)$, Thurstone's LCJ Case V can be written as

$$p(a, b) = \Phi[u(a) - u(b)] \tag{1}$$

where Φ is the standard normal distribution function and $u(a)$ the utility of object a. The BTL model is usually presented as

$$p(a, b) = \frac{v(a)}{v(a) + v(b)} \tag{2}$$

with v a positive real-valued function on $\{a, b, \cdots\}$. Defining $u(a) = \log[v(a)]$, eq. (2) becomes

$$p(a, b) = \Psi[u(a) - u(b)] \tag{3}$$

where Ψ denotes the standard logistic distribution function. As is apparent from (1) and (3), Thurstone's LCJ Case V and the BTL model are very similar (in fact, the standard normal and logistic distribution functions are so similar in form that enormous numbers of trials are required to distinguish between the two models statistically). Both models imply the following condition commonly referred to as strong stochastic transitivity:

$$\text{if } p(a, b) \geqslant 1/2 \text{ and } p(b, c) \geqslant 1/2, \\ \text{then } p(a, c) \geqslant \max\,[p(a, b), p(b, c)]\,. \tag{4}$$

Unfortunately, there is ample experimental evidence (Becker, DeGroot, & Marschak, 1963; Krantz, 1967; Rumelhart & Greeno, 1971; Sjöberg, 1975, 1977, 1980; Sjöberg & Capozza, 1975; Tversky & Russo, 1969; Tversky & Sattath, 1979) that empirical choice proportions often violate strong stochastic transitivity in a systematic way. Empirical choice proportions seem to be influenced not only by differences in utility between the choice objects, but also, to some extent, by the similarity or comparability of the choice alternatives. Subjects tend to be somewhat indifferent between very dissimilar alternatives, even when the objects differ considerably in utility. Similar choice objects, on the contrary, tend to evoke more extreme choice proportions, even when they do not differ very much in utility.

Although empirical choice proportions were often found to violate (4), they usually satisfied a less restrictive transitivity condition known as moderate stochastic transitivity. Moderate stochastic transitivity states that

$$\text{if } p(a, b) \geqslant 1/2 \text{ and } p(b, c) \geqslant 1/2, \\ \text{then } p(a, c) \geqslant \min\,[p(a, b), p(b, c)]\,. \tag{5}$$

It can be proved that any model of the form

$$p(a, b) = F\left(\frac{[u(a) - u(b)]}{d(a, b)}\right), \tag{6}$$

where F is monotonically increasing with $F(x) = 1 - F(-x)$, u a real-valued function on $\{a, b, \cdots\}$, and d a (semi-)metric, implies (5) but not necessarily (4) (Halff, 1976). A model of the form (6) is called a moderate utility model. Contrary to models implying (4), moderate utility models can account for the empirically observed similarity effects. Consequently, in order to be realistic, a pairwise choice model should be a moderate utility model. In this paper, we concentrate on probabilistic multidimensional choice models for representing paired comparisons data that are moderate utility models.

2.3 No a priori psychological information about the choice objects may be required

Because of the similarity effects mentioned above, various moderate utility models have been proposed in the psychological literature in recent years (e.g., Carroll, 1980; Edgell & Geisler, 1980; Heiser & de Leeuw, 1981; Marley, 1981; Strauss, 1981; Takane, 1980; Tversky, 1972; Tversky & Sattath, 1979). Several of these models are inspired by Restle's (1961) set-theoretic choice model and require a priori information about the psychological feature structure of the choice objects in order to be applicable. However, in many applied settings this a priori information is not available and the models cannot be utilized. In order to be routinely applicable in a variety of marketing research situations, a choice model should not require any a priori psychological information about the choice objects.

In the next two sections, two recently developed multidimensional choice models are described. These models take into account the three considerations discussed above: they are probabilistic, imply only moderate stochastic transitivity, and do not require any a priori psychological information about the choice objects. In addition, they can in principle be applied to three-way paired comparisons data.

3. THE WANDERING VECTOR MODEL

The wandering vector (WV) model is a probabilistic version of the well-known vector model (Slater, 1960; Tucker, 1960) for analyzing choice data. It was introduced by Carroll (1980) and generalized by De Soete and Carroll (1983). An even more general formulation of the WV model is given in De Soete and Carroll (in press) and is presented in the next section.

3.1 General formulation

The WV model seeks a joint representation of the subjects (or groups of subjects) and the choice objects in an r-dimensional space. Each subject i ($i = 1, N$) is represented by a vector emanating from the origin, with a terminus $\mathbf{Y}_i$ that follows a multivariate normal distribution:

$$\mathbf{Y}_i \sim N(\boldsymbol{\mu}_i, \Sigma_i) . \tag{7}$$

It is assumed that the distributions of the N subject vectors are independent of each other. I.e.,

$$\text{Covar}\,(\mathbf{Y}_i, \mathbf{Y}_{i'}) = \mathbf{0}$$

for $i, i' = 1, N$ and $i \neq i'$. Each choice object j ($j = 1, M$) is represented as a fixed point $\mathbf{x}_j$ in the same multidimensional space.

According to the model, a subject i samples a $\mathbf{y}_i$ from $\mathbf{Y}_i$ each time he or she is presented a pair of choice objects (j, k). Alternative j is preferred to k whenever the orthogonal projection of $\mathbf{x}_j$ on the vector from the origin to $\mathbf{y}_i$ exceeds the orthogonal projection of $\mathbf{x}_k$ on that vector, i.e., whenever

$$\frac{\mathbf{x}_j'\mathbf{y}_i}{\|\mathbf{y}_i\|} > \frac{\mathbf{x}_k'\mathbf{y}_i}{\|\mathbf{y}_i\|} . \tag{8}$$

This is illustrated in Figure 1.

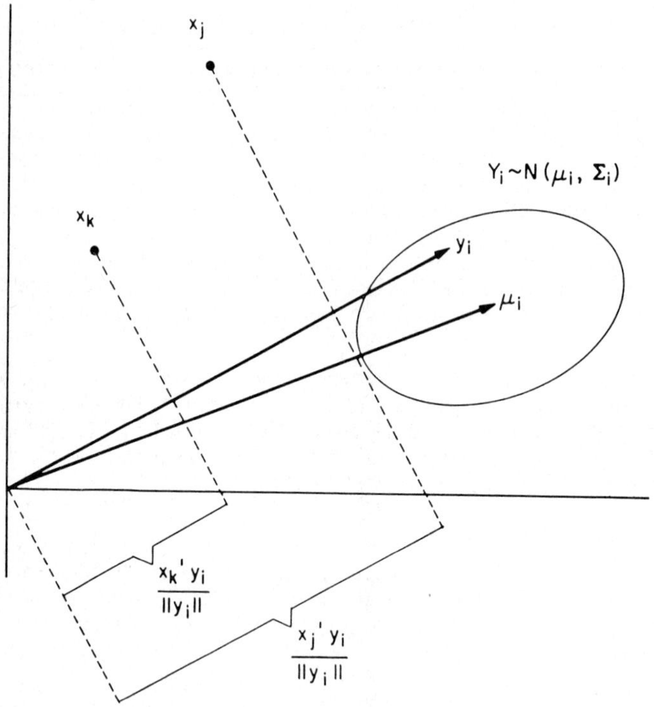

Figure 1. Illustration of the wandering vector model. The ellipse represents the random terminus of the subject vector.

In the figure, the orthogonal projection of $\mathbf{x}_j$ on the sampled vector is larger than that of

$\mathbf{x}_k$. Therefore, the subject would prefer j to k. Since each time the subject is presented a pair of choice objects a new $\mathbf{y}_i$ is sampled, a subject does not always respond consistently on repeated presentations of the same object pair.

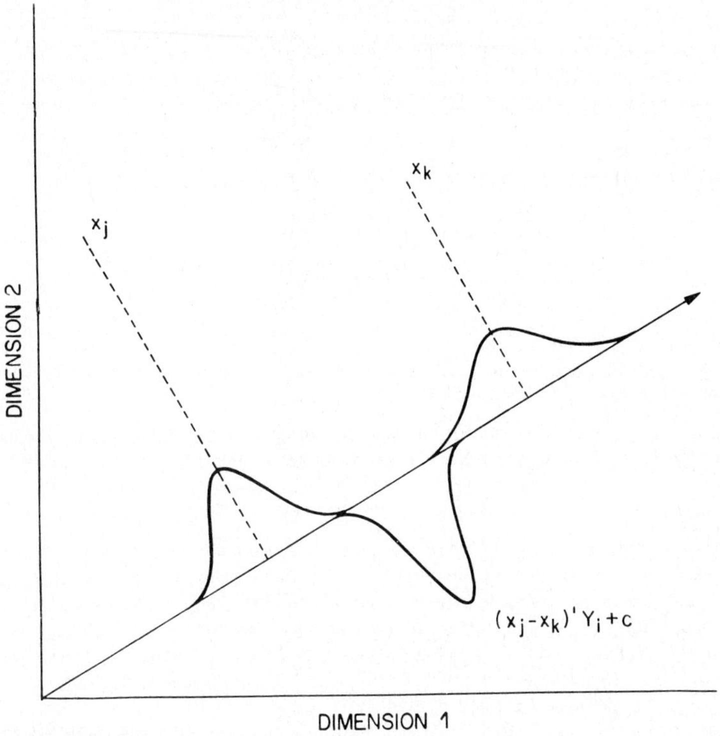

Figure 2. Two objects and their utility distributions on the subject vector. The distribution of $\mathbf{x}_j - \mathbf{x}_k)'\mathbf{Y}_i$ is translated such that its expectation coincides with $(\mathbf{x}_j'\boldsymbol{\mu}_i + \mathbf{x}_k'\boldsymbol{\mu}_i)/2$ (i.e., $\mathbf{c} = (3\mathbf{x}_k - x_j)'\boldsymbol{\mu}_i/2$).

Multiplying both sides of (8) by $||\mathbf{y}_i||$ shows that subject i prefers j to k whenever

$$\mathbf{x}_j'\mathbf{y}_i > \mathbf{x}_k'\mathbf{y}_i \,.$$

Consequently, the probability that subject i prefers choice object j to k, written as p_{ijk}, is

$$\begin{aligned} p_{ijk} &= \text{Prob}\,\{\mathbf{x}_j'\mathbf{Y}_i > \mathbf{x}_k'\mathbf{Y}_i\} \\ &= \text{Prob}\,\{(\mathbf{x}_j - \mathbf{x}_k)'\mathbf{Y}_i > 0\}\,. \end{aligned} \tag{9}$$

Since it follows from (7) that

$$(\mathbf{x}_j - \mathbf{x}_k)'\mathbf{Y}_i \sim N((\mathbf{x}_j - \mathbf{x}_k)'\boldsymbol{\mu}_i,\ \delta^2_{ijk}) \tag{10}$$

with

$$\delta_{ijk}^2 = (\mathbf{x}_j - \mathbf{x}_k)' \, \boldsymbol{\Sigma}_i (\mathbf{x}_j - \mathbf{x}_k) \,, \tag{11}$$

eq. (9) becomes

$$p_{ijk} = \phi \left[\frac{(\mathbf{x}_j - \mathbf{x}_k)' \boldsymbol{\mu}_i}{\delta_{ijk}} \right] , \tag{12}$$

which constitutes the general formulation of the WV model. When $N = 1$, the WV model can be shown to be a special case of a Thurstonian choice model proposed by Heiser and de Leeuw (1981) and Takane (1980) (cf. De Soete, 1983).

3.2 Properties

To show that the WV model is a moderate utility model, we can rewrite (12) as

$$p_{ijk} = \phi \left[\frac{u_{ij} - u_{ik}}{\delta_{ijk}} \right] \tag{13}$$

with

$$u_{ij} = \mathbf{x}_j' \boldsymbol{\mu}_i \,. \tag{14}$$

Since as a variance-covariance matrix $\boldsymbol{\Sigma}_i$ is always positive (semi-)definite, δ_{ijk} is a (semi-) metric and (13) is of the form (6).

The following counterexample illustrates that the choice probabilities generated by the WV model do not necessarily satisfy strong stochastic transitivity. Let $\boldsymbol{\mu}_i = (1, 1/3)'$, $\boldsymbol{\Sigma}_i = \mathbf{I}$, $\mathbf{x}_j = (1/3, 2)'$, $\mathbf{x}_k = (2/3, 1)'$, $\mathbf{x}_l = (1/3, 2/3)$, then $p_{ijk} = 0.50$ and $p_{ikl} = 0.83$, but $p_{ijl} = 0.63$.

To demonstrate that a choice probability p_{ijk} is influenced not only by the difference in mean utilities $u_{ij} - u_{ik}$, but also by the distance δ_{ijk} between the two choice objects, compare Figures 2 and 3. In these figures, two choice objects j and k are shown as well as the utility distributions $\mathbf{x}_j'\mathbf{Y}_i$ and $\mathbf{x}_k'\mathbf{Y}_i$. On the other side of the subject vector, the distribution of $(\mathbf{x}_j - \mathbf{x}_k)'\mathbf{Y}_i$ is drawn translated such that its expectation coincides with $(\mathbf{x}_j'\boldsymbol{\mu}_i + \mathbf{x}_k'\boldsymbol{\mu}_i)/2$. The distributions of $\mathbf{x}_j'\mathbf{Y}_i$ and $\mathbf{x}_k'\mathbf{Y}_i$ are the same in both figures. However, because the distance between $\mathbf{x}_j$ and $\mathbf{x}_k$ is larger in Figure 3 than in Figure 2, the variance of $(\mathbf{x}_j - \mathbf{x}_k)'\mathbf{Y}_i$ is larger in Figure 3 than in Figure 2. This will give rise to less extreme choice proportions.

3.3 Degrees of freedom of the WV model

In its general form, the WV model has

$$Mr + N[r + r(r+1)/2] \tag{15}$$

parameters. However, the model does not determine all these parameters uniquely. More specifically, the choice probabilities defined by (12) remain invariant under the following transformations of the parameters:

a) *Dilation of the subject vectors*
Replacing

$$\mathbf{Y}_i \sim N(\boldsymbol{\mu}_i, \boldsymbol{\Sigma}_i)$$

by

$$\alpha_i \mathbf{Y}_i \sim N(\alpha_i \boldsymbol{\mu}_i, \alpha_i^2 \boldsymbol{\Sigma}_i) \,,$$

where α_i is an arbitrary positive constant, does not affect the choice probabilities.

b) *Translation of the object points*
Applying to every object point $\mathbf{x}_j$ ($j = 1, M$) the transformation

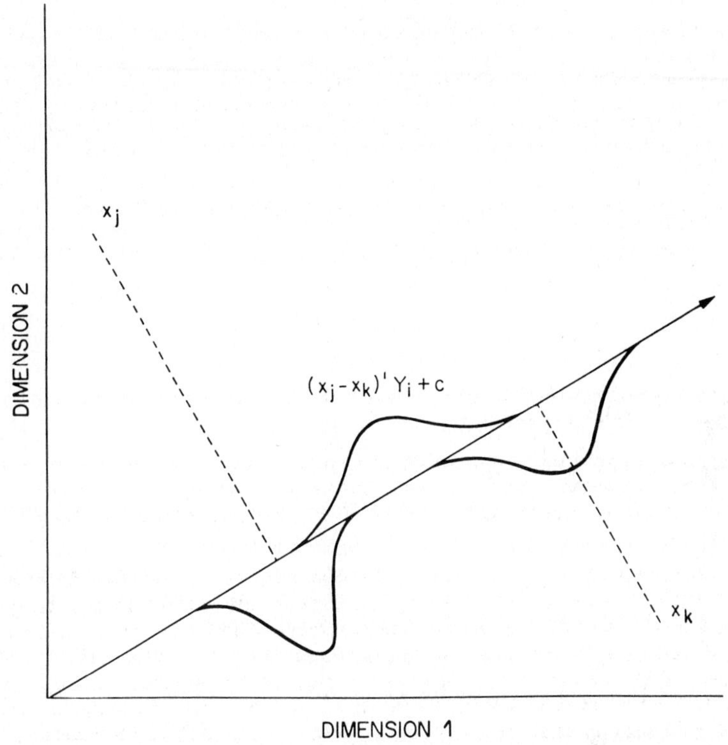

Figure 3. Two objects with identical utility distributions as in Figure 2, but the distribution of $\mathbf{x}_j - \mathbf{x}_k)'\mathbf{Y}_i$ has a larger variance because $\mathbf{x}_j$ and $\mathbf{x}_k$ are farther apart.

$$\mathbf{x}_j \rightarrow \boldsymbol{\beta} + \mathbf{x}_j \, ,$$

where $\boldsymbol{\beta}$ is an arbitrary r-component vector, leaves the choice probabilities invariant.

c) *Nonsingular transformation of the object points and the subject vectors*
If $\mathbf{Q}$ is an arbitrary r by r matrix such that $\mathbf{Q}^{-1}$ exists, then the choice probabilities are not affected by the following transformations

$$\mathbf{x}_j \rightarrow \mathbf{Q}'\mathbf{x}_j \tag{16a}$$
$$\mathbf{Y}_i \rightarrow \mathbf{Q}^{-1}\mathbf{Y}_i \, , \tag{16b}$$

where

$$\mathbf{Q}^{-1}\mathbf{Y}_i \sim N(\mathbf{Q}^{-1}\boldsymbol{\mu}_i, \;\; \mathbf{Q}^{-1}\boldsymbol{\Sigma}_i(\mathbf{Q}^{-1})') \, .$$

Because of these indeterminacies, the degrees of freedom of the general WV model reduce to

$$Mr + N[r + r(r+1)/2] - N - r - r^2 =$$
$$Mr + N[r(r+3)/2 - 1] - r - r^2 \,. \tag{17}$$

3.4 Special cases

In empirical applications, it might be interesting to impose restrictions on the general model to verify specific hypotheses. The validity of a hypothesis can then be tested by statistically comparing the fit of the restricted model with the fit of the general model.

De Soete and Carroll (1983) consider the special case where the variance-covariance matrices of the subject vector termini are restricted to be proportional to each other, i.e.,

$$\boldsymbol{\Sigma}_i = c_i \boldsymbol{\Sigma} \quad (c_i > 0) \,. \tag{18}$$

Because of the indeterminacy expressed in (16), $\boldsymbol{\Sigma}$ can without loss of generality be set equal to an identity matrix. Defining $\boldsymbol{\mu}_i^*$ as $\boldsymbol{\mu}_i / \sqrt{c_i}$, this restricted model can be written as

$$p_{ijk} = \Phi\left(\frac{(\mathbf{x}_j - \mathbf{x}_k)'\boldsymbol{\mu}_i^*}{d(\mathbf{x}_j, \mathbf{x}_k)}\right), \tag{19}$$

where $d(.,.)$ denotes the Euclidean distance function, i.e.,

$$d^2(\mathbf{x}_j, \mathbf{x}_k) = (\mathbf{x}_j - \mathbf{x}_k)'(\mathbf{x}_j - \mathbf{x}_k) \,. \tag{20}$$

The degrees of freedom of this restricted model are

$$(M + N)r - r(r+1)/2 - 1 \,. \tag{21}$$

When $N = 1$, (21) equals (17) showing that (18) only imposes real constraints on the general model when $N > 1$. One could also constrain the $\boldsymbol{\Sigma}_i$ to be diagonal. Because of the indeterminacy stated in (16), this case only imposes real constraints on the general model when $N > 2$. The degrees of freedom for this model are

$$Mr + N(2r - 1) - 2r \,. \tag{22}$$

Note that (17) exceeds (22) only when $N > 2$.

Besides, or in addition to, constraining the variance-covariance matrices $\boldsymbol{\Sigma}_i$, various linear constraints could be imposed on the coordinates of the object points in order to relate these coordinates to known characteristics of the choice objects. Similarly, the directions of the subject vectors can be related to background information on the subjects by imposing appropriate linear restrictions on the $\boldsymbol{\mu}_i$.

De Soete and Carroll (1983) introduced in the general model an additional error term to account for response variability due to unaccounted variance associated with dimensions not present in the model. This extended model can, however, be shown to be a special case of the general model. Suppose, in analogy to the factor analysis model, that the M objects have r ($< M$) dimensions in common and that in addition there is a specific dimension for each object. The object coordinates can therefore be written as

$$\mathbf{X}^* = (\mathbf{X} \;\; \mathbf{I}_M) \,, \tag{23}$$

where $\mathbf{X} = (\mathbf{x}_1, \ldots, \mathbf{x}_M)'$ contains the coordinates of the M objects on the r common dimensions and $\mathbf{I}_M$ is an identity matrix of order M. Assume that $\mathbf{Y}_i^*$, the $r + M$ dimensional random variable representing the terminus of the i'th subject vector, is distributed as follows

$$\mathbf{Y}_i^* \sim N\left(\begin{bmatrix} \boldsymbol{\mu}_i \\ {}_M\mathbf{0}_1 \end{bmatrix}, \begin{bmatrix} \boldsymbol{\Sigma}_i & {}_r\mathbf{0}_M \\ {}_M\mathbf{0}_r & \gamma_i^2 \mathbf{I}_M \end{bmatrix}\right) \tag{24}$$

where ${}_s\mathbf{0}_t$ denotes an s by t matrix filled with zeros. I.e., $\mathbf{Y}_i^*$ is assumed to have variance γ_i^2 along each specific dimension. Now, since

$$(\mathbf{x}_j^* - \mathbf{x}_k^*)'\mathbf{Y}_i^* \sim N((\mathbf{x}_j - \mathbf{x}_k)'\boldsymbol{\mu}_i,\ \ \delta_{ijk}^2 + 2\gamma_i^2), \tag{25}$$

the model becomes

$$p_{ijk} = \Phi\left(\frac{(\mathbf{x}_j - \mathbf{x}_k)'\boldsymbol{\mu}_i}{\sqrt{\delta_{ijk}^2 + 2\gamma_i^2}}\right). \tag{25}$$

For a more extensive discussion of this model in connection with restriction (18), the reader is referred to De Soete and Carroll (1983). Note that (24) could be generalized by allowing for non-zero expectations on the specific dimensions as well as for differential variances along the specific dimensions. The latter extension is closely related to a general model for proximity data proposed by Winsberg and Carroll (1985) called the extended two-way Euclidean model, and generalized to the three-way (or individual differences) case by Carroll and Winsberg (1985). The three-way generalization has been called the extended INDSCAL model by Carroll and Winsberg.

4. THE WANDERING IDEAL POINT MODEL

The geometric representation on which the WV model is based assumes that all subjects have a monotone preference function on each dimension. I.e., it is assumed that the larger the value of an object on a dimension, the more the object will be preferred. This is true for a lot of object attributes, but certainly not for all attributes. If, for example, the objects are job compensation plans characterized by the amount of salary and the number of vacation days, it is very likely that when the number of vacation days is constant, the subject will prefer the job with the largest salary and vice versa. But, if the objects are coffee brands varying in bitterness, a subject probably has a most preferred bitterness level. The farther away the bitterness of a coffee brand is from this optimum level, the less it will be preferred. This implies that the subject does not have a monotone, but a single-peaked preference function on the bitterness dimension.

A model that allows for single-peaked preference functions is Coombs' (1964) unfolding model. In the next section, we present a probabilistic unfolding model, called the wandering ideal point (WIP) model, which was recently proposed by De Soete, Carroll, and DeSarbo (1986).

4.1 General formulation

In the WIP model, both the subjects and the choice objects are presented as points in a joint r-dimensional Euclidean space. Whereas the choice objects $1, 2, \ldots, M$ are represented by fixed points $\mathbf{x}_1, \mathbf{x}_2, \ldots, \mathbf{x}_M$, the subjects are represented by random points. More specifically, each subject i ($i = 1, N$) is represented by a random point $\mathbf{Y}_i$ which follows a multivariate normal distribution:

$$\mathbf{Y}_i \sim N(\boldsymbol{\mu}_i, \Sigma_i)\ . \tag{26}$$

As with the WV model, it is assumed that the distributions of the N subject points are independent of each other, i.e.,

$$\text{Covar}\,(\mathbf{Y}_i, \mathbf{Y}_{i'}) = \mathbf{0}$$

for $i, i' = 1, N$ and $i \neq i'$.

Each time a pair of choice objects (j, k) is presented to a subject i, he or she samples a point $\mathbf{y}_i$ from $\mathbf{Y}_i$. In accordance with Coombs' (1964) unfolding theory, the subject prefers j to k whenever

$$d(\mathbf{y}_i, \mathbf{x}_j) < d(\mathbf{y}_i, \mathbf{x}_k) \tag{27}$$

with $d(\cdot,\cdot)$ defined in eq. (20). An illustration of the WIP model is given in Figure 4.

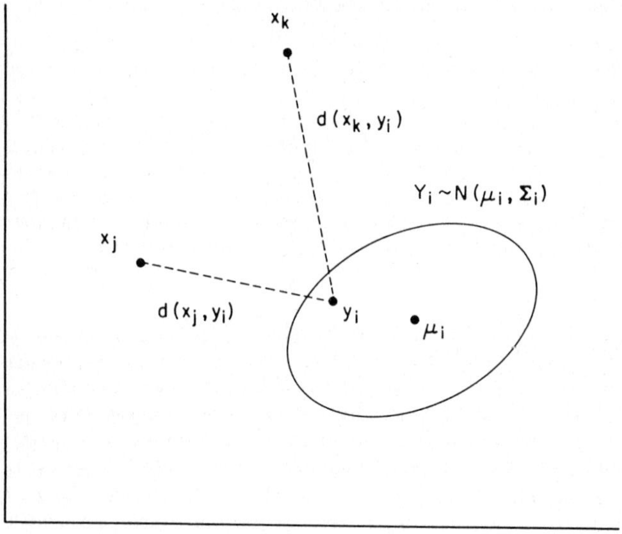

Figure 4. Illustration of the wandering ideal point model. The ellipse represents the random subject point.

In the figure, $\mathbf{y}_i$ is closer to $\mathbf{x}_j$ than to $\mathbf{x}_k$. Therefore, subject i would prefer j to k. Because a subject always chooses the object that is closest to $\mathbf{y}_i$, $\mathbf{y}_i$ can be considered as subject i's ideal point. However, since each time a pair of objects is presented, a new $\mathbf{y}_i$ is sampled, the subject's ideal point is not fixed but "wanders" from trial to trial. Hence the name the wandering ideal point model.

By squaring both sides of (27) and rearranging terms, we obtain that subject i prefers j to k whenever

$$(\mathbf{x}_k - \mathbf{x}_j)'\mathbf{y}_i < (\mathbf{x}_k'\mathbf{x}_k - \mathbf{x}_j'\mathbf{x}_j)/2 .$$

Therefore, the probability that subject i prefers object j to k is

$$p_{ijk} = \text{Prob}\,\{(\mathbf{x}_k - \mathbf{x}_j)'\mathbf{Y}_i < (\mathbf{x}_k'\mathbf{x}_k - \mathbf{x}_j'\mathbf{x}_j)/2\} . \tag{28}$$

It follows from (26) that

$$(\mathbf{x}_k - \mathbf{x}_j)'\mathbf{Y}_j \sim N((\mathbf{x}_k - \mathbf{x}_j)'\boldsymbol{\mu}_i,\ \delta^2_{ijk}) \tag{29}$$

where

$$\delta^2_{ijk} = (\mathbf{x}_k - \mathbf{x}_j)'\boldsymbol{\Sigma}_i(\mathbf{x}_k - \mathbf{x}_j) . \tag{30}$$

Consequently, eq. (28) becomes

$$p_{ijk} = \Phi\left(\frac{(\mathbf{x}_k'\mathbf{x}_k - \mathbf{x}_j'\mathbf{x}_j) - 2(\mathbf{x}_k - \mathbf{x}_j)'\boldsymbol{\mu}_i}{2\delta_{ijk}}\right). \tag{31}$$

Equation (31) provides the general formulation of the WIP model.

4.2 Properties

Again it is straightforward to prove that the WIP model is a moderate utility model. By defining

$$u_{ij} = \mathbf{x}_j'\boldsymbol{\mu}_i - \mathbf{x}_j'\mathbf{x}_j/2, \tag{32}$$

eq. (31) can be written as

$$p_{ijk} = \phi\left(\frac{u_{ij} - u_{ik}}{\delta_{ijk}}\right). \tag{33}$$

By means of a simple counterexample we show that the choice probabilities defined by the WIP model do not necessarily satisfy strong stochastic transitivity. Let $\boldsymbol{\mu}_i = (0, 0)'$, $\boldsymbol{\Sigma}_i = I$, $\mathbf{x}_j = (3, 1)'$, $\mathbf{x}_k = (4, 0)'$, $\mathbf{x}_l = (-5, 0)'$, then $p_{ijk} = 0.98$ and $p_{ikl} = 0.69$, but $p_{ijl} = 0.82$.

Figure 5, taken from De Soete, Carroll, and DeSarbo (1986), visualizes some of the properties of the WIP model. When the distance between $\mathbf{x}_j$ and $\boldsymbol{\mu}_i$ and between $\mathbf{x}_k$ and $\boldsymbol{\mu}_i$ (in the figure indicated as d_{ij} and d_{ik} respectively) are fixed, the probability of preferring j to k varies as a function of the distance between $\mathbf{x}_j$ and $\mathbf{x}_k$. The figure clearly illustrates that extreme choice proportions are likely to occur when the choice objects are close, while distant choice objects are likely to induce more moderate choice proportions.

4.3 Degrees of freedom of the WIP model

Just like the WV model, the WIP model has in its general form

$$Mr + N[r + r(r+1)/2]$$

parameters. Again, all these parameters are not uniquely determined by the model. More precisely, the choice probabilities defined by (33) are invariant under the following transformations of the parameters:

a) *Translation of the subject and object points*
Adding the same arbitrary r-component vector to all subject and object points, does not affect the choice probabilities.

b) *Central dilation of the subject and the object points*
The transformations

$$\mathbf{x}_j \rightarrow \alpha\mathbf{x}_j \quad (i = 1, M)$$
$$\mathbf{Y}_i \rightarrow \alpha\mathbf{Y}_i \quad (i = 1, N),$$

where α is an arbitrary positive constant, leaves the choice probabilities invariant. Note that

$$\alpha\mathbf{Y}_i \sim N(\alpha\boldsymbol{\mu}_i, \alpha^2\boldsymbol{\Sigma}_i).$$

c) *Orthogonal rotation of the subject and object points*
Applying the same orthogonal rotation $\mathbf{T}$ to all object and subject points does not affect the choice probabilities predicted by the model. Note that the distribution of $\mathbf{TY}_i$ is

$$\mathbf{TY}_i \sim N(\mathbf{T}\boldsymbol{\mu}_i, \mathbf{T}\boldsymbol{\Sigma}_i\mathbf{T}').$$

By subtracting these indeterminacies from the total number of parameters, we obtain the degrees of freedom of the general WIP model:

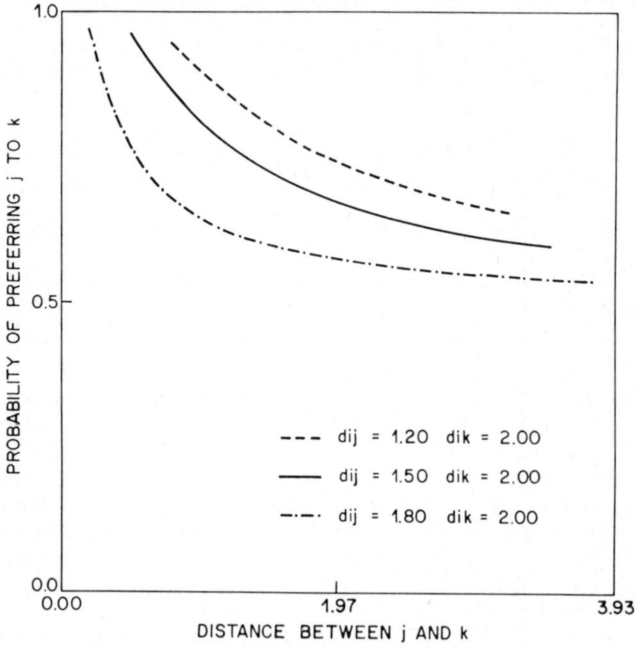

Figure 5. Probability of preferring j to k as a function of the distance between j and k for fixed d_{ij} and d_{ik} (Taken from De Soete, Carroll, & DeSarbo, 1986).

$$(M + N)r + Nr(r + 1)/2 - r(r + 1)/2 - 1 . \tag{34}$$

4.4 Special cases

As with the WV model, it might be interesting to impose certain restrictions on the general model. First of all, various kinds of restrictions can be imposed on the variance-covariance matrices of the subject points. The Σ_i can for instance be constrained to be diagonal. Due to the rotational indeterminacy of the general model, setting the off-diagonal elements of the variance-covariance matrices equal to zero only imposes real constraints on the general model when $N > 1$. The degrees of freedom for this model are

$$(M + 2N)r - r - 1 .$$

A more restrictive case constrains all Σ_i to be identity matrices. In this case, the model has

$$(M + N)r - r(r + 1)/2$$

degrees of freedom.

Analogously to the last special case discussed in the context of the WV model, one could assume besides r ($< M$) common dimensions, a specific dimension for each choice object. If the object coordinates can be written as in (23) and if the subject points are distributed as in (24), the WIP model becomes

$$p_{ijk} = \Phi\left[\frac{(\mathbf{x}_k'\mathbf{x}_k - \mathbf{x}_j'\mathbf{x}_j) - 2(\mathbf{x}_k - \mathbf{x}_j)'\mu_i}{2\sqrt{\delta_{ijk}^2 + 2\gamma_i^2}}\right]. \tag{35}$$

5. ILLUSTRATIVE APPLICATION

In the previous sections, two probabilistic multidimensional choice models have been described. In order to apply the models, one must dispose of replicated paired comparisons data from one or more subjects (or groups of subjects). On the basis of these data, the model parameters need to be estimated. This can be done in several ways. Maximum likelihood estimation using Fisher's scoring method has been found quite satisfactory for estimating the parameters of certain special cases of the WV and the WIP model (cf. De Soete & Carroll, 1983; De Soete et al., 1986). Maximum likelihood estimating has the advantage of enabling model selection by means of likelihood ratio tests. However, likelihood ratio statistics can only be constructed for comparing the fit of nested models. When non-nested models are to be compared, one can use an information criterion such as those proposed by Akaike (1977), Chow (1981), and Schwarz (1978). For illustrative purposes, this approach is applied to a data set collected by Rumelhart and Greeno (1971).

Rumelhart and Greeno (1971) obtained pairwise preference judgments from 234 undergraduates about nine celebrities. The nine celebrities consisted of three politicians (L. B. Johnson, Harold Wilson, Charles De Gaulle), three athletes (A. J. Foyt, Johnny Unitas, Carl Yastrzemski), and three movie stars (Brigitte Bardot, Sophia Loren, Elizabeth Taylor). The subjects were treated as replications of each other (as might often be assumed in practical marketing applications where, for example, balanced incomplete block or other designs may be used in which each subject judges only a subset of the complete set of pairs), yielding a single replicated paired comparison matrix. These data were analyzed according to a variety of models using maximum likelihood estimation. The results of these analyses are summarized in Table 1. The null model refers to a completely saturated model in which no structural constraints are imposed on the choice probabilities. In addition to the wandering vector model and the wandering ideal point model, two other models are considered: Thurstone's (1927) model of comparative judgment Case V (defined in eq. (1)) and the factorial model of comparative judgment proposed by Takane (1980) and Heiser and de Leeuw (1981). This model amounts to Thurstone's general model of comparative judgment in which the matrix of covariances between the stimulus utilities is constrained to have a prescribed rank. De Soete (1983) showed that when $N = 1$, the WV model is a special case of the factorial model. The goodness-of-fit of the different models is assessed by a likelihood ratio test against the null model. As can be inferred from column 5 of Table 1, the factorial model, the WV model and the WIP model all give a good account of the data, while Thurstone's Law of Comparative Judgment Case V must be rejected. Contrary to the multidimensional models, Thurstone's Case V implies strong stochastic transitivity. Two versions of the WIP model were applied: the general model with a diagonal variance-covariance matrix (which is in the case of $N = 1$ equivalent to using an unconstrained variance-covariance matrix) and the WIP model with an identity matrix as variance-covariance matrix. The chi-square statistic for comparing both representations has 2 degrees of freedom and amounts to 0.3, which is clearly nonsignificant. This implies that the ideal point wanders to an equal degree in all directions of the space. Of all models considered here, the WIP model with an identity variance-covariance matrix gives the best representation of the Rumelhart and Greeno (1971) data, according to Akaike's (1977) AIC criterion. This

Table 1. Summary of the analyses on the Rumelhart and Greeno (1971) data

model	log L (+5000)	effective no. of parameters	AIC (−10000)	Likelihood ratio test against null model χ^2	d.f.
null model	-310.6	36	693.3		
Thurstone's model of comparative judgment, case V	-351.4[a]	8	718.7	81.6[d]	28
factorial model of comparative judgment, 2 dim.	-314.1[a]	22	672.1	7.0	14
wandering vector model, 2 dim.	-318.4[b]	16	668.8	15.4	20
wandering ideal point model, 2 dim., diag. Σ_1	-315.5[c]	19	669.1	9.8	17
wandering ideal point model, 2 dim., $\Sigma_1 = I$	-315.7[c]	17	665.1	10.1	19

[a]Obtained by Takane. These values are not identical to those reported in Table 2 of Takane (1980) due to an error in the data on which the latter table is based.
[b]Obtained by De Soete (1983).
[c]Obtained by De Soete, Carroll, and DeSarbo (1986).
[d]$p < 0.001$

two-dimensional solution is presented in Figure 6. As is apparent from the figure, the politicians, the athletes, and the movie stars clearly show up as separate clusters. The politicians constitute the most preferred group of celebrities, while the movie stars are generally preferred to the athletes.

Although the data analyzed in the application discussed above are not real marketing data, the analyses should shed some light on the possible usefulness of the wandering vector model and the wandering ideal point model in marketing research. In fact, such analyses of data on celebrities could be useful in a marketing context in which, for example, one were interested in selecting one or more celebrities for endorsing a product in an advertising campaign.

6. APPLICATIONS TO TELECOMMUNICATIONS DEMAND MODELING

A potential application of the wandering vector and/or wandering ideal point models (and the associated maximum likelihood estimation (MLE) methods for fitting these models) to telecommunications demand modeling might take the following form:

Subjects, who might be either a random or selectively stratified sample of consumers of telecommunications services, could be presented with a number of potential pricing options for such services, in which various combinations of pricing "features" (e.g., discounts on service, say for calling during off-peak hours or for calls to specific geographic areas, options involving special charge card services, or other specialized pricing or price related attributes) could be varied systematically. This would result in a

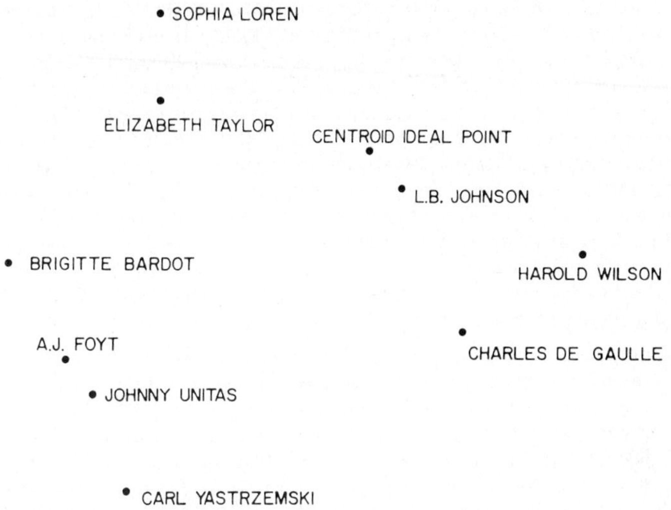

Figure 6. Representation of the Rumelhart and Greeno (1971) data according to the wandering ideal point model with identity covariance matrix.

complete or fractional factorial design, such as is often used in "conjoint analysis" approaches, entailing pricing "packages" defined in terms of profiles comprising presence or absence of various of these features, or of combinations of levels of multi-level features. In cases in which the pricing options have been so designed as to allow an overall cost factor to vary relatively independently of other factors, one might expect the wandering vector model (in which subjective utility varies monotonically with dimensions) to provide optimal fit, while in cases where the options have been so designed as to build in price-value tradeoffs in an economically realistic manner, the wandering ideal point model might be more appropriate. Since fitting either the WV or the WIP model does not require prespecification of the dimensions underlying the stimuli (the profiles defining the pricing "packages" in the present example), these dimensions may be related in a simple linear fashion to the features in terms of which the stimuli (pricing alternatives) are defined, or may, on the other hand, be related in a much more complex non-linear fashion. If an initial (unconstrained) exploratory analysis suggests a fairly simple (e.g., linear) relationship, the constrained versions of the MLE fitting procedures could then be used in a confirmatory statistical mode, to fit models with explicit constraints relating the stimulus (pricing alternative) dimensions to these a priori features. Such explicit constraints might be quite useful in connection with designing new products (pricing plans) comprising different combinations of the feature sets. Such "new product design" could be based on seeking combinations of features (or feature levels) corresponding to theoretical stimulus positions predicted, based on extrapolation from the fitted model(s), to be optimal from a marketing point of view (i.e., to optimize market share, profitability, or some other well defined objective).

A word would seem appropriate concerning data collection for purposes of fitting these models. While in a laboratory setting it may well be feasible to collect complete paired comparisons data from each subject, even on all pairs of a reasonably large set of alternatives (say 10-20), in field trials or telephone interview situations this probably is not feasible. If, however, one is willing to treat different subjects as replications (as we have done in the analyses reported above on the "celebrities" data, even though complete paired comparisons data were in fact collected from each subject) it should be quite feasible to carry out analyses in terms of these models even with such relatively sparse data from each respondent. For many kinds of stimuli this assumption (of interchangeability of subjects, implying lack of systematic individual differences) appears quite reasonable; even in cases in which it might not be readily justified, either theoretically or empirically, the models seem to be sufficiently robust that the results are most probably not seriously affected. In such cases the distributions of vector termini or ideal points are presumably composites of systematic between subject variation (of centroids of vectors or of ideal points, as the case may be) and within subject variability (of the type posited in the wandering vector or ideal point models, respectively).

In large scale field trials or telephone interviews it should be possible to segment respondents on the basis of calling patterns, demographic or "psychographic" information, and/or by use of some data-based clustering procedure, and then to define subsets of respondents corresponding to these segments, which could then be treated as "pseudo-subjects" in analyses allowing fitting of individual differences parameters (vector or ideal point centroids and/or covariance matrices) for these segmental "pseudo-subjects". Combined with appropriate computational algorithms for predicting probabilities of first choice from among larger sets of alternatives (ideally, separately within "pseudo-subject" segments) and with appropriate use of the statistical tests and procedures for imposing constraints on the solution configurations, these models and associated methods should provide a very powerful methodology for both exploratory and confirmatory analysis of preferential choice data in marketing studies related to pricing policy, as well as to more standard marketing situations concerned with new product design and promotion (e.g., in the areas of new telecommunications technology, computer hardware and software, or integrated telecommunications/computer network systems).

In the specific case of telecommunications demand analysis, synthesis of modeling of choice behavior based on the WV and WIP models with other stochastic choice models based on econometric principles, and incorporating standard economic variables, should lead to especially powerful new techniques for measurement and prediction of consumer response to a wide variety of telecommunications products. The strength of such a synthesis of psychometric and econometric modeling and methodology lies in the potential for encompasssing both subjective factors related to consumer perception of alternatives and directly measurable objective economic and demographic variables in composite models/methods more realistic, powerful, and accurate than those based on either approach (the psychometric or the econometric) alone. This should lead to a concomitant increase in the precision of prediction of consumer choice among complex pricing plans (or other alternatives in which price is a significant factor) and thus of forecasting of demand, anticipation of effects of introduction of new pricing plans or other products by competitors, or, most critically, development of new pricing plans or other products designed to optimize strategically important marketing objectives. Such new methodology, simultaneously sensitive to essential economic, psychological, sociological and technological factors, and capable of rapid adaptation to the dynamically evolving competitive environment in which the entire telecommunications industry now finds itself, would seem to be the conditio sine qua non for effectively navigating the exceedingly convoluted terrain our society in general, and our industry in particular, must traverse for the foreseeable future. Perhaps the general approach to analysis of consumer choice provided by the models presented in this paper can provide at least a part of the solution to this momentous and intellectually challenging problem.

REFERENCES

[1] Akaike, H. (1977). On entropy maximization principle. In P. R. Krishnaiah (Ed.), *Applications of statistics* (pp. 27-41). Amsterdam: North-Holland.

[2] Becker, G. M., DeGroot, M. H., & Marschak, J. (1963). Probabilities of choice among very similar objects. *Behavioral Science, 8*, 306-311.

[3] Bradley, R. A., & Terry, M. E. (1952). Rank analysis of incomplete block designs. I. The method of paired comparisons. *Biometrika, 39*, 324-345.

[4] Carroll, J. D. (1980). Models and methods for multidimensional analysis of preferential choice (or other dominance) data. In E. D. Lantermann & H. Feger (Eds.), *Similarity and choice* (pp. 234-289). Bern: Huber.

[5] Carroll, J. D., & Winsberg, S. (1985). Maximum likelihood procedures for metric and quasi-nonmetric fitting of an extended INDSCAL model assuming both common and specific dimensions. Paper presented at the Multidimensional Data Analysis Workshop, Cambridge, England. (Abstract to be published in *Proceedings of Multidimensional Data Analysis Workshop*. Leiden: DSWO Press)

[6] Chow, G. C. (1981). A comparison of information and posterior probability criteria for model selection. *Journal of Econometrics, 16*, 21-33.

[7] Coombs, C. H. (1964). *A theory of data*. New York: Wiley.

[8] De Soete, G. (1983). On the relation between two generalized cases of Thurstone's Law of Comparative Judgment. *Mathématiques et Sciences humaines, 81*, 47-57.

[9] De Soete, G., & Carroll, J. D. (1983). A maximum likelihood method for fitting the wandering vector model. *Psychometrika, 48*, 553-566.

[10] De Soete, G., & Carroll, J. D. (in press). Probabilistic multidimensional choice models for representing paired comparisons data. In E. Diday et al. (Eds.), *Data analysis and information IV*. Amsterdam: North-Holland.

[11] De Soete, G., Carroll, J. D., & DeSarbo, W. S. (1986). The wandering ideal point model: A probabilistic multidimensional unfolding model for paired comparisons data. *Journal of Mathematical Psychology, 30*, 28-41.

[12] Edgell, S. E., & Geisler, W. S. (1980). A set-theoretic, random utility model of choice behavior. *Journal of Mathematical Psychology, 21*, 265-278.

[13] Halff, H. M. (1976). Choice theories for differentially comparable alternatives. *Journal of Mathematical Psychology, 14*, 244-246.

[14] Harshman, R. A., & Lundy, M. E. (1987). Multidimensional analysis of preference structures. This volume.

[15] Heiser, W. J., & de Leeuw, J. (1981). Multidimensional mapping of preference data. *Mathématiques et Sciences humaines, 73*, 39-96.

[16] Krantz, D. H. (1967). Rational distance function for multidimensional scaling. *Journal of Mathematical Psychology, 4*, 226-245.

[17] Luce, R. D. (1959). *Individual choice behavior. A theoretical analysis.* New York: Wiley.

[18] Marley, A. A. (1981). Multivariate stochastic processes compatible with "aspect" models of similarity and choice. *Psychometrika, 46*, 421-428.

[19] Restle, F. (1961). *Psychology of judgment and choice: A theoretical essay.* New York: Wiley.

[20] Rumelhart, D. L., & Greeno, J. G. (1971). Similarity between stimuli: An experimental test of the Luce and Restle choice models. *Journal of Mathematical Psychology, 8*, 370-381.

[21] Schwarz, G. (1978). Estimating the dimensions of a model. *Annals of Statistics, 6*, 461-464.

[22] Sjöberg, L. (1975). Uncertainty of comparative judgment and multidimensional structure. *Multivariate Behavioral Research, 10*, 207-218.

[23] Sjöberg, L. (1977). Choice frequency and similarity. *Scandinavian Journal of Psychology, 18*, 103-115.

[24] Sjöberg, L. (1980). Similarity and correlation. In E. D. Lantermann & H. Feger (Eds.), *Similarity and choice* (pp. 70-87). Bern: Huber.

[25] Sjöberg, L., & Cappoza, D. (1975). Preference and cognitive structure of Italian political parties. *Italian Journal of Psychology, 2*, 391-402.

[26] Slater, P. (1960). The analysis of personal preferences. *British Journal of Statistical Psychology, 13*, 119-135.

[27] Strauss, D. (1981). Choice by features: An extension of Luce's choice model to account for similarities. *British Journal of Mathematical and Statistical Psychology, 22*, 188-196.

[28] Takane, Y. (1980). Maximum likelihood estimation in the generalized case of Thurstone's model of comparative judgment. *Japanese Psychological Research, 22*, 188-196.

[29] Thurstone, L. L. (1927). A law of comparative judgment. *Psychological Review, 34*, 273-286.

[30] Tucker, L. R. (1960). Intra-individual and inter-individual multidimensionality; In H. Gulliksen & S. Messick (Eds.), *Psychological scaling: Theory and applications* (pp. 155-167). New York: Wiley.

[31] Tversky, A. (1972). Elimination by aspects: A theory of choice. *Psychological Review, 79*, 281-299.

[32] Tversky, A., & Russo, J. E. (1969). Substitutability and similarity in binary choices. *Journal of Mathematical Psychology, 6*, 1-12.

[33] Tversky, A., & Sattath, S. (1979). Preference trees. *Psychological Review, 86*, 542-573.

[34] Winsberg, S., & Carroll, J. D. (1985). A metric and quasi-nonmetric method for fitting an extended Euclidean model postulating both common and specific dimensions. Submitted to *Psychometrika*.

TELECOMMUNICATIONS DEMAND MODELLING
An Integrated View
A. de Fontenay, M.H. Shugard, D.S. Sibley (Editors)

MULTIDIMENSIONAL ANALYSIS OF PREFERENCE STRUCTURES*

Richard A. Harshman and Margaret E. Lundy

Psychology Department, University of Western Ontario, London, Ontario, Canada N6A 5C2**

A novel approach is proposed for analysis of paired-comparison preference data. Instead of fitting a specific psychological model, an exploratory analysis method called DEDICOM can be used to obtain a multidimensional representation of preferences much like the representation that factor analysis or multidimensional scaling provides for other kinds of data. To demonstrate the method, it is applied to an 8 by 8 matrix giving strength of preferences among foods, and a 9 by 9 matrix of preference choice frequencies among celebrities. Both analyses yield interpretable "dimensions" of stimulus preference or utility. They also demonstrate how DEDICOM can provide (a) simple linear or nonlinear preference scales, when appropriate; (b) multiple "dimensions" of preference when required by the data; (c) scale values for the stimuli on each scale; and (d) increments in the variance accounted for resulting from each additional "dimension" of preference or stimulus utility. Exploratory and psychological-model-based approaches are compared, and two possible explanations for the multidimensionality of preferences are discussed.

1. INTRODUCTION

It is often assumed that preference relationships among a set of objects can be represented by ordering the objects along a single line or dimension, corresponding to the "utility" of the stimuli. One can force a subject's data to have this type of structure by asking him/her to order the objects from most to least preferred. But if instead the subject is asked to independently assess all the pairwise relationships among the stimuli (e.g., by the method of "graded paired comparisons"), the resulting data often seem too complex to be produced by any ordering along a single dimension.

One of two approaches might be taken to investigating these complexities. The first is theory-based, and involves construction of special purpose models which incorporate psychological effects such as modulation of preference strength by stimulus similarity, or hierarchical choice. By fitting these models to observed preference data, one can test how well such

---

* Preparation of this manuscript was supported in part by Bell Communications Research, Morristown, NJ, and by a Natural Sciences and Engineering Research Council of Canada grant to the first author.
** Portions of this research were carried out while the first author was a consultant at Bell Communications Research, Morristown, NJ.

effects account for the complexities, and one can also estimate the "underlying" preferences that the stimuli would elicit if the extra psychological effects were not present. This is the approach taken by most current investigators (e.g., [1]; [2]; [3], section III; [4]; [5]; cf. [6]) and is well exemplified by DeSoete and Carroll's contribution to this volume [7].

The second approach is more descriptive and "data-analytic". It attempts to decompose the complexities by some general purpose analytical procedure similar to multidimensional scaling, rather than theoretically "accounting for" and possibly eliminating them. Recall, for example, how theoretical models often try to account for the "multidimensionality" of preference by first constructing a multidimensional "attribute space" that is evaluatively neutral, and then defining a preference rule which operates on that space. In contrast, DEDICOM constructs no neutral attribute space. It decomposes the preferences themselves, and constructs evaluative "dimensions" which span a preference structure that is only quasi-spatial. One can then examine the obtained preference dimensions to see what kind of psychological theory they might suggest. In more applied situations, one can also use the extracted dimensions or patterns for marketing and forecasting purposes (e.g., to suggest important patterns of judgment underlying complex product evaluations, to infer which stimulus attributes affect preference, or to predict preferences for objects not in the stimulus set).

While interesting data-analytic methods had been developed for certain other kinds of data such as preference rankings (see [1] and [3] for excellent discussions), until recently the proper mathematical tools did not exist for matrices of paired comparisons. Traditional methods of uncovering latent multidimensional structure in a table of relationships, such as factor analysis or multidimensional scaling, all require that the relations be symmetric (as with "correlation" or "similarity"); preferences, on the other hand, are strongly asymmetric. Indeed, preference relationships tend to be anti- or "skew-symmetric" (i.e., pref(a over b) = -pref(b over a)). Recently, several methods for scaling asymmetric relationships have been proposed (e.g., [8]; [9]; [10]; [11]; [12]; [13]); the most general of these is DEDICOM (for DEcomposition into DIrectional COMponents; see [14] and [15]). The following sections illustrate the DEDICOM data-analytic multidimensional approach for discovering the structure of paired comparison preference data.

2. THE DEDICOM PROCEDURE

DEDICOM is a procedure for multidimensional analysis of matrices of relationships. It can be applied to real-valued matrices of any form: symmetric (where $x_{ij}=x_{ji}$); general asymmetric (where x_{ij} might not equal x_{ji}); or skew-symmetric (where $x_{ij}=-x_{ji}$). We restrict discussion here to the special case of skew-symmetric data, since this is the form taken by pairwise preference relationships. (The two-way DEDICOM representation of general asymmetric data is briefly described in Appendix A.) We show how the DEDICOM model directly decomposes a table of pairwise preferences into a few basic preference patterns, each of which can be attributed to one "dimension" or aspect of utility (instead of first finding a stimulus attribute space and then deriving utilities from directions or locations in this space, as most other models do). To facilitate interpretation, we present the component patterns graphically. "Scale values" for particular stimuli can also be determined by taking projections onto straight or curved lines, but because of space restrictions this is not shown.

2.1. The DEDICOM Model for Skew-symmetric Data

Let **X** be a matrix of pairwise preference relationships, with the entry x_{ij} describing the amount that alternative i is preferred over alternative j (a negative preference value is used if j is actually preferred over i). In such a matrix, x_{ij} is usually (approximately) equal to $-x_{ji}$, that is, **X** is usually (approximately) skew-symmetric. Here we assume that any deviations from skew-symmetry are due to measurement error; if the data are not perfectly skew-symmetric, we compute their skew-symmetric part by taking $0.5(\mathbf{X} - \mathbf{X}')$.

The DEDICOM representation of skew-symmetric **X** has the form

$$(1)\quad \mathbf{X} = \mathbf{A}\,\mathbf{R}\,\mathbf{A}' + \mathbf{E}$$

where **A** is an n by s matrix of "loadings" of the n stimuli on s dimensions (for even s only), **R** is a skew-symmetric s by s matrix of relationships among the dimensions, and **E** is an n by n matrix of residual error terms. The model is fit by least squares. For skew-symmetric **X**, this is accomplished by taking the singular value decomposition of X and transforming it to appropriate form (see Appendix B for the details).

Note that we specified for (1) that the number of dimensions be even. This is a consequence of the well-known "even rank" property of skew-symmetric matrices (see, e.g., [16]). That is, <u>two</u> vectors or "dimensions" are required to generate even the simplest skew-symmetric matrix; more complex matrices may require four, six, or more component vectors, but the number required is never odd. Thus, any complex skew-symmetric matrix is represented as the sum of elementary rank-2 skew-symmetric matrices rather than rank-1 components as in factor analysis. Taken pairwise, DEDICOM dimensions (i.e., columns of **A**) generate such rank-2 matrices. **X** is approximated by weighting each of these matrices by the appropriate value from **R** and then summing the weighted matrices.

For example, consider a four-dimensional DEDICOM solution. Let $\mathbf{a}_1$, $\mathbf{a}_2$, $\mathbf{a}_3$ and $\mathbf{a}_4$ be the columns of **A**. Taking them pairwise, we can generate two rank-2 skew-symmetric matrices $\mathbf{K}_1$ and $\mathbf{K}_2$ such that $\mathbf{K}_1 = \mathbf{a}_1\mathbf{a}_2' - \mathbf{a}_2\mathbf{a}_1'$ and $\mathbf{K}_2 = \mathbf{a}_3\mathbf{a}_4' - \mathbf{a}_4\mathbf{a}_3'$. Weighting $\mathbf{K}_1$ and $\mathbf{K}_2$ by elements of **R**, the "orthogonal" (explained below) representation is then $X = r_{12}\mathbf{K}_1 + r_{34}\mathbf{K}_2 + \mathbf{E}$.

The even-rank property of skew-symmetric matrices has important consequences for DEDICOM analysis. First, DEDICOM must deal with pairs of dimensions in much the same way that traditional factor analysis deals with single axes or dimensions. For example, DEDICOM rotates two <u>pairs</u> of dimensions at a time rather than two axes. Second, we do not interpret each individual dimension but rather the plane defined by each pair of dimensions. We view the plane as representing the preference pattern contributed by one kind of utility underlying the stimulus preferences. (Interpretation of single dimensions is not prohibited, but is usually not done for the skew-symmetric case. What we are interested in is dominance patterns, which are revealed by the planes; this is discussed below.) Finally, we use the term "bimension" to refer to the plane defined by two dimensions (i.e., the four-<u>d</u>imensional solution above is said to be two-<u>b</u>imensional). Hereafter, "bimension" will be used to mean the plane defined by a pair of DEDICOM dimensions.

2.2. Transformations or "Rotations"

Immediately after the singular value decomposition of X (see Appendix B), the bimensions are like principal axes, mutually orthogonal and oriented so that each successive bimension accounts for as much of the remaining variance as possible. These initial "unrotated" bimensions usually represent linear combinations of the preference patterns arising from several different kinds of utility. More interpretable bimensions, corresponding to the (partial) preferences due to individual "kinds of utility" or aspects of preference, can often be obtained by means of axis "rotations" similar to those used in factor analysis. Modified versions of Varimax and Orthoblique transformations (e.g., see [17] and [18]) have proven generally useful. The main difference between DEDICOM rotation and standard factor analytic rotation is that DEDICOM transformations are carried out on two planes or bimensions at a time, rather than two axes or dimensions. To accomplish this, two pairs of columns from **A** are postmultiplied by a 4 by 4 transformation matrix T which is composed of four 2 by 2 blocks of the form

$$\begin{matrix} d & c \\ -c & d \end{matrix}$$

This type of constrained transformation is said to "preserve bimensional form" (see [19], pp. 10-12, 32-33). As in factor analysis, the variance is redistributed among bimensions but the fitted sum for each x_{ij} is not changed. Thus, **R** must absorb the inverse transformations; that is, if transformed $\tilde{\mathbf{A}}=\mathbf{AT}$, then transformed $\tilde{\mathbf{R}}=(\mathbf{T}^{-1})\mathbf{R}(\mathbf{T}^{-1})'$.

When **X** is skew-symmetric, so is the corresponding **R** matrix. In the initial, "unrotated" solution, the only nonzero values in **R** are adjacent to the main diagonal; thus **R** is the skew-symmetric equivalent of a diagonal matrix (what we call "skew-diagonal"). The nonzero values correspond to the singular values of **X** (i.e., $d_{11}=r_{12}=-r_{21}$, see Appendix B). After orthogonal planewise transformations, the matrix is still skew-diagonal, but the variance is redistributed among the planes, and so the sizes of r_{12}, r_{34}, etc. are altered. After oblique planar transformations, however, additional cells of **R** are nonzero, and are organized into 2 by 2 blocks as described above. These blocks describe the strength of cross-dominances between elements in one bimensional plane and those in another (i.e., they provide an index of "obliqueness" of the bimensional planes). For preference data, this "obliqueness" indicates the extent to which preference relationships hold across bimensions. (Space limitations prevent us from discussing the algebra of bimension "rotation", or the associated changes in **R** and their interpretation; for details, see [19].)

2.3. Interpretation of Solutions

A bimension can be represented and interpreted geometrically. Each successive pair of DEDICOM axes (i.e., pair of columns from **A**) defines a plane. The stimulus coordinates on those axes can be used to plot a two-dimensional configuration of points which gives the locations of the stimuli in (their projections onto) each bimension. Gower [9] gives a geometric method by which one can interpret the resulting stimulus configuration in terms of dominance (here, preference) relationships. He points out that the dominance of stimulus y over z is proportional to the (directed) area of the triangle whose vertices are y, z and the origin. If the path from the origin to y, then to z, and back to the origin corresponds to clockwise rotation, the area--and associated dominance of y over z--is taken as positive; if counterclockwise, negative. In this way a bimension's contribution to the dominance relationship between any two stimuli can be presented visually. We call such plots of bimensions, with the associated geometric interpretation, "Gower diagrams". In the two examples presented below, we restrict ourselves to interpreting the **A** loading matrix and associated Gower diagrams, but it is also possible, and enlightening, to look at **R**.

3. APPLICATIONS

3.1. Food Preferences

Our first example involves the analysis of an 8 by 8 set of food preferences obtained by the method of "graded paired comparisons". One person rated how much each of eight foods (four entrees and four desserts) was preferred over each of the others, using a scale ranging from -12 to +12, with 0 meaning "no preference" (Table 1). The data were then skew-symmetrized (new $\mathbf{X}=.05(\mathbf{X}-\mathbf{X}')$), and several bimensions extracted.

TABLE 1
Food preference ratings for one subject*

	1	2	3	4	5	6	7	8
1. Beef Burgundy	0	-4	-2	+4	+3	+3	+4	+5
2. Steak	+1	0	+3	+6	+1	+1	+5	+8
3. Pork Chops	0	-2	0	+4	+1	+1	+3	+5
4. Hamburger	-4	-6	-4	0	0	0	+1	+4
5. Chocolate Pie	+1	-3	0	+4	0	+1	+4	+8
6. Lemon Pie	0	-2	+1	+4	-1	0	+4	+8
7. Berry Pie	-4	-6	-4	-3	-6	-5	0	+4
8. Peach Pie	-7	-9	-7	-5	-10	-8	-6	0

*Each rating is amount row food is preferred over column food.

Unrotated solution. Gower diagrams of the first "unrotated" bimension are depicted in Figure 1. To illustrate how such a plot can be interpreted in terms of areas of triangles, we have drawn and shaded two such triangles in Figure 1a. Here (and in the other figures) the most preferred things are placed at the top of the plot so that the direction of decreasing preference corresponds to clockwise rotation about the origin. Thus, the area of the upper triangle represents the amount that Steak is preferred over Beef Burgundy Stew, and the area of the lower triangle the amount that Beef Burgundy Stew is preferred over Hamburger. Note that these two areas approximately add up to the area of the larger triangle that would be formed by connecting the origin, Steak and Hamburger. This illustrates an additive preference hierarchy, that is, a hierarchy where pref(a over b) + pref(b over c) = pref(a over c). (Collinear points represent additive hierarchies because they define a set of triangles sharing a common base --the line-- and a common vertex --the origin-- thus insuring the additivity [9]).

The first unrotated bimension provides an overall summary of the food preference relationships. The four entrees fall approximately on a straight line, as do the four desserts, but the two lines are obliquely oriented

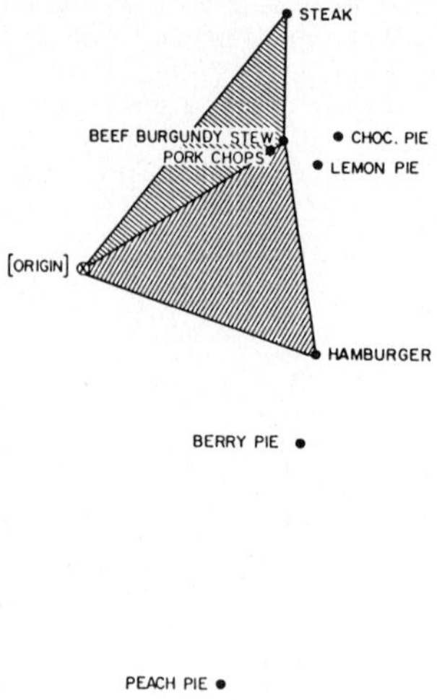

FIGURE 1a
Unrotated one-bimensional solution of food preferences data, showing geometric representation of preferences as areas of triangles.

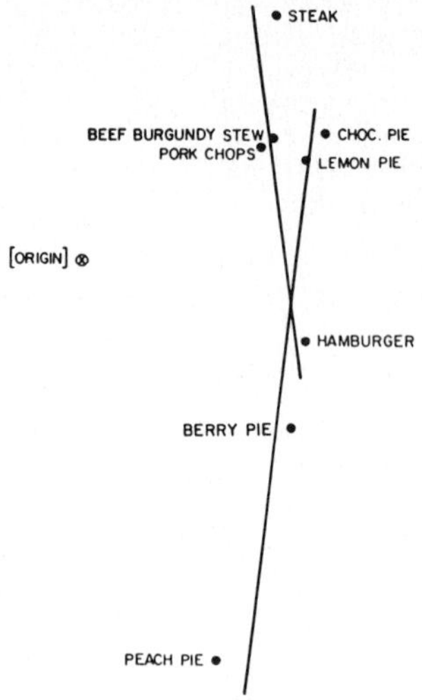

FIGURE 1b
Unrotated one-bimensional solution of food preferences data, showing linear preference hierarchies.

towards one another (Figure 1b). This suggests that the observed preference ratings may involve two separable components. To better resolve them, we obliquely rotated a two-bimensional solution, using planewise Orthoblique transformation (see [18] for a description of the Orthoblique method and [19] for the planewise transformation).

Rotated solution. The first and second rotated bimensions are shown in Figures 2a and 2b, respectively. The first bimension shows a preference hierarchy which holds mainly among desserts, while the second one shows a preference hierarchy among the entrees. Each hierarchy involves a curved rather than straight line, which indicates a modest nonadditivity in the preference relationships. This might only mean that the subject tends to underrate strong preferences relative to weak ones, or it might reflect attenuation of preference by stimulus dissimilarity, as postulated by some current theories (we say more about this later). The desserts cluster relatively close to the origin in Figure 2b and the entrees, except for Steak, do the same in Figure 2a; this suggests that the "kind of utility" represented by each bimension contributes only a small amount to the preference ratings of foods falling off the line. In terms of the subject's taste preference, it is difficult to label what the utility is that each bimension represents. Perhaps it is sufficient to say that the observed preferences can be decomposed into two simpler preference patterns, one for the desserts (and Steak), and one for the entrees (plus inter-group preferences arising from "obliqueness" of bimensions). It is also interesting to note that Steak enjoys two kinds of utility: the kind associated with other entrees, and (to a lesser degree) the special utility

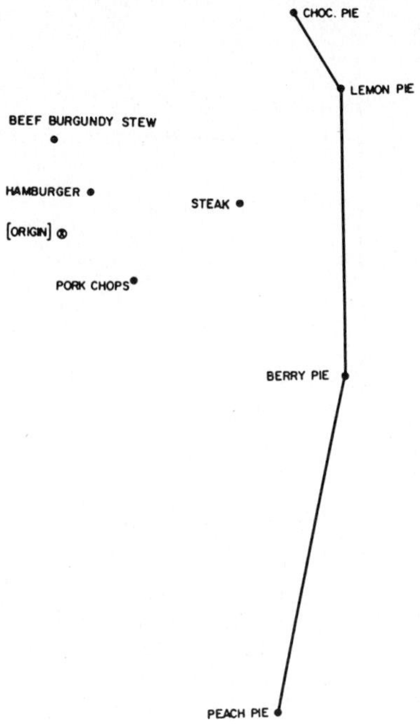

FIGURE 2a
Food preferences data, two-bimensional solution: First obliquely rotated bimension.

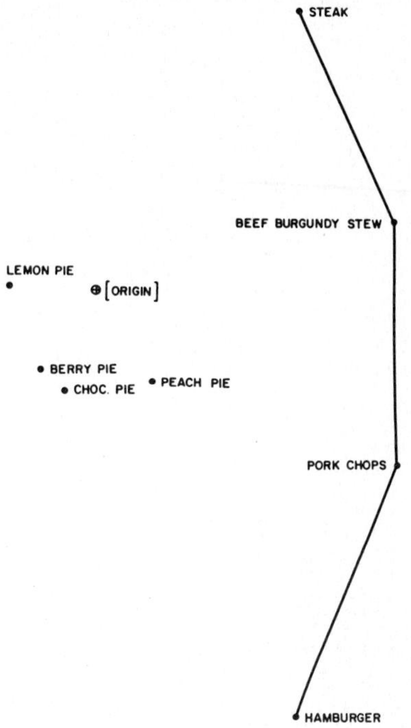

FIGURE 2b
Food preferences data, two-bimensional solution: Second obliquely rotated bimension.

associated with desserts (perhaps "enjoyment of an extravagant treat"?).

3.2. Celebrity Preferences

Our second application of DEDICOM is to a more typical kind of paired-comparison data: z-scores derived from choice proportions. The data, taken from [20], have been used by several other investigators to study similarity effects on preference strength ([4]; [7]; [21], p. 371). Thus our DEDICOM analysis can be compared to these theory-based solutions, and, in particular, to the "wandering ideal point" solution in this volume [7].

The stimuli were nine celebrities (listed in the Figure 3 legend). All possible pairs were presented, and subjects selected from each pair the celebrity with whom they would rather spend one hour, discussing a topic of their choice. The x_{ij} values for the DEDICOM analysis were obtained by converting the proportion of the sample preferring celebrity i over celebrity j to z-scores, by taking the z which cuts off the corresponding proportion of area under the normal curve.

Unrotated solution. The first unrotated plane (Figure 3) shows that Lyndon

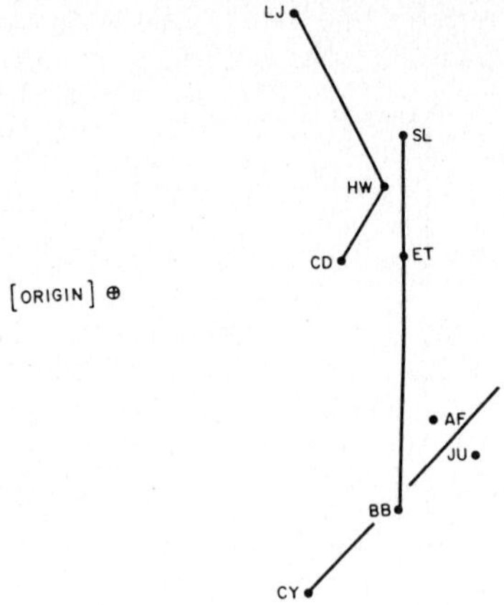

FIGURE 3
Unrotated one-bimensional solution of celebrity preferences data.
LJ=Lyndon Johnson, HW=Harold Wilson, CD=Charles DeGaulle, SL=Sophia Loren, ET=Elizabeth Taylor, BB=Brigitte Bardot, AF=A. J. Foyt, JU=Johnny Unitas, CY=Carl Yastrzemski.

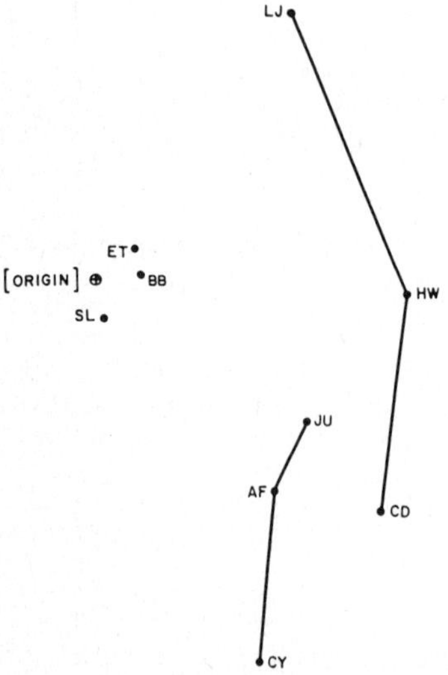

FIGURE 4a
Celebrity preferences data, two-bimensional solution: First obliquely rotated bimension.

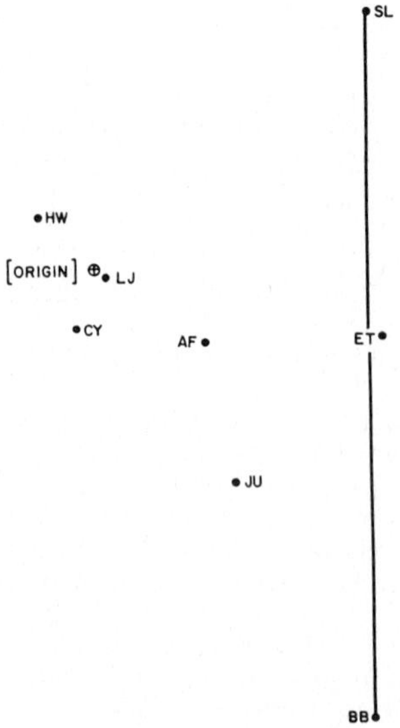

FIGURE 4b
Celebrity preferences data, two-bimensional solution: Second obliquely rotated bimension.

Johnson was by far the most preferred and Carl Yastrzemski the least preferred of the celebrities. Again there is some curvature, so that the extreme preference values are not as great as the sum of the intermediate preferences, but this may be due to several straight subsets of points, rather than a general curvature. There are indications in Figure 3 that a single overall preference hierarchy may not adequately describe these data; the politicians, movie stars, and athletes may fall on separate lines, much as the two food groups did in the prior example. (However, there seems to be something atypical about the position of Charles DeGaulle.)

Rotated solution. An oblique rotation of two bimensions was performed to see if interpretable stimulus subsets and thus identifiable "aspects of preference" could be better identified. In the rotated solution, preference relationships among actresses are shifted into the second bimension (Figure 4b), which might be labelled "glamour". Rotation also clarifies the structure of the main preference hierarchy (Figure 4a), which now shows what might be called an "importance" or "general interest" ordering for the rest of the stimuli. However, there is still some indication of two distinct groups in the first bimension, and the athletes are still nonzero in the second bimension. Thus, an oblique rotation of the three-bimensional solution was performed. (This may be an over-extraction, but it is presented for illustrative purposes). As Figure 5 shows, this solution approx-

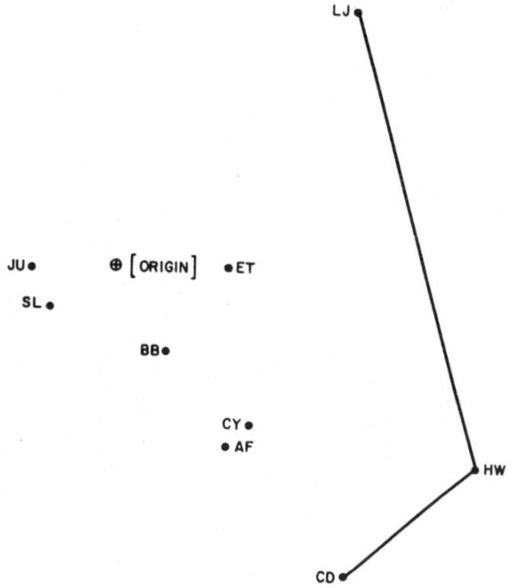

FIGURE 5a
Celebrity preferences data, three-bimensional solution: First obliquely rotated bimension.

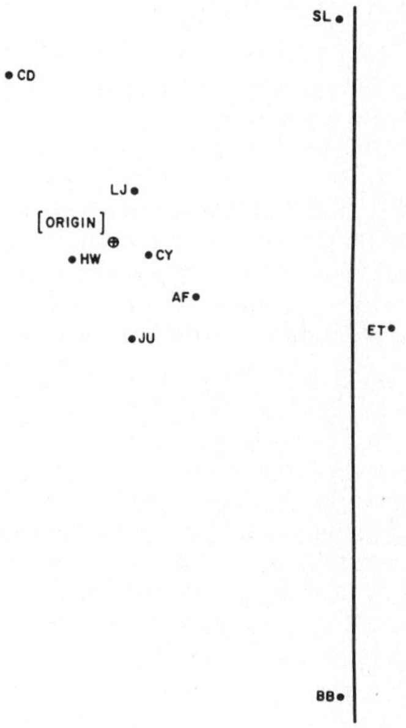

FIGURE 5b
Celebrity preferences data, three-bimensional solution: Second obliquely rotated bimension.

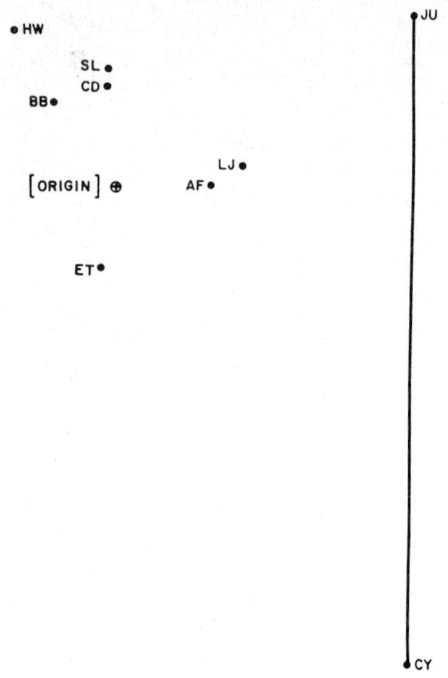

FIGURE 5c
Celebrity preferences data, three-bimensional solution: Third obliquely rotated bimension.

imately identifies a separate preference hierarchy for each of the three stimulus groups in the data. This suggests that there are three different kinds or "aspects" of utility which contribute to the overall preference judgments. One could speculate that they are the conversational interest associated with "power", "glamour" and "sports achievement".

There are still some anomalies in the patterns, however, and these might reflect subtleties of the subjects' perceptions of these stimuli. For example, Charles DeGaulle seems to take on less involvement with the first "power" bimension and more with the second "glamour" bimension than would be expected (mainly with respect to Sophia Loren). Could this have something to do with his "glamorous" historical role, or his linkage with the "glory of France"? Also, A. J. Foyt does not have the expected position on the third bimension. Of course, before taking such anomalies too seriously, it is important to establish that they are reliable (e.g., by split-half analysis, "jack-knifing" or "bootstrapping", or by comparing different samples).

4. DISCUSSION

4.1. Interpretation of DEDICOM Solutions

Information provided. The representation of preference structures in terms of component bimensions, and their depiction in Gower diagrams, reveals several kinds of useful information. The examples above show how it can uncover (a) approximations to simple linear or additive preference scales, when appropriate; (b) nonadditive or curved preference scales, when present; (c) multiple hierarchies, if present, and their mutual obliqueness; (d) scale values for stimuli, or relative distances between stimuli on each scale or preference hierarchy; and (e) increments in variance contributed by each additional hierarchy (discussed below). It can also provide information on systematic violations of moderate or weak stochastic transitivity, and can represent circular or intransitive preferences (not illustrated here, but demonstrated in other applications).

With the food preference data, for example, a single overall scale of desirability or utility (Figure 1) captured 97% of the variance, and so a one-scale approach might be useful for most purposes. Since the points were roughly collinear, scale values for these stimuli could be approximated by taking their projections onto a single best-fitting line. Only an additional 2.7% of the variance could be fit by taking into account the split between the desserts and entrees (Figures 2a, 2b), but the discovery that this distinction affects preferences might be theoretically useful (and might have practical value in applications such as marketing). Moreover, the two-bimensional solution may provide improved scale values for the stimuli, since the locations of the entrees are no longer directly constrained by relationships to the desserts. (The entrees in Figure 2b lie on a curved line, but one way to obtain utility scale values that reflect relative preferences among them is to take their projections onto the least-squares best-fitting straight line passing near them.) Preferences between the groups are accounted for by obliqueness of the bimensions. The resulting cross-group preference strengths can be computed using the **R** matrix.

Similar results were obtained in the celebrities example, where the variance accounted for by one- through three-bimensional solutions was 96%, 98.8% and 99.9%, respectively. Again the groupings appeared meaningful, and the small size of the increments in fit after the first bimension can be attributed to the nonorthogonality of the planes.

Evaluating fit values. These fit values are quite high, but we are fitting a large number of parameters in proportion to the number of data points. (Indeed, both example data sets were fit perfectly at four bimensions.) Consequently, we should ask whether the fits of 96% and 98.7% are really higher than one should expect by chance, that is, higher than would occur if there were no structure in the data. As yet, no distributional theory allows us to analytically answer this question, and so we must use more empirical methods.

We have adopted a permutation test approach to evaluating fit values [15]. The directions of asymmetry are scrambled by exchanging the x_{ij} and x_{ji} values for a randomly determined half of the pairs in the matrix **X**. The scrambled data are then analyzed with DEDICOM and the resulting fit values obtained. The process is performed for 19 different permutations of the data. The original (unpermuted) **X** provides the 20th fit value, and all 20 are then ranked. Under the null hypothesis, all 20 datasets are equivalent and all rankings are equally likely, and so the probability that the unpermuted data would have the highest fit is .05. If it does, the hypothesis of no structure can be rejected at the .05 level.

The results of applying this test to the foods and celebrities data sets are shown in Table 2. Clearly, the fit is better than would be expected by chance at the .05 level of probability. Nonetheless, with real applications it would be preferable to have a larger data set, so as to increase the ratio of data points to model parameters. (Unfortunately, this type of test does not directly tell us how many dimensions are needed. We note, however, that the three-bimensional solutions provide uncomfortably high fit to random data.)

TABLE 2
DEDICOM Fit Values: Original and Randomly Permuted Datasets

	Food Preferences Data Bimensions Extracted			Celebrity Preferences Data Bimensions Extracted		
	1	2	3	1	2	3
Original Data						
Obtained R-SQ	.9711	.9985	.9999	.9603	.9879	.9996
Randomized Data						
Mean R-SQ	.7487	.9419	.9980	.6183	.8782	.9784
S.D.	.0560	.0256	.0026	.0844	.0626	.0221
Highest R-SQ	.8987	.9858	1.0000	.8176	.9580	.9995
Lowest R-SQ	.6420	.8838	.9888	.4995	.7388	.9250

4.2. Psychological Implications of DEDICOM Structure

The mathematical structure of a DEDICOM representation has at least two features which distinguish it from other models of paired comparison preference data: (a) the preference structure is decomposed into simple rank-2 component patterns; and (b) intransitive and weakly transitive relationships can be represented. We will now discuss what useful psychological interpretation can be given to the structures that these features permit.

Additive components as reflecting incommensurable utilities. When DEDICOM analysis reveals more than one systematic bimension, the pairwise preference relationships in the data must be too complex to be approximated by a single rank-2 preference pattern. As we have said, each basic component pattern of preferences is represented by a separate ("rotated") bimension and is generated primarily by preference relations among that subset of the stimuli that have high loadings on the associated bimension. Between subsets, preference relations are either not generated (if bimensions are orthogonal) or are attenuated (if bimensions are oblique). How might such multi-component preference structures arise? One possible psychological interpretation is in terms of incommensurable (or only partly commensurable) aspects of utility. If two stimuli are perceived to possess different kinds of utility, these utilities may be incommensurable (or partly so) and comparisons between them will be difficult. As a result, judged preferences will be attenuated. On the other hand, if two stimuli have the same kind of utility, but different amounts of it, then they are commensurable with respect to that aspect. Comparisons between them are easy and so judged preferences will not be attenuated. For example, subjects may easily compare two stimuli in terms of aesthetic value, but may find it difficult to say which they prefer if two properties are placed in conflict (e.g., if one stimulus is very aesthetic but inconvenient and the other is highly convenient but not at all aesthetic.) Under this interpretation, then, each utility dimension would generate a distinct pattern of pairwise preferences (e.g., one pattern for "aesthetic" preferences, the other for "convenience" preferences), and each such pattern would give rise to a separate bimension in the DEDICOM analysis.

An alternative: stimulus clusters. The emergence of several bimensions need not imply incommensurable dimensions of preference. An alternative interpretation is possible, one which postulates a general attenuation of preference by dissimilarity (as in the "wandering vector" and related theoretical models to be discussed later). This general attenuation could appear to operate more selectively--between, but not within certain groups of stimuli--if the stimuli happen to be arranged in tight clusters in the "similarity space." In this case, the overall preference structure would be the sum of several separate patterns of preference simply because judgments of preference between members of different clusters would be strongly attenuated relative to those within clusters; each cluster would appear as one bimension in the DEDICOM solution.

Observable differences between predictions. It would seem, then, that modulation of preferences by either dissimilarity or incommensurability gives rise to a similar DEDICOM representation. Nonetheless, the effects of the two processes are distinguishable, because incommensurability theory and dissimilarity theory make different predictions about the expected configuration of points within a bimension.

Incommensurability theory implies that attenuation of preferences occurs when dissimilarity is one of kind, but not when it is simply one of degree. Now note that only differences of degree are found within a given bimension, since the various locations of points reflect (commensurable) differences in the amount of whatever kind of utility the bimension represents. Consequently, there should be no differential attenuation of preferences within a bimension; if we adopt the usual presumption that the underlying utility is additive, the points loading strongly on the bimension should lie along a straight line. In contrast, the general dissimilarity theory predicts attenuation whenever stimuli differ in any respect. Thus even within a bimension, the preferences should be underestimated more for points falling further apart on the bimension (points that are more dissimilar) than for points lying closer together. Consequently, the set of strongly loading points should be "bent" into a curve that is concave toward the origin.

The two bimensions for the food preference example show this type of curvature, and so one might say that these data provide support for a theory of general similarity effects. (This example demonstrates how the descriptive information provided by a DEDICOM analysis can complement a "wandering vector" or related theory-based analysis.) However, there is another possible explanation. The food preference data were collected using a rating scale, and the subject may simply have avoided the endpoints of the scale; strong preferences might be underrated because of this response style.

The celebrities preference values were obtained without the use of rating scales, however. Here the evidence for curvature within bimensions is much more ambiguous. In particular, the movie stars seem to lie on an almost straight line. This suggests at least partial support for an incommensurability between the utilities resulting from interest in power and interest in glamour.

It is possible, of course, that general dissimilarity effects and specific incommensurability effects could each be found, depending on the type of stimuli. For example, data presented by Sjoberg [22, Table 5] might be interpreted as indicating both a general dissimilarity effect which attenuates preferences within dimensions, and attenuation due to partial incommensurability across dimensions.

"Factorial" models. One qualification should be noted here. Strictly speaking, the more general "factorial" models of Takane [4] and Heiser and DeLeeuw [3] need not imply a curved within-bimension preference hierarchy. Even though these models postulate the same attenuation of preference by dissimilarity as the wandering vector or wandering ideal point models, they do not imply that stimuli showing a greater difference in utility should also be more dissimilar. This is because they do not require that the similarity space contain a direction, or set of distances, onto which one can map the utilities. Instead, the similarity space is simply determined by the pattern of inferred preference attenuations which best account for the deviations from simple additivity. These "factorial" models would fit data containing linear, additive preference relationships by specifying that all stimuli are equally dissimilar to one another (lie on vertices of a simplex), and so all preferences are attenuated equally.

Although these more general dissimilarity models can account for linear, additive preference hierarchies, they do so by implying relationships that might be psychologically questionable. In particular, they have to allow stimuli with quite different values on a utility dimension to be just as similar as those with the same value (even in the absence of any "counter-balancing" differences on other dimensions). In DEDICOM terms, stimuli which were widely separated within a bimension could be just as similar as adjacent stimuli within a bimension.

Generality of DEDICOM representation. The second feature that distinguishes DEDICOM from the other models is its greater generality. It can represent data that obey only weak stochastic transitivity, and it can even handle intransitive relationships (discussed in more detail later). For example, DEDICOM can easily represent circular dominance relationships as points arranged in a circular pattern around the origin. While this generality can be quite important for other kinds of data (e.g., [23]), it is less clear how it applies to preferences. Nonetheless, circularities have been observed (e.g., [22]) and so DEDICOM's generality might be useful, at the very least, to detect and display such puzzling phenomena when they arise. While some authors have proposed intransitive utility theories (e.g., [24]) and others have offered psychological explanations for moderate intransitivity (e.g., [25]), this area of research is beyond the scope of this paper. Consequently, we will not speculate here on what psychological theories might be compatible with such preference structures. We simply note that DEDICOM will reveal such patterns, should they be present in the data.

4.3. Comparison of Theoretical and Data-Analytic Approaches

When the theory is appropriate. The descriptive information provided by a data-analytic approach can be viewed as complementary to that provided by a theory-based approach. If the theory is appropriate, the theory-based analysis gives a more illuminating representation of the underlying process that generated the data, including estimates of underlying utility values for the stimuli being fitted, and stimulus similarities that might be useful for, say, marketing purposes. On the other hand, a data-analytic approach like DEDICOM gives a more detailed description of the preference structure that actually exists in the data. Some of these descriptive details can also be valuable for marketing and forecasting purposes.

A DEDICOM analysis reveals both regularities and anomalies in the preference structure, and both can be used by the applied researcher. For example, the occurrence of any stimuli in anomalous or unexpected positions (as Charles DeGaulle in our analysis) signals unusual preference patterns involving those stimuli. The researcher might thus be alerted to important characteristics of the stimuli requiring further study. On the other hand, regular patterns, as indicated by orderly stimulus arrangements on one or more bimensions, may suggest distinct aspects of preference that need to be considered when designing or marketing new products.

When a DEDICOM bimension reveals linear or simple curvilinear preference hierarchies (e.g., Figures 2, 4b), one interpretation would be in terms of stimulus differences in the amount of some specific attribute. Product designers might use such information to estimate how varying that attribute in a new product would change the preference for the product. On the other hand, suppose two groups of points fall on differently oriented lines within a given bimension, or on different bimensions. Then the researcher would have reason to believe that more complex relationships were involved, and a product change that would strongly alter the product's preference relationships with respect to one group would probably not produce such large changes with respect to the other group.

<u>When the theory is questionable.</u> When it is unclear which psychological theory is appropriate, or when the data contain theoretically troublesome preference relationships, DEDICOM analysis might be useful precisely because its method of representation is much more general than those based on specific models. Thus, it can deal with a much wider range of preference patterns and provide a useful description of the data structure, even in the absence of a theory explaining <u>why</u> the structure would have a given form. Good descriptive information about that form is valuable to those involved in marketing, etc., whether or not the "wandering vector" or some other psychological model is appropriate. In addition, by studying the structure obtained, one might gain insights that would suggest a general theory that could account for the observed patterns.

We can compare the generality of various models by considering the different restrictions they would place on the arrangement of points in a DEDICOM representation. Figure 6 shows the range of relationships that can be represented by several psychological models of preference, compared to what DEDICOM can represent. Points in the figure are arranged in a circular arc around the origin of a DEDICOM bimension. A highly restricted utility model such as Thurstone Case V [26] would require that all points fall on the straight line, since it implies that the fitted parts of the data obey a linear additive preference hierarchy. "Strong" utility models such as Bradley-Terry-Luce ([27]; [28]) would not require linear additivity, but would require that the relations obey strong stochastic transitivity (i.e., when a is preferred over b and c, and b is preferred over c, pref(a over c) is greater than or equal to the larger of pref(a over b) and pref(b over c)). In the circular bimension shown in Figure 6, this would correspond to the relationships among points within a 90 degree arc, such as the points to the right of the vertical line.

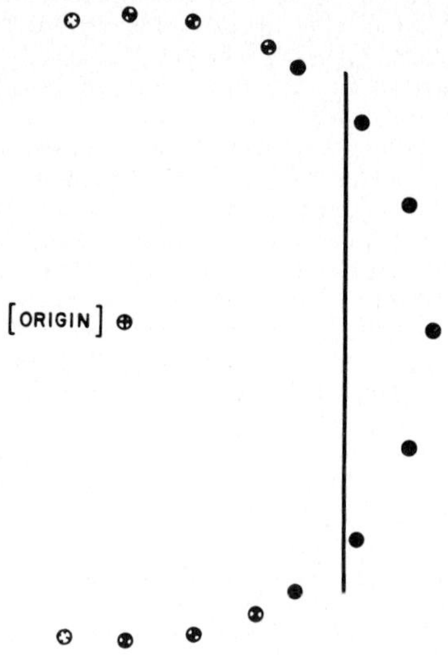

FIGURE 6
Diagrammatic representation of relationships that conform to various kinds of transitivity.

Weaker restrictions are also depicted in Figure 6. Solid points in the figure lie on a 120 degree arc, where moderate stochastic transitivity holds (i.e., pref(a over c) is greater than or equal to the smaller of pref(a over b) and pref(b over c)). Within 180 degrees of arc, the relationships obey weak stochastic transitivity (i.e., a preferred over b, and b over c, implies a is preferred over c), and beyond 180 degrees, the relationships begin to show intransitivities (circular triads). Only relationships that obey strong or moderate stochastic transitivity (i.e., involving the solid points in the figure) can be represented by current theory-based methods such as "wandering vector" or "factorial" models. In contrast, DEDICOM can represent all the different kinds of relationships shown in Figure 6, and even completely intransitive ones.

4.4. Other Applications

The relevance of the methods used above, and their possible application to telecommunications problems, should be apparent. Potential customers could be asked to rate pairwise preferences between possible options for telecommunications service, such as different bandwidths, toll options, etc. Or, different hardware configurations could be compared. The analysis would then be directed at uncovering the underlying dimensions of preference and inferring those aspects or features of the products which determine relative preference, as well as quantifying the relative positions of the product on each preference dimension and the relative contributions of each dimension to total overall preference. Careful construction of the response scales (e.g., by asking the subjects to state how much they would be willing to pay to have one option over another) might make it possible to calibrate the preferences in terms of actual dollar values. Then the areas of the triangles in each bimension would represent the difference in price needed to produce indifference between alternatives, when the contributions of only that aspect of preference is extracted from the total.

The general DEDICOM model can also be applied to a matrix in which x_{ij} represents the amount of telephone traffic between points i and j. If the skew-symmetric part of the data is extracted, so that x_{ij} represents the imbalance of telephone traffic between different points, the bimensional representation used above would apply. To explore this area of application, we performed a DEDICOM analysis of the skew-symmetric part of a data set giving the telephone traffic between 16 regions in the U.S.A. One result of the analysis is shown in Figure 7. As before, clockwise rotation represents dominance. For example, the size of the triangle between the

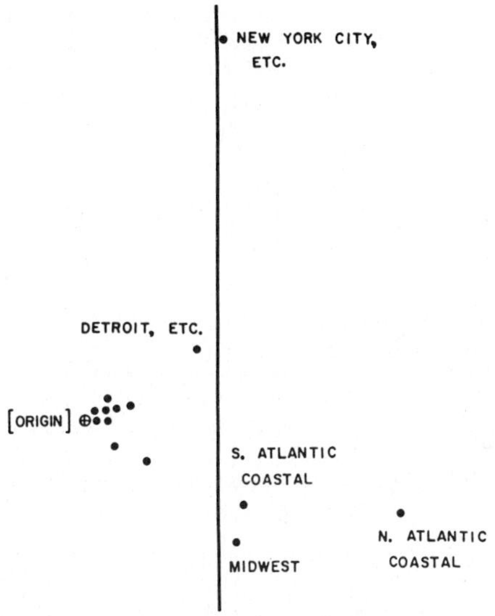

FIGURE 7
Skew-symmetric part of telephone traffic data: First bimension of a DEDICOM two-bimensional solution.

origin, "New York City, etc." (including the New York City, New Jersey and Washington, D. C. areas) and "Detroit, etc." (including the Detroit, Chicago and Cleveland-Pittsburgh metropolitan areas) represents the imbalance in calls between these two regions (more calls from NY to Detroit).

Overall, this bimension shows a linear imbalance hierarchy that seems to order the regions according to urbanization and population. The main exception to the linear pattern is the North Atlantic Coastal region (which includes Maine, Vermont, New Hampshire, part of New York state, part of

Pennsylvania, Virginia and southern Florida). Falling outside the hierarchy, it reflects an extra large imbalance with New York City, etc. This anomaly might be due in part to the proximity of most of this region to New York City, etc. In general, this bimension reveals a very orderly (linear and additive) tendency for the difference in urbanization level of two regions to be reflected in the imbalance of telephone traffic between them. We see a pattern originating in the eastern USA for a relatively more urbanized region to generate more calls to a less urbanized region than vice versa. The second bimension (not shown here) describes patterns of imbalance between other areas; more detailed analyses may suggest patterns due to factors other than urbanization.

4.5. Conclusion

We have presented an exploratory or "data-analytic" approach to studying the structure underlying a paired comparison preference matrix, based on the DEDICOM generalization of multidimensional scaling or factor analysis. It can reveal salient dimensions useful for product design and marketing, and can also highlight peculiarities of preferences involving particular stimuli deserving of further study. DEDICOM analysis is viewed as complementary to the more theoretically based analysis methods such as the "wandering vector", which fit specific psychological models to the preferences. Both approaches are useful for gaining a deeper understanding of the detailed structure underlying a pattern of pairwise preferences among several stimuli.

5. APPENDIX A

5.1. The General DEDICOM Model

Consider an n by n data matrix **X**, with the rows and columns representing the same set of entities. The entry x_{ij} represents the directed relationship (e.g., amount of telephone traffic, brand switching, etc.) from entity i to entity j. Often, such directed relationships are asymmetric; that is, $x_{ij} \neq x_{ji}$. DEDICOM represents the underlying structure in X in a way that sheds light on systematic asymmetric as well as symmetric aspects.

The (single domain, two-way) DEDICOM representation of X has the form $\mathbf{X} = \mathbf{ARA}' + \mathbf{E}$, where **A** is an n by s matrix giving "loadings" of the n stimuli on s dimensions, **R** is an s by s matrix of relationships among the dimensions, and E is a matrix of residual error terms. Asymmetries in **R** reflect systematic asymmetries in X, and R can be interpreted in the same terms as **X**; it gives a "miniature" version of **X**, but at a higher level of description. For example, if X were a matrix describing how many persons switched from the row brand to the column brand in some period of time, then the columns of **A** would give types or aspects of products, and **R** would give a description of switching between these different types or aspects.

As in the skew-symmetric case, solutions are "rotated" to maximize interpretability. However, in the general asymmetric case, two dimensions are rotated at a time, rather than two planes at a time. The interpretation of the model for general asymmetric **X** is discussed in more detail elsewhere ([14]; [15]).

6. APPENDIX B

6.1. Least-squares Fitting of DEDICOM in the Skew-Symmetric Case

Let the singular value decomposition (SVD) of **X** be written

$$(2) \qquad \mathbf{X}=\overset{*}{\mathbf{U}}\overset{*}{\mathbf{D}}\overset{*}{\mathbf{V}}' \quad ,$$

with $\overset{*}{\mathbf{D}}$ diagonal and $\overset{*}{\mathbf{U}}$, $\overset{*}{\mathbf{V}}$ orthogonal. The s-dimensional (s/2 bimensional) "unrotated" DEDICOM representation can be obtained from (2) by defining **A**=**U** and **R**=**DP**, where U is the orthonormal section composed of the first s columns of $\overset{*}{\mathbf{U}}$, **D** is composed of the first s rows and columns of $\overset{*}{\mathbf{D}}$, and **P** is an s by s permutation (and sign changing) matrix, of the form

$$\begin{matrix} 0 & 1 & 0 & 0 & \cdots \\ -1 & 0 & 0 & 0 & \cdots \\ 0 & 0 & 0 & 1 & \cdots \\ 0 & 0 & -1 & 0 & \cdots \\ . & . & . & . & \cdots \\ . & . & . & . & \cdots \end{matrix}$$

In the SVD of any skew-symmetric matrix, **U**'=**P**'**V**'. Consequently, we can obtain the DEDICOM representation from the SVD by taking advantage of the fact that (**PP**'=**I**) and inserting it into the rank-s SVD approximation of **X** as follows:

$$\mathbf{X} = \mathbf{UD}(\mathbf{PP}')\mathbf{V}' = \mathbf{U}(\mathbf{DP})(\mathbf{P}'\mathbf{V}') = \mathbf{URU}' = \mathbf{ARA}'$$

After this transformation, the **R** matrix is skew-symmetric, with the singular values adjacent to the main diagonal (e.g., $d_{11}=r_{12}=-r_{21}$) and all other values zero. This DEDICOM solution has orthogonal bimensions, with maximum variance accounted for by the first bimension, and maximum portions of residual variance accounted for by each successive bimension. This "principal planes" solution is analogous to the unrotated rank-s principal components approximation of a symmetric positive definite or semidefinite matrix [19, p. 13]. As noted earlier, it is usually linearly transformed into an alternative form before interpretation.

REFERENCES

[1] Carroll, J. D. (1980). Models and methods for multidimensional analysis of preferential choice (or other dominance) data. In E. D. Lantermann & H. Feger (Eds.), Similarity and choice (pp. 234-289). Bern: Huber.

[2] Edgell, S. E. & Geisler, W. S. (1980). A set theoretic random utility model of choice behavior. Journal of Mathematical Psychology, 21, 265-278.

[3] Heiser, W. J., & De Leeuw, J. (1981). Multidimensional mapping of preference data. Mathematiques et sciences humaines, 73, 39-96.

[4] Takane, Y. (1980). Maximum likelihood estimation in the generalized case of Thurstone's model of comparative judgment. Japanese Psychological Research, 22, 188-196.

[5] Tversky, A. (1969). Intransitivity of preferences. Psychological Review, 76, 31-48.

[6] Bockenholt, I., & Gaul, W. (1984). A multidimensional analysis of consumer preference judgments related to print ads. Diskussionspapier

Nr. 64, Institut fur Entscheidungstheorie und Unternehmensforschung, Universitat Karlsruhe (TH), Karlsruhe, Germany.

[7] DeSoete, G. & Carroll, J. D. (1987). Probabilistic multidimensional choice models for marketing research. This volume.

[8] Chino, N. (1978). A graphical technique in representing the asymmetric relationships between n objects. Behaviormetrika, 5, 23-40.

[9] Gower, J. C. (1977). The analysis of asymmetry and orthogonality. In J. R. Barra, F. Brodeau, G. Romier, and B. Van Cutsem (Eds.), Recent developments in statistics, (pp. 109-123). Amsterdam: North-Holland.

[10] Gower, J. C. (1984). Multivariate analysis: Ordination, multidimensional scaling, and allied topics. In E. Lloyd (Vol. Ed.) Handbook of applicable mathematics: Vol. 6. Statistics. (W. Ledermann, Series Ed.). Chichester: J. Wiley.

[11] Tobler, W. (1979). Estimation of attractiveness from interactions. Environment and Planning, Series A, 11, 121-127.

[12] Weeks, D. G & Bentler, P. M. (1982). Restricted multidimensional scaling models for asymmetric proximities. Psychometrika, 47, 201-208.

[13] Young, F. W. (1974, August). An asymmetric euclidean model for multiprocess asymmetric data. In Theory, methods, and applications of multidimensional scaling and related techniques, Proceedings of the U.S.-Japan Joint Seminar, University of California at San Diego.

[14] Harshman, R. A. (1978, August). Models for the analysis of asymmetrical relationships among N objects or stimuli. Paper presented at the first joint meeting of the Psychometric Society and the Society for Mathematical Psychology, McMaster University, Hamilton, Ontario.

[15] Harshman, R. A., Green, P. E., Wind, Y. & Lundy, M. E. (1982). A model for the analysis of asymmetric data in marketing research. Marketing Science, 1, 205-242.

[16] Hoffman, K. & Kunze, R. (1971). Linear algebra (2nd ed.). Englewood Cliffs, New Jersey: Prentice-Hall.

[17] Kaiser, H. F. (1958). The varimax rotation for analytic rotation in factor analysis. Psychometrika, 23, 187-200.

[18] Harris, C. W. & Kaiser, H. F. (1964). Oblique factor analytic solutions by orthogonal transformations. Psychometrika, 29, 347-362.

[19] Harshman, R. A. (1981). DEDICOM multidimensional analysis of skew-symmetric data. Part I: Theory. Unpublished Bell Laboratories Technical Memorandum.

[20] Rumelhart, D. L., & Greeno, J. G. (1971). Similarity between stimuli: An experimental test of the Luce and Restle choice models. Journal of Mathematical Psychology, 8, 370-381.

[21] Restle, F., & Greeno, J. G. (1970). Introduction to mathematical psychology. Reading, Mass.: Addison-Wesley.

[22] Sjoberg, L. (1975). Uncertainty of comparative judgments and multidimensional structure. Multivariate Behavioral Research, 10, 207-218.

[23] Shepard, R. N. (1964). Circularity in judgments of relative pitch. Journal of the Acoustical Society of America, 36, 2346-2353.

[24] Fishburn, P. C. (1981). Nontransitive measurable utility. Unpublished Bell Laboratories Technical Memorandum.

[25] Tversky, A. & Sattath, S. (1979). Preference trees. Psychological Review, 86, 542-573.

[26] Thurstone, L. L. (1927). A law of comparative judgment. Psychological Review, 34, 273-286.

[27] Bradley, R. A., & Terry, M. R. (1952). Rank analysis of incomplete block designs. Biometrika, 39, 324-335.

[28] Luce, R. D. (1959). Individual choice behavior. New York: Wiley.

TELECOMMUNICATIONS DEMAND MODELLING
An Integrated View
A. de Fontenay, M.H. Shugard, D.S. Sibley (Editors)

ORDINAL NETWORK SCALING OF COMMUNICATION NETWORKS

Hubert Feger

Department of Psychology
University of Hamburg

Overview

Little is known about the structure of communication networks which form outside the laboratory. Who communicates what to whom how often? One of the reasons for this scarcity of information is the lack of methods to analyze communication behavior from a structural point of view. Ordinal Network Scaling (ONS) is introduced in this paper and applied to matrices describing communication behavior between several members of a group. The analysis reveals simple and basic regularities governing communication exchange.

1. Basic concepts of Ordinal Network Scaling (example 1)

Ordinal Network Scaling (ONS, Feger & Bien 1982) is a procedure for representing proximities - the rank order of distances - between objects by means of an undirected or directed weighted graph (for an introduction into the formal model and some examples see Droge & Feger 1983, Feger & Droge 1984, Orth 1985). The objects which, e.g., may be persons or institutions, correspond to points of the graph. The tied or untied proximities are represented as geodetic, i.e., shortest paths connecting these points. The lines in the graph and their weights are determined such that a strictly monotone relation exists between the proximities and the sum of the weights of the geodesics.

Example 1 is used as an introduction. The observations are conversation frequencies between four persons during office work. The frequencies are transformed into ranks (see Tab. 1) because the reliable properties of data such as these are not the absolute size or the size of the differences but the rank order.

The analysis of the ranks proceeds by applying two rules.
Rule 1: Of the three distances in a triple, the largest could be the sum of the two smaller distances. Let ^{1}BC denote the rank of the distance BC which is - according to Tab. 1 - rank 1. Rule 1 results in:

Triple:		Additivity:
A B C	(1)	$^{4}AC = {}^{1}BC + {}^{3}AB$,
A B D	(2)	$^{6}AD = {}^{3}AB + {}^{5}BD$,
A C D	(3)	$^{6}AD = {}^{4}AC + {}^{2}CD$,
B C D	(4)	$^{5}BD = {}^{1}BC + {}^{2}CD$.

Table 1: Example 1
Conversation frequences

	A	B	C	D
A	-	3	4	6
B	46	-	1	5
C	42	69	-	2
D	20	30	53	-

Legend: Lower half of the matrix: Conversation frequencies between persons A,B,C,D, observed for two weeks during their office work (from Gullahorn 1960, his Tab. II, Row III). Upper half: Ranks of the frequencies, 1 = maximal frequency = shortest interaction distance.

Rule 2: For every pair of additivities, e.g.:

(1) AC = AB + BC ,
and (2) DF = DE + EF ,

if AC > DF then not: AB + BC < DE + EF

DE + EF > AB + BC if DE > AB and EF > BC or DE > BC and EF > AB.

In example 1, rule 2 is violated for equations (1) and (4). Therefore, (1) is deleted from the set of admissible additivities. Then the three additivities (2), (3), (4) remain, specifying sums to replace the lines BD and AD. The remaining lines form the qualitative solution:

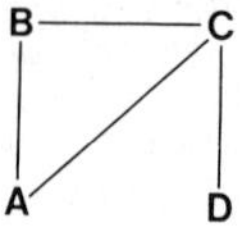

Quantitative solution: For every set of admissible additivities for the same line, e.g.:

1st set	(m)	AC = AB + BC ,
	(m+1)	AC = AD + CD ,
	(m+2)	AC = AE + CE ,
	.	
	.	
	.	
2nd set	(n)	DF = DE + EF ,
	(n+1)	DF = DG + FG ,
	(n+2)	DF = DH + FH ,
	.	
	.	
	.	

if AC > DF, then the smallest sum for DF has to be smaller than the smallest sum for AC.

A quantitative solution for example 1 (a metric space in the form of a weighted graph) is:

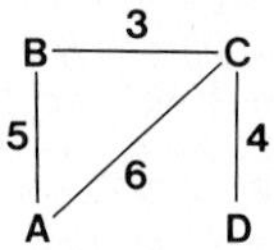

As Droge (1984) has shown, a representation of proximities as an undirected weighted graph (without loops from one point to itself) is always possible if there are no cycles (inconsistencies in the comparisons of sizes between three or more elements) in the empirical relation, and if all distances between an element and itself are minimal and equal. Further, he has proven that those lines which can not be replaced by rules 1 and 2 are a necessary part of every solution graph. Thus, one basic result with respect to structural uniqueness is that for a graph with path length metric all necessary lines constituting this connected graph can be determined by the two rules. If the necessary lines are sufficient then the set of necessary lines is minimal, i.e., no line has to be added to reproduce the data without error.

In example 1, additivity (1) has to be removed from the set of admissible additivities. It should be pointed out that the set of nonadmissible additivities is a finer structural description of a path length metric than the set of lines of a graph because these additivities imply the lines but not vice versa.

A natural characterization of a graph, called its simplicity, is the number of admissible additivities for this graph (Droge 1984, chap. 3). It has been shown that a chain - a "one dimensional" or simplex representation - is the most simple graph. A star with one center is the least simple of all tree graphs.

In order to interpret a solution as the one given in Tab. 1, the fact that the set of lines is minimal will be used. The following assumptions are made: A communication network has to be build between the participants. The costs of this network should be minimal. Every direct connection between two participants increases the costs, while indirect connections are possible and provide no additional costs. Costs increase proportionally with an increase in channel capacity.

The quantitative ONS solution has the property to determine the minimal number of direct channels and their (relative) capacity which should be inversely proportional to the weights in the solution. The resulting connected network is optimal as far as it provides the most direct connections with the highest capacities for the most frequent interactions.

2. The uniqueness of solutions (example 2)

Rules 1 and 2 determine all necessary lines. It is possible that the set of necessary lines is not sufficient to reproduce the data without error. This is the case if quantitative implications of admissible additivities lead to contradictions. This is shown by example 2 (Tab. 2). In Tab. 3, rows 5 and 8 imply $^{5}BE < {}^{2}BC + {}^{4}CD$, while rows 10 and 11 imply $^{2}BC + {}^{4}CD < {}^{5}BE$.

In such case, at least one line has to be added to the set of necessary lines to reproduce the ranks perfectly from a quantitative solution. The solution then is no longer structurally unique, and more than one minimal solution exists in general. Droge (working paper) has conceived an algorithm and written a computer program to find all quantitative solutions, explicitly listing minimal solutions also if the necessary lines are not sufficient. Radtke (1985) by referring to a theorem of Motzkin (1936) stated the necessary and sufficient conditions for the existence of a solution for a system of linear inequalities to which the information in the additivities may be transformed. Radtke also provided a computer program for unconditional similarities with or without ties. Monte Carlo studies (Droge 1984, chap. 10) have shown that with random data the necessary lines as found by rules 1 and 2 constitute 90 %, usually very close to 100 % of the sufficient lines. In applications to real data, no instance was found until now demanding more than 2 additional lines.

Table 2: Example 2
Necessary lines are not sufficient

	A	B	C	D	E
A	-	1	3	9	10
B		-	2	8	5
C			-	4	7
D				-	6
E					-

Legend: Ficitious data taken from Droge (1984, pp. 73).
Rank 1 = shortest distance.

Admissible additivities:

(1) $^{3}AC = {}^{1}AB + {}^{2}BC$,
(2) $^{8}BD = {}^{2}BC + {}^{4}CD$,
(3) $^{8}BD = {}^{5}BE + {}^{6}DE$,
(4) $^{9}AD = {}^{3}AC + {}^{4}CD$,
(5) $^{10}AE = {}^{1}AB + {}^{5}BE$,
(6) $^{10}AE = {}^{3}AC + {}^{7}CE$,
(7) $^{10}AE = {}^{9}AC + {}^{6}DE$.

This completes the treatment of structural uniqueness. The scale level of the weights is characterized by the following equivalences (Droge 1984, Radtke 1985):

If x_1 and x_2 are two quantitative solutions then

(a) cx_1 , $c \in IR > 0$,
(b) $x_1 + x_2$,
(c) $cx_1 + (1-c)x_2$, $c \in 0,1$,

are quantitative solutions as well. This type of measurement has been called (Coombs 1964) a "higher order metric scale". It is conjectured that interval scale measurement is approached with increasing simplicity of the graph and increasing number of elements.

Table 3: Quantitative implications of example 2

ROW		${}^{1}AB$	${}^{2}BC$	${}^{4}CD$	${}^{5}BE$	${}^{6}DE$	${}^{7}CE$
1	${}^{1}AB$	1					
2	${}^{2}BC$		1				
3	${}^{3}AC$	1	1				
4	${}^{4}CD$			1			
5	${}^{5}BE$				1		
6	${}^{6}DE$					1	
7	${}^{7}CE$						1
8	${}^{8}BD$		1	1			
9	${}^{8}BD$				1	1	
10	${}^{9}AD$	1	1	1			
11	${}^{10}AE$	1			1		
12	${}^{10}AE$	1	1				1
13	${}^{10}AE$	1	1	1		1	

Legend: The columns represent all necessary lines. The rows show these lines and the sums of all admissible additivities. The "ones" in the cells indicate which column elements constitute the row elements.

3. An example with ties

Nothing new in principle is to be considered when analyzing observations with tied ranks. Example 3 in Tab. 4 reports the percentages of 801 exchanges between 5 groups participating in a computer conference system (Pieper 1982). There are two blocks of ties (5,5,5) and (9.5, 9.5) in the ranks of the percentages.

Table 4: Treatment of tied observations (example 3)

City:		D	W	K	A	M
Duisburg	D	-	18	9	9	3
Wuppertal	W	1	-	10	9	3
Kiel	K	5	3	-	14	5
Aachen	A	5	5	2	-	7
München	M	9.5	9.5	8	7	-

Legend: Upper half of the matrix: Percentages of 801 communications. Lower half: Ranks, with 1 = highest frequency.

Table 4 (contin.)

Additivities		Implications of tie-breaking
(1)	${}^{5}DK = {}^{1}DW + {}^{3}KW$	
(2)	${}^{5}DK = {}^{5}AD + {}^{2}AK$	DK > AD
(3)	${}^{5}AD = {}^{2}AK + {}^{5}DK$	AD > DK
(4)	${}^{5}AD = {}^{5}AW + {}^{1}DW$	AD > AW
(5)	${}^{5}AW = {}^{2}AK + {}^{3}KW$	
(6)	${}^{5}AW = {}^{5}AD + {}^{1}DW$	AW > AD
(7)	${}^{8}KM = {}^{2}AK + {}^{7}AM$	
(8)	${}^{9.5}DM = {}^{5}DK + {}^{8}KM$	
(9)	${}^{9.5}DM = {}^{1}DW + {}^{9.5}MW$	DM > MW
(10)	${}^{9.5}DM = {}^{5}AD + {}^{7}AM$	
(11)	${}^{9.5}MW = {}^{1}DW + {}^{9.5}DM$	MW > DM
(12)	${}^{9.5}MW = {}^{3}KW + {}^{8}KM$	
(13)	${}^{9.5}MW = {}^{5}AW + {}^{7}AM$	

There exist 8 different solutions with only one of every pair of additivities:

(2) or (3),
(4) or (6),
(9) or (11).

A solution with highest simplicity is

$$D \overset{1}{-} W \overset{3}{-} K \overset{2}{-} A \overset{7}{-} M$$

not satisfying additivities (2), (6), and (11).

With tied observations, the number of additivities identified by rules 1 and 2 may be larger than the number of triples because there may be more than one admissible additivity per triple. E.g.:

${}^{5}DK = {}^{5}AD + {}^{2}AK$, implying DK > AD, or:

${}^{5}AD = {}^{2}AK + {}^{5}DK$, implying AD > DK .

Especially with the weak approach to ties (tied ranks may be untied in determining admissible additivities) the structural and quantitative uniqueness suffers from ties. All possible breakings of ties and their combinations lead to 8 different solutions for example 3. Usually a solution with maximal simplicity will be chosen, such as the simplex $D \overset{1}{-} W \overset{3}{-} K \overset{2}{-} A \overset{7}{-} M$.

4. Symmetrization (example 4)

The example in Tab. 5 contains information frequencies describing how often information was given from one department to an other. As can be seen there are asymmetries in the observations, e.g., department A as a sender initiates 70 messages to B as a receiver, while B as a sender provides 100 messages to A as receiver. Quite often, as in Müller-Merbach (1963, see also Dichtl & Schobert 1979) from whom these data are taken, such observations are rendered symmetric, usually by averaging. From a practical viewpoint, this may be justified if the communication network to be established or maintained will provide channels that are open in both directions anyhow.

Table 5: Communication frequency between seven departments (from Müller-Merbach 1973, his Tab. 1; see also Dichtl & Schobert 1979) = Example 4

	Department						
	A	B	C	D	E	F	G
A	-	100	120	10	70	60	90
B	70	-	30	40	110	70	10
C	80	10	-	0	20	30	50
D	0	20	20	-	70	20	10
E	40	90	30	40	-	20	60
F	40	0	50	10	0	-	70
G	60	20	30	0	40	100	-

Legend: rows = departments as senders
columns = as receivers

From a formal viewpoint, the symmetrization implies the assumption that - except for possible unreliabilities of the observations -

$$AB = BA$$

with the sender written in the first, the receiver in the second position. The assumption has a testable consequence: The combined rank order for all distances has to be without cycles.

By Monte-Carlo studies, Droge (1985) has shown that for random data it is very unlikely that row conditional data exhibit no cycles if symmetry is imposed. But the data in Tab. 5 show no cycles, a result with very low probability if the data were random. So there are no violations and no loss of information generated by them in symmetrization.

If one wants to find a symmetric representation, there are several possibilities. One that takes into account the appearance of rounded estimations of the data in Tab. 5 is to assign a larger rank to a distance AB than to a distance CD if neither quasi-distance AB or BA represent fewer interactions than CD or DC. Applying this criterion to the data in Tab. 5, the partial order in Tab. 6 results.

Table 6: Partial order of the quasi-distances in Tab. 5 after symmetrisation

	A	B	C	D	E	F	G
A	-						
B	2	-					
C	1	9	-				
D	12	7	11	-			
E	4	1	8	4	-		
F	5	11	6	10	11	-	
G	3	10	6	12	5	2	-

Legend: Ties, e.g.: ^{1}AC and ^{1}BE indicate the absense of order information between the tied distances.

For this partial order, a qualitative ONS solution is given in Fig. 1. As Monte-Carlo studies show (see Droge & Feger 1985, with several tables for variing numbers of points) it is very unlikely for random data that they can be represented by so few lines relative to all possible lines. The probability of the reduction observed for this graph (8 lines out of 21 are replaced) is below the 1 % level of chance.

Figure 1: Qualitative ONS solution for the partial order of ranks in Tab.6

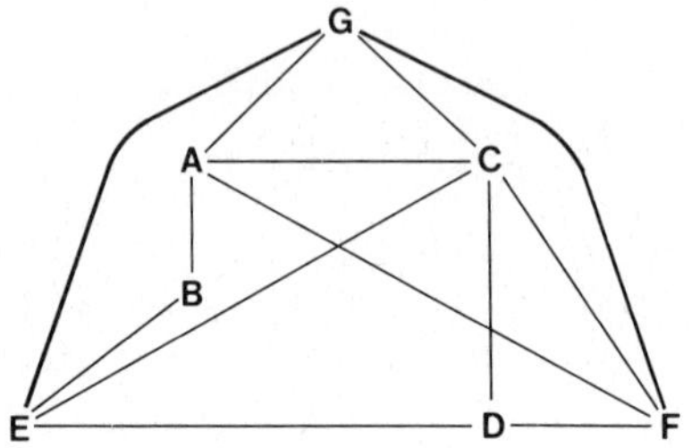

Legend: Letters indicate departments, see. Tab. 5

5. Directed graphs (example 6)

Even if a symmetrization is possible without a violation of the cycle criterion, quantitative information may be lost. This loss can be avoided by using a representation by a directed graph. This is also an alternative if violations should exist. In example 6 the well-known data of Bales (1970, his Appendix 4) are used. Tab. 7 shows the interaction frequencies and their ranks in a six person group. A detailed analysis is reported in Feger (1985 a,b).

Table 7: Interaction frequencies in a six-person-group (after Bales 1970, his appendix 4) = example 5

		receiver					
		A	B	C	D	E	F
sender							
	A	-	1238	961	545	445	317
	B	1748	-	443	310	175	102
	C	1371	415	-	305	125	69
	D	952	310	282	-	83	49
	E	662	224	144	83	-	28
	F	470	126	114	65	44	-

Legend: Aggregate matrix for 18 sessions of a problem solving laboratory group

		RANKS (1 = largest frequency)					
		A	B	C	D	E	F
sender							
	A	-	3	4	7	9	12
	B	1	-	10	13.5	18	23
	C	2	11	-	15	21	26
	D	5	13.5	16	-	24.5	28
	E	6	17	19	24.5	-	30
	F	8	20	22	27	29	-

All results on undirected graphs as reported above are valid for directed graph solutions except that the "distances" are not metric in general because the symmetry axiom may be violated. Rule 1 has to be slightly changed for digraphs: For every tripel of points A,B,C, there exist 6 permutations. For every permutation: if the largest "directed distance" is between the first and third point, it could be replaced by the sum of the two other distances. Fig. 2 reports a quantitative solution for the ranks in Tab. 7.

While a detailed theoretical interpretation is given elsewhere (Feger 1985 a,b), a few points will be mentioned:

(1) This ONS solutions reproduces the ranks without errors and requires fewer parameters than models using one sender and one receiver score per person, while those using only one parameter per person show a "poor fit to the data", see Goetsch & McFarland (1980).

(2) The network for this six persons example - as well as for the solutions with 3,4,5,7 and 8 Persons, not reported here - are not minimal. A connected minimal digraph with N points requires only N arrows arranged as a circle in one direction. But such a solution implies maximal asymmetry in every pair of persons, i.e., if a sender communicates very much with a specific receiver, this receiver-person as a sender would communicate very little with his partner. The social norm of reciprocity would be severely violated.

Figure 2: An ONS-solution as a valued digraph for the ranks in Tab. 7

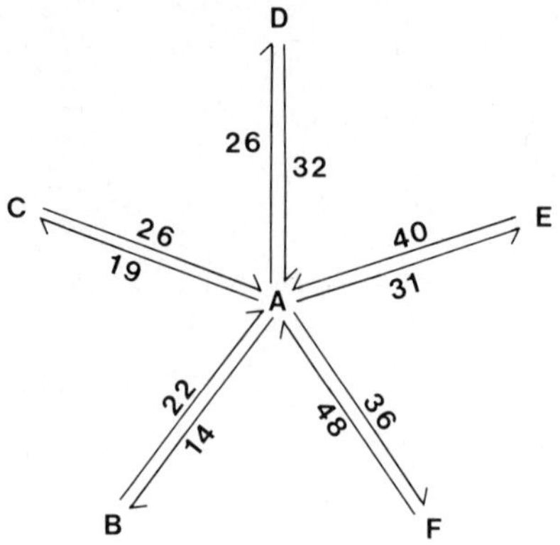

Legend: Arrows are not drawn proportionally to their lengths, indicated by the weights.

(3) The solution in its form of a star reveals a hierarchical structure: If one knows the relative intensity of communication from the central person to all others and vice versa, then all other intensities can be derived.

(4) If an ONS solution is in the form of a star or a double star as in Fig. 2, the ranks of communication intensity might be reproduced by a model assuming intensity to be a function of the sum of sender activity and receiver attractivity. The sender activities, denoted a,...,f, and the receiver attractivities, A,...,F, can be determined by solving a system of inequalities. For every pair of adjacent ranks (with rank 1 corresponding to the lowest frequency), one inequality is derived, e.g., from Tab. 7:
$e + F > f + E$,
$f + E > d + F$,
$d + F > f + D$, etc.
An errorfree solution for the ranks in Tab. 5 is:

A = 132	D = 48	a = 99	d = 31
B = 72	E = 28	b = 58	e = 16
C = 66	F = 8	c = 50	f = 8

showing more variability in attractivity than in activity.

(5) Three tendencies describe this communication structure:
(a) The number of messages sent is a decreasing function of sender centrality.
(b) The number of messages received is a decreasing function of sender centrality.
(c) For every pair of persons, the person with less total activity sends more frequently to the one with higher total activity than is sent in the other direction. Thus, the asymmetries are not random in their direction - an information which would be lost by symmetrization.

6. Confirmatory Ordinal Network Scaling (example 6, desire for communication)

Riley et al. (1954, see also Homans 1974, pp. 188) asked 1500 girl students to name those in their grade considered "popular", and derived the sociometric status from these nominations. Every girl also answered a questionnaire to find out how much she wanted to talk to every other girl in her grade on several topics. Tab. 8 shows the ranks of the frequencies which were derived by averaging over all girls and all communication topics.

Table 8: Frequency ranks for desire to communicate to different status groups (after Riley et al. 1954)

		Communication choice received					
	STATUS	A	B	C	D	E	F
Communi-	A	1	6	15	21	26	30
cation	B	2	3	10	19	24	29
choice	C	3	4	7	17	25	28
given	D	5	9	13	16	23	28
	E	8	12	18	20	20	27
	F	11	14	20	22	20	28

Legend: Rank 1 = strongest desire to communicate; ties are indicated by identical numbers.
A = highest sociometric status.

One may note that the observations are not only asymmetrical but that the main diagonal of Tab. 8 contains values which are neither all equal nor all minimal. Thus, to represent a status category by a single point would violate the first metric axiom of minimality. Analyzing Tab. 8 for overall trends, one may realize:
(1) The intention to communicate with members of one's own status category decreases with decreasing status.
(2) The less similar two status categories are, the lower the intention to communicate.

In order to incorporate these trends into an ONS solution, every status category is represented by two points, one showing the category in its role as a sender (a,...,f), the other in the receiver role (A,...,F). The overall trends in Tab. 8 are represented in Fig. 3, together with an ONS solution.

A possible qualitative solution is theoretically derived from the overall trends in the data: the upper part of the matrix in Tab. 8 shows the order structure of a simplex. This form is represented as a similarity dimension of the status categories in their role as senders. The lower part of the matrix shows the pattern of a star with A as the central element. The ONS solution represents the receivers in this form and combines it with the simplex. Choosing a confirmatory approach, the additivities can be read off from the hypothesized qualitative solution.

Figure 3:
(1) Overall trends in Tab. 8

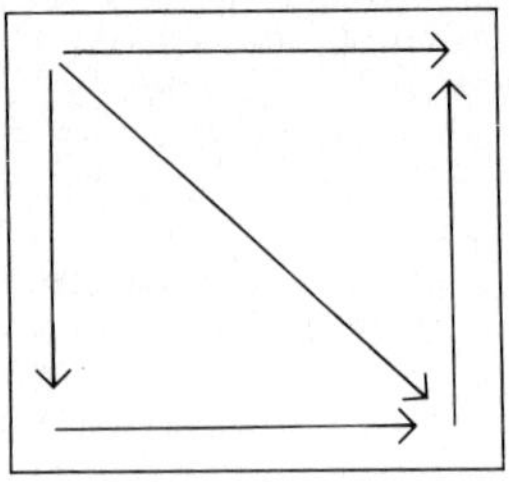

(2) A quantitative solution to the data in Tab. 8

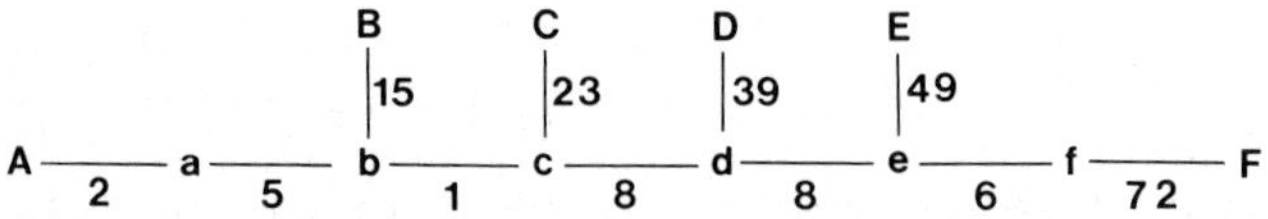

Legend: Lines are not drawn in proportion to their length as indicated by the weights.

The solution permits the following statements:
(1) In order to determine the strength of a desire to communicate two parameters have to be known: the self-attractiveness of each status category, and the distance or dissimilarity of the two categories on the status dimension. An additive combination of attractiveness and dissimilarity describes the intention strength.
(2) The attractiveness weights are much larger than the values describing status dissimilarity. It is not so much the possibility of an inhibition created by status differences which determines communication intentions but the positive aspects which a status provides - and provides this to the same extent for its own members as well as for others.

7. Concluding remarks

In closing, it should be mentioned that due to space limitations this presentation has been incomplete in several respects. A formal mathematical treatment had to be omitted as well as the treatment of some kinds of data, e.g., row conditional data with the same or with different elements defining the rows and columns of a data matrix. Several topics like interindividual difference scaling, approximate solutions, and the relationships between MDS, ONS, and other graph theory approaches have not been mentioned (see Carroll 1976, Carroll & Wish 1974).

Not only the analytical methods could merely be sketched but also only few types of communication networks were considered. As described elsewhere (Feger 1986), three basic types of networks may be differentiated. A network is called a participant network if it is based on information about who sends to whom, what, how much, how often, when and how. All empirical examples used in this paper are of this kind. A structure is called a message network if it traces the paths the messages take, how fast they propagate, how a message is changed, etc. And it is called a channel network if qualitative and quantitative differences in characteristics of channels are analyzed such as the use of channels, their costs, capacities, alternatives, etc.

Finally, for most social scientists, a theory to explain regularities in communication networks - some of which may be revealed by ONS - is the ultimate goal. There is work in progress to specify optimal networks as defined by a knowledge of the participants goals in communicating, their perception of the obstacles and possibilities of communication, and the strategies and tactics they use to reach their goals despite the barriers. One purpose of such a theory is to derive predictions about the form different networks will take. ONS will be used to test these predictions.

References:

Bales, R.F. (1970), Personality and interpersonal behavior. New York: Holt, Rinehart & Winston

Carroll, J.D. (1976), Spatial, non-spatial and hybrid models for scaling. Psychometrika, 41, 439-463

Carroll, J.D. & Wish, M. (1974), Models and methods for three-way multidimensional scaling. In: Krantz, D.H., Atkinson, R.C., Luce, R.D. & Suppes, P. (eds.), Contemporary developments in mathematical psychology. Vol. 2. San Francisco: Freeman, 57-105

Coombs, C.H. (1964), A theory of data. New York: Wiley

Dichtl, E. & Schobert, R. (1979), Mehrdimensionale Skalierung. München: Vahlen

Droge, U. (1984), Ordinale Netzwerkskalierung. Ph.-D.-Thesis, University of Hamburg (mimeo.)

Droge, U. (1985), Symmetrische Modelle und zeilenkonditionale Daten: eine Monte-Carlo-Studie und eine neue Skala. Arbeitsbericht. Psychologisches Institut I, Universität Hamburg

Droge, U. & Feger, H. (1983), Ordinal Network Scaling. Paper presented at the joint meeting of the Classification Society and the Psychometric Society, Paris

Droge, U. & Feger, H. (1985), Reduktionsindices bei Zufallsdaten: eine Monte-Carlo-Studie. In: Arbeitsbericht, Psychologisches Institut I, Universität Hamburg

Feger, H. (1985a), Kommunikationsstrukturen in Kleingruppen. In: Grewe-Partsch, M. & Groebel, J. (Hrsg.), Mensch und Medien. Festschrift für Hertha Sturm. Mainz: v.Hase & Koehler Verlag

Feger, H. (1985b), Structures of interaction in small groups. International Journal of Small Group Research, 1, 104-121

Feger, H. (1986), Netzwerkanalyse in Kleingruppen: Datenarten, Strukturregeln und Strukturmodelle. In: Pappi, F.U. (ed.), Methoden der Netzwerkanalyse

Feger, H. & Bien, W. (1982), Network unfolding. Social Networks, 4, 257-283

Feger, H. & Droge, U. (1984), Repräsentation von Ordinaldaten durch Graphen: Ordinale Netzwerkskalierung. Kölner Zeitschrift für Soziologie und Sozialpsychologie, 36, 494-510

Goetsch, G.G. & McFarland, D.D. (1980), Models of the distribution of acts in small discussion groups. Social Psychology Quarterly, 43, 173-183

Gullahorn, J.T. (1960), Distance and friendship as factors in the gross interaction matrix. In: Moreno, J.L. (ed.), Sociometry reader. Glencoe, Ill.: The Free Press, 506-517

Homans, C.G. (1961), Social behavior: Its elementary forms. New York: Harcourt, Brace, Jovanovich, rev. edition 1974

Motzkin, T. (1936), Beiträge zur Theorie der linearen Ungleichungen. Ph.-D. Thesis, Basel

Müller-Merbach, H. (1973), OR-Ansätze zur optimalen Abteilungsgliederung in Institutionen. In: Kirsch, W. (Hrsg.), Unternehmensführung und Organisation. Wiesbaden, 91-124

Orth, B. (1985), ONSCAL: Ein Verfahren für die "Ordinale Netzwerkskalierung". Teil I: Unkonditionale Ähnlichkeitsdaten (working paper). Department of Psychology, University of Hamburg

Pieper, M. (1982), 'Computerkonferenzsysteme': Nachrichtenübermittlungssysteme oder Kommunikationsmedien zur Unterstützung sozialer Interaktion. Gesellschaft für Mathematik und Datenverarbeitung, (KOMEX-Abschlußbericht), Bonn/Birlinghoven

Radtke, B. (1985), Ein Verfahren zur Konstruktion ordinaler Netzwerkskalen: OSDS, Ph.-D. Thesis, Department of Psychology, University of Hamburg

Riley, M.W., Cohn, R., Toby, J. & Riley, J.W. Jr. (1954), Interpersonal orientations in small groups: A consideration of the questionnaire approach. American Sociological Review, 19, 715-724

SECTION II: ECONOMETRICS OF MARKET BEHAVIOR

II.1. Theory

TELECOMMUNICATIONS DEMAND MODELLING
An Integrated View
A. de Fontenay, M.H. Shugard, D.S. Sibley (Editors)

TESTING THE SPECIFICATION OF ECONOMETRIC MODELS IN REGRESSION AND NON-REGRESSION DIRECTIONS

Russell DAVIDSON and James G. MACKINNON

Department of Economics, Queen's University, Kingston, Ontario, Canada, K7L 3N6.*

The asymptotic power of a statistical test depends on the model being tested, the (implicit) alternative against which the test is constructed, and the process which actually generated the data. The exact way in which it does so is examined for several classes of models and tests. First, we analyse the power of tests of nonlinear regression models in regression directions, i.e. tests which are equivalent to tests for omitted variables. Next, we consider the power of heteroskedasticity-robust variants of these tests. Finally, we examine the power of very general tests in the context of a very general class of models.

1. INTRODUCTION

In any area of applied econometric research, and especially in the usual situation where the experimental design cannot be controlled, it is essential to subject any model to a great many statistical tests before even tentatively accepting it as valid. When that is done routinely, especially early in the model-building process, it is inevitable that most models will fail many tests. The applied econometrician must then attempt to infer from those failures what is wrong with the model. The interpretation of test statistics, both those which reject and those which do not reject the null hypothesis, is thus an important part of the econometrician's job. There is however surprisingly little literature on the subject, a recent exception being Davidson and MacKinnon [6].

The vast majority of the models that econometricians estimate are regression models of some sort, linear or nonlinear, univariate or multivariate. The tests to which such models may be subjected fall into three broad categories, according to a scheme discussed in [6]. First of all, there are tests in "regression directions", in which the (possibly implicit) alternative model is also a regression model, at least locally. Secondly, there are tests in "higher moment directions", which are only concerned with the properties of the error terms; these might include tests for heteroskedasticity, skewness and kurtosis. Finally, there are tests in "mixed directions", which combine both regression and higher moment components.

This paper will deal primarily with tests in regression directions, although there will be some discussion of tests in mixed directions.

---

* This research was supported, in part, by grants from the Social Sciences and Humanities Research Council of Canada.

Tests in regression directions form the lion's share of the tests that are commonly employed by econometricians when testing regression models. The asymptotic analysis of such tests is reasonably easy, and turns out to be remarkably similar to the asymptotic analysis of much more general families of tests; see Davidson and MacKinnon [7] and section 5 below. In contrast to [6], which deals only with linear regression models, we will focus on the case of univariate, nonlinear regression models. Because we are dealing with nonlinear models, all of the analysis is asymptotic; this is probably an advantage rather than otherwise, since even in the linear case many inessential complexities can be eliminated by focussing on what happens as the sample size tends to infinity.

In section 2, we discuss nonlinear regression models and tests in regression directions. In section 3, we analyse the asymptotic power of these tests when the data generating process (or DGP) is not the alternative against which the test is constructed. In section 4, we discuss heteroskedasticity-robust variants of tests in regression directions. Finally, in section 5, we discuss tests in mixed directions.

2. NONLINEAR REGRESSION MODELS AND TESTS IN REGRESSION DIRECTIONS

The model of interest is the nonlinear regression model

$$\mathbf{y} = \mathbf{f}(\beta,\gamma) + \mathbf{u}\ , \qquad E(\mathbf{u}) = 0\ , \qquad E(\mathbf{uu}') = \sigma^2\mathbf{I}\ , \tag{1}$$

where $\mathbf{y}$ and $\mathbf{u}$ are n-vectors and $\mathbf{f}$ denotes a vector of twice continuously differentiable functions $f_t(\beta,\gamma)$ which depend on β, a k-vector, and γ, an r-vector, of (generally unknown) parameters. The matrices of derivatives of the elements of $\mathbf{f}$ with respect to the elements of β and γ will be denoted $\mathbf{F}_\beta(\beta,\gamma)$ and $\mathbf{F}_\gamma(\beta,\gamma)$; these matrices are $n \times k$ and $n \times r$ respectively. The quantities $\mathbf{f}(\beta,\gamma)$, $\mathbf{F}_\beta(\beta,\gamma)$ and $\mathbf{F}_\gamma(\beta,\gamma)$ are all functions of β and γ, as are several other quantities to be introduced below. When one of these, $\mathbf{f}(\beta,\gamma)$ for example, is evaluated at the true values, say β_0 and γ_0, it will simply be denoted $\mathbf{f}$.

The functions f_t may depend on past values of y_t, but not on current or future values, since otherwise (1) would not be a regression model and least squares would no longer be an appropriate estimating technique. In order to ensure that all estimators and test statistics behave sensibly as $n \to \infty$, it is assumed that $\mathbf{F}_\beta'\mathbf{F}_\beta/n$, $\mathbf{F}_\gamma'\mathbf{F}_\gamma/n$ and $\mathbf{F}_\beta'\mathbf{F}_\gamma/n$ all tend to finite limiting matrices with ranks k, r and $\min(k,r)$ respectively as $n \to \infty$, while the matrix $[\mathbf{F}_\beta\ \ \mathbf{F}_\gamma]$ always has rank $k + r$ for large n .

It is remarkably easy to test hypotheses about β and γ by means of artificial linear regressions, without ever estimating the unrestricted model; see, among others, Engle [11] and Davidson and MacKinnon [5]. Suppose that we wish to test the null hypothesis $\gamma = 0$. If we estimate (1) by least squares imposing this restriction, we will obtain restricted estimates $\tilde{\beta}$ and $\tilde{\sigma}^2$. Evaluating $\mathbf{f}(\beta,\gamma)$ and $\mathbf{F}_\beta(\beta,\gamma)$ and $\mathbf{F}_\gamma(\beta,\gamma)$

at $(\tilde{\beta},0)$, we obtain $\tilde{\mathbf{f}}$, $\tilde{\mathbf{F}}_\beta$ and $\tilde{\mathbf{F}}_\gamma$, and can thus construct the artificial linear regression

$$(\mathbf{y} - \tilde{\mathbf{f}})/\tilde{\sigma} = \tilde{\mathbf{F}}_\beta\mathbf{b} + \tilde{\mathbf{F}}_\gamma\mathbf{c} + \text{errors} \quad . \tag{2}$$

The explained sum of squares from this regression (which is also n times the uncentred R^2) is

$$(\mathbf{y} - \tilde{\mathbf{f}})'\tilde{\mathbf{F}}_\gamma(\tilde{\mathbf{F}}_\gamma'\tilde{\mathbf{M}}_\beta\tilde{\mathbf{F}}_\gamma)^{-1}\tilde{\mathbf{F}}_\gamma'(\mathbf{y} - \tilde{\mathbf{f}})/\tilde{\sigma}^2 \quad , \tag{3}$$

where $\tilde{\mathbf{M}}_\beta = \mathbf{I} - \tilde{\mathbf{F}}_\beta(\tilde{\mathbf{F}}_\beta'\tilde{\mathbf{F}}_\beta)^{-1}\tilde{\mathbf{F}}_\beta'$. It is straightforward to show that this test statistic is asymptotically distributed as chi-squared with r degrees of freedom under the null hypothesis. The proof makes use of the facts that, asymptotically, $\mathbf{y} - \tilde{\mathbf{f}} = \mathbf{M}_\beta\mathbf{u}$, and that a central limit theorem can be applied to the $r \times 1$ vector $\mathbf{F}_\gamma'\mathbf{M}_\beta\mathbf{u}$. Note that the numerator of (3) is also the numerator of the ordinary F statistic for a test of $\mathbf{c} = 0$ in (2), so that (3) is asymptotically equivalent to an F test.

The test statistic (3) may be thought of as a generic test in what [6] refers to as "regression directions". It is generic because the fact that it is based solely on restricted estimates does not prevent it from having the same asymptotic properties, under the null and under local alternatives, as other standard tests. If we had assumed that the u_t's were normally distributed, (3) could have been derived as a Lagrange Multiplier test, and standard results on the equivalence of LM, Wald and Likelihood Ratio tests could have been invoked; see Engle [12]. It is obvious that this equivalence does not depend on normality.

By a regression direction is meant any direction from the null hypothesis, in the space of likelihood functions, which corresponds, at least locally, to a regression model. It is clear that (3) is testing in regression directions because (1) is a regression model whether or not $\gamma = 0$. But a test in regression directions need not be explicitly derived from an alternative hypothesis which is a regression model, although of course the null must be such a model. If we were simply to replace the matrix $\tilde{\mathbf{F}}_\gamma$ in (2) and (3) by an arbitrary matrix $\mathbf{Z}$, asymptotically uncorrelated with $\mathbf{u}$ under the null hypothesis, we would obtain the asymptotically valid test statistic

$$(\mathbf{y} - \tilde{\mathbf{f}})'\mathbf{Z}(\mathbf{Z}'\tilde{\mathbf{M}}_\beta\mathbf{Z})^{-1}\mathbf{Z}'(\mathbf{y} - \tilde{\mathbf{f}})/\tilde{\sigma}^2 \quad . \tag{4}$$

Using this device, specification tests (in the sense of Hausman [14] and Holly [15]) can be computed in the same way as other tests in regression directions, when the null is a regression model; see Davidson and MacKinnon [8]. So can non-nested hypothesis tests, encompassing tests and differencing tests; see Davidson and MacKinnon [3], Mizon and Richard [18], and Davidson, Godfrey and MacKinnon [10] respectively. In the next section, we shall ignore where the matrix $\mathbf{Z}$ and associated test statistic (4) came from, and analyse what determines the power of all tests in regression directions.

3. THE LOCAL POWER OF TESTS IN REGRESSION DIRECTIONS

In order to say anything about the power of a test statistic, one must specify how the data are actually generated. Since we are concerned with tests in regression directions, we shall restrict our attention to DGP's which differ from the null hypothesis only in such directions. This restriction is of course by no means innocuous, as will be made clear in section 5 below. The null hypothesis will be (1) with $\gamma = 0$. Since the alternative that $\gamma \neq 0$ is only one of many alternatives against which we may wish to test the null, we shall usually suppress γ and write $\mathbf{f}(\beta, 0)$ as $\mathbf{f}(\beta)$.

Suppose that the data are generated by the sequence of local DGP's

$$\mathbf{y} = \mathbf{f}(\beta_0) + \alpha n^{-\frac{1}{2}}\mathbf{a} + \mathbf{u}, \qquad E(\mathbf{u}) = \mathbf{0}, \qquad E(\mathbf{uu}') = \sigma_0^2\mathbf{I}, \tag{5}$$

where β_0 and σ_0 denote particular values of β and σ^2, $\mathbf{a}$ is an $n \times 1$ vector which may depend on exogenous variables, β_0 and past values of y, and α is a parameter which determines how far the DGP is from the <u>simple</u> null hypothesis

$$\mathbf{y} = \mathbf{f}(\beta_0) + \mathbf{u}, \qquad E(\mathbf{u}) = \mathbf{0}, \qquad E(\mathbf{uu}') = \sigma_0^2\mathbf{I}. \tag{6}$$

We assume that $\mathbf{a}'\mathbf{a}/n$, $\mathbf{a}'\mathbf{F}_\beta/n$ and $\mathbf{a}'\mathbf{Z}/n$ all tend to finite limiting matrices.

The notion of a sequence of local DGP's requires some discussion. Following Pitman [20] and many subsequent authors, we adopt it because it seems the most reasonable way to deal with power in the context of asymptotic theory. The sequence (5) approaches the simple null (6) at a rate of $n^{-\frac{1}{2}}$. This rate is chosen so that the test statistic (4), and all asymptotically equivalent test statistics, will be of order unity as $n \to \infty$. If, on the contrary, the DGP were held fixed as the sample size was increased, the test statistic would normally tend to blow up, and it would be impossible to talk about its asymptotic distribution.

Sequences like (5) have not been widely used in econometrics. Most authors who investigate the asymptotic power of test statistics have been content to conduct their analysis on the assumption that the sequence of local DGP's actually lies within the compound alternative against which the test is constructed; see for example Engle [12]. But this makes it impossible to study how the power of a test depends on the relations among the null, the alternative and the DGP, so that the ensuing analysis can shed no light on the difficult question of how to interpret significant test statistics. One paper which uses a sequence similar to (5) is Davidson and MacKinnon [4], who study the power of various non-nested hypothesis tests. They conclude that several of the tests are asymptotically equivalent under all sequences of local DGP's, a stronger conclusion than would be possible using the conventional assumption.

The sequence (5) provides a perfectly general *local* representation of any regression model which is sufficiently close to the simple null (6). For example, suppose that we wanted to see how a certain test performed when the data were generated by an alternative like (1), with $\gamma \neq 0$. We could simply specify the sequence of local DGP's as

$$\mathbf{y} = \mathbf{f}(\beta_0,\ \alpha n^{-\frac{1}{2}}\gamma_0) + \mathbf{u}\ , \tag{7}$$

where γ_0 is fixed and α determines how far (7) is from the simple null (6). Because (7) approaches (6) as $n^{-\frac{1}{2}} \to 0$, a first-order Taylor approximation to (7) around $\alpha = 0$ must yield exactly the same results, in an asymptotic analysis, as (7) itself. This approximation is

$$\mathbf{y} = \mathbf{f}(\beta_0, 0) + \alpha n^{-\frac{1}{2}}\ \mathbf{F}_\gamma(\beta_0, 0)\gamma_0 + \mathbf{u}\ . \tag{8}$$

Defining $\mathbf{f}(\beta_0)$ as $\mathbf{f}(\beta_0, 0)$ and $\mathbf{a}$ as $\mathbf{F}_\gamma(\beta_0, 0)\gamma_0$, we see immediately that (8) is simply a particular case of (5).

We now wish to find the asymptotic distribution of the test statistic (4) under the sequence of local DGP's (5). We first rewrite (4) so that all factors are O(1):

$$[(\mathbf{y}-\tilde{\mathbf{f}})'\mathbf{Z}/n^{\frac{1}{2}}][(\mathbf{Z}'\tilde{\mathbf{M}}_\beta\mathbf{Z})/n]^{-1}[\mathbf{Z}'(\mathbf{y}-\tilde{\mathbf{f}})/n^{\frac{1}{2}}]/\tilde{\sigma}^2\ . \tag{9}$$

The $n^{-\frac{1}{2}}$ factor in (5) ensures that $\tilde{\sigma}^2 \to \sigma_0^2$ as $n \to \infty$, and it is obvious that

$$[(\mathbf{Z}'\tilde{\mathbf{M}}_\beta\mathbf{Z})/n]^{-1} \to [\text{plim}(\mathbf{Z}'\mathbf{M}_\beta\mathbf{Z}/n)]^{-1} \quad \text{as} \quad n \to \infty\ . \tag{10}$$

Now recall that

$$\mathbf{f}(\tilde{\beta}) \overset{a}{=} \mathbf{f}(\beta_0) + \mathbf{F}_\beta(\mathbf{F}_\beta'\mathbf{F}_\beta)^{-1}\mathbf{F}_\beta'[\mathbf{y}-\mathbf{f}(\beta_0)]\ , \tag{11}$$

where "$\overset{a}{=}$" means "is asymptotically equal to". It follows from (11) that

$$\begin{aligned}\mathbf{y} - \mathbf{f}(\tilde{\beta}) &\overset{a}{=} \alpha n^{-\frac{1}{2}}\mathbf{a} + \mathbf{u} - \mathbf{F}_\beta(\mathbf{F}_\beta'\mathbf{F}_\beta)^{-1}\mathbf{F}_\beta'(\alpha n^{-\frac{1}{2}}\mathbf{a} + \mathbf{u}) \\ &= \mathbf{M}_\beta(\alpha n^{-\frac{1}{2}}\mathbf{a} + \mathbf{u})\ .\end{aligned} \tag{12}$$

The test statistic (4) is thus asymptotically equal to

$$(\alpha n^{-1}\mathbf{a} + n^{-\frac{1}{2}}\mathbf{u})'\mathbf{M}_\beta\mathbf{Z}[\text{plim}(\mathbf{Z}'\mathbf{M}_\beta\mathbf{Z}/n)]^{-1}\mathbf{Z}'\mathbf{M}_\beta(\alpha n^{-1}\mathbf{a} + n^{-\frac{1}{2}}\mathbf{u})/\sigma_0^2\ . \tag{13}$$

It is easy to find the asymptotic distribution of (13). First, define $\mathbf{P}_0$ as an $r \times r$ triangular matrix such that

$$\mathbf{P}_0\mathbf{P}_0' = [\text{plim}(\mathbf{Z}'\mathbf{M}_\beta\mathbf{Z}/n)]^{-1}\ , \tag{14}$$

and then define the $r \times 1$ vector η as

$$\eta = \mathbf{P}_0'\mathbf{Z}'\mathbf{M}_\beta(\alpha n^{-1}\mathbf{a} + n^{-\frac{1}{2}}\mathbf{u})/\sigma_0\ . \tag{15}$$

The test statistic (13) now takes the very simple form $\eta'\eta$; it is just the sum of r squared random variables which are the elements of the vector η . It is clear that, asymptotically, the mean of η is the vector

$$\operatorname{plim}(\alpha\sigma_0^{-1}\ \mathbf{P}_0'\mathbf{Z}'\mathbf{M}_\beta\mathbf{a}/n) \quad , \tag{16}$$

and its variance-covariance matrix is

$$\operatorname{plim}(\sigma_0^{-2}\mathbf{P}_0'\mathbf{Z}'\mathbf{M}_\beta E(\mathbf{uu}')\mathbf{M}_\beta\mathbf{ZP}_0/n) \tag{17}$$

$$= \mathbf{P}_0'\operatorname{plim}(\mathbf{Z}'\mathbf{M}_\beta\mathbf{Z}/n)\mathbf{P}_0 = \mathbf{I}_r \quad . \tag{18}$$

Since η is equal to $n^{-\frac{1}{2}}$ times a weighted sum of random variables with mean zero and finite variance, and since our assumptions keep those weights bounded from above and below, a central limit theorem can be applied to it. The test statistic (9) is thus asymptotically equal to a sum of r independent squared normal random variates, each with variance unity, and with means given by the vector (16). Such a sum has the non-central chi-squared distribution with r degrees of freedom and non-centrality parameter, or NCP, given by the squared norm of the mean vector, which in this case is equal to

$$\alpha^2[\operatorname{plim}(\mathbf{a}'\mathbf{M}_\beta\mathbf{Z}/n)][\operatorname{plim}(\mathbf{Z}'\mathbf{M}_\beta\mathbf{Z}/n)]^{-1}[\operatorname{plim}(\mathbf{Z}'\mathbf{M}_\beta\mathbf{a}/n)]/\sigma_0^2 \quad . \tag{19}$$

The NCP (19) can be rewritten in a more illuminating way. Consider the vector $\alpha n^{-\frac{1}{2}}\mathbf{M}_\beta\mathbf{a}$, the length of which, asymptotically, is

$$\alpha^2 \operatorname{plim}(\mathbf{a}'\mathbf{M}_\beta\mathbf{a}/n) \quad . \tag{20}$$

The quantity (20) is a measure of the distance between the DGP (5) and a linear approximation to the null hypothesis around the simple null (6); in a sense, it tells us how "wrong" the model being tested is. Now consider the artificial regression

$$\alpha n^{-\frac{1}{2}}\mathbf{M}_\beta\mathbf{a}/\sigma_0 = \mathbf{M}_\beta\mathbf{Zd} + \text{errors} \quad . \tag{21}$$

The total sum of squares for this regression, asymptotically, is expression (20) divided by σ_0^2 . The explained sum of squares, asymptotically, is the NCP (19). Thus the asymptotic uncentred R^2 from regression (21) is

$$\{\operatorname{plim}(\mathbf{a}'\mathbf{M}_\beta\mathbf{Z}/n)[\operatorname{plim}(\mathbf{Z}'\mathbf{M}_\beta\mathbf{Z}/n)]^{-1}\operatorname{plim}(\mathbf{Z}'\mathbf{M}_\beta\mathbf{a}/n)\}/[\operatorname{plim}(\mathbf{a}'\mathbf{M}_\beta\mathbf{a}/n)]. \tag{22}$$

Expression (22) has an alternative interpretation. Consider the asymptotic projection of $\alpha n^{-\frac{1}{2}}\mathbf{M}_\beta\mathbf{a}$ onto the space spanned by $\mathbf{F}_\beta$ and $\mathbf{Z}$ jointly. This projection is

$$\alpha n^{-\frac{1}{2}}\mathbf{M}_\beta\mathbf{Z}\ [\operatorname{plim}(\mathbf{Z}'\mathbf{M}_\beta\mathbf{Z}/n)]^{-1}[\operatorname{plim}(\mathbf{Z}'\mathbf{M}_\beta\mathbf{a}/n)] \quad . \tag{23}$$

Now let ϕ be the angle between $\alpha n^{-\frac{1}{2}}\mathbf{M}_\beta\mathbf{a}$ and the projection (23). By the definition of a cosine, it is easily seen that $\cos^2\phi$ is equal to the R^2 (22). Thus we may rewrite the NCP (19) as

$$\alpha^2 \sigma_0^{-2}[\text{plim}(\mathbf{a}'\mathbf{M}_\beta\mathbf{a}/n)] \cos^2\phi \quad , \tag{24}$$

or simply as

$$\alpha^2 \sigma_0^{-2} \cos^2\phi \tag{25}$$

if we normalize the vector $\mathbf{a}$ so that $\text{plim}(\mathbf{a}'\mathbf{M}_\beta\mathbf{a}/n) = 1$, and rescale α appropriately.

If a test statistic which has the non-central chi-squared distribution with r degrees of freedom is compared with critical values from the corresponding central chi-squared distribution, the power of the test increases monotonically with the NCP. Expression (25) writes this NCP as the product of three factors. The first factor, α^2 , measures the distance between the DGP and the closest point on a linear approximation to the null hypothesis. Note that this distance in no way depends on $\mathbf{Z}$; the greater α^2 , the more powerful *any* test will be. Like the other two factors, this first factor is independent of n . For the first factor, that comes about because the DGP approaches the null hypothesis at a rate of $n^{-\frac{1}{2}}$; if, on the contrary, the DGP were fixed as n increased, this factor would have to be proportional to n . Of course, the asymptotic analysis we have done would not be valid if the DGP were fixed, although it would provide a useful approximation in most cases.

The second factor in expression (25) is σ_0^{-2} . This tells us that the NCP is inversely proportional to the variance of the DGP, which makes sense because as the DGP becomes noisier it should become harder to reject any null hypothesis. What affects the NCP is α^2/σ_0^2 , the ratio of the systematic discrepancy between the null and the DGP to the noise in the latter. Note that this ratio does not depend on $\mathbf{Z}$. It will be the same for all tests in regression directions of any given null hypothesis with any given data set.

The most interesting factor in expression (25) is the third one, $\cos^2\phi$. It is only through this factor that the choice of $\mathbf{Z}$ affects the NCP. A test will have maximal power, for a given number of degrees of freedom, when $\cos^2\phi$ is one, that is, when the artificial regression (21) has an R^2 of one. That will be the case whenever the vector $\mathbf{a}$ is a linear combination of the vectors in $\mathbf{F}_\beta$ and $\mathbf{Z}$, which will occur whenever the DGP lies within the alternative against which the test is constructed. For example, if the null hypothesis were $\mathbf{y} = \mathbf{f}(\beta,0)$, the alternative were $\mathbf{y} = \mathbf{f}(\beta,\gamma)$ and the DGP were (8), $\mathbf{Z}$ would then be $\mathbf{F}_\gamma$ and $\mathbf{a}$ would be a linear combination of the columns in $\mathbf{F}_\gamma$ (see (8)). Thus the power of a test is maximized when we test against the truth.

On the other hand, a test will have no power at all when $\cos^2\phi$ is zero. This would occur if $\mathbf{M}_\beta\mathbf{a}$ were asymptotically orthogonal to $\mathbf{M}_\beta\mathbf{Z}$, something which in general would seem to be highly unlikely. However, special features of a model, or of the sample design, may make such a situation less uncommon than one might think. Nevertheless, it is probably not very misleading to assert that, when a null hypothesis is false in a regression direction, almost any test in regression directions can be expected to have *some* power, although perhaps not very much.

These results make it clear that there is always a tradeoff when we choose what regression directions to test against. By increasing the number of columns in Z, we can always increase $\cos^2\phi$, or at worst leave it unchanged, which by itself will increase the power of the test. But doing so also increases r, the number of degrees of freedom, which by itself reduces the power of the test. This tradeoff is at the heart of a number of controversies in the literature on hypothesis testing, such as the debate over non-nested hypothesis tests versus encompassing tests (Dastoor [2], Mizon and Richard [18]), and the literature on specification tests versus classical tests (Hausman [14], Holly [15]).

The tradeoff between $\cos^2\phi$ and degrees of freedom is affected by the sample size. As n increases, the NCP can be expected to increase (since in reality the DGP is not approaching the null as $n \to \infty$), so that a given change in $\cos^2\phi$ will have a larger effect on power the larger is n. On the other hand, the effect of r on the critical value is independent of sample size. Thus when the sample size is small, it is particularly important to use tests with few degrees of freedom, while when the sample size is large it becomes feasible to look in many directions at once so as to maximize $\cos^2\phi$.

If we were confident that the null could only be false in a single direction (i.e. if we knew exactly what the vector $\mathbf{a}$ might be) the optimal procedure would be to have only one column in Z, that column being proportional to $\mathbf{a}$. In practice, we are rarely in that happy position. There are normally a number of things which we suspect might be wrong with our model, and hence a large number of regression directions in which to test. Faced with this situation, there are at least two ways to proceed.

One approach is to test against each type of potential misspecification separately, with each test having only one or a few degrees of freedom. If the model is in fact wrong in one or a few of the regression directions in which these tests are carried out, such a procedure is as likely as any to inform us of that fact. However, the investigator must be careful to control the overall size of the test (since when one does, say, ten different tests each at the .05 level, the overall size could be as high as .40), and should avoid jumping to the conclusion that the model is wrong in a particular way just because a certain test statistic is significant. Remember that $\cos^2\phi$ will often be well above zero for *many* tests, even if only one thing is wrong with the model.

Alternatively, it is possible to test for a great many types of misspecification at once by putting all the regression directions we want to test against into one big Z matrix. This maximizes $\cos^2\phi$, and hence maximizes the chance that the test is consistent, and it also makes it easy to control the size of the test. But such a test will have many degrees of freedom, so that power may be poor when the sample size is small. Moreover, if such a test rejects the null, that gives us very little information as to what may be wrong with the model.

It may be possible to make some tentative inferences about the true model by looking at the values of several test statistics. Suppose that we test a model against several sets of regression directions, represented by regressor matrices Z_1, Z_2 and so on, and thus generate test statistics T_1, T_2 .... Each of the test statistics T_i can be used to estimate the corresponding NCP, NCP_i. Since the mean of a non-central chi-squared random variable with r degrees of freedom is r plus the NCP, the obvious estimate of NCP_i is $T_i - r_i$. It is far from certain that the Z_i with the highest estimated NCP_i, say Z_i^*, actually represents truly omitted directions. Nevertheless, modifying the model in the directions represented by Z_i^* would seem to be a reasonable thing to do in many cases, especially when the number of columns in Z_i^* is small. It might be useful to perform a test in all the interesting regression directions one can think of, thus obtaining a test statistic with the largest NCP obtainable. If that test statistic is not much larger than T_i^*, then one might feel reasonably confident that the directions represented by Z_i^* adequately capture the discrepancy between the null and the DGP.

4. TESTS THAT ARE ROBUST TO HETEROSKEDASTICITY OF UNKNOWN FORM

The distinguishing feature of regression models is that the error term is simply added to a regression function which determines the mean of the dependent variable. This greatly simplifies the analysis of such models. In particular, it means that test statistics such as (4) are asymptotically valid regardless of how the error terms u_t are distributed, provided only that, for all t, $E(u_t) = 0$, $E(u_t u_s) = 0$ for all $s \neq t$, and $E(u_t^2) = \sigma^2$. Without normality, least squares will not be asymptotically efficient, and tests based on least squares will not be most powerful, but least squares estimates will be consistent and tests based on them will be asymptotically valid.

When the error terms u_t display heteroskedasticity of unknown form, least squares estimates remain consistent, but test statistics such as (4) are no longer asymptotically valid. However, the results of White [23] make it clear that asymptotically valid test statistics can be constructed in this case. Davidson and MacKinnon [8] show how to compute such tests by means of artificial linear regressions, and provide some results on their finite-sample properties. These heteroskedasticity-robust tests are likely to be very useful when analysing cross-section data. In this section we consider the power properties of such tests.

Under the null hypothesis, the test statistic (4) tends to the random variable

$$[\mathbf{u}'M_\beta Z/n^{\frac{1}{2}}][\text{plim}(Z'M_\beta Z/n)]^{-1}[Z'M_\beta \mathbf{u}/n^{\frac{1}{2}}] \qquad (26)$$

as $n \to \infty$. Thus it is evident that (4) is really testing the hypothesis that

$$\lim_{n\to\infty} E(n^{-\frac{1}{2}}\mathbf{u}'M_\beta Z) = 0 \quad . \tag{27}$$

Now suppose that $E(\mathbf{uu}') = \Omega$, where Ω is a diagonal matrix with diagonal elements σ_t^2 bounded from above. The asymptotic variance-covariance matrix of $n^{-\frac{1}{2}}\mathbf{u}'M_\beta Z$ is

$$\text{plim}[Z'M_\beta \Omega M_\beta Z/n] \quad . \tag{28}$$

Even though Ω is unknown and has as many unknown elements as there are observations, the matrix (28) may be estimated consistently in a number of different ways. One of the simplest is to replace Ω by $\tilde{\Omega}$, where Ω is a diagonal matrix with diagonal elements $\tilde{u}_t^2$, the $\tilde{u}_t$'s being the residuals from least squares estimation of the null hypothesis. Other estimators, which may have better finite-sample properties, are discussed by MacKinnon and White [17].

It is now straightforward to derive a heteroskedasticity-robust test statistic. Written in the same form as (9), so that all factors are O(1), it is

$$[(y - \tilde{f})'Z/n^{\frac{1}{2}}][(Z'\tilde{M}_\beta \tilde{\Omega}\tilde{M}_\beta Z)/n]^{-1}[Z'(y - \tilde{f})/n^{\frac{1}{2}}] \quad . \tag{29}$$

This statistic is simply n minus the sum of squared residuals from the artificial regression

$$\iota = \tilde{U}\tilde{M}_\beta Zc + \text{errors} \quad , \tag{30}$$

where ι is an n-vector of ones and $\tilde{U} = \text{diag}(\tilde{u}_t)$. That (29) can be calculated in this simple way follows from the facts that

$$\iota'\tilde{U}\tilde{M}_\beta Z = (y - \tilde{f})'\tilde{M}_\beta Z = (y - \tilde{f})'Z \tag{31}$$

and

$$Z'\tilde{M}_\beta \tilde{U}'\tilde{U}\tilde{M}_\beta Z = Z'\tilde{M}_\beta \tilde{\Omega}\tilde{M}_\beta Z \quad . \tag{32}$$

Finding the asymptotic distribution of the heteroskedasticity-robust test statistic (29) is very similar to finding that of the ordinary test statistic (9). Under a sequence of local DGP's like (5), but with $E(\mathbf{uu}') = \Omega_0$, it is easy to show that (29) is asymptotically equal to

$$(\alpha n^{-1}\mathbf{a} + n^{-\frac{1}{2}}\mathbf{u})'M_\beta Z[\text{plim}(Z'M_\beta \Omega_0 M_\beta Z/n)]^{-1}Z'M_\beta(\alpha n^{-1}\mathbf{a} + n^{-\frac{1}{2}}\mathbf{u}) \quad , \tag{33}$$

where Ω_0 is the covariance matrix of the error terms in a sequence of local DGP's similar to (5) except for the heteroskedasticity. An argument very similar to that used earlier can then be employed to show that, asymptotically, (33) has the non-central chi-squared distribution with r degrees of freedom and NCP

$$\alpha^2[\text{plim}(\mathbf{a}'\mathbf{M}_\beta Z/n)][\text{plim}(Z'\mathbf{M}_\beta\Omega_0\mathbf{M}_\beta Z/n)]^{-1}[\text{plim}(Z'\mathbf{M}_\beta\mathbf{a}/n)] \quad . \tag{34}$$

Like the earlier NCP (19), expression (34) can also be interpreted as the explained sum of squares from a certain artificial regression. In this case, the tth element of the regressand is $\alpha n^{-\frac{1}{2}}(\mathbf{M}_\beta\mathbf{a})_t\sigma_t$ and the tth row of the regressor matrix is $\sigma_t(\mathbf{M}_\beta Z)_t$. The NCP may then be written as

$$\alpha^2 \ \text{plim}(\mathbf{a}'\mathbf{M}_\beta\Omega_0^{-1}\mathbf{M}_\beta\mathbf{a}/n)\psi \quad , \tag{35}$$

where ψ denotes the uncentred, asymptotic R^2 from the artificial regression.

Expression (35) resembles expression (24), but differs from it in two important respects. First of all, the factor which measures the distance between the DGP and the null hypothesis relative to the noisiness of the DGP is now $\alpha^2 \ \text{plim}(\mathbf{a}'\mathbf{M}_\beta\Omega_0^{-1}\mathbf{M}_\beta\mathbf{a}/n)$ rather than $\alpha^2 \ \text{plim}(\mathbf{a}'\mathbf{M}_\beta\mathbf{a}/n)/\sigma_0^2$. Secondly, although ψ plays the same role as $\cos^2\phi$, and is the only factor which is affected by the choice of Z , ψ does not have quite the same properties as $\cos^2\phi$. It is possible to make ψ zero by choosing Z appropriately, but it is usually not possible to make ψ unity even by choosing Z so that $\mathbf{a}$ lies in the span of the columns of $\mathbf{F}_\beta$ and Z . That would of course be possible if Ω_0 were proportional to the identity matrix, in which case (35) and (24) would be identical. Thus when there is no heteroskedasticity, the heteroskedasticity-robust test statistic (29) is asymptotically equivalent, under all sequences of local DGP's, to the ordinary test statistic (9). There will however be some loss of power in finite samples; see Davidson and MacKinnon [8].

It is clear from expression (35) that when there is in fact heteroskedasticity, the pattern of the error variances will affect the power of the test. Multiplying all elements of Ω_0 by a factor λ will of course reduce the NCP by a factor $1/\lambda$, as in the homoskedastic case. But changes in the pattern of heteroskedasticity, even if they do not affect the average value of σ_t^2 , may well affect $\text{plim}(Z'\mathbf{M}_\beta\Omega_0\mathbf{M}_\beta Z/n)$, and hence the power of the test.

As a result of this, the interpretation of heteroskedasticity-robust tests is even harder than the interpretation of ordinary tests in regression directions. As discussed in the last section, we know in the ordinary case that the NCP will be highest when we test against the truth, and so looking at several test statistics can provide some guidance as to where the truth lies. In the heteroskedasticity-robust case, however, things are not so simple. It is entirely possible that $\mathbf{a}$ may lie in the span of the columns of $\mathbf{F}_\beta$ and Z_1 , so that a test against the directions represented by Z_1 is in effect a test against the truth, and yet the NCP may be substantially higher when testing against some quite different set of directions represented by Z_2 . Thus in the common situation where several different tests reject the null hypothesis, it may be far from obvious how the model should be modified.

5. TESTS IN MIXED NON-REGRESSION DIRECTIONS

The popularity of regression models is easy to understand. They have evolved naturally from the classical problem of estimating a mean; they are easy to write down and interpret; and they are usually quite easy to estimate. The regression specification is, however, very restrictive. By forcing the error term to be additive, it greatly limits the way in which random events outside the model can affect the dependent variable. In order to see whether this is in fact a severe restriction, careful applied workers will usually wish to test their models in non-regression as well as regression directions.

As we saw above, regression directions are those which correspond, at least locally, to a more general regression model in which the null is nested. Non-regression directions, then, are those which correspond either to a regression model with a different error structure (as in tests for heteroskedasticity, skewness and kurtosis), or to a more general non-regression model. We shall refer to the latter as "mixed" directions, since they typically affect both the mean and the higher moments of the dependent variable.

One non-regression model which has been widely used in applied econometrics is the Box-Cox regression model; see, among others, Zarembka [25] and Savin and White [21]. This model can be written as

$$y_t(\lambda) = \sum_{i=1}^{k} \beta_i X^1_{ti}(\lambda) + \sum_{j=1}^{m} \gamma_j X^2_{tj} + u_t \quad , \tag{36}$$

where $x(\lambda)$ denotes the Box-Cox transformation,

$$x(\lambda) = (x^\lambda - 1)/\lambda \quad \text{if} \quad \lambda \neq 0 \ ; \quad x(\lambda) = \log(x) \quad \text{if} \quad \lambda = 0 \quad . \tag{37}$$

Here the X^1_{ti} are regressors which may sensibly be subjected to the Box-Cox transformation, while the X^2_{tj} are ones which cannot sensibly be thus transformed, such as the constant term, dummy variables and variables which can take on non-positive values.

Conditional on λ, the Box-Cox model (36) is a regression model. When $\lambda = 1$, it is a linear model, and when $\lambda = 0$, it is a loglinear one. Either of these null hypotheses may be tested against the Box-Cox alternative (36), and numerous procedures exist for doing so. These are examples of tests in mixed directions, because (36) is not a regression model when λ is a parameter to be estimated. The reason is that the regression function in (36) determines the mean of $y_t(\lambda)$ rather than the mean of the actual dependent variable. If we were to rewrite (36) so that y_t was on the left-hand side, it would be clear that y_t is actually a nonlinear function of the X^1_{ti}'s, X^2_{tj}'s and u_t.

The obvious way to test both linear and loglinear null hypotheses against the Box-Cox alternative (36) is to use some form of the Lagrange

Multiplier test. Several such tests have been proposed. In particular, Godfrey and Wickens [13] suggest using the "outer product of the gradient" (or OPG) variant of the LM test, while Davidson and MacKinnon [9] suggest using the "double-length regression" (or DLR) variant. Both these variants can be computed by means of a single artificial linear regression, and both can handle a wide variety of tests in mixed non-regression directions. In the remainder of this section, we shall discuss the OPG variant of the LM test. We shall not discuss the DLR variant, which was originally proposed by Davidson and MacKinnon [5], even though it appears to have substantially better finite-sample properties than the OPG variant. Discussion of the asymptotic power properties of the DLR variant may be found in Davidson and MacKinnon [6,9], and there would be no point in repeating that discussion here. Moreover, the DLR variant is applicable to a narrower class of models than the OPG variant, and is somewhat more complicated to analyse. Note that our results do not apply merely to the OPG form of the LM test; they will be equally valid for any form of the LM test, and for asymptotically equivalent Wald and Likelihood Ratio tests as well.

The OPG variant of the LM test is applicable to any model for which the loglikelihood function may be written as a sum of the contributions from all the observations. Thus if $\ell_t(y_t,\theta_1,\theta_2)$ denotes the contribution from the tth observation (possibly conditional on previous observations), the loglikelihood function is

$$L = \sum_{t=1}^{n} \ell_t(y_t,\theta_1,\theta_2) \quad . \tag{38}$$

The derivatives of $\ell_t(\cdot)$ with respect to the ith element of θ_1 and the jth element of θ_2 will be denoted by $G^1_{ti}(\cdot)$ and $G^2_{tj}(\cdot)$ respectively. These may then be formed into the $n \times k$ and $n \times r$ matrices $\mathbf{G}_1(\cdot)$ and $\mathbf{G}_2(\cdot)$. The null hypothesis is that $\theta_2 = 0$, and for the computation of the LM test, all quantities will be evaluated at the restricted ML estimates $(\tilde{\theta}_1,0)$

In the particular case of testing the null hypothesis of loglinearity against the Box-Cox alternative (36), θ_1 would be a vector of the β_i's, the γ_j's and σ (or σ^2) , and θ_2 would be the scalar λ . Assuming normality, it is easy to write down the loglikelihood function, and we see that

$$\ell_t = -\tfrac{1}{2}\log(2\pi) - \log(\sigma) - \tfrac{1}{2}u_t^2/\sigma^2 + (\lambda - 1)\log(y_t) \quad , \tag{39}$$

where u_t is implicitly defined by (36). It is now easy to calculate the elements of the matrix $\mathbf{G}(\cdot)$ and to evaluate them under the null hypothesis that $\lambda = 0$, which simply requires that one estimate the loglinear null and obtain estimates $\tilde{\beta}$, $\tilde{\gamma}$ and $\tilde{\sigma}$; see Godfrey and Wickens [13].

The OPG form of the LM test is remarkably easy to calculate. It is simply the explained sum of squares (or n minus the sum of squared residuals) from the artificial regression

$$\iota = \tilde{\mathbf{G}}_1\mathbf{b} + \tilde{\mathbf{G}}_2\mathbf{c} + \text{errors} \quad , \tag{40}$$

where ι is an n-vector of ones, $\tilde{\mathbf{G}}_1 = \mathbf{G}_1(\tilde{\theta}_1, 0)$ and $\tilde{\mathbf{G}}_2 = \mathbf{G}_2(\tilde{\theta}_1, 0)$. The explained sum of squares from regression (40) is

$$\iota'\tilde{\mathbf{G}}(\tilde{\mathbf{G}}'\tilde{\mathbf{G}})^{-1}\tilde{\mathbf{G}}'\iota = \iota'\tilde{\mathbf{G}}_2(\tilde{\mathbf{G}}_2'\tilde{\mathbf{M}}_1\tilde{\mathbf{G}}_2)^{-1}\tilde{\mathbf{G}}_2'\iota \quad , \tag{41}$$

where $\tilde{\mathbf{M}}_1 = \mathbf{I} - \tilde{\mathbf{G}}_1(\tilde{\mathbf{G}}_1'\tilde{\mathbf{G}}_1)^{-1}\tilde{\mathbf{G}}_1'$. The equality in (41) follows from the facts that the sum of squared residuals from regression (40) is identical to that from the regression

$$\tilde{\mathbf{M}}_1\iota = \tilde{\mathbf{M}}_1\tilde{\mathbf{G}}_2\mathbf{c} + \text{errors} \quad , \tag{42}$$

and that $\iota'\tilde{\mathbf{G}}_1 = \mathbf{0}$ by the first-order conditions for $\tilde{\theta}_1$.

Just as a test in regression directions may look in any such directions, not merely those which are suggested by an explicit alternative hypothesis, so may a test in non-regression directions. It is clearly valid to replace $\tilde{\mathbf{G}}_2$ in regression (40) by any $n \times r$ matrix $\mathbf{Z}$, provided that the elements of $\mathbf{Z}$ are $O(1)$, and that the asymptotic expectation of the mean of every column of $\mathbf{Z}$ is zero under the null hypothesis. Newey [19] and Tauchen [22] exploit this fact to propose families of specification tests, while Lancaster [16] uses it to provide à simple way to compute the information matrix test of White [24].

What we are interested in, then, is the asymptotic distribution of the test statistic

$$(\iota'\mathbf{Z}/n^{\frac{1}{2}})[(\mathbf{Z}'\tilde{\mathbf{M}}_1\mathbf{Z})/n]^{-1}(\mathbf{Z}'\iota/n^{\frac{1}{2}}) \quad , \tag{43}$$

which is the explained sum of squares from the artificial regression

$$\iota = \tilde{\mathbf{G}}_1\mathbf{b} + \mathbf{Z}\mathbf{c} + \text{errors}, \tag{44}$$

written in such a way that all factors are $O(1)$. Following Davidson and MacKinnon [7], we shall suppose that the data are generated by a process which can be described by the loglikelihood

$$L' = \sum_{t=1}^{n} [\ell_t(y_t, \theta_1^0, 0) + \alpha n^{-\frac{1}{2}} a_t(y_t)] \quad . \tag{45}$$

Here θ_1^0 is a vector of fixed parameters which determines the simple null hypothesis to which the sequence of DGP's (45) tends as $n \to \infty$, α is a parameter which determines how far (45) is from that simple null, and the $a_t(y_t)$ are random variables which are $O(1)$ and have mean zero under the null. The sequence of local DGP's (45) plays the same role here as the sequence (5) did in our earlier analysis. The major difference between the two is that (45) is written in terms of loglikelihoods, so that the DGP may differ from the null hypothesis in *any* direction in likelihood space, while (5) was written in terms of regression functions, so that the DGP could only differ from the null in regression directions.

Using results of Davidson and MacKinnon [7], it is possible to show that, under the sequence of local DGP's (45), the statistic (43) is asymptotically distributed as non-central chi-squared with r degrees of freedom and non-centrality parameter

$$\alpha^2[\mathrm{plim}(\mathbf{a}'\mathbf{M}_1 Z/n)][\mathrm{plim}(Z'\mathbf{M}_1 Z/n)]^{-1}[\mathrm{plim}(Z'\mathbf{M}_1\mathbf{a}/n)] \quad , \tag{46}$$

where $\mathbf{M}_1 = \mathbf{I} - \mathbf{G}_1(\mathbf{G}_1'\mathbf{G}_1)^{-1}\mathbf{G}_1'$ and $\mathbf{G}_1 = \mathbf{G}_1(\theta_1^0, 0)$. The similarity between expressions (46) and (19) is striking, and by no means coincidental. Note that $\mathbf{M}_1$ plays exactly the same role here that $\mathbf{M}_\beta$ did previously, that $\mathbf{a}$ and Z play the same roles as before, although of course their interpretation is different, and that σ_0^2 has no place in (46), because the variance parameters (if any) are subsumed in θ_1 and/or θ_2 .

Consider the vector $\alpha n^{-\frac{1}{2}}\mathbf{M}_1\mathbf{a}$. Its asymptotic projection onto the space spanned by $\mathbf{G}_1$ and Z jointly is

$$\alpha n^{-\frac{1}{2}}\mathbf{M}_1 Z[\mathrm{plim}(Z'\mathbf{M}_1 Z/n)]^{-1}[\mathrm{plim}(Z'\mathbf{M}_1\mathbf{a}/n)] \quad . \tag{47}$$

If ϕ denotes the angle between $\alpha n^{-\frac{1}{2}}\mathbf{M}_1\mathbf{a}$ and the projection (47), then

$$\cos^2\phi = \mathrm{plim}(\mathbf{a}'\mathbf{M}_1 Z/n)[\mathrm{plim}(Z'\mathbf{M}_1 Z/n)]^{-1}\mathrm{plim}(Z'\mathbf{M}_1\mathbf{a}/n)/\mathrm{plim}(\mathbf{a}'\mathbf{M}_1\mathbf{a}/n), \tag{48}$$

which is the uncentred asymptotic R^2 from the artificial regression

$$\alpha n^{-\frac{1}{2}}\mathbf{M}_1\mathbf{a} = \mathbf{M}_1 Z\mathbf{b} + \text{errors} \quad . \tag{49}$$

Thus the NCP (46) may be rewritten as

$$\alpha^2[\mathrm{plim}(\mathbf{a}'\mathbf{M}_1\mathbf{a}/n)]\cos^2\phi \quad . \tag{50}$$

The interpretation of expression (50) is almost exactly the same as the interpretation of expression (24). The first two factors measure the distance between the DGP and the closest point on a linear approximation to the null hypothesis. The larger these factors, the greater will be the power of *any* test statistic like (43), or of any asymptotically equivalent test statistic. The choice of Z only affects the NCP through $\cos^2\phi$, and a test will have maximal power, for a given number of degrees of freedom, when $\cos^2\phi = 1$. This will be the case if $\mathbf{a}$ is a linear combination of the vectors in $\mathbf{G}_1$ and Z , which will happen whenever the DGP lies within the alternative against which the test is constructed. Thus once again, the power of a test is maximized when we test against the truth.

When $\cos^2\phi = 0$, a test will have no power at all asymptotically. This is a situation which is likely to arise quite often when using tests in non-regression directions. For example, it can be shown that $\cos^2\phi = 0$ whenever the DGP is in a higher moment direction and we test in regression directions, or *vice versa*. This is true for essentially the same reason that the information matrix for a regression model is block-diagonal between the parameters which determine the regression function and those which determine the higher moments of the error terms. Notice that when Z has only one column, expression (48) for $\cos^2\phi$ is symmetrical in α and Z ; thus if a test against alternative 1 has power when alternative 2 is true, a test against alternative 2 must have power when alternative 1 is true.

The artificial regression (49) may actually be used to compute NCPs, and values of $\cos^2\phi$, for models and tests where it is too difficult to work them out analytically. This requires a computer simulation, in which n is allowed to become large enough so that the probability limits in expression (46) are calculated with reasonable accuracy. Such a procedure may tell us quite a lot about the ability of certain test statistics to pick up various types of misspecification.

As an illustration of this technique, we consider certain tests for functional form. The null hypothesis will be the linear regression model which emerges from the Box-Cox model (36) when $\lambda = 1$:

$$y_t = \sum_{i=1}^{k} \beta_i (X_{ti}^1 - 1) + \sum_{j=1}^{m} \gamma_j X_{tj}^2 + u_t \quad . \tag{51}$$

The data will be assumed to be generated by the Box-Cox model, or, more precisely, by a sequence of local approximations to that model having the form of expression (45). The artificial regression (49) will be used to compute $\cos^2\phi$ for three tests of the model (51), none of which is a classical test of (51) against (36).

First of all, we will consider the LM test of (51) against the model

$$y_t(\lambda) = \sum_{i=1}^{k} \beta_i X_{ti}^1 + \sum_{j=1}^{m} \gamma_j X_{tj}^2 + u_t \quad , \tag{52}$$

in which the Box-Cox transformation is applied only to the dependent variable. This model is often just as plausible as (36), and would seem to be a reasonable alternative to test against in many cases.

Secondly, we will consider a test originally proposed by Andrews [1] and later extended by Godfrey and Wickens [13]. The basic idea of the Andrews test is to replace the non-regression direction in which a classical test of the linear or loglinear null against (36) would look with a regression direction which approximates it. One first takes a first-order Taylor approximation to the Box-Cox model (36) around $\lambda = 0$ or $\lambda = 1$. The term which multiplies λ in the Taylor approximation necessarily involves y_t , which is replaced by the fitted value of y_t from estimation under the null. This yields an OLS regression which looks exactly the same as the original regression, with the addition of one extra regressor. The t-statistic on that regressor is the test statistic, and, under the usual conditions for t-tests to be exact, it actually has the Student's t distribution with $n-k-m-1$ degrees of freedom. As Davidson and MacKinnon [9] showed, and as we shall see shortly, the test regressor often provides a poor approximation to the true non-regression direction of the Box-Cox model, so that the Andrews test can be seriously lacking in power.

Finally, we will consider a test against a particular form of heteroskedasticity, namely that associated with the model

$$y_t = X_t\beta + (1 + \alpha X_t\beta)u_t \quad , \qquad u_t \quad NID(0,\sigma^2) \quad . \tag{53}$$

When α is greater than zero, this model has heteroskedastic errors with variance proportional to $(1 + \alpha X_t\beta)^2$. It is well-known to users of the Box-Cox transformation that estimates of models like (36) are very sensitive to heteroskedasticity, and so it seems likely that a test of $\alpha = 0$ in (53) will have power when the DGP is actually the Box-Cox model.

The contribution of the tth observation to the loglikelihood function for the Box-Cox model (36) is given by expression (39). In order to approximate this expression around $\lambda = 1$ by a sequence like (45), we must set

$$a_t(y_t) = \log(y_t) - (u_t/\sigma^2)[h(y_t) - \sum_{i=1}^{k} \beta_i h(X_{ti}^1)] \quad , \tag{54}$$

where

$$h(x) = x \log(x) - x + 1 \quad . \tag{55}$$

The vector $\mathbf{a}$ depends only on the DGP, and will be the same for any test we wish to analyse. If we were interested in the power of an LM test of (51) against (36), Z would be a vector and would be identical to the vector $\mathbf{a}$, so that $\cos^2\phi$ would necessarily be unity. In fact, we are interested in testing (51) against the alternative Box-Cox model (52) and the heteroskedastic model (53), and in the Andrews test. For the first, we find that

$$Z_t = \log(y_t) - (u_t/\sigma^2)\, h(y_t) \quad , \tag{56}$$

for the second that

$$Z_t = X_t\beta(u_t^2/\sigma^2 - 1) \tag{57}$$

and for the third that

$$Z_t = (u_t/\sigma^2)[h(\bar{y}_t) - \sum_{i=1}^{k} \beta_i h(X_{ti}^1)] \quad , \tag{58}$$

where $\bar{y}_t$ is the non-stochastic part of y_t . The matrix $\mathbf{G}_1$ will be the same for all the tests and is easily derived.

In order actually to compute $\cos^2\phi$ by regression (49), we must specify the model and DGP more concretely than has been done so far. For simplicity, we will examine a model with only one regressor in addition to the constant term: constant dollar quarterly GNP for Canada for the period 1955:1 to 1979:4 (100 observations). The constant term was chosen to be 1000, and the coefficient of the other regressor unity. Based on results in [9], we expect $\cos^2\phi$ to be very sensitive to the choice of σ^2 , so the calculations were performed for a range of values. In order to obtain reasonably accurate approximations to probability limits, n was set to 5000; this involved repeating the actual 100 observations on the regressor fifty times. The results presented in Table 1 are averages over 200 replications, and are quite accurate.

TABLE 1

Calculations of $\cos^2\phi$

σ	Alternative Box-Cox Test	Andrews Test	Heteroskedasticity Test
10	.9784	.9913	.0095
20	.9508	.9661	.0349
50	.7907	.8198	.1767
100	.4767	.5318	.4574
200	.1439	.2207	.7621
500	.0025	.0431	.9342
1000	.1220	.0113	.9530
1500	.2958	.0057	.9377
2000	.4596	.0033	.9115
2500	.5932	.0025	.8765

All figures were calculated numerically using $n = 5000$ and 200 replications. Standard errors never exceed .0022, and are usually much smaller.

The results in Table 1 are quite striking. When σ is small, $\cos^2\phi$ is very close to unity for the Andrews test and the alternative Box-Cox test, and very close to zero for the heteroskedasticity test. Note that $\sigma = 10$ is very small indeed, since the mean value of the dependent variable is 21394. For the Andrews test, $\cos^2\phi$ then declines monotonically towards zero as σ increases, a result previously noted by Davidson and MacKinnon [9]. This could have been predicted by looking at expressions (54) and (58) for a_t and Z_t. The behavior of $\cos^2\phi$ for the other two tests is more interesting. For the alternative Box-Cox test, it initially declines, essentially to zero, but then begins to increase again as σ is increased beyond 500. By examining expressions (54) and (56), we can see the reason for this: when σ is large, the second terms in these expressions become small, and the first terms, which are $\log(y_t)$ in both cases, become dominant. For the heteroskedasticity test, $\cos^2\phi$ initially rises as σ increases from zero, but reaches a maximum around $\sigma = 1000$ and then falls somewhat thereafter. The reason for this is not entirely clear; possibly the fact that (57) has no $\log(y_t)$ term begins to matter as σ gets large.

This example illustrates that, once we leave the realm of regression models, the power of a test may depend in quite a complicated way on the parameters of the DGP, as well as on the structure of the null, the alternative and the DGP. Thus techniques for computing $\cos^2\phi$ may be quite useful in practice. The technique we have used here is very widely applicable and quite easy to use, but may be computationally inefficient in many cases. When LM tests can be computed by means of double-length regressions (Davidson and MacKinnon [5]), a more efficient but basically similar technique is available; see [6] and [9].

This example also shows that approximating a mixed non-regression direction by a regression direction may yield a test with adequate power, as in the case of the Andrews test with σ small, but may also yield a test with very low power, as in the case of σ large. Despite the possibly large loss of power, there may sometimes be a reason to do this. Tests in regression directions are asymptotically insensitive to misspecification of the error process, such as normality. Moreover, the techniques of section 4 can be used to make tests in regression directions robust to heteroskedasticity. Thus by applying the artificial regression (30) to the Andrews test regression, one could obtain a heteroskedasticity-robust test of linear and loglinear models against Box-Cox alternatives. If such a test rejected the null hypothesis, and the sample was reasonably large, one could be quite confident that rejection was justified.

6. CONCLUSION

Any test of an econometric model can be thought of as a test in certain directions in likelihood space. If the null is a regression model, these may be regression directions, higher moment directions, or mixed non-regression directions. The power of a test will depend on the model being tested, the process that generated the data, and the directions in which the test is looking. Section 3 provided a detailed analysis of what determines power when the null hypothesis is a univariate nonlinear regression model, the DGP is also a regression model, and we are testing in regression directions. Section 4 extended this analysis to the case of heteroskedasticity-robust tests, and obtained the surprising result that a test may not have highest power when looking in the direction of the truth. Section 5 then considered a much more general case, in which the null and the DGP are merely described by loglikelihood functions, and tests may look in any direction. The results are remarkably similar to those for the regression case, and are concrete enough to allow one to compute the power of test statistics in a variety of cases.

REFERENCES

[1] Andrews, D.F., "A Note on the selection of data transformations," Biometrika 58 (1971), 249-254.

[2] Dastoor, N.K., "Some aspects of testing non-nested hypotheses," Journal of Econometrics, 21 (1983), 213-228.

[3] Davidson, R., and J.G. MacKinnon, "Several tests for model specification in the presence of alternative hypotheses," Econometrica, 49 (1981), 781-793.

[4] Davidson, R., and J.G. MacKinnon, "Some non-nested hypothesis tests and the relations among them," Review of Economic Studies, 49 (1982), 551-565.

[5] Davidson, R., and J.G. MacKinnon, "Model specification tests based on artificial linear regressions," International Economic Review, 25 (1984), 485-502.

[6] Davidson, R., and J.G. MacKinnon, "The interpretation of test statistics," Canadian Journal of Economics, 18 (1985), 38-57.

[7] Davidson, R., and J.G. MacKinnon, "Implicit alternatives and the local power of test statistics," CORE Discussion Paper No. 8525, 1985.
[8] Davidson, R., and J.G. MacKinnon, "Heteroskedasticity-robust tests in regression directions," Annales de l'INSEE, 59/60 (1985), 183-218.
[9] Davidson, R. and J.G. MacKinnon, "Testing linear and loglinear regressions against Box-Cox alternatives," Canadian Journal of Economics, 25 (1985), 499-517.
[10] Davidson, R., L.G. Godfrey and J.G. MacKinnon, "A simplified version of the differencing test," International Economic Review, 26 (1985), 639-647.
[11] Engle, R. F., "A general approach to Lagrange Multiplier model diagnostics," Journal of Econometrics, 20 (1982), 83-104.
[12] Engle, R. F., "Wald, Likelihood Ratio and Lagrange Multiplier tests in econometrics," in: Z. Griliches and M. Intriligator (eds.), Handbook of Econometrics, (North-Holland, Amsterdam, 1984), 775-826.
[13] Godfrey, L.G., and M.R. Wickens, "Testing linear and log-linear regressions for functional form," Review of Economic Studies, 48 (1981), 487-496.
[14] Hausman, J. A., "Specification tests in econometrics," Econometrica, 46 (1978), 1251-1272.
[15] Holly, A., "A remark on Hausman's specification test," Econometrica, 50 (1982), 749-759.
[16] Lancaster, T., "The covariance matrix of the information matrix test," Econometrica, 52 (1984), 1051-1053.
[17] MacKinnon, J.G., and H. White, "Some heteroskedasticity consistent covariance matrix estimators with improved finite sample properties," Journal of Econometrics, 29 (1985), 305-325.
[18] Mizon, G.E., and J.-F. Richard, "The encompassing principle and its application to testing non-nested hypotheses," Econometrica, 54 (1986), 657-678.
[19] Newey, W.K., 1985, "Maximum likelihood specification testing and conditional moment tests," Econometrica, 53 (1985), 1047-1070.
[20] Pitman, E.J.G., "Notes on non-parametric statistical inference," Columbia University, mimeographed, 1949.
[21] Savin, N.E., and K.J. White, "Estimation and testing for functional form and autocorrelation: a simultaneous approach," Journal of Econometrics, 8 (1978), 1-12.
[22] Tauchen, G., "Diagnostic testing and evaluation of maximum likelihood models," Journal of Econometrics, 30 (1985), 415-443.
[23] White, H., "A heteroskedasticity-consistent covariance matrix estimator and a direct test for heteroskedasticity," Econometrica, 48 (1980), 817-838.
[24] White, H., "Maximum likelihood estimation of misspecified models," Econometrica, 50 (1982), 1-25.
[25] Zarembka, P., "Transformation of variables in econometrics," in: P. Zarembka (ed.), Frontiers in Econometrics, (Academic Press, New York, 1974).

TELECOMMUNICATIONS DEMAND MODELLING
An Integrated View
A. de Fontenay, M.H. Shugard, D.S. Sibley (Editors)

CONSISTENT ESTIMATION OF LIMITED DEPENDENT VARIABLE MODELS DESPITE MISSPECIFICATION OF DISTRIBUTION

Paul A. RUUD

Department of Economics
University of California, Berkeley
Berkeley, California 94720*

By exploiting conditions on the behavior of regressors, estimators are developed which are consistent, up to a factor of proportionality, for the slope coefficients of the regression function of a imperfectly observed dependent variable. Limited dependent variables are a special case of interest. The consistency is robust to misspecification of the distribution of the latent dependent variable. The estimators are weighted pseudo maximum likelihood estimators which can be readily computed using popular methods and software.

1. INTRODUCTION

Econometric parameterizations of economic models rely heavily on functions which are linear in parameters. The ordinary regression model is the most familiar example:

$$y = \alpha + \beta_1 x_1 + \ldots + \beta_K x_K + u$$
$$E(y|x) = \alpha + \beta_1 x_1 + \ldots + \beta_K x_K + u$$

where the unknown coefficients, $\beta_1, \ldots, \beta_K$, appear as a weighted sum. Such parameterizations do not require that a variable enter linearly, of course, and therein lies the ability of these parameterizations to approximate many functions over a closed domain.

Two characteristics of the linear specification contribute to its popularity. The coefficients are the partial derivatives of the conditional expectation of the dependent variable with respect to each explanatory variable and can be interpreted as *ceteris paribus* effects. Second, this specification is statistically convenient. The estimation of the unknown parameters and the sampling distributions of the estimators is inexpensive in researchers' and computers' time.

The interest a researcher holds for the slope coefficients varies widely with the application. When the primary purpose is to fit the data well for prediction, then the values of the parameters have secondary importance. The role of point estimation is often diagnostic, flagging relationships that may be unreliable. However, researchers also examine th performance of theories with actual data and, in that case, the parameter estimates are often primary. At the simplest level of specification, theories may predict that a set of coefficients is zero. In macroeconomic

* The research described in this paper was supported by the National Science Foundation. This paper is an abridged version of Ruud (1986).

models with rational expectations, for example, unanticipated future values of explanatory variables have coefficients equal to zero in equations describing behavior at a moment in time. A more restrictive specification gives the sign of some slope parameters, as in several theories of labor economists which claim that wages of workers should rise with their level of education. Still more ambitious models focus interest on the ratios of a subset of coefficients. For example, simple life cycle theory of consumption and saving predicts that the shape of the wealthage profile rises and then falls; when wealth is regressed on age and agesquared, the second slope should not only be negative but should also be large enough to imply a declining path of wealth in later years. Other examples might include the earnings-experience profile of labor economics and the shape of Engel curves.

Interest in parameter values is not restricted, of course, to testing theories. Empirical research is also a valuable tool for informing both theoretical inquiry and policy making and for describing actual economic phenomena.

Econometric methods are designed to respond to these interests and the robustness of these methods against errors of specification is an important research question. In this paper, we focus on estimation when observation of the true dependent variable, y, is imperfect. For such reasons as those just given, the researcher maintains the linear specification, but a variety of circumstances can obscure the dependent variable. Limited dependent variable (LDV) models and their applications provide many examples. The discrete choice model is one that we will give special attention. The ratios of the slope parameters have the interpretation of marginal rates of substitution and hold interest despite our inability to observe the consumer's preference relation. Popular methods of estimation for LDV models specify the error distribution and their robustness against misspecification is an open question.

Another class of problems we will examine are characterized by observation of y up to an unknown, but monotonic increasing, transformation. Such transformations as the Box-Cox are often used to cope with this difficulty. When the transformation is differentiable and the observed dependent variable has a finite conditional expectation, the slope coefficients retain their relevance to the questions mentioned above. The partial derivatives of the conditional expectation of the observed dependent variable are proportional to the β, retaining their signs and relative values. Again, the robustness of procedures which treat the transformation as known up to a finite parameterization is poorly understood although, in this case, the literature on robust estimation methods suggests that there is reason for concern.

In this paper, we propose a weighted M-estimator (WME) for these estimation problems. This estimator is consistent, up to an unknown scaling factor, for the slope parameters in the regression function of the imperfectly observed dependent variable. Because of this characteristic, our estimator is applicable in its present form to a discrete choice estimation problem where the distribution function is either unknown or computationally unattractive. The estimation method makes weak demands on the relationship between the observed dependent variable and its latent counterpart, however, suggesting to us that exploitation of known characteristics of the relationship can further extend the usefulness of the method. Because the consistency only of the estimator is established, this research is a first step toward inference in the problems described above.

2. THE PROBLEM

Suppose an unobserved vector of M dependent variables, y^*, is generated by a seemingly unrelated system of equations

$$y^* = \alpha_0 + (I_M \otimes x')\beta_0 + u \tag{1}$$

where α_0 is a column vector of M unknown constants, I_M is an $M \times M$ identity matrix, x is a column vector of K explanatory variables, β_0 is a vector of MK unknown coefficients, and u is a column vector of M unobserved, random error terms. We will denote the distribution function of u, conditional on x, by $F(\cdot|x)$ or simply F. F is also unknown.

One observes a sample of N independent observations of the vector y that is a transformation of the unobserved y^*:

$$y_n = \tau(y_n^*) \quad , \quad (n=1,\ldots,N) \ . \tag{2}$$

The transformation $\tau(\cdot)$, which is often called an observation rule, is a function from $\mathbb{R}^M$ onto $\mathbb{R}^M$. The u_n are independently distributed for all n.

Many LDV models are represented in this way. If, for example, $M = 1$ and $\tau(y^*) = 1(y^* > 0)$, then one has the familiar binary dependent variable of probit, logit, and linear probability models.[1] Alternatively, when $M = 1$ and $\tau(y^*) = 1(y^* > 0)\cdot y^*$, we obtain a censored regression model. Finally, $M = 2$, $y_1 = 1(y_1^* > 0)$, and $y_2 = y_1 \cdot y_2^*$ yields the nonrandom sample selection model. In all of these models, τ is known but it is not one-to-one, and it may be discontinuous.

Note, however, that τ can be unknown. Some *characteristics* of τ might be known. τ chould belong to such families of functions as the Box-Cox or the monotonic, differentiable functions. Given such information, one may still be able to relate the population parameters β_0 to the characteristics of the sample, and hence, have interest in their values.

Now, we wish to estimate the unknown parameters (α_0, β_0), but due to ignorance about the disytribution of the observed y, and in particular the expectation of y conditional on x, ordinary least squares (OLS) yields misleading estimators. When the distribution function F and the transformation function τ are known up to a finite number of parameters, the maximum likelihood estimator (MLE) is a popular estimator. This estimator is often unique, easy to compute, consistent, and asymptotically normal. Even though researchers specify F and τ, these functions are frequently unknown, and probably misspecified, so that the desirable properties of the MLE are not obtained in practice.

Econometricians may hedge against misspecification of F and τ, and hence the distribution of y, by giving these functions flexible parametric forms. If the family of functions is rich, then one can approximate the true distribution function of the observed y and obtain reasonable estimators for the parameters of interest (α, β). Duncan (1982), Gallant (1985), Lee (1981, 1983), and Ruud (1981) all employ this method in various forms. This method suffers, of course, from the original flaw. Despite the flexibility of the chosen forms, the researcher ultimately continues to misspecify the distribution of y and the significance of such errors for inference is unknown. A second potential drawback of flexible parameterizations of the distribution of y is onerous computational costs. The added dimensions of the estimation problem may introduce unbounded likelihood functions on parameter space boundaries, multiple local maxima, and nonconcavities that make numerical maximization of the log-likelihood function difficult.

Semi-parametric methods of estimation provide an alternative approach for specific τ. For binary discrete data, Cosslett (1983) developed the MLE over distribution functions F. Manski (1976) describes a maximum score method for multinomial choice models. To the censored and truncated regression case, Powell (1981) applied least absolute deviations (LAD) methods. All of these distribution-free estimators for specific types of data, like the parametric estimators mentioned above, are costly to compute. In addition, only Powell's LAD estimators are known to have asymptotically normal distributions.

In this paper, we propose another distribution-free method for estimation of these latent dependent variable models. The method is motivated by research that has examined the robustness of the pseudo-MLE to misspecification of the distribution of y, due to misspecification of the underlying distribution function F. Because misspecification of the transformation function τ is an alternative cause of misspecification of the distribution of the observed y, the proposed method of estimation also applies to estimation in that case.

Several researchers have noted that sufficient conditions exist to ensure the consistency, up to a scalar, of the pseudo-ML estimators of the slope coefficients β. Brillinger (1977) apparently first observed that bivariate normality for (y^*,x), where $M = K = 1$, assures that for many τ

$$Cov(y,x) = Cov(y^*,x)\cdot Cov(y,y^*)/Var(y^*) .$$

Brillinger (1982) applied this simple result to a nonlinear multiple regression case,

$$y = \tau(\alpha_0 + x'\beta_0) + u$$

and showed that ordinary least squares estimates may provide an estimate of β up to an unknown constant of proportionality, if the regressors are normally distributed and u is independent of the regressors with finite variance. Note that Brillinger's formulation differs from ours in that we do not assume the independence of $y - E(y|x)$ and x, although their covariance is implied to be zero.

Goldberger (1981) showed independently that when the explanatory variables are multivariate normally distributed and the dependent variable y^* is a truncated normal random variable, the OLS regression of y on X yields an estimator that converges to a scalar multiple of the slope vector. He also showed that a similar result failed to hold up under nonrandom sample selection. In an extension of Goldberger's work, Greene (1981) also suggested an estimator for censored regression that corrects the OLS estimator for the scalar.

Ruud (1983) demonstrated that for discrete data and a broad class of assumed and underlying distribution functions F, the pseudo-MLE provides consistent (up to a scalar) estimates of the slopes if

$$E(x|x'\beta_0) = \theta_0 + \theta_1 x'\beta_0 \tag{3}$$

where θ_0 and θ_1 are column vectors of K constants. Multivariate normal x obviously satisfy (3). Ruud's proof extends immediately to more general kinds of data than purely discrete, as well as to such estimators as method-of-moments or minimum distance estimators. Thus, as we will see, Goldberger's result for OLS on truncated data is a special case, and furthermore, a similar result can be established for the sample selection model. Chung and Goldberger (1984) have since derived similar results for the estimation by OLS of the *population projection* of y^* on x .

The proposed estimator of this paper builds on the consistency result of Ruud (1983) by reweighting the sample data points so that the regressors appear to satisfy (3), even when they do not. Given the weights, such standard estimators as OLS regression of y on x and an intercept yield consistent estimators of β , up to an unknown constant of proportionality. Two broad implications are that many popular estimators can be modified in a simple way to obtain estimators that are robust to a general form of misspecification and that many LDV models can be estimated consistently in the presence of misspecified distribution functions.

The rest of the paper begins with the generalization of Ruud (1983) to a broader class of estimation problems than discrete choice models. Second, consistent estimation in cases where sufficient condition (3) does not hold is explained. The last section discusses further research and concludes the paper.

3. SUFFICIENT CONDITIONS FOR CONSISTENCY

We begin by restricting the argument to the single equation case, $M = 1$. In this case, the matrix β is a column vector of K elements. Consider the estimator $\hat{\theta} = (\hat{\alpha},\hat{\beta})$ that is the unique solution, at least asymptotically in sample size N , to the implicit set of equations:

$$(\hat{\alpha},\hat{\beta}) : \quad N^{-1} [\iota \; X]' \, \lambda(y,\iota\hat{\alpha}+X\hat{\beta}) = 0 \tag{4}$$

where X is the $N \times K$ matrix of the regressors, ι is a column vector of N ones, and $\lambda(\cdot,\cdot)$ is a function from $\mathbb{R}^N \times \mathbb{R}^N$ onto $\mathbb{R}^N$, continuous and infinitely differentiable in the second set of arguments. For example, $\lambda = y - \iota\hat{\alpha} - X\hat{\beta}$ yielding OLS for $\hat{\beta}$ or, in the case of the pseudo-MLE

$$\lambda = \partial[\Sigma \, L(y_n,\hat{\alpha}+x_n'\hat{\beta})]/\partial(\iota\hat{\alpha}+X\hat{\beta}) \tag{5}$$

where $L(\cdot,\cdot)$ is the chosen log-likelihood function for each observation. Another important example is

$$\lambda = [1(y_n > 0)\cdot(y_n - \hat{\alpha} - x_n'\hat{\beta}) \; ; \; n=1,\ldots,N]$$

which leads to OLS estimates for a truncated sample. This example illustrates how our framework can accomodate sample selection. The transformation τ is defined over the entire support of y and does not accommodate "unobserved" (or, undefined) values.

Solving for such estimators as those in (4) is equivalent to solving a maximization problem [see Hansen (1982), Burgete et al (1981), and Manski

(1983)]. Therefore, we will call estimators that solve such normal equations as (4), pseudo MLE's and call a corresponding maximand the pseudo log-likelihood function, ΣL_n. The function L is obviously not unique.

We assume

Assumption 1. $(\alpha_0, \beta_0) = \theta_0$ *is an interior point of the compact, convex parameter space* Θ.

Assumption 2. u_n *is independently and identically distributed with distribution function* F *which is continuously differentiable.*

Assumption 3. x_n *are independently and identically distributed with distribution function* H. x_n *is independent of* u_n.

Assumption 4. $N^{-1} \Sigma L_n$ *and* $N^{-1} [\iota\ X]'\lambda$ *are uniformly integrable with respect to* u *and* x *and converge a.e. uniformly in* $\theta \in \Theta$ *to* $E(L)$ *and* $E(N^{-1} [\iota\ X]'\lambda)$, *which both exist.* $\partial E(N^{-1} \Sigma L_n)/\partial\theta \equiv E(N^{-1} [\iota\ X]'\lambda)$.

Assumption 5. $E(L) = 0$ *has a unique maximum at* $\theta = \theta^*$, *an interior point of* Θ.

Assumption 6: λ *and* $E(\lambda|X)$ *are infinitely differentiable functions of* $\theta \in \Theta$. $\partial^2 L/\partial\theta\partial\theta'$ *is uniformly bounded.*

Assumption 3 restricts the behavior of the regressors unduly and will be relaxed in the next section. Assumptions 1, 4, 5, and 6 correspond to standard regularity conditions which ensure that $\hat{\theta}$ converges almost surely to $\theta^* \equiv (\alpha^*, \beta^*)$.

Following section 3 in Ruud (1983), we have

Theorem 1. Given assumptions 1-5, and given the model in equations (1) and (2) with $M = 1$, *if (3) holds then* β^* *is a scalar multiple of* β.[2]

One key element of the proof of this theorem is the observation that

$$\lambda^*(\iota\alpha + X\beta, \iota\hat{\alpha} + X\hat{\beta}) = E[\lambda(y, \iota\hat{\alpha} + X\hat{\beta})|X] .$$

That is, the expectation of the "residual" λ is a function of X only through the underlying regression function $\iota\alpha + X\beta$. Although the unknown transformation is treated as fixed here, such behavior can also hold if τ is stochastic. In other words, some forms of errors-in-variables in the observed y do not vitiate the consistency of estimators, just like the familiar regression case.

Essentially the same point explains the relationship between β and the observed dependent variable y. Its expectation (when it exists) is also a function of X only through $X\beta$ and therefore the partial derivatives of $E(y|X)$ with respect to X are proportional to β. When τ is differentiable and that derivative has an expectation, we can explicitly write $E\{\partial E[y|x]/\partial x\} = \beta\ E[\tau'(y^*)]$ so that the factor of proportionality is the average slope of the transformation from y^* to the observed y. If in addition, x and u are normally distributed, this expression is also the almost sure limit of the OLS estimator $(X'X)^{-1}X'y$, but β^*

does not generally have this simple interpretation.[3] Although normally distributed x is an important case, multivariate distributions that imply condition (3) comprise a larger class than one might think. For specific values of β, it is simple to construct appropriate density functions. If (3) must hold for virtually all β, which will be the cases of interest, such constructions are harder. One example, however, is the class of elliptically symmetric distributions which have densities proportional to

$$h[(x - \mu)'\Omega (x - \mu)]$$

where h is a valid univariate probability density function (pdf), Ω is a positive definite symmetric matrix, and μ is a vector of K constants.[4] The same mathematical argument as for the normal density proves that, if expectations exist, (3) holds for random variables with this pdf for all values of β.[5] Thus, distributions with closed support can satisfy (3).

Goldberger's (1981) analogous result for the application of OLS to truncated normal dependent variables with normally distributed regressors is a special case of Theorem 1. OLS estimators are a special case of (4) and assumptions 1-6 certainly hold. However, Theorem 1 extends this special case to more general transformation functions τ. For example, one need not have continuous information about the underlying y^* ; sign information is sufficient. Alternatively, one could observe the logarithm of y^*, even its square, and Theorem 1 would still apply. In addition, neither the estimator nor the distribution of the errors need be associated with the normal distribution, as in Goldberger's case. Similar results hold if the errors have a Student t distribution and one uses LAD estimators.

Theorem 1 is different, however, from that found in Chung and Goldberger (1984) where the emphasis is OLS estimation and the analogous sufficient condition for consistency is

$$E(x|y^*) = \theta_0 + \theta_1 y^*$$

Here the conditioning random variable is y^*, whose distribution is a convolution of those of x and u. As will become apparent, our approach to estimation in more general settings rests heavily on the fact that only the behavior of x is restricted by Theorem 1. It does not appear to us, therefore, that one can capitalize on this alternative formulation.

We can extend Theorem 1 straight-forwardly to multiple equation systems ($M > 1$) by replacing x with $I_M \otimes x$ in equations (3) and (4) and re-dimensioning θ_1 to be a $K \times M$ matrix and λ to be a vector of length $N \cdot M$. We must also add identification restrictions that enable us to distinguish one regression equation from another:

Assumption 7. (i) Let $\beta = (\beta_1,\ldots,\beta_M)$ where β_m $(m=1,\ldots,M)$ is a $K \times 1$ vector of the slope coefficients in the m-th equation. There exists at least one a priori linear restriction for each β_m, $R_m\beta_m - r = 0$ say, such that the matrix of all restrictions, $[R_m]$, is full rank. (ii) Furthermore, let the variance-covariance matrix $V(\lambda)$ be full rank for every finite N.

More general restrictions will also suffice for identification. The linear restrictions on the coefficients, (i), is simply familiar and illustrates the potential identification problems that arise in our models. A simple example of assumption 7 is a different regressor excluded from each regression. In the sample selection model, this particular restriction is made to distinguish the selection equation from the regression equation of the experimental outcome. We speak of system identification here, but one could easily limit the discussion to single equation identification.

Corollary 1. Given the conditions of Theorem 1 but allowing $M > 1$, and given assumption 7, the slope coefficients β_m for each regression equation are a scalar multiple of the corresponding β_m^* if

$$E[x|(I_M \otimes x')\beta] = \theta_0 + \theta_1(I_M \otimes x')\beta \qquad (m=1,\ldots,M)$$

where θ_0 is a column vector of K constants and θ_1 is a $K \times M$ matrix of constants.

Using this theorem we can examine as an example a sample selection problem related to that Goldberger (1981) analyzed, and obtain a further consistency result. In a simple sample selection model, there are two equations ($M = 2$) and $y_1 = 1(y_1^* > 0)$, (y_2^*, x_2) observed only if $y_1 = 1$; the observable dependent variables have expectations such that

$$E(y_{1n}|x_{1n},x_{2n}) = f_1(x_{1n}'\beta_1), \qquad (n=1,\ldots,N)$$
$$E(y_{2n}|x_{1n},x_{2n}) = f_2(x_{1n}'\beta_1, x_{2n}'\beta_2), \quad (y_{1n}=1, n=1,\ldots,N).$$

If we estimate linear regressions for y_1 on x_1 and y_2 on both x_1 and x_2 (for the observed y_2 only) by least-squares,

$$y_{1n} = a_1 + x_{1n}b_1 + \epsilon_{1n},$$
$$y_{2n} = a_2 + c\cdot(x_{1n}b_1) + x_{2n}b_2 + \epsilon_{2n}, \qquad (y_{1n} = 1)$$

maintaining the proportionality of the coefficients on x_1 in both equations by including the scalar c, then Theorem 2 indicates that the estimates of b_1 and b_2 will converge to constants proportional to β_1 and β_2, respectively. Because a scale multiple of β_1 can be estimated from the regression for y_1 alone, a convenient two-step estimator for a scale multiple of β_2 is contained in a "two stage least-squares" estimator:

$$[\bar{a}_1 \;\; \bar{b}_1']' = (X_1'X_1)^{-1} X_1'y_1,$$
$$[\bar{a}_2 \;\; \bar{c} \;\; \bar{b}_2']' = (\bar{X}_2'\bar{X}_2)^{-1} \bar{X}_2'y_2,$$

where $X_1 = [1, x_{1n}]$, $\bar{X}_2 = [1, x_{1n}'\bar{b}_1, x_{2n}; y_{1n}= 1]$, $y_2 = [y_{2n}; y_{1n}= 1]$, and $\bar{b}_1 = (X_1'X_1)^{-1} X_1'y_1$ (from a regression using the entire sample). This estimator is analogous to the two-step estimator proposed by Heckman (1976) and Lee (1976) for this model when the y^*'s are known to be normally distributed.

If we followed an even simpler route, and regressed y_2 linearly on both x_1 and x_2 several cases need to be considered. If x_1 and x_2 share no variables, proportional estimates of both β_1 and β_2 are obtained. If, as is common, some explanatory variables appear in both sets of regressors, then only the coefficients of the exclusive regressors are identifiable and estimable up to a proportionality constant using OLS.

Clearly, no elements of β_1 or β_2 can be estimated in this way when $x_1 = x_2$.[6] If, however, $x_1'b_1$ enters the estimated regression function non-linearly, the non-linearity will generally help to distinguish b_1 from b_2 in non-linear least squares (NLS) estimation. Such specification is, in fact, the practice when one assumes bivariate normality for the y^*'s and therefore specifies the expectation of y_2 to be

$$E(y_2|x_1,x_2,y_1=1) = \alpha_2 + x_2'\beta_2 + \rho\cdot\phi(\alpha_1+x_1'\beta_1)/\Phi(\alpha_1+x_1'\beta_1)$$

where ϕ and Φ are the pdf and cdf of the standard normal distribution. Thus, two devices assist in estimating both β_1 and β_2 : such non-linearities as the one immediately above and such additional estimation equations as that for y_1 discussed earlier.

Returning to Theorem 1 and its Corollary 1, note that the proportionality factor between β_m^* and β_m might be zero. This would be an unusual and unfortunate event, but it is possible. Because we do not observe the underlying dependent variable y^* or know the transformation function τ , we cannot, in general, identify the scale of y^* . Therefore, the scale of β is also unidentifiable. Furthermore, the precision of the estimators is potentially poor. The noise introduced by the unknown τ , low variation in the explanatory variables, and a poor choice of λ presumably cloud perceptions of β . Such propositions must await the exploration of asymptotic approximations to the distribution of the pseudo-MLE and questions of relative efficiency.

4. CONSISTENT ESTIMATION IN THE GENERAL CASE

Although we have a great deal of freedom in the models to which we can apply these consistency results, in actual application we are restricted to regressors that satisfy equation (3) and assumption 3 -- unrealistic conditions. It is possible, however, to "trick" the pseudo-MLE into treating the regressors as though they were such special regressors. This is done by reweighting the sample points so that the induced distribution function for the regressors satisfies the linear expectation condition in (3) and the regressors appear to be independent and identically distributed.

In this section, first we explain heuristically why failures in condition (3) may lead to inconsistent estimators. This explanation suggests that weighting the data can, in principle, remove this source of inconsistency. Weighted M-estimation, the resultant method, is similar to methods of robust estimation. We compare the two approaches and then present a feasible estimator. The theoretically appropriate weights depend on the unknown distribution of the explanatory variables, but alternative weights that are feasible are discussed.

4.1 Sources of Estimator Inconsistency

To understand why weighting works, let us consider how failures of (3) lead to inconsistent estimation. When the x's fail the linear expectation condition, we run the risk of a sample of regression functions, $x_n'\beta$ $(n=1,\ldots,N)$, that associates particular values disproportionately with individual explanatory variables. Such asymetries may cause the ceteris paribus effect of one variable on the regression function to be exaggerated while the effect of another is discounted, depending on τ and F .

For example, suppose the underlying regression model is the familiar log form with two continuous explanatory variables

$$ln(y) = \alpha + x_1\beta + x_2\beta + u$$

but one mistakenly runs the regression using y, instead of $ln(y)$, as the dependent variable: $\tau(.) = exp(.)$. Let x_1 and x_2 behave in the following, peculiar, way:

$$x_1 \cdot x_2 = 0 , \quad x_1 \text{ or } x_2 > 0 .$$

As a result, one regressor determines the value of the regression function for each observation; and a regression on both regressors is equivalent to separate regressions using only nonzero observations for each regressor. In this case, if β_1 and β_2 are equal in magnitude, but opposite in sign, the misspecified regression will find a relatively larger effect for the positive slope coefficient.

Several points bear emphasis. First, note that although the regressors themselves behave idiosyncratically, the regression function itself is simply uniformly distributed, $U(-2,2)$. This fact ensures the second point, that the population projection approximates the population regression function quite well. Despite this good fit, we also note that the relative slopes will be misestimated: the coefficient on x_2 will appear to be smaller in magnitude than the coefficient on x_1, even though these are equal in the underlying data generating process. If goodness of fit is the estimation criterion of the researcher, then further analysis is unnecessary. However, interest in the parameters of the regression function in the data generating process will require better performance than OLS.

Similar distortions occur in LDV models when a distribution function is misspecified and truncation or censoring points depend unevenly on different explanatory variables. If only $1(y>0)$ were observed in the previous example, and if the distribution of u were skewed to the right but assumed to be symmetric (as in probit, logit, or linear probability models), the same effects occur. In this case the regression function is

$$E[1(y>0)] = P(x_1 + x_2 + u > 0)$$

Small regression, or truncation, points will depend on the large values of x_1 which are, in turn, associated with relatively steep parts of the cumulative distribution function, P. A failure to attribute this large derivative to the underlying skewness of the distribution function will cause the coefficient on x_1 to capture this effect: that coefficient will be increased in size relative to the other.

These distortions apparently do not occur if the expectation of each explanatory variable, conditional on the regression function, is a linear function of the regression function. Intuitively, such a restriction on the distribution of the regressors forces the distribution of each explanatory variable to be similar for each regression value. At least asymetries like those above are ruled out. On the other hand, symmetry of the distribution of the regressors obviously satisfies the restriction. Therefore, we might try to weight sample points to remove assymetries.

4.2 Weighted M-Estimation

Theoretically such weighting is possible. If we introduce a scalar weight function, $a(x)$, from $\mathbb{R}^K$ onto $\mathbb{R}$, of the explanatory variables and weight each observation, Theorem 1 takes a slightly modified form. Returning to the single equation case ($M = 1$) for simplicity, condition (3) is replaced with

$$E[a(x)x|x'\beta] = (\theta_0 + \theta_1 x'\beta)\cdot E[a(x)|x'\beta] \tag{6}$$

and the definition of the estimator (4) becomes:

Definition. A weighted M-estimator (WME), $\hat{\beta}$, of a vector proportional to β_0 is a solution $(\hat{\alpha},\hat{\beta})$ to the maximization

$$\max_{\alpha,\beta} \; N^{-1} \Sigma_{n=1}^{N} a_n(x_n)\cdot L(y_n, \alpha + x_n'\beta) ,$$

given by sovlving the first-order conditions

$$N^{-1} [\iota \; X]' \; A \; \lambda(y, \iota\hat{\alpha}+X\hat{\beta}) = 0 \tag{7}$$

where L is the chosen pseudo log-likelihood function for each observation and $A = diag[a(x_n)]$ is an $N \times N$ diagonal matrix of the weights for each observation.

To state the new form of Theorem 1 explicitly:

Corollary 2. Given assumptions 1-5 with $a\cdot L$ replacing L and $[\iota \; X] \; A$ replacing $[\iota \; X]$, and given the model in equations (1) and (2) with $M = 1$, if (6) holds then the WME $\hat{\beta} \xrightarrow{\text{a.s.}} \beta^*$, which is a scalar multiple of β.

While Corollary 2 establishes an important statistical property of the WME, one should also note the computational conveniences of the WME. Given the weights, modification of popular maximum likelihood estimators simply involves re-weighting the log-likelihood function and its derivatives. Furthermore, if the original log-likelihood function is globablly concave, the weighted log-likelihood function of the WME is also globally concave. As a result, the actual computation of a WME for many non-linear estimators is familiar and straight-forward.

Now given that we can formally introduce weights, we need to find weights satisfying (6). There are many potential weight functions. An important family of weight functions are those that effectively "replace" one distribution function with another in all expectations. Let us illustrate this by "replacing" the actual distribution function of the x with the normal distribution, a distribution that has linear conditional expectations and that is closed under convolutions. Suppose the domain of H, the distribution function of x, is $\mathbb{R}^K$ and H is continuously differentiable with density h. Let $\phi_K(.,\mu,\Sigma)$ denote the K-variate normal density with expectation vector μ and covariance matrix Σ and let

$$a(x) = \phi_K(x,\mu,\Sigma)/h(x) .$$

Also, factor h and ϕ into conditional and marginal densities $h(x) = h_1(z)\cdot h_2(x|x'\beta=z)$ and $\phi(x)=\phi_1(z)\phi_2(x|x'\beta=z)$.[7] Then

$$\begin{aligned} E[a(x)x|x'\beta] &= \int [\phi_K(x)/h(x)] \;\; x \; h_2(x|x'\beta)\, dx \\ &= [\phi_1(x'\beta)/h(x'\beta)] \;\; \int \phi_2(x|x'\beta)\; x \; dx \\ &= E[a(x)|x'\beta)] \; [\theta_0 + \theta_1\cdot x'\beta] \end{aligned}$$

for some θ_0 and θ_1 , by the linearity of conditional expectations of the multivariate normal distribution and the multivariate normality of x and $x'\beta$.

Our choice of the normal distribution is purposeful. Note that for the normal density (9), and therefore (6), hold for all β . This is an important characteristic because we do not want the weights to depend on unknown parameters when we try to use this procedure to form estimators. Since some explanatory variables are discrete and cannot be reweighted by a normal density, note also that we could reweight a subset of regressors only. Several obstacles to such uses remain however: one must specify μ , Σ , and the density h . Before discussing these, we consider several interpretations of the weighted pseudo-MLE in (7).

Our weight function, $a(x)$, plays a similar role to the trimming functions of robust estimation. In robust estimators, the influence of outlying observations is reduced by putting less weight, or no weight, on outlying observations. The resultant estimator is less sensitive to misleading data. The weights in (7) also reduce the importance of misleading observations.

There is a fundamental difference, too. Influential observations in the robust estimation problem are those with large residual errors. On the other hand, our concern is with down-weighting observations based on the distribution of the explanatory variables, not the error terms. This in no way implies that outliers in explanatory variables are down-weighted.

The weighted pseudo MLE can also be interpreted as an instrumental variables (IV) or weighted least squares (WLS) estimator. The difference between estimators (4) and (7) is that the explanatory variables matrix is replaced with an "instrumental variables" matrix composed of transformed explanatory variables. Interpreted in this way, we view the problem with (4) as a covariance between x and λ caused by the misspecification of the distribution of y . Consequently, (4) fails to be an orthogonality condition and yields inconsistent estimators. Instead, one uses (7), an instrumental variables orthogonality condition, where the chosen instruments are clearly correlated with the explanatory variables.

Finally we note that if the weight function a were completely known, the WME is a member of the class of estimators considered by White (1982). The consistency and asymptotic normality of such estimators is therefore already established.

4.3 The Calculation of Weights

The suggested weight function, $a(x)$, depends on the unknown distribution function of the explanatory variables. It appears that we have only succeeded in replacing the problem of estimating the density of the u with the estimation of the multivariate density of the x . Given our uncertainty about the effects of parametric distributional assumptions on estimators, we do not wish to fall back on parametric density estimation at this point. Nor are nonparametric methods appealing, because such point estimators are required for every observation. One might defend such procedures, however, on the grounds that estimating the density of such observed random variables as x may prove more reliable than estimating the density of an unobserved random variable like u. The ability to do the latter will require correct specification of other parts of the model, the regression function and the transformation τ.

Recognizing that our objective is to weight the data so that the x appears to have a desired distribution leads to another approach: form an empirical measure for the data whose distribution function converges asymptotically to the desired probability measure. If, in addition, integrals with respect to such empirical measures also converge to integrals with respect to the desired probability measure, the weights will serve our purpose.

Consider first a scalar random variable with the sampling space of the real line. Given a sample, $X_N = (x_1, \ldots, x_N)$ say, we can calculate the sample distribution function (sdf) as[8]

$$H_N(X) = \Sigma_{n=1}^{N} N^{-1} 1(x_n \leq X) . \tag{11}$$

The generalized derivative of H_N is an empirical measure that assigns N^{-1} to each sample point and zero to the rest of the real line. Integration with respect to this measure yields expectations that correspond to sample statistics, for example the sample mean

$$\bar{x} = \Sigma_{n=1}^{N} N^{-1} x_n = \int x \, dH_N .$$

The Glivenko-Cantelli Lemma establishes that H_N converges asymptotically to H. Laws of large numbers establish when such sample expectations as x converge to their population values. Although the convergence of H_N generally does not imply the convergence of sample expectations, their coincidence is common in econometric settings.

Other empirical measures can be formed by replacing the weights of the sdf with others that are also positive and sum to one. In particular, we might choose a set of weights $(a_{N1}, \ldots, a_{NN})$ so that the induced edf "is close to" a desired distribution function. The edf "induced" by a set of weights is defined to be

$$G_N(X) = \Sigma_{n=1}^{N} N^{-1} a_{Nn} 1(x_n \leq X) , \tag{12}$$

where $a_{Nn} > 0$ $(n=1, \ldots, N)$, and $\Sigma_n a_{Nn} = N$. Note that G_N and H_N are bounded and monotonic step functions such that everywhere G_N has a step, so does H_N. Given a desired distribution function G, let us choose the weights that minimize a measure of the distance between G_N and G:

$$(a_{N1}, \ldots, a_{NN}) : \min_{a_{Nn}} \max_{X_N} |G_N(x) - G(x)| \tag{13}$$

One solution to (13) in this scalar case is quite simple and appealing. For large samples, the weights are given approximately by

$$N^{-1} a_{N(n)} = G(x_{N(n)}) - G(x_{N(n-1)}) , \quad (n=1, \ldots, N) , \tag{14}$$

where $G(x_{N(0)}) = 0$ and $G(x_{N(N)}) = 1$, and $x_{N(1)} \leq \ldots \leq x_{N(N)}$ are the order statistics of the sample of size N. This solution is interesting because it illustrates how weights based only on explicit knowledge of G succeed in "replacing" the unknown density h. By a simple Taylor series expansion,

$$N^{-1} a_{N(n)} = [H(x_{N(n)}) - H(x_{N(n-1)})] \cdot [g(x^*)/h(x^*)]$$

where $x_{N(n-1)} \leq x^* \leq x_{N(n)}$. The leading term on the right hand side is the difference between adjacent order statistics of the uniform distribution and is therefore $O_p(N^{-1})$; its expectation is $(N+1)^{-1}$. On average, then, the weight for each x is approximately $N^{-1}g(x)/h(x)$, the very weight we desire.

It is straightforward to see that $G_N(X)$ converges uniformly in X to $G(X)$. Equation (14) implies that $G_N(x_n) = G(x_n)$ (except at the extremes of the sample). The largest deviation between G_N and G is the largest value of the weights a_{Nn} divided by the sample size, because G is continuous and monotonic. Since the a_{Nn} shrink monotonically with N, though discontinuously, the largest deviation will generally converge to zero and, hence, G_N converges to G.

Our hope is that in addition, the weights given in (14) yield the desired expectations as the sample size grows. To see why this is reasonable, consider the expectation of a function of x, say $w(x)$:

$$\begin{aligned} \int w(x)\, dG_N &= \Sigma_{n=1}^{N} w(x_{N(n)}) a_{N(n)}/N \qquad (15) \\ &= \Sigma_{n=1}^{N} w(x_{N(n)}) \cdot [G(x_{N(n)}) - G(x_{N(n-1)})] \end{aligned}$$

The latter expression is a familiar Stieltjes sum. One anticipates that if $w(x)$ is Stieltjes integrable, the expressions above will converge to that integral as N goes to infinity. Asymptotically, each interval will go to zero as the sample populates the sampling space and the limit of the sum, the Stieltjes integral, will be obtained.

5. ASYMPTOTIC DISTRIBUTION THEORY

The components of the formal argument that an induced empirical distribution function provides the appropriate measure are three. First, we show that the induced edf for the regressors converges asymptotically to the desired distribution function. Second, we show that if we extend the empirical measure to a joint distribution function (df) for the regressors and the stochastic error, the induced edf still converges to the desired joint df. Third, we show that typical regularity conditions ensure the convergence of expectations measured with respect to this induced joint edf. Thus, these conditions guarantee the consistency of estimators based on the feasible weighted pseudo MLE.

Although the scalar case just described is suggestive, our intuitive discussion does not extend to the multi-regressor case. Deeper arguments must be mustered to obtain the convergence of G_N to G in the general case. That this occurs still follows from the notions that the sample points eventually cover the entire sample space as the number of observations grows and, as a result, G_N gets closer to G. Difficulties in several dimensions arise from the fact that discontinuities in G_N no longer occur at sample points alone. As a result, bounds on the distance between G_N and G over a sequence of length N do not imply bounds over the entire sample space. Despite this problem, we have the following result:

Theorem 2. Let H and G be continuous df's with densities h and g respectively. Let $\{x_n\}$ be an iid sequence of random variables with df H. If $E\{[g(x)/h(x)]^2\}$ exists then $G_N(X)$ converges uniformly in X to $G(X)$ almost surely.

This theorem is analogous to the Glivenko-Cantelli theorem for sdf's. Having extended interest to more general edf's than the sdf, Theorem 2 establishes the same uniform convergence, subject, however, to a caveat that f and g cannot be extremely different. That $E[(g/f)^2]$ must exist does not require that the ratio of the densities be bounded, but it does ensure that large values have low probability.

Theorem 2 can be weakened in several ways. First, it is unnecessary that H and G be continuous over $\mathbb{R}^K$: a finite set of atoms and smaller supports both can be accommodated. However, the set of atoms of G and the support of G must be subsets of their counterparts for H. Second, the x's need not be identically or independently distributed. All that our proof of Theorem 2 requires is that x satisfy a strong law of large numbers. Admitting non-stationary processes would simply require additional restrictions on the behavior of g/f. Finally, other criteria for the approximation to G in (13) could be considered. One might, for example, minimize the supremum norm over the entire support of G. This would lead to a stronger result than Theorem 2, but the computational cost would be substantial. Alternatively, the least absolute deviations and the least squares criterion functions would be useful choices.

The functions that we seek to integrate with respect to the induced empirical distribution are functions of the random variable u as well as the regressors. We cannot be content, therefore, with the convergence of G_N to G. Instead, the convergence of the stochastic distribution function

$$D_N(U,X) = \Sigma_{n=1}^{N} 1(u_n < U)\cdot 1(x_n < X) a_{Nn}/N$$

to $F(U)\cdot G(X)$ must be established. It is possible that G_N does not converge quickly enough to prevent the introduction of noise into the marginal distribution of u so that expressions like (7) fail to converge at all. Indeed, an additional constraint on the G_N is introduced in the next theorem to obtain the required convergence.

Theorem 3. Take the conditions of Theorem 2 and in addition, let $\{u_n\}$ be an iid sequence of variables with df F; $\{u_n\}$ is independent of $\{x_n\}$. If there is a Δ such that $a_{Nn} \leq \Delta N^{\delta/2}$ for all N, n, and $\delta < 1$, then $\Sigma_n 1(u_n \leq U)\cdot 1(x_n \leq X)\cdot a_{Nn}/N$ converges uniformly in U and X to $F(U)\cdot G(X)$ almost surely.

Moving to the measurement of the joint distribution function of u and x has lead us to modify our definition of G_N. Theorem 3 adds a smoothing condition that guarantees that the mass assigned to any sample point shrinks asymptotically. Because we do not know that this will occur in general, the researcher typically must either assume such behavior for G_N or impose it by choosing a conservative value for Δ, perhaps based on a nonparametric estimator for h. Note that this theorem, like the previous one, still admits an unbounded g/h.

Note also that the independence of u and x plays a central role in our proof of Theorem 3. While it may not be a necessary condition, we feel that independence is a natural assumption given that interest focusses on the β parameters. Otherwise, the identification and interpretation of β may be problematic. However, the assumption that the regressors are iid (independent and identically distributed) is unnecessary.

Finally, we must establish that functions integrated with respect to our empirical measure converge to the desired expectation. If this holds, then asymptotically our estimators will be the solution to normal equations of our own design. The following lemma contains a weak version of such results.

Lemma. Take the conditions of Theorem 3. If w *is continuous and* $\int w\, dG$ *is uniformly integrable,* $G_N \longrightarrow G$ *implies that* $\int w\, (dG_N - dG)$ *converges to zero if there is a* $\Delta < \infty$ *such that* $a_{Nn} < \Delta \cdot g(x_n)/h(x_n)$ *for all* N *and* n .

Like the previous theorem, this lemma places a restriction on the values of the weights, though this restriction involves the ratio of densities and not the sample size. Neither restriction implies the other and they play similar roles: these constraints on the weights of the empirical measure prevent large weights from appearing "too frequently" as the sample size grows. In both cases, it is difficult to provide a more fundamental circumstance that would imply the restriction. While it is possible to impose these conditions as constraints on the solution for the weights, the verity of the constraints will be difficult to assess in finite samples. On the other hand, potential violations are easy to detect, either by examining the extreme values of the calculated weights or by nonparametric estimation of the actual density of the regressors. We therefore view restrictions on the weights as regularity conditions and assume

Assumption 8. There exist Δ_1 , $\Delta_2 < \infty$ *and* $\delta < 1$ *such that* $a_{Nn} < \min[\Delta_1 N^{\delta/2}, \Delta_2 \cdot g(x_n)/h(x_n)]$.

Theorem 4. Given assumptions 1-8 and given that the data are generated according to equations (1) and (2), the WME converges almost surely to $\gamma\beta$ *for some scalar* γ .

6. SIMPLE ILLUSTRATIONS OF THE WEIGHTED M-ESTIMATOR

In this section, we illustrate the use of the weighted M-estimator with a simple example. Our purpose is to suggest that these estimators can be feasible and can perform well in finite samples. This example is purely illustrative; we leave serious Monte Carlo experimentation and empirical application to future research.

Our example parallels the earlier illustration of sources of estimator inconsistency using the exponential function as the unknown τ . The data were generated as follows. Two explanatory variables were drawn from a mixture of normal distributions:

$$h(x_1,x_2) = 1/2\ [\phi(x_1 - 1/2)\phi(2\cdot x_2) + + \phi(2\cdot x_1)\phi(x_2 + 1/2)] .$$

where ϕ is the standard normal pdf. In this way, positive x_1 tend to coincide with small x_2 and negative x_2 tend to coincide with small

x_1 . This distribution is a continuous version of the simple mixed distribution used before.

The dependent variable was generated by

$$y = exp[x_1 + x_2 + .2(u-.5)]$$

where u had a standard uniform distribution. Because the exponential function is convex, the OLS estimator for the linear regression of y on x_1 , x_2 , and a constant will overstate the relative effect of x_1 compared to the effect of x_2 .

The weights of the weighted M-estimator were calculated for a simpler and less flexible class than discussed above. The weights were restricted to be proportional to

$$\phi(x_{1n})\cdot\phi(x_{2n})/\{\Sigma_m \phi[(x_{1n}- x_{1m})/\theta]\cdot\phi[(x_{2n}- x_{2m})/\theta]\} . \tag{16}$$

The numerator corresponds to the choice of the desired distribution for the x : iid standard normal. This choice is probably not best since the bulk of the sample will fall in one orthant of $\mathbb{R}^2$. The performance of the WME might improve if the means and the correlation were nonzero.

The denominator is proportional to a kernel density estimator for the density of the x , where the window width of the estimator is θ . This specification reflects the notion that the weights are proxies for the ratio containing the actual density in the denominator. Because the kernel estimator is consistent, we can be sure that this functional form is sufficiently flexible asymptotically to enable the induced edf to converge to the chosen df, the bivariate normal.

After a few trials, the smoothing parameter θ was set permanently to 2 for every replication of the Monte Carlo. According to the theory, θ should be chosen to solve the optimization problem

$$\begin{aligned} &\min_\theta \max_{X_N} |G_N - G| \quad subject\ to \\ &G_N(X) = \Sigma_n a_{Nn} I(x_n \le X)/N \\ &G(X) = \Phi(X_1)\Phi(X_2) \\ &\Sigma_n a_{Nn} = N \end{aligned}$$

and equation (16), where Φ is the standard normal cumulative distribution function (cdf). Indeed, casual experimentation suggested that such optimization of the weights leads to estimator improvement.

The example was replicated one thousand times with fifty observations. In Table 1, we give the sample means and sample standard deviations of the OLS and WM estimators. The WM estimator is simply weighted least squares (WLS). The OLS estimators estimate the ratio of the coefficients on the continuous regressors with the predicted bias. The WM estimators reduce this bias substantially. Remarkably, the WM estimator also has smaller sampling errors than the OLS estimator. One should not expect this to be a general phenomenon.

We find these rough experiments encouraging. With a small sample size and crude weights, the WME performs quite well. Future research will investigate such complex models as the two-equation sample selection and the multinomial choice models, as well as consider more sophisticated weighting functions.

Table 1: Monte Carlo Results

Sample Mean of Estimated Coefficient
(Sample Standard Deviation)

Estimator	β_1	β_2	Intercept	β_2/β_1
OLS	2.50	1.46	1.75	0.64
	(1.07)	(0.51)	(0.29)	(0.23)
WM (WLS)	1.46	1.28	1.36	0.88
	(0.19)	(0.17)	(0.09)	(0.12)

7. CONCLUSION

This paper describes a weighted M-estimator that provides consistent estimates (up to a scalar) of the slope parameters of the regression function of an imperfectly observed dependent variable. The WME is attractive because it has weak requirements for knowledge of the relationship between the observed dependent variable and its latent counterpart and because the computation is a modification of popular estimators.

Besides those mentioned previously, many questions remain for future research. An immediate concern is the approximate distribution of the proposed estimators. We anticipate that asymptotic normality holds because the weighted estimation criterion function remains continuous in the estimator. Many cases of other nonparametric estimators which do not have established asymptotic normal distributions violate this characteristic.

Exploitation of information about τ , the transformation from the latent to the observed dependent variable, remains another important area of research. The method presented in this work capitalizes only on the stability of τ and the variation in the observed dependent variable that can be attributed to variation in the regressors. LDV models frequently postulate that τ is homogeneous, for example, and that information might be brought to bear on estimating the unknown scale of the slope parameters. Conversely, it may be possible to uncover τ , or the distribution of error terms, given consistent estimates of the slope parameters.

Progress in these two areas would substantially extend the usefulness of the weighted M-estimator and raise a host of other questions. Tests about the significance, signs, and ratios of parameters could be constructed that would be robust to a potentially important specification error. Efficiency within the class of weighted M-estimators is probably very difficult to characterize but two-step, or iterative, procedures may be found to improve the asymptotic sampling variance. Many applications to specification testing by comparing a pseudo MLE with the WME, or comparing alternative WME's, also exist.

Finally, since this research first appeared an approach similar in spirit, but very different in substance, has been suggested by Stoker (1985). The relationship between the method suggested here and that of Stoker requires study. Both of the methods exploit the distribution of the regressors, which is usually ancillary to estimation. The understanding of this feature may be revealed in the commonality of the methods.

FOOTNOTES

1. $1(\cdot)$ is the indicator function, which equals one when the argument is true and zero otherwise.

2. Proofs for the theorems are given in an extended version of this paper, Ruud (1986).

3. See Brillinger (1982).

4. I am grateful to Gary Chamberlain for pointing this out.

5. Brillinger (1982) suggests that normal regressors will prove most useful. Thus, our formulation implies a broader set of regressors than he indicates.

6. Goldberger (1981) also discussed the case of $x_1 = x_2$ in this estimation approach. Our discussion shows that the failure of proportionality in this situation is due to lack of identification.

7. Note that $h_2(x|x'\beta=z)$ and $\phi_2(x|x'\beta=z)$ are singular densities defined only on the hyperplane $z = x'\beta$. Integration with respect to these densities is defined only on this hyperplane.

8. The "sample distribution function" is often called the "empiricial distribution function." Because we speak of empirical measures in general, and their corresponding df's, we adopt this alternative label.

REFERENCES

Amemiya, Takeshi, 1973, "Regression Analysis When the Dependent Variable is Truncated Normal," *Econometrica*, 41(6), 997-1016.

Brillinger, David R., 1977, "The Identification of a Particular Nonlinear Time Series System," *Biometrika*, 64, 509-515.

Brillinger, David R., 1982, "A Generalized Linear Model with 'Gaussian' Regressor Variables," *A Festschrift for Erich L. Lehmann* (Peter J. Bickel, Kjell A. Doksum, and J. L. Hodges, eds.), Woodsworth International Group: Belmont.

Burguete, J. F., A. R. Gallant, and G. Souza, 1982, "On the Unification of the Asymptotic Theory of Nonlinear Econometric Models," *Econometric Reviews*, 1(2), 151-190.

Chung, Ching-Fan and Arthur S. Goldberger, 1984, "Proportional Projections in Limited Dependent Variable Models," *Econometrica*, 52(2), 531-534.

Cosslett, Stephen R., 1983, "Distribution-Free Maximum Likelihood Estimator of the Binary Choice Model," *Econometrica*, 51(3), 765-782.

Duncan, Gregory M., 1981, "A Relatively Distribution Robust Censored Regression Estimator," Washington State Univ. Working Paper No. 581-3.

Goldberger, Arthur S., 1981, "Linear Regression After Selection," *Journal of Econometrics*, 15, 357-366.

Greene, William H., 1981, "On the Asymptotic Bias of the Ordinary Least Squares Estimator of the Tobit Model," *Econometrica*, 49(2), 505-514.

Hansen, Lars Peter, 1982, "Large Sample Properties of Generalized Method of Moments Estimators," *Econometrica*, 50(4), 1029-1054.

Heckman, James J, 1976, "The Common Structure of Statistical Models of Truncation, Sample Selection, and Limited Dependent Variables and a Simple Estimator for Such Models," *Annals of Economic and Social Measurement*, 5, 475-492.

Lee, Lung-Fei, 1976, "Two Stage Estimations of Limited Dependent Variables Models," Ph.D. dissertation, Department of Economics, Univ. of Rochester.

Lee, Lung-Fei, 1981, "Estimation of Some Non-Normal Dependent Variable Models," Center for Economic Research Discussion Paper No. 81-148, Department of Economics, Univ. of Minnesota.

Lee, Lung-Fei, 1983, "Generalized Econometric Models of Selectivity," *Econometrica*, 51(2), 507-512.

Manski, Charles F., 1975, "Maximum Score Estimation of the Stochastic Utility Model of Choice," *Journal of Econometrics*, 3(3), 205-228.

Manski, Charles F., 1983, "Closest Empirical Distribution Estimation," *Econometrica*, 51(2), 305-320.

Powell, James L., 1981, "Least Absolute Deviations Estimation of Censored and Truncated Regression Models," Institute for Mathematical Studies in the Social Sciences, The Economics Series, Technical Report No. 356, Stanford Univ.

Rao, R. Ranga, 1963, "Relations between Weak and Uniform Convergence of Measures with Applications," *Annals of Mathematical Statistics*, 659-680.

Ruud, Paul A., 1981, "Misspecification Error in Limited Dependent Variable Models," Ph.D. Thesis, Massachusetts Institute of Technology.

Ruud, Paul A., 1983, "Sufficient Conditions for the Consistency of Maximum Likelihood Estimation Despite Misspecification of Distribution," *Econometrica*, 51(1), 225-228.

Ruud, Paul A., 1986, "Consistent Estimation of Limited Dependent Variable Models Despite Misspecification of Distribution," *Journal of Econometrics*, 32(1), 157-187.

White, H., 1982, "Maximum Likelihood Estimation of Misspecified Models," *Econometrica*, 50(1), 1-26.

TELECOMMUNICATIONS DEMAND MODELLING
An Integrated View
A. de Fontenay, M.H. Shugard, D.S. Sibley (Editors)

ON ESTIMATING DYNAMIC MODELS WITH SEASONALITY

Eric Ghysels*

1. INTRODUCTION

Most econometricians would recommend the use of seasonally unadjusted time series. This recommendation is usually not put into practice for several reasons. First, because it regularly argued that economists have no "theory" for dealing with seasonal fluctuations so that the non-seasonal part is filtered out to test economic theories. Second, because the use of seasonally adjusted time series is quite often justified by results on the treatment of seasonality in linear regression (see e.g. Sims (1974)). Third, because econometricians do not really know how to take account of seasonality in a model. In this paper it is explained how to deal with seasonality in a dynamic model. It is shown that it is, in the first place, a problem of defining the proper decomposition of seasonal time series. Traditionally one decomposes economic time series exhibiting seasonality in a cyclical component, a seasonal component and possible other components. The fundamental assumptions almost exclusively made about this decomposition are that (1) the components are mutually orthogonal and (2) the seasonal component has power only at the seasonal frequency and its harmonics. It is believed then that removing the peaks of the spectrum of the observed series yields the spectrum of the non-seasonal part. Such a decomposition of an economic time series is usually inappropriate. In dynamic economic models, the seasonality in exogenous variables may induce power at all frequencies of the spectrum of endogenous variables. This point has been made several times by economists, notably by Sargent (1978), and has been endorsed by the author, in Ghysels (1986a), with the analysis of a dynamic economic model with closed-form solutions. It should be noted that the inability to locate the seasonality in endogenous variables in a dynamic economic model induced by the seasonality in the exogenous variables at the so-called seasonal frequencies has far-reaching implications. Almost all economic time series are the outcome of a complexity of dynamic interactions. In Ghysels (1986a) it is noted that any attempt to base the component structure of one isolated series on arbitrary identifications assumptions, namely those mentioned above which are almost exclusively made by statisticians, is bound to give false interpretations of co-movements of economic time series. In the same paper, the author also reconsidered the treatment of seasonality in the standard linear regression model. This analysis lead to some positive results on seasonality in regression, results which differ substantially from the recommendations made by Sims (1974) and others. In Ghysels (1986b) the concept of Granger causality was used to identify exogeneity. The Granger exogeneity results help to identify the pattern of induced

* Université de Montréal, Department of Economics and C.R.D.E., P.O. Box 6128, Station A, Montreal H3C 3J7, Canada.

seasonality from exogenous to endogenous series. Such procedures could make "seasonal adjustments without too much a priori economic theory" in a multivariate context. The analysis in this paper goes beyond exogeneity tests based on unidirectional Granger causality and beyond the analysis of a standard linear regression. There are several reasons for this. The most important one is that a necessary condition for the identification of induced seasonality is unidirectional causality. This condition may not necessarily be met by the data. If so, an explicit dynamic model has to be formulated and estimated. This is the task focused on in this paper.

In section 2 the dynamic model presented in Ghysels (1986a) is introduced. It is a simple market equilibrium model with three alternative specifications for the cost function. The models lead to three possible decompositions of the endogenous variables along the line of seasonality, including the traditional decomposition. In section 3 the properties of the decompositions for the alternative cost function specifications are discussed. Section 4 digresses on the estimation of the decompositions. Conclusions follow in the last section.

2. A MARKET EQUILIBRIUM OF PRODUCTION

In this section a market equilibrium model of production, introduced in Ghysels (1986a), is reviewed and extended. A market with seasonally fluctuating demand is modeled via the following two components.

(2.1) $D_t^s = \theta_1 D_{12}^s + \mu_t^s$

μ_t^s i.i.d. $N(0, \sigma_s^2)$

(2.2) $D_t^c = a_0 - a_1 p_t + \mu_t^c$

μ_t^c i.i.d. $N(0, \sigma_c^2)$

with $|\theta_1| \leqslant i$ for $i = 0, 1$. The first component, denoted D_t^s, is a purely exogenous monthly seasonal autoregressive process. A seasonal AR process was selected in order to facilitate analytical derivations. The second component, denoted by D_t^c, corresponds to demand responding to market price p_t. Observed market demand is defined as the sum of the two components:

(2.3) $D_t = D_t^c + D_t^s$

with $\{D_{t-k}, P_{t-k}\}_{k=0}^{+00}$ observed at time t. Using (2.1), (2.2) and (2.3), a standard inverse demand function is obtained as follows:

(2.4) $P_t = A_0 - A_1 D_t + \mu_t$

with $A_0 = a_0 a_1^{-1}$

$A_1 = a_1^{-1}$

$\mu_t = a_1^{-1}(1-\theta_1 L^{12})^{-1} \mu_t^s + a_1^{-1} \mu_t^c$

The stochastic shock to demand μ_t will normally have an ARMA (12,12) structure (see Granger and Morris (1976)).

The supply side of the market is modeled via a representative firm standing in as a typical firm producing the particular product. Instead of looking at one specific specification of the representative firm, three model specifications will be considered. The first being a special case of the second and the latter a special case of the third. The first model specification is a completely time separable essentially static model with linear cost functions:

Model 1

$$(2.5) \quad \max_{\{q_{t+j}\}} E_0 \lim_{T\to\infty} \sum_{j=0}^{T} \beta^j (p_{t+j}\, q_{t+j} - dq_{t+j})$$

with output at time t being denoted q_t and $0 < \beta < 1$, β the discount factor and $d > 0$. Hence the firm plans future production, starting from $t = 0$, with the objective to maximize the expected value, given information at zero, of the infinite horizon stream of discounted profits. The specification in (2.5) is a special case of the next model with quadratic cost function:

Model 2

$$(2.6) \quad \max_{\{q_{t+j}\}} E_0 \lim_{T\to\infty} \sum_{j=0}^{T} \beta^j (p_{t+j} q_{t+j} - dq_{t+j} - (e/2) q_{t+j}^2)$$

with $e > 0$. Similarly, model 2 is a special case of a genuine dynamic model where the entrepreneur faces an intertemporal maximization problem characterized by the fact that changing production from one period to another is costly:

Model 3

$$(2.7) \quad \max_{\{q_{t+j}\}} E_0 \lim_{T\to\infty} \sum_{j=0}^{T} \beta^j (p_{t+j} q_{t+j} - dq_{t+j} - (e/2) q_{t+j}^2 - (f/2)\,(q_{t+j} - q_{t+j-1})^2)$$

With the same market demand, as given in (2.4), equilibrium output will be computed, assuming $D_t = Q_t = Nq_t \;\forall\, t$ with N the number of firms producing. For simplicity, but without loss of generality, however, N will be set equal to one.

3. THREE DECOMPOSITIONS AND THEIR PROPERTIES

Let us suppose that there is a sample of equilibrium output $\{Q_t\}_{t=0}^{T}$ for each of the three cost specifications. Hence, for one of the endogenous variables of the model, namely output, there are three sets of observations, each time with the same market demand but different technologies of production which yield different equilibrium outcomes. The equilibrium outcomes will be discussed by examining three decompositions along the lines of seasonality. They are:

Decomposition 1. $Q_t = D_t = D_t^c + D_t^s$

which is an economically meaningful decomposition since it corresponds to the building blocks of our market demand specification.

Decomposition 2. $Q_t = Q_t(\mu_t^c) + Q_t(\mu_t^s)$

with, $Q_t(\mu_t^c)$ standing for the part of Q_t originating from the shocks $\{\mu_t^c\}$ and similarly $Q_t(\mu_t^s)$ as the part originating from $\{\mu_t^s\}$. Since $\{\mu_t^c\}$ and $\{\mu_t^s\}$ are assumed to be orthogonal, an orthogonal decomposition of Q_t is obtained. This decomposition is built on the fact that μ_t^c and μ_t^s are the two exogenous shocks of cyclical and seasonal nature, respectively, driving the model and producing the spectral power of Q_t.

Decomposition 3. $Q_t = Q_t^c + Q_t^s$

which stand for the traditional decomposition based on the following assumptions:

1. $Q_t = Q_t^c + Q_t^s$
2. $\{Q_t^c\}$ and $\{Q_t^s\}$ are independent of each other
3. Q_t follows a known ARMA - process model $\phi(L)Q_t = \Omega(L)a_t$
4. Q_t^c follows an unknown ARMA - process model $\phi_c(L)Q_t^c = \Omega_c(L)a_t^c$
5. Q_t^s follows an unknown ARMA - process model $Q_s(L)Q_t^s = \Omega_s(L)a_t^s$

For Model 1 all three decompositions coincide. On the contrary, for Model 3 each decomposition is different. To show this model 1 is first analyzed. The computational details of the rational expectations market equilibrium outcome are deferred to the appendix. The first model yields:

(3.1) $Q_t = (a_0 - a_1 d + \mu_t^c) + (1-\theta_1 L^{12})^{-1}\mu_t^s$

From (3.1) one obtains that:

(3.2) $D_t^c = Q_t(\mu_t^c) = Q_t^c = a_0 - a_1 d + \mu_t^c$

(3.3) $D_t^s = Q_t(\mu_t^s) = Q_t^s - (1-\theta_1 L^{12})^{-1}\mu_t^s$

The result in (3.2) is easy to explain. Model 1 is essentially a sequence of static models with linear cost functions. Under perfect competition profit maximization corresponds to setting the marginal cost equal to the market price, i.e. $p_t = d$. Consequently, using (2.2) one obtains (3.2) immediately.

The equilibrium output law of motion for the second model is considered next:

(3.4) $Q_t = (1+a_1 e)^{-1}(a_0 - a_1 d + \mu_t^c + (1-\theta_1 L^{12})^{-1}\mu_t^s)$

The equilibrium market price p_t is not constant anymore in the second model since the cost structure is quadratic and hence the marginal cost depends on the volume of output with the latter fluctuating over the season. Consequently, equilibrium prices exhibit seasonal fluctuations, or more precisely:

$$(3.5)\quad p_t = (1+a_1e)^{-1}[d - ea_0 + a_1ed - e\mu_t^c + e(1-\theta_1L^{12})^{-1}\mu_t^s]$$

The latter implies that the first decomposition is not orthogonal, indeed one can write:

$$(3.6)\quad D_t^c = (1+a_1e)^{-1}[a_0 - a_1d + 2a_0a_1e + 2a_1^2de + (1+2a_1e)\mu_t^c + a_1e(1-\theta_1L^{12})^{-1}\mu_t^s$$

The cross-covariance generating function between D_t^c and D_t^s can be obtained from (3.6) as:

$$(3.7)\quad g_{cs}^D(z) = \sigma_1^2a_1e(1+a_1e)^{-1}\,[(1-\theta_1z^{12})^{-1}(1-\theta_1z^{-12})^{-1}]$$

Contrary to the $D_t = D_t^c + D_t^s$ decomposition the two other ones remain orthogonal and equal to:

$$(3.8)\quad Q_t(\mu_t^c) = Q_t^c = (1-a_1e)^{-1}\,(a_0-a_1d+\mu_t^c)$$

$$(3.9)\quad Q_t(\mu_t^s) = Q_t^s = (1+a_1e)^{-1}\,(1-\theta_1L^{12})^{-1}\mu_t^s$$

Notice that neither $Q_t(\mu_t^s)$ nor Q_t^s coincide with D_t^s since the seasonality induced by equilibrium prices into D_t^c is captured by $Q_t(\mu_t^s)$ and Q_t^s.

Since the third model has a quadratic cost structure we expect there will be a difference between (D_t^c, D_t^c) and the two other decompositions. There is more to the model, however, when one derives equilibrium output:

$$(3.10)\quad Q_t = -(1-\rho_1L)^{-1}(1-\rho_2^{-1}L^{-1})^{-1}\,\rho_1f^{-1}[\theta(L)-L^{-1}\rho_2^{-1}\theta(\rho_2^{-1})]\,(\theta(L))^{-1}\mu_t$$

where ρ_1 and ρ_2 are the roots of the characteristic polynomial obtained from (A.9) while $\theta(L)$ is defined in (A.11) as the polynomial of the fundamental moving average representation of the μ_t process. In deriving (3.10) it was assumed that the representative entrepreneur does not observe any series which Granger causes the μ_t process. It is reasonable, however, to assume that entrepreneur observes series which Granger cause μ_t^s in particular, in which case equilibrium outcome becomes:

$$(3.11)\quad Q_t = -(1-\rho_1L)^{-1}(1-\rho_2^{-1}L^{-1})^{-1}\,\rho_1f^{-1}[\delta(L) - L^{-1}\rho_2^{-1} \times \delta(\rho_2^{-1})](\delta(L))^{-1}v_t^x$$

where V_t^x is the vector of fundamental shocks of the multivariate moving average representation of the vector process containing μ_t and all other series which help predict μ_t. Finally, $\delta(L)$ is the matrix polynomial of that particular joint process. The distinction between (3.10) and (3.11) is important when the estimation of relevant decompositions will be discussed in the next section. The three decompositions of (3.10) have the following properties:

$$(3.12)\quad D_t^c = -\,[1+(1-\rho_1 L)^{-1}(1-\rho_2^{-1}L^{-1})^{-1}\rho_1 f^{-1}[\theta(L) - L^{-1}\rho_2^{-1}\theta(\rho_2^{-1})]]\,(\theta(L))^{-1}(1-\theta_1 L^{12})^{-1}\mu_t^s - [2+(1-\rho_1 L)^{-1}(1-\rho_2^{-1}L^{-1})^{-1}\rho_1 f^{-1}[\theta(L)-L^{-1}\rho_2^{-1}\theta(\rho_2^{-1})]]\,(\theta(L))^{-1}\mu_t^c$$

Equation (3.12) represents the first decomposition with D_t^c. It shows that D_t^c has again seasonal properties like in model 2. The second decomposition is considered next.

$$(3.13)\quad Q_t(\mu_t^c) = -a_1^{-1}(1-\rho_1 L)^{-1}(1-\rho_2^{-1}L^{-1})^{-1}\rho_1 f^{-1}\mu_t^c + a_1^{-1}(1-\rho_1 L)^{-1}(1-\rho_2^{-1}L^{-1})^{-1}L^{-1}\rho_2^{-1}\theta(\rho_2^{-1}) \times (\theta(L))^{-1}\mu_t^c$$

$$(3.14)\quad Q_t(\mu_t^s) = -a_1^{-1}(1-\rho_1 L)^{-1}(1-\rho_2^{-1}L^{-1})^{-1}\rho_1 f^{-1}(1-\theta_1 L^{12})^{-1}\mu_t^s + a_1^{-1}(1-\rho_1 L)^{-1}(1-\rho_2^{-1}L^{-1})^{-1}L^{-1} \times \rho_2^{-1}\theta(\rho_2^{-1})(\theta(L))^{-1}(1-\theta_1 L^{12})^{-1}\mu_t^s$$

where (3.13) has power at all frequencies, including possibly the seasonal frequency and harmonics, and (3.14) has obviously power at the seasonal frequency and its harmonics, but also at all other frequencies. Hence there are "spillover" effects among the frequencies caused by the fact that the model is characterized by intertemporal adjustment costs. This can easily be explained intuitively. Let there be at any particular time a high seasonal shock to demand. In case there is no adjustment cost the firm can produce a high output instantaneously without affecting the cost of production in the next period. With adjustment costs, however, it is expensive to change production activity rapidly in both directions, i.e. increase or decrease production rapidly.

Some numerical examples can be found in Ghysels (1986a) which highlight the differences between a model with relatively low adjustment costs, which is essentially like model 1, and a model with high adjustment costs. These numerical examples clearly illustrate how the power of $Q(\mu_t^s)$, defined in (3.14), is present at all frequencies, including the low frequencies where substantial power may be accumulated. The latter is due to the adjustment costs which spread shocks, whatever their nature may be, over "the long run".

Finally from the discussion of the second decomposition $Q_t = Q_t(\mu_t^c) + Q_t(\mu_t^s)$ it follows that:

(3.15) $Q_t^c = \Omega_c^1(L)\mu_t^c + \Omega_s^2(L)\mu_t^s$

i.e. by removing peaks at seasonal frequencies one obtains a spectrum composed of power generated by μ_t^c at all other frequencies as well as μ_t^s.

4. MAXIMUM LIKELIHOOD ESTIMATION OF DECOMPOSITIONS

The analysis in section 3 raises several issues about estimating dynamic models with seasonality. After having read the third section the question should arise whether estimating dynamic economic models with seasonally adjusted time series can be salvaged. The answer is negative because traditional seasonal adjustment procedures identify seasonality in the endogenous series via ad hoc identification assumptions which are incompatible with the proper identification of co-movements of the exogenous and endogenous variables of the model. The question then arises how to estimate dynamic models of seasonality and interpret the results in the context of the decompositions defined in the preceding section. The discussion presented here certainly does not constitute a full analysis of all aspects of estimation. First, it will be implicitely assumed that the econometrician is interested in finding empirical estimates of the decompositions which were defined. Second, it will also be assumed that the econometrician is interested in estimating structural parameters of dynamic models. These two restrictions narrow down the focus of the discussion considerably. When there is no interest in estimating actual decompositions then it would not be necessary to apply the analysis which will be presented. Indeed, it would be sufficient to apply instrumental variable estimation techniques to Euler conditions emerging from the dynamic economic models, using seasonally unadjusted data. This technique is known to yield consistent parameter estimates for dynamic models with rational expectations, without having to specify explicitly the distributional assumptions of the environments in which economic agents operate (see Hansen (1982) and Hansen and Singleton (1982)). An example of such an approach, with seasonality being taking fully into account, appeared in Miron (1986). Furthermore, if the econometrician is not interested in structural parameter estimates he can apply the Granger causality-based analysis presented in Ghysels (1986a) and Ghysels (1986b) to reduced form equations. If, the econometrician is interested in characterizing the sample path of endogenous variables and their decompositions along the lines of seasonality and is also interested in estimating structural parameters, then the following discussion should be a guidance to accomplishing such goals. The discussion will focus on the ML estimation of the second decomposition. The first decomposition is of lesser interest in the context of a demand model. In related work, however, the author has shown that non-orthogonal decomposition of dynamic models may be very important (see Ghysels (1986c)). The analysis presented here will not elaborate on this particular extension.

Two examples, very similar to the models of section 3, will be used here to analyse the problem of estimation. In the first example a dynamic economic model is considered with exogenous processes observable to the econometrician. The second example does not have this property. It turns out that this distinction is important since it makes the task of estimation very different. The two examples are models of inventory behavior and labor demand. Output which was defined in the previous section as endogenous variable is being decomposed into labor requirements and inventory laws of motion. This was done in order to construct the two separate examples. In both examples a representative form makes labor demand L_t and inventory decisions I_t with a cost function depending on the levels of L_t and I_t and intertemporal changes of the variables.

Example 1: Inventory and labor demand model with exogenous demand and prices

Representative firm

$$(4.1) \quad \max_{\{L_{t+j}, I_{t+j}\}} E_0 \sum_{j=0}^{T} \beta^j \{P_{t+j} D_{t+j} - w_{t+j} L_{t+j}$$
$$- (c_1/2)(I_{t+j} - \theta D_{t+j+1})^2 - (c_2/2) L_{t+j} (L_{t+j} + \phi^b_{t+j})$$
$$- (c_3/2)\, I_{t+j}(I_{t+j} + \phi^c_{t+j}) - (c_4/2)\,(L_{t+j} - L_{t+j-1})^2$$
$$-(c_5/2)\,(I_{t+j} - I_{t+j-1})^2\}$$

subject to the constraint

$$D_{t+j} = I_{t+j} - I_{t+j+1} + bL_{t+j} + \phi^a_{t+j}$$

with

$$\gamma(L) X_t = V^x_t$$
$$X_t' = (P_t, D_t, w_t, \phi^a_t, \phi^b_t, \phi^c_t, \ldots)$$

V^x_t fundamental to X_t and $\gamma(L)$ a matrix polynomial.

The second example is almost identical in structure except for the exogeneity assumptions, namely:

Example 2: Market equilibrium inventory and production model

Representative firm

$$(4.2) \quad \max_{\{I_{t+j}, L_{t+j}\}} \lim_{T\to\infty} E_0 \sum_{j=0}^{T} \beta^j \{P_{t+j}(I_{t+j} - I_{t+j+1} + bL_{t+j} + \phi^a_{t+j})$$
$$- w_{t+j} L_{t+j} - (c_1/2)\,(I_{t+j} - \theta D_{t+j+1})^2 - (c_2/2)\, L_{t+j}(L_{t+j} + \phi^b_{t+j})$$
$$- (c_3/2)\, I_{t+j}(I_{t+j} + \phi^c_{t+j}) - (c_4/2)\,(L_{t+j} - L_{t+j-1})^2$$
$$- (c_5/2)\,(I_{t+j} - I_{t+j-1})^2\}$$

subject to the constraints

(4.3) $D_t^s = \theta_1 D_{t-12}^s + \mu_t^s$

(4.4) $D_t^c = a_0 - a_1 P_t + \mu_t^c$

$D_t = D_t^c + D_t^s = I_t - I_{t+1} + bL_t + \phi_t^a$

Here $\{D_{t+j}\}$ and $\{P_{t+j}\}$ are not assumed exogenous but instead computed endogenously as the outcome of a full equilibrium solution similar to the example considered in the previous section.

The exogenous processes can only be decomposed according to the principles of the third decomposition. Any alternative decomposition would be arbitrary. Hence $\{P_{t+j}\}$ and $\{D_{t+j}\}$, which are in the first example exogenous, can only be decomposed along the traditional lines into a cyclical and seasonal component. Such a decomposition will be denoted as:

(4.5) $D_t = S_t = S_t^c + S_t^s$

(4.6) $P_t = p_t^c + p_t^s$

where S_t stands for supply since $D_t = Q_t = Q_t^c + Q_t^s$ can no longer be used because output Q_t and supply don't coincide when inventories are present. Given such a decomposition one can consider a decomposition of the second type:

(4.7) $L_t = L_t(S_t^c, p_t^c) + L_t(S_t^s, p_t^s)$

(4.8) $I_t = I_t(S_t^c, p_t^c) + I_t(S_t^s, p_t^s)$

In the second example, however, D_{t+j} and P_{t+j} no longer belong to the primitive structure of the model so that one can for instance define:

(4.9) $D_t = D_t(\mu_t^c) + D_t(\mu_t^s)$

(4.10) $L_t = L_t(\mu_t^c) + L_t(\mu_t^s)$

(4.11) $I_t = I_t(\mu_t^c) + I_t(\mu_t^s)$

The differences between (4.7-8) and (4.9-11) are important. Decompositions (4.7) and (4.8) are defined on cyclical and seasonal components of observed exogenous series. The Euler conditions for model (4.1) are:

(4.12) $\beta D_1' E_t Y(t+1) + D_0 Y(t) + D_1 Y(t-1) = E_t R(t)$

with $Y'(t) = (L_t, I_t)$

$R'(t) = (w_t - bp_t + (C_2/2)\,\phi_t^b,\ p_t + (c_3/2)\phi_t^c - \beta p_{t+1} - \beta c_1 D_{t+1})$

The matrices D_i depend on b, c; , β and θ. The solution to the Euler equations which satisfy the transversality conditions (cfr Hansen and Sargent (1981)) is:

(4.13) $Y(t) = -C_0^{-1}C_1Y(t-1) + E_t[C_0'C_0 + C_1'C_0\beta L^{-1}]^{-1}R(t)$

with

$$D_1 = C_0'C_1 \text{ and } D_0 = C_0'C_0 + \beta C_1'C_1$$

Equation (4.13) is a multivariate version of (A.10). It implies that current (L_t, I_t) decisions depend on the previous period state of labor and inventories and the conditional expectation of R(t), R(t+1), R(t+2) ... given current information. The R(t) process contains elements of the process X_t defined in (4.1). Let $X_t' = (p_t, D_t, w_t, \phi_t^a, \phi_t^b, \phi_t^c, \ldots)$ containing R(t) as well as observations the entrepreneur makes of series which Granger cause the elements of R(t) (cfr (3.11)). The X_t process has the AR representation:

(4.14) $\gamma(L)X_t = V_t^X$

with $\gamma(L)$ a polynomial lag matrix of order r. Based on (4.13) and (4.14) a final solution can be obtained using Wiener-Kolmogorov prediction formula for $E_tR(t+j)$:

(4.15) $Y(t) = -C_0^{-1}C_1Y(t-1) - \sum_{j=1}^{k} z_j^{-1}N_j\gamma(z_j^{-1}\beta)^{-1}\{I + \sum_{s=1}^{r-1}\sum_{i=s+1}^{r}(z_j^{-1}\beta)^{i-s}\gamma_iL^s\}X_t$

with z_j the roots of $\det[C_0'C_0 + C_1'C_0\beta L^{-1}]$ and N_j are obtained from the matrix partial fraction expansion of the inverse of the matrix (see Hansen and Sargent (1981) for details). Let us define the decomposition for the exogenous process:

$$X_t = X_t^c + X_t^s$$

(4.16) $\gamma_c(L)X_t^c = V_t^c$

(4.17) $\gamma_s(L)X_t^s = V_t^s$

Which implies that

$$I_t = I_t(X_t^c) + I_t(X_t^s)$$

$$L_t = L_t(X_t^c) + L_t(X_t^s)$$

It is conceptually feasible to estimate equations (4.15), (4.16) and (4.17) jointly to yield FIML estimates of the model but it is computationally tedious and impractical. Instead, one might estimate on the one hand (4.14) and (4.15) jointly to capture the cross-equation restrictions between I_t, L_t, p_t, D_t, ... imposed by the model and estimate separately (4.16) and (4.17). The procedure to estimate (4.14) and (4.15) is described by Hansen and Sargent (1981) and applied by Blanchard (1983) to the automobile sector for a model similar to (4.1). A methodology to estimate an unobserved component multivariate seasonal model with extraction of X_t^c and X_t^s has to our knowledge not yet been worked out. The current literature on model-based seasonal deals with univariate models (see Hillmer and Bell (1984) and the literature surveyed there). Bell (1980) considered estimation of a multivariate model consisting of signal contaminated with white noise. Hence one might at this stage only consider single variable model-based extraction for the elements of X_t separately. It should parenthetically be noted that one can introduce seasonal dummies to capture deterministic seasonality when one models the X_t process.

The second example is considerably more difficult. In the event that economic agents do not observe any series which Granger cause the exogenous μ_t process the estimation problem is quite similar to the one discussed above. The μ_t process is an exogenous error process exhibiting seasonal features. One can solve for the simultaneous equilibrium of market and production decisions. Eichenbaum (1984) discusses this solution method in the context of an inventory and production model which ignores seasonality and was estimated with seasonally adjusted series. When economic agents do observe series which Granger cause the exogenous μ_t process then the quasi-maximum likelihood estimation procedure becomes infeasible. In particular when the econometrician omits variables from the model which Granger cause μ_t (since the latter is unobserved by the econometrician this can never be verified) the endogenous variables may Granger cause the exogenous variables in the model (see Granger (1969) and Hansen and Sargent (1981)). The second example clearly illustrates that estimating such models and their decompositions looks prohibitely tedious and prone to identification problems.

CONCLUSIONS

In this paper, the author extended his previous work on seasonality in the context of dynamic economic models by discussing the problem of estimating structural models with closed-form solutions. The major difficulty is that the decomposition of a time series into unobserved components usually suffers from an inherent arbitrariness. The work of Wallis (1978) and Plosser (1978) suggested that an important contribution of economists was to model the interrelation between variables which would help to clarify the interrelations between the seasonality in the different series involved. Simon Kuznets (1933) was already concerned with how seasonality is introduced in some markets by seasonality in other markets. It was shown here and elsewhere that a first requirement is a clear definition and separation of exogenous versus endogenous variables in a model since this forms the basis of defining induced seasonality. In dynamic models, the seasonality in exogenous variables may induce power at all frequencies of the spectrum of the endogenous variables. Hence, it is improper to estimate dynamic models with series which have been filtered a priori by an adjustment procedure which eliminates power only at seasonal frequencies. The task of specifying properly defined decompositions and estimating them in the context of a dynamic model is tedious but feasible, as the paper shows. Quite often easier estimation procedures are applicable (see Ghysels (1986a) and (1986b)). There are, however, many cases where there is no choice but to follow the approach discussed in this paper.

APPENDIX

This appendix briefly reviews and extends material, presented in Ghysels (1986a), on the computations of the equilibrium in the models discussed in the text. The first model consists of equations (2.4) and (2.5). Lucas and Prescott (1971) pointed out that a rational expectations competitive equilibrium for a model of this type solves a particular type of social planning problem. It consists of considering the area under the demand curve:

$$(A.1)\quad \int_0^{Q_t}[A_0 - A_1X_t + \mu_t]dX = [A_0 + \mu_t]Q_t - (A_1/2)Q_t^2 = [A_0 + \mu_t]Nq_t - (A_1/2)N^2Q_t^2$$

The social planning problem corresponding to the competitive equilibrium consists of maximizing the expected discounted area (A.1) minus the social cost of producing the output which implies:

$$(A.2)\quad \underset{\{q_{t+j}\}}{\text{Max}} \lim_{T\to\infty} E_0 \sum_{t=0}^{T} \beta^t [A_0 + \mu_{t+j}) q_{t+j} - (A_1/2)q_{t+j}^2 - dq_{t+j}]$$

The first order necessary conditions for (A.2) yield the Euler equation (in particular for j = 0)

$$(A.3)\quad (A_0+\mu_t) - A_1q_t - d = 0$$

where $A_0 = a_0a_1^{-1}$ and $A_1 = a_1^{-1}$. With N = 1, we can set q_t equal to Q_t so that (A.3) is equivalent to:

$$(A.4)\quad Q_t = (a_0 - a_1d) + (1-\theta_1L^{12})^{-1}\mu_t^s + \mu_t^c$$

which is obtained after substituting the different parameters and variables.

The equilibrium output law of motion for the second model is obtained in the same way. The rational expectations market equilibrium is equivalent to a social planners problem:

$$(A.5)\quad \underset{\{q_{t+j}\}}{\text{Max}} \lim_{T\to\infty} E_0 \sum_{t=0}^{T} \beta^t [(A_0+\mu_{t+j}) q_{t+j} - (A_1/2)q_{t+j}^2 - dq_{t+j} - (e/2)q_{t+j}^2]$$

The first order necessary condition for (A.5) is:

$$(A.6)\quad Q_t = (1+a_1e)^{-1}(a_0 - a_1d + \mu_t^c + (1-\theta_1L^{12})^{-1}\mu_t^s)$$

The law of motion for the equilirium price can be derived by using (2.4), realizing that $Q_t = D_t$, and (A.6) which yields:

$$(A.7)\quad p_t = (1+a_1e)^{-1}[d - ea_0 + a_1ed - e\mu_t^c + e(1-\theta_1L^{12})^{-1}\mu_t^s]$$

Finally, the equilibrium solution for the third model is computed according to the same principles as the previous two. The objective function is formulated:

$$\text{(A.8)}\quad \underset{\{q_{t+j}\}}{\text{Max}} \lim_{T\to\infty} E_0 \sum_{t=0}^{T} \beta^t[(A_0+\mu_{t+j})\, q_{t+j} - (A_1/2)q_{t+j}^2 - dq_{t+j} - (e/2)q_{t+j}^2 - (f/2)\,(q_{t+j}-q_{t+j-1})^2]$$

The above expression leads to the Euler equation, for any j:

$$\text{(A.9)}\quad \beta E_{t+j}Q_{t+j+1} + \phi Q_{t+j} + Q_{t+j-1} = f^{-1}(a_0 a_1^{-1} + \mu_{t+j} + d)$$

with $\phi = -[1 + \beta + (a_1^{-1}+e)f^{-1}]$.

Equilibrium output satisfying the Euler conditions (A.9) and the transversality condition (cfr. Hansen and Sargent (1981)) can then be expressed as (up to a constant):

$$\text{(A.10)}\quad Q_t = \rho_1 Q_{t-1} - \rho_1 f^{-1} \sum_{j=0}^{+\infty} \rho_2^{-j} E_t \mu_{t+j}$$

with ρ_1 and ρ_2 the roots of the characteristic polynomial obtained from the Euler equation (A.9). The conditional expectations appearing in (A.10) can be solved for by using the Wiener-Kolmogorov prediction formula. Based on the assumption that the fundamental MA representation of the $\{\mu_t\}$ process can be written as:

$$\text{(A.11)}\quad \mu_t = \theta(L)\varepsilon_t$$

and assuming that entrepreneurs do not observe any series which Granger causes μ_t the closed form solution for equilibrium output can be written as:

$$\text{(A.12)}\quad Q_t = -(1-\rho_1 L)^{-1}(1-\rho_2^{-1}L^{-1})^{-1}\ \rho_1 f^{-1}[\theta(L)-L^{-1}\rho_2^{-1}\theta(\rho_2^{-1})]\ (\theta(L))^{-1}\mu_t$$

as Hansen and Sargent (1981) have shown. In the event that entrepreneurs do observe series which Granger cause μ_t, they will use them to help predict μ_{t+j} for $j > 0$ to make their optimal decisions as expressed by (A.10). Let us define the vector X_t as a vector with the first element being μ_t and all others the variables Granger causing μ_t with X_t having the fundamental MA representation:

$$\text{(A.13)}\quad X_t = \delta(L)V_t^x$$

Then equilibrium output becomes:

$$\text{(A.14)}\quad Q_t = -(1-\rho_1 L)^{-1}(1-\rho_2^{-1}L^{-1})^{-1}\ \rho_1 f^{-1}[\delta(L) - L^{-1}\rho_2^{-1}\ \delta(\rho_2^{-1})](\delta(L))^{-1}V_t^x$$

REFERENCES

Bell, W.R., 1980, Multivariate Time Series: Smoothing and Backward Models, Doctoral Dissertation, Department of Statistics, University of Wisconsin-Madison.

Bell, W.R., and S.C. Hillmer, 1984, Issues Involved with the Seasonal Adjustment of Economic Time Series, Journal of Business and Economic Statistics, 2, 291-320.

Blanchard, O.J., 1983, The Production and Inventory Behavior of the American Automobile Industry, Journal of Political Economy, 91, 365-400.

Eichenbaum, M.S., 1984, Rational Expectations and the Smoothing Properties of Inventories of Finished Goods, Journal of Monetary Economics, 14, 71-96.

Ghysels, E., 1986a, A Study towards a Dynamic Theory of Seasonality for Economic Time Series, Discussion Paper, Université de Montréal.

Ghysels, E., 1986b, Seasonal Adjustment without Too Much a Priori Economic Theory, Proceedings of the American Statistical Association, Business and Economic Statistics Section (forthcoming).

Ghysels, E., 1986c, Seasonal Extraction in the Presence of Feedback, Journal of Business and Economic Statistics (forthcoming).

Granger, C.W.J., 1969, Investigating Causal Relations by Econometric Models and Cross-Spectral Methods, Econometrica, 37, 424-438.

Granger, C.W.J., and M.J. Morris, 1976, Time Series Modelling and Interpretation, Journal of the Royal Statistical Society, Series A., 139, 246-257.

Hansen, L.P., 1982, Large Sample Properties of Generalized Methods of Moments, Econometrica 50, 1029-1054.

Hansen, L.P., and T.J. Sargent, 1980, Formulating and Estimating Dynamic Linear Rational Expectations Models, Journal of Economic Dynamics and Control 2, reprinted in R.E. Lucas, Jr. and T.J. Sargent (ed.), 1981, Rational Expectations and Econometric Practice (The University of Minneapolis Press, Minneapolis), 91-126.

Hansen, L.P., and T.J. Sargent, 1981b, Linear Rational Expectations Models for Dynamically Interrelated Variables, in R.E. Lucas, Jr., and T.J. Sargent (ed.), Rational Expectations and Econometric Practice (The University of Minneapolis Press, Minneapolis, 127-156.

Hansen, L.P., and K.J. Singleton, 1982, Generalized Instrumental Variables Estimation of Nonlinear Rational Expectations Models, Econometrica, 50, 1269-1298.

Kuznets, S., 1933, Seasonal Variations in Industry and Trade (N.B.E.R., New York).

Lucas, R.E., and E. Prescott, 1971, Investment Under Uncertainty, Econometrica, 39, 659-681.

Miron, J.A., 1986, Seasonal Fluctuations and the Life Cycle-Permanent Income Model of Consumption, Journal of Political Economy (forthcoming).

Plosser, C.I., 1978, A Time Series Analysis of Seasonality in Econometric Models, in A. Zellner (1978), 365-397.

Sargent, T.J., 1978, Comments on "Seasonal Adjustment and Multiple Time Series Analysis", by K.F. Wallis, in A. Zellner (1978), 361-364.

Wallis, K.F., 1978, Seasonal Adjustment and Multiple Time Series Analysis, in A. Zellner (1978), 347-357.

Zellner, A. (ed.), 1978, Seasonal Analysis of Economic Time Series (U.S. Department of Commerce, Bureau of Census, Washington, D.C.).

SECTION II: ECONOMETRICS OF MARKET BEHAVIOR

II.2. Application

TELECOMMUNICATIONS DEMAND MODELLING
An Integrated View
A. de Fontenay, M.H. Shugard, D.S. Sibley (Editors)

SAMPLE DESIGN CONSIDERATIONS FOR TELEPHONE TIME-OF-USE PRICING EXPERIMENTS WITH AN APPLICATION TO OTC-AUSTRALIA

by

Dennis J. Aigner

University of Southern California, U.S.A.

and

Denzil G. Fiebig

University of Sydney, Australia

1. INTRODUCTION

All telephone traffic is subject to time-of-use (TOU) variation. In particular, distinct peaks in traffic occur, usually during day-time hours. As a consequence the marginal cost of providing calls during these peak periods is higher than in the off-peak periods. The theory of peak-load pricing suggests that prices should vary over the day (or, the week or season in the case of weekly or seasonal variation in demand) in order to reflect these cost differences.[1]

Trunk calls in Australia are subject to time-of-use price variation but as yet no such tariff structure is permanently in place for international calls. The Overseas Telecommunications Commission of Australia, OTC, has recently expressed interest in the implementation of TOU rates for international calls. In fact off-peak discounts for a selection of streams were implemented for a trial period of six months starting in February 1985 and have been subsequently extended. Unfortunately, for planning purposes evidence from such an exercise will necessarily be of limited generalizability. Many questions will remain regarding the design of tariff structures appropriate for full-scale implementation. For example, it will be impossible to infer anything about tariff structures that use different peak time periods or a different scale of discounts than those included in the trial tariff.

Given a commitment to TOU pricing, a process of learning-by-doing should be avoided. (In other words, to implement a tariff structure and then revise it on the basis of observed responses and non-responses. This process continues until it eventually iterates to an appropriate tariff structure.) This may be costly in terms of revenue losses, may cause uncertainty and disruption amongst consumers and provides no guarantee that the iterative process will reach anything desirable within a reasonable period of time.

The purpose of this paper is to discuss some of the modelling procedures that are available to planners when designing new tariff structures. In particular we provide an argument for the use of a carefully designed experiment on a randomly selected group of customers. This discussion is illustrated by OTC's particular situation but it is relevant to many other situations, such as redesigning existing TOU tariffs and evaluating local measured service options.

Designed experiments recently have been utilized extensively in the U.S.A. to analyze the impact of TOU pricing in the electricity industry. This provides an extensive body of literature concerned with experimentation in economics and in particular how it relates to TOU pricing. We find that these experiences can usefully be employed to supply guidance in the design, implementation and analysis of similar experiments in telecommunications.

The next section includes a discussion of some alternative modelling procedures that can be used to analyze the likely response to tariff changes. This is followed by an extended discussion of the designed experiment, together with a stylized example for TOU telephone demand.

2. MODELLING THE LIKELY RESPONSE TO TOU PRICES

The theory of public-sector economics provides clear guidance on the setting of optimal prices. Economic efficiency can be increased by implementing prices equal to marginal costs. If such prices lead to a financial deficit then these prices will need to be modified to cover the costs of the public enterprise. As is well known this may be achieved by increasing prices above marginal costs in inverse proportion to the individual price elasticities of demand. (See for example Ramsey [18] and Baumol and Bradford [4].)

The potential benefits from such a pricing scheme need to be compared with the additional costs associated with new metering and billing procedures. In the case of telephone calls these costs are not likely to be large, although it will depend on the existing price structure and hence existing metering and billing procedures. For example, there is currently much discussion in the U.S.A. on the appropriate method for the pricing of local telephone services. Many customers face a flat rate tariff where a single fee is paid irrespective of usage. In this case the telephone company only needs to know whether a customer is connected to the system or not. An alternative is usage-sensitive pricing where there is a charge per call in addition to the connection fee. This is a two-part tariff structure. Further one could consider varying the charge per call by the time of the day that the call was made. The additional costs involved in metering the time of the call are minimal once the equipment for metering usage is available.

Both Mathewson and Quirin [13] and Mitchell [14] suggest that benefits net of metering costs will usually accrue when moving from flat to usage-sensitive pricing. In addition Mitchell suggests that: "... peak-load pricing applied to residential customers in metropolitan areas would double or triple the welfare gains achieved from a two-part tariff alone." These analyses do not incorporate the added dimensions associated with international telephone demand, but the qualitative conclusions are likely to be the same.

Given a commitment to the introduction of TOU prices, the planning question of most interest is the actual structure of these prices. It would be desirable to have information on the likely response of subscribers to alternative TOU price schedules. The development of formal demand models to measure such responses is difficult because of the lack of historical data relating to such structures. (At least, for the particular telephone company in question.) Nevertheless, the following options can be considered:

(i) sample surveys,
(ii) test market,
(iii) transferability,
(iv) designed experiment.

Sample surveys will provide relatively inexpensive information with a very short time delay. Unfortunately, there is a great problem in obtaining an unbiased response from the consumers surveyed. There is no strong incentive for them to reveal their preferences accurately. (Nor may they be able to, faced with a purely hypothetical situation.) On balance, market data is superior to survey data. Market situations provide obvious incentives for the consumer. Moreover, actual market behavior is not subject to the confusion or implicit assumptions that exist in survey responses.

One method of obtaining market data is a test market study. Examples of such procedures are the GTE Measured Service Experiment that was conducted in the Jacksonville, Clinton and Tuscola exchanges in Illinois (for further details see Park, Mitchell, Wetzel and Alleman [16], Park, Wetzel and Mitchell [17] and Wilkinson [24]) and the ISD Experiment conducted by OTC. The latter ran for a period of six months, starting in February 1985, with an off-peak discount for 6 streams of outgoing ISD calls. The 6 streams chosen were USA, Hawaii, Canada, Hong Kong, Singapore and Malaysia.

There are several factors which may militate against the usefulness of test market procedures. Most important is the difficulty associated with the identification of a market segment that is truly representative of the total market. This in turn makes inferences regarding total demand somewhat difficult. In the ISD Experiment, how relevant are the changes in demand for the Hong Kong and U.S.A. streams to the likely changes in demand for say Great Britain? In fact there are no European streams amongst those chosen and consequently an important time zone is not represented in the ISD Experiment. Moreover, there are limitations on the number of different tariff structures that can be applied and, because of the absence of a control group, the analysis of data from test markets is problematic.

The techniques discussed above involve the collection of sample information from customers of the telephone company of interest. Another source of information is other telephone companies which have histories of TOU prices. In OTC's case, trunk calls within Australia have been subject to TOU prices for some time. Probably more appropriate are the histories of other countries that have TOU prices for international calls.

Here the key notion is one of transferability. Is it possible to transfer the findings from locations where TOU prices have been implemented to areas where they have not? Although such transfers are likely to occur on an informal basis, it is interesting to note that there has been some recent work investigating this question more formally. In particular, there have been a number of designed experiments in the U.S.A. aimed at determining customer response to TOU pricing of electricity. Several studies including Aigner and Leamer [3], Caves, Christensen and Herriges [5] and Kohler and Mitchell [11], have statistically analyzed whether it is possible to transfer these findings to areas where experiments have not been carried out. On balance they suggest that it is possible. However, more experience is required with this methodology before a complete evaluation is available.[2]

It is these sorts of deficiencies that suggest the use of a designed experiment on a carefully selected random sample of customers. These customers would be subjected to a variety of alternative tariffs and their responses compared with those of a control group. In this way the results can be generalized to situations not specifically tested in the experiment. A detailed discussion of this general procedure together with associated advantages and disadvantages are provided in the following sections.

3. BASIC PRINCIPLES OF OPTIMAL DESIGN

Social experiments tend to be expensive, and for that reason it is recommended that the experimental design be optimal in the sense that it provide for the most precise estimates possible of relevant parameters in the family of models that would be used to analyze the data were it available. In this regard we present a very brief exposition of the methodology of optimal design at this point.

Suppose a single-equation framework is appropriate and, following standard notation, let the regression model of interest be:[3]

$$\underset{(n \times 1)}{Y} = \underset{[n \times (K+1)]}{X} \; \underset{[(K+1) \times 1]}{\beta} + \underset{(n \times 1)}{u} \tag{1}$$

with $E(u) = 0$ and $E(uu') = \sigma^2 I_n$. The regressor matrix, X, depends, row-for-row, on a matrix of design variables, Z, of dimension (n × L). This specification appropriately focusses attention on the design variables at the stage of sample design and the regressor variables at the analysis stage. It is noted that while (1) concerns parameters that appear linearly, the design variables can be subject to various non-linear (but known) transformations in arriving at the regressors in X. No presumption is made on the additivity of regressor effects. Models that are inherently non-linear in parameters are not included, however.[4]

The design problem is to choose the rows of the design matrix, Z, and hence the rows of X, in an optimal way. To simplify matters, it is assumed that over the relevant column partition of X (the partition of variables subject to experimental control) row selection is limited to a set of m admissible rows ($m < n$). In the row space of the design matrix these admissible rows of X correspond to design points or cells, selected in advance so as to give adequate coverage to the experimental region. Actual determination of these points (how many of them and their exact specification) is an artful matter that depends on a variety of things, among them a prior knowledge of the relevant range of variation for policy purposes of each design variable, interest in particular combinations of values to be taken on by design variables, etc. In any event, the design problem is thus reduced to the problem of selecting m non-negative integers, $n_1,\ldots,n_m$, where $\Sigma_{h=1}^{m} n_h$, corresponding to the selection of admissible row one of X (denoted X_1') n_1 times, row two n_2 times, etc., so that $X'X = \Sigma_{h=1}^{m} n_h X_h X_h'$.[5] Next, an objective function must be formulated which adequately reflects the goals of analysis as functions of $n_1,\ldots,n_m$. We take the primary goal of analysis to be precise estimation of linear functions of basic model parameters, the elements of β. Thus, encompassed in this specification is prediction at one or more specific points and estimation of one or more linear functions of β which may be of interest for purposes of hypothesis testing. Let the set of such linear functions of interest be denoted by $P\beta$, where P is a $[H \times (K + 1)]$ matrix whose rows reflect particular linear functions of β. To represent interest in all individual coefficients, P would be specified as a $[(K + 1) \times (K + 1)]$ identity matrix.

Because (1) is written as a classical linear regression model, OLS will yield best linear unbiased estimates of the elements of β. Their variance-covariance matrix is given by the familiar expression $\sigma^2(X'X)^{-1}$. Likewise, the best linear unbiased estimator for $P\beta$ is $P\beta$, with $(H \times H)$ variance-covariance matrix $V(P\beta) = \sigma^2 P(X'X)^{-1}P'$.

Finally, suppose W is given as a diagonal matrix (of dimension $(H \times H)$) of weights to be applied to the functions of $P\beta$, reflecting their importance to the experimenter, and that we adopt as our objective function the weighted trace of $V(P\beta)$; that is,

$$
\begin{aligned}
(2) \quad \phi(n_1,\ldots,n_m) &= \sigma^{-2}\mathrm{tr}WV(P\beta) \\
&= \sigma^{-2}\mathrm{tr}P'WPV(\beta) \\
&= \mathrm{tr}\Big[D\Big(\sum_{h=1}^{m} n_h X_h X_h'\Big)^{-1}\Big],
\end{aligned}
$$

where $D = P'WP$

$\phi(n_1,\ldots,n_m)$ is now to be minimized subject to a cost constraint. [The model (1) is already an implicit constraint on ϕ.] Letting c_h be the unit cost of an observation at the hth design point, and denoting by C the total budget, we have $\Sigma_{h=1}^{m} c_h n_h \leqslant C$. As may be the case in any actual application, the c_h's can themselves depend on values taken on by variables at each design point.

Given (2) and the cost constraint, our problem is as follows:

$$(3) \quad \min \phi(n_1,\ldots,n_m) = \mathrm{tr}\,[D(\sum_{h=1}^{m} n_h X_h X_h')^{-1}],$$

subject to

$$\sum_{h=1}^{m} c_h n_h \leq C, \quad h = 1,\ldots,m, \qquad n_h \geq 0,$$

which, strictly speaking, is a non-linear integer programming problem. The nature of a solution to (3) will be such that the budget constraint holds at equality. Thus, in solution all design points included must have the same marginal effectiveness (per dollar of cost) in reducing the criterion function. Moreover, since ϕ is homogeneous of degree minus one in the n_h's, the optimal proportional sample allocations, $p_h^* = n_h^*/n$, are independent of C (although a larger value of C implies as larger total sample size).[6] Also from Elfving's [9] work we know that there will be at most $1/2(K + 1) \times (K + 2)$ design points ultimately chosen with $n_h > 0$.

Optimal designs based on the weighted trace criterion (3) generally produce unbalanced designs, and some (perhaps many) cells will receive no observations in the optimal allocation. Subject to the qualification that 'optimality' is defined over an assumed 'true' model, we are merely discovering by this that on the margin there is no information in such cells.

Certain problems may arise, however, because of the fact that some admissible rows of Z do not appear in the final design. There is no guarantee, for instance, that the rank of X'X for the optimal design will be sufficient to allow even for estimation of all parameters in the assumed response model. Clearly we require sufficient degrees of freedom for judging the fit of that model and to consider other alternatives as well. But if a certain subgroup of the population is eliminated, that presents the difficulty that our inferences will be necessarily limited, were they to be based on a sample exhibiting such characteristics.

To guard against these difficulties, one may add constraints in (3) to ensure that a minimum number of observations appears in each major population subgroup (which normally involves several cells). By so doing the convenient homogeneity property ascribed to the optimal solution is destroyed, so that a new calculation must be done for each sample size. Moreover, the computational costs appear to be higher for problems that incorporate such constraints, but the advantages are compelling.

A persuasive advantage of this allocation framework is the case by which particular linear functions of parameters can be singled out for attention in the criterion function. Moreover, each can be given a specific weighting through the matrix W. But the scaling of independent variables can have a pronounced effect on the final outcome, acting in the same way as the elements in W. In practice it is therefore appropriate to scale the regressors in order to make their numerical values roughly equal in order of magnitude. The weighting matrix can then clearly influence the design in the desired direction. In a related way, if a variable is to be coded, that coding should be selected so as to be compatible with the magnitudes of other regressors.

There are a few other guidelines to mention that relate to the formation of the experimental region. One is that the optimal design is heavily influenced by extreme values of the regressors. Therefore, rather beneficial results will be achieved by spreading out the endpoints of the experimental region, whereas including more intermediate points is not likely to have much effect.[7] Controlling the extent of multicollinearity that may be induced in the final design by this means is also a matter of concern because it likewise influences the anticipated and actual levels of precision attached to parameter estimates. Care is to be taken, in as much as possible, to anticipate other collinearity problems not only in the assumed response function but in competing specifications, by recognizing the unfortunate fact that it may not be possible as a practical matter to break down correlation among concomitant variables that exists naturally in the population.

Still very problematic is that fact that the 'optimal' design is aptly characterized that way only with reference to a particular model and a particular criterion function. And, unfortunately, these design results are not very robust to variations in either direction. One approach to guard against severe mistakes in model specification is to add appropriate cell minima constraints in specifying the allocation problem. Another supposes that a small number of alternative models with known prior probabilities can adequately represent the 'truth', in which case a composite weighted trace function is suggested.

Extension of the design problem (3) to the multivariate regression case is not difficult conceptually, but it does put on additional demands at the point of implementation. Nevertheless this extension is especially relevant for telephone demand. Taylor [21,22] emphasizes the need to distinguish between <u>access</u> to the network and <u>usage</u>, the latter involving both number of calls and duration. When considering international calls, the demand for access is much less important than the demand for use. A consumer contemplating making overseas calls is likely to have access to the telephone network and be making local calls.

Given access, now the consumer needs to decide on the total number of calls to be made and the duration of these calls. The distinction between calls and minutes is similar to that between access and use. Consumers decide whether to make a call just as they decide whether to subscribe to the system. Then they decide on the level of usage that is represented by the length of the call. These distinct aspects of demand will often have separate prices. There may be an initial surcharge or a minimum call duration; this is the "access" price of making a call. The price of extra minutes is then the "usage" price. Importantly, both the demand for calls and for minutes will depend on the two prices; access and usage. A decrease in the access price will lead to more calls being made and in turn more minutes and more calls are likely to be made. Here one would expect overall call duration to increase.

Previous studies of international telecommunications demand (Lago [12],Yatrakis [25], Craver [6], Rea and Lage [19], Craver and Neckowitz [7] and Schultz and Triantis [20] model either calls or minutes but not both. For example, Rea and Lage [19] argue that because the two are highly correlated, results are not materially affected by the choice of one in favour of the other. This does not follow. Calls and minutes are both important aspects of telephone demand that may react quite differently to price changes.

4. AN EXAMPLE EXPERIMENTAL DESIGN FOR TOU TELEPHONE DEMAND

In order to make these ideas concrete, consider the problem of designing an experiment to measure the response of customers to off-peak discounts in the price of overseas telephone calls, a situation now under study in Australia.

As in most statistical sampling, at the outset we attempt to identify the set of potentially important causal variables to help explain telephone demand by time-of-use, for use as stratifying variables. The efficiency of a stratified design over simple random sampling is well-known, and the case for its use in applications like this is compelling because of the high cost involved in any large-scale social experiment.

The most important such variables or dimensions of stratification in the OTC situation are these:

- Customer "size"
- Business vs. Residential account
- Availability of alternative services (e.g., telex)
- Ethnic background of residential customers
- Extent of overseas trade for business customers

In order for a variable to qualify as a stratifier for sampling purposes, however, it must be observable on the entire population. Moreover, if the individual account is to be our primary sampling unit, these stratifiers will have to be measurable at that level of disaggregation. This presents problems in the case of several items on the above list.

To keep things simple, we model telephone demand by time-of-day with paid minutes as the basic consumption variable. As we discussed previously, access and use are aspects of demand that should be modelled separately. But since in the prevailing OTC tariff the call-minute is what customers pay for and, moreover, because access to the local network is a prerequisite to making an overseas call, we will ignore access as a dependent variable for the design exercise. For non-operator-assisted calls there is no minimum duration and hence no "price" associated with placing a call apart from its duration. Although there is such a minimum for operator-assisted calls, we ignore its possible influence in what follows.

There is a strong historical precedent for aggregate telephone demand to be represented by a double-log function (of price) but in a recent paper deFontenay and Lee [8] argue that this formulation is based primarily on convention without a firm grounding in demand theory. They propose (and use) a translog model in their work. Since the translog demand system has been employed extensively in the analysis work on electricity demand by TOU, it certainly is a candidate for driving a model-based design. Another would be the CES function.

For example, consider the homothetic translog system

$$(4)\quad w_1 = \beta_{10} + \beta_{11}\log(p_1/z) + \beta_{12}\log(p_2/z) + \ldots + u_1$$

$$w_2 = \beta_{20} + \beta_{21}\log(p_1/z) + \beta_{22}\log(p_2/z) + \ldots + u_2$$

where:

w_i = expenditure share on telephone consumption in period i,

p_i = price per paid-minute in period i,

x_i = number of paid-minutes in period i,

$z = \sum_{i=1}^{2} p_i x_i$, total expenditure on telephone calls in the billing period (say, one month),

u_i = a random term with mean zero and constant variance within each pricing period, which is also independent over observations within the period, but is correlated over periods (i.e., $\mathrm{Cov}(u_1,u_2) \neq 0$).

Other stratifiers than customer size (measured by z) may appear in each equation, including those represented by indicator (or "dummy") variables.

Now since $w_1 + w_2 = 1$ for every observation, it turns out that one need only estimate the parameters of one of the two equations. We take the peak period equation (w_1) as the one to be estimated. However, if homogeneity <u>within</u> each equation is imposed (i.e., $\beta_{12} = -\beta_{11}$), then the equation of interest reduces to

$$w_1 = \beta_{10} + \beta_{11}\log(p_1/p_2)$$

It is useful to note that from the optimal design point-of-view this is indistinguishable from the simple CES model.

For the optimal design exercise we expand the equation to include an income term (proxied for by the logarithm of annual expenditure on overseas calls) and the interaction of this variable with $\log(p_1/p_2)$. The complete design model is therefore

$$(5)\quad w_1 = \beta_{10} + \beta_{11}\log(p_1/p_2) + \beta_{13}\log(z) + \beta_{14}\log(z)*\log(p_1/p_2) + u_1$$

As to the tariff structure to be studied, Table 1 presents the current OTC tariffs for both short- and long-haul calling. There was a six-month "demonstration" TOU tariff in place from February through July 1985 involving an off-peak discount on ISD calls for six long-haul countries: U.S.A., Hawaii, Canada, Hong Kong, Singapore and

Malaysia. For these countries T_1=\$1.50 during the periods midnight to 6.00 a.m. Monday to Saturday and midday to midnight Saturday. So, we pattern our experimental traiffs after it. These are presented in Table 2.

Note that there is variation in the peak period rate, but only in one direction, reflecting the anticipation that the nominal price of long-distance calls will continue to fall over time. Not only will the experiment be capable of generating estimates of demand response by time-of-day, it will also be useful in estimating the response of total demand to a falling average price per paid-minute.

The Operator Assisted and ISD tariffs are constructed in a parallel fashion, both with respect to the TOU peak to off-peak ratios and the relative declines in average price involved, both of which range from 1:1 to 2:1 (1:1 to 0.5:1). Sixteen price pairs are thereby specified. As to values for z, a convenient breakdown consists of four values, being the mid-points of an equal frequency stratification of customers by annual numbers of calls. This subsumes within it business and social (residential) callers.[8] Here we have assumed no allowances for special services (collect calls, person-to-person) and use the "long-haul" rates, which is consistent with the experimental prices themselves.

The results of our optimal design calculations are contained in Tables 3ab and 4ab for the CES/translog model. Because the relevant price variable in the CES/translog case is $\log(p_1/p_2)$, the design problem revolves around a total of 16 cells (four price ratios times four values of z).

With equal weight attached to the policy interest of all parameters (through the W-matrix) and equal cost associated with observations taken at each design point, the optimal allocation is one of extremes, as shown in Table 3a. Only four cells receive positive allocations, those corresponding to the categories of least and most annual expense on overseas calls combined with the smallest and largest values of (p_1/p_2). A measure of the effect of an optimal allocation for this model is to compare the value of the objective function [equation (3)] under an equal allocation to the optimal allocation. For our design matrix the objective function goes from 314.42 (equal allocation) to 93.78 (optimal allocation), a reduction of 70%.

Considering the fact that no observations are, however, allocated in this design to the two intermediate levels of annual expenditure, we also computed a design on this model that constrains there to be an equal allocation of observations over the four expenditure categories in order to force them to be "representative" of the underlying population. This results in the allocation shown in Table 4a. The same tendency toward a concentration in the four "extreme" cells is evident, but now there are equal allocations called for within the cells corresponding to each intermediate expenditure level. The value of the objective function for this design is 208.42, a reduction of 34% from the completely equal allocation.

How large should the experiment be? Tables 3b and 4b contain the anticipated variances for regression coefficients (apart from the term

σ_1^2/n) for the CES/translog designs we have been discussing.

Since estimating the price elasticity of demand for either peak or off-peak consumption is of primary interest (or, ES in the CES model), the experiment can be "sized" accordingly. The key relationship needed is

$$(6) \quad V(\hat{\eta}_{11}) = [V(\hat{\beta}_{11}) + \log^2(z)\, V(\hat{\beta}_{14}) + 2\log(z)\, C(\hat{\beta}_{11},\hat{\beta}_{14})]/w_1^{\,2},$$

which is the variance of the estimated own-price elasticity in the translog model, namely

$$\hat{\eta}_{11} = -1 + [\hat{\beta}_{11} + \hat{\beta}_{14}\log(z)]/w_1$$

Using $w_1 = \sigma_1 = 0.5$[9] and observing that $\bar{z} = 88.71$ for ISD calls, we

find

$$V(\hat{\eta}_{11}) = 14.05/n$$

for the unconstrained optimal design.

With this admittedly conservative formula, sample size requirements are exceedingly large. For instance, in order to insure a margin of error of no more than .05 (which is about 10% of the anticipated value of the own-price elasticity)[10] and a 95% confidence level, we would need a sample size of over 21,500. Settling for a larger margin of error, a lower confidence level or a less conservative assumption about σ will reduce the sample size requirments. Imposing the illustrative constraint on the sample distribution over z will obviously increase them. A 90% confidence level with $\sigma = .30$ and a margin of error of .10 implies a sample size of just under 2300, an impressive reduction in the required size of the experiment and one that seems much more realistic.[11]

The primary reason for why these large sample sizes occur is that in the telephone experiment under consideration, TOU price variation is very limited. By way of contrast, sample size requirements are approximately 1000 (as compared to 21,500) under the conservative assumptions used herein for a TOU electricity pricing experiment where peak to off-peak price ratios range to 9:1.[12]

To complete the analysis, we also computed an optimal design for the quadratic model

$$(7) \quad w_1 = \gamma_{10} + \gamma_{11}p_1 + \gamma_{12}p_1^{\,2} + \gamma_{13}p_2 + \gamma_{14}p_2^{\,2} + \gamma_{15}z$$
$$+ \gamma_{16}z^*p_1 + \gamma_1 z_1{}^*p_2 + v_1$$

In the interest of space, detailed results are not presented. To summarize, the allocation is more balanced and ranges over the full 64 cells. The relative allocation over intervals of annual expenditure, while favoring the extremes somewhat, also is approximately balanced. As to sample size requirements, we find, consistent with the results reported in Aigner [2], that approximately three times more observations would be needed to produce an estimate of η_{11} with comparable precision under the quadratic model than under the CES/translog design.

5. FINAL REMARKS

Designed experiments have been used successfully to facilitate the measurement of customer response to time-differentiated prices for electricity. There is no reason to think this strategy would be any less successful when applied to telecommunications. Indeed, in more than one important respect the TOU-pricing experiment for telephone service will be easier: the service itself is much more homogeneous than electricity and the cost of metering, such an important constraint in the sizing and ultimate evaluation of the electricity experiments, is essentially free.[13] That the sample size requirements for the OTC experiment as currently conceived are large is therefore not of overriding concern, although some economies in total sample size requirements can be had by TOU metering of selected customers prior to the implementation of experimental rates, thereby lessening the need for as large a cross-section control group.

As to the duration of such an experiment, a full 12 months is the absolute minimum. This would be long enough, probably, to see the most important behavioral changes but it would not be long enough to allow customers to alter their stock of telecommunications "appliances", something that may be very important if the experiment is to yield relevant information for the future.

For example, owing to the large time differences between Australia and many of its major calling countries, a facsimile machine may become economically feasible even for small firms that do substantial overseas business by taking advantage of lower off-peak rates.[14]

A major problem with the designed experiment for telephone (or, more generally, telecommunications) pricing is that telephone usage is not a unilateral thing: Unlike electricity usage, which for the customer is something that affects only him/herself or the household, the telephone call involves another party who "benefits" from it. Telephone demand has a reciprocity feature that electricity demand generally does not. This must be represented in any econometric model used for purposes of analysis by focussing on the point-to-point nature of telephone demand.

For a pricing experiment one can imagine that a favourable off-peak rate in Australia will affect incoming traffic as well as outgoing traffic. U.S. callers, for example, may merely signal their intention to communicate and let the Australian party initiate the real call. Obviously a transfer of funds must take place between the two parties in order to complete the transaction, which may act as a deterrant for some or even most social callers, but it would not in business situations where the facility for sending and collecting payments to other countries already exists.

If our orientation is limited to Australian originated calls and the substitution and stimulation effects on them that an off-peak tariff may have, the scenario just described could be viewed as a serious biasing element in the estimation of telephone demand by time-of-day from experimental data. Viewed properly, it is not. With the facility to measure the effect both on outgoing and incoming traffic of an off-peak tariff for OTC, the experiment will yield exactly what it should: information on the responses of customers at both ends of the stream to changes in OTC prices and price strucutres.

A final consideration is whether sample participation in such an experiment ought to be mandatory or voluntary. Clearly if the purpose of experimentation is to learn something about a mandatory tariff, sample participation ought to be mandatory. In some of the electricity-pricing experiments mandatory sample participation was not possible since it involved forcing randomly selected households to participate in something that may have made them worse off. (In the OTC design this is not a problem because experimental customers can only be better off.) The fact that volunteers were used in such instances, often with compensation allowances or participation incentives to induce them to volunteer and remain in the experiment for its duration, introduced an element of selection bias into these experiments and, while there are econometric techniques available to deal with the selection bias problem, they are not as effective as eliminating it in the experimental design directly.

While TOU-pricing for OTC is an example of a desirable mandatory tariff on grounds of economic efficiency, there are other examples of tariff alternatives, especially now under consideration in the U.S., that may be desirable as options for customers. A voluntary sample is appropriate for learning about an optional program. But unlike the case of volunteers who are protected and induced, in this they must be free to opt in, opt out or attrit. These are all features of a real optional tariff that need to be measured. The sample design we have presented, which relates only to the measurement of customer response to a changed price structure for those in and remaining in the experiment, would have to be expanded in order to satisfy these requirements.

ACKNOWLEDGEMENTS

Aigner would like to acknowledge the good offices of Instituut voor Actuariaat en Econometrie, University of Amsterdam, and Department d'Econometrie, Universite de Geneve, where much of the writing of this paper took place while on a brief study leave during April and May, 1985, and the support of NSF Grant SES-8319129.
The optimal design computations reported in Section 4 were supervised by our colleague Joseph G. Hirschberg at USC. To him we owe special thanks. Bridger Mitchell contributed some useful insights in his discussion of the paper which we have incorporated in this revision.

FOOTNOTES :

1. Mitchell [15] gives a good synthesis of the peak-load pricing issue applied to telephone services. von Weizsacker [23] emphasizes cost-based rather than value-based prices as

appropriate in telecommunications in general, in order to foster an efficient industry organization.

2. Most importantly, how sensitive is the analysis to misspecification of the formal models used? This point is stressed by Aigner and Leamer [3].

3. In this presentation we have borrowed liberally from Aigner [1], to which source the reader is referred for additional material and references. We include it here in order for the paper to be self-contained.

4. The essential difference in results for non-linear models and those for linear ones is that in the non-linear case the optimal design depends on the unknown parameters.

5. In what follows, we are assuming for expository convenience that all the variables in Z are controlled. This will not be the normal case in practice, in which case the 'optimality' of sample design applies only to that subset of variables under experimental control.

6. Little is lost, therefore, by treating the n_h's as continuous variables in actual computation and rounding.

7. This assumes there are already enough intermediate points specified to allow for estimation of all model parameters.

8. Obviously we cannot observe whether a call is for business or social purposes. But we can identify business and residential accounts. The frequency distribution is based on sylized facts and not on hard data, although OTC does monitor incoming overseas calls directly and therefore could produce the needed information. On this same basis customer "size" determined by incoming calls would be available as a stratifying variable.

9. That is, we assume the peak expenditure share is approximately 0.5 and take σ=0.5, which is as large as it can be given that the dependent variable in the translog model is a proportion.

10. We are using .50 as our guesstimate of the peak period own-price elasticity. This may be on the high side, in which event the margin of error of .05 is perhaps larger than 10%.

11. On the other hand, since metering costs are usually not a serious constraint on the size of a TOU experiment for telephone usage, even a sample of 21,500 may be completely feasible.

12. Aigner [2].

13. It should be mentioned that not all exchanges in Australia are yet automated to such a degree that TOU billing is readily available and so this statement needs some qualification.

14. In an experimental setting it may not be possible to observe such changes being made unless there is some assurance that the

experimental tariff will be available after the experiment itself is over.

REFERENCES :

[1] Aigner, D.J., A Brief Introduction to the Methodology of Optimal Experimental Design, Journal of Econometrics, Vol. 11 (1979) 7-26.

[2] Aigner, D.J., Sample Design for Electricity Pricing Experiments, Journal of Econometrics, Vol. 11 (1979) 192-205.

[3] Aigner, D.J. and Leamer, E.E., Estimation of Time-of-Use Pricing Response in the Absence of Experimental Data: An Application of the Methodology of Data Transferability, Journal of Econometrics, Vol. 26 (1984), 205-228.

[4] Baumol, W.J. and Bradford, F.F., Optimal Departures from Marginal Cost Pricing, American Economic Review, Vol. 60 (1970) 265-283.

[5] Caves, D.W., Christensen, L.R., and Herriges, J.A., Consistency of Residential Customer Response in Time-of-Use Electricity Pricing Experiments, Journal of Econometrics, Vol. 26 (1984) 179-204.

[6] Craver, R.F., An Estimate of The Price Elasticity of Demand for International Telecommunications, Telecommunication Journal, Vol. 43 (1976) 671-675.

[7] Craver, R.F. and Neckowitz, H., International Telecommunications: The Evolution of Demand Analysis, Telecommunication Journal, Vol. 47 (1980) 217-223.

[8] de Fontenay, A., and Lee, J.T.M., B.C./Alberta Long Distance Calling, in Courville, L., et al. (eds) Economic Analysis of Telecomunications, (North-Holland Publishing Co, 1983), 199-230.

[9] Elfving, G., Optimum Allocation in Linear Regression Theory, Annals of Mathematical Statistics, Vol. 23 (1952) 255-62.

[10] Keifer, J., Optimum Experimental Designs, Journal of the Royal Statistical Society, B, Vol.21 (1959) 272-304.

[11] Kohler, D.F. and Mitchell, B.M., Response to Residential Time-of-Use Electricity Rate: How Transferable Are the Findings?, Journal of Econometrics, Vol. 26 (1984) 141-178.

[12] Lago, A.M., Demand Forecasting Model of International Telecommunications and Their Policy Implications, Journal of Industrial Economics, Vol. 19 (1970) 6-21.

[13] Mathewson, G.F. and Quirin, G.D., Metering Costs and Marginal Cost Pricing in Public Utilities, Bell Journal of Economics and Management Science, Vol. 3 (1972) 335-339.

[14] Mitchell, B., Optimal Pricing of Local Telephone Service, American Economic Review, Vol. 68 (1978) 517-537.

[15] Mitchell, B., Local Telephone Costs and The Design of Rate Structures, in Courville, L., et al. (eds.), Economic Analysis of Telecommunications, (North-Holland Publishing Co. 1983) 293-304.

[16] Park, R.E., Mitchell, B.M., Wetzel, B.M., and Alleman, J.H., Charging for Local Telephone Calls: How Household Characteristics Affect the Distribution of Calls in the GTE Illinois Experiment, Journal of Econometrics, Vol. 22 (1983) 339-364.

[17] Park, R.E., Wetzel, B.M., and Mitchell, B.M., Price Elasticities for Local Telephone Calls, Econometrica, Vol. 51 (1983) 1699-1730.

[18] Ramsey, F., A Contribution to the Theory of Taxation, Economic Journal, Vol. 37 (1927) 47-61.

[19] Rea, J.D. and Lage, G.M., Estimates of Demand Elasticities for International Telecommunications Services, Journal of Industrial Economics, Vol. 26 (1978) 363-381.

[20] Schultz, W.R. and Triantis, J.E., An International Telephone Demand Study Using Pooled Estimation Techniques, Proceedings of the American Statistical Association, Business and Economic Statistics Section, (1982) 537-542.

[21] Taylor, L.D., Telecommunications Demand: A Survey and Critique., (Ballinger, Cambridge, MA, 1980).

[22] Taylor, L.D., Problems and Issues in Modelling Telecommunications Demand, in Courville, L., et al. (eds.), Economic Analysis of Telecommunications, (North-Holland Publishing Co 1983), 183-98.

[23] von Weizsacker, Christian, C., Free Entry into Telecommunications?, Information Economics and Policy, Vol. 1 (1984) 197-216.

[24] Wilkinson, G.F., The Estimation of Usage Repression under local Measured Service: Empirical Evidence from the GTE Experiment, in Courville, L., et al. (eds.), Economic Analysis of Telecommunications, (North-Holland Publishing Co 1983), 253-62.

[25] Yatrakis, P.G., Determinants of the Demand for International Telecommunications, Telecommunication Journal, Vol. 39 (1972) 732-746.
Courville, L., et al. (eds.), Economic Analysis of Telecommunications, (North-Holland Publishing Co 1983), 253-62.

Table 1: Current OTC Tariffs

Services: Four services are available, ISD, person-to-person, station-to-station and collect.

Destination: Charges vary with the destination of the call. Two regions are defined:

(i) Short haul or South Pacific: includes Fiji, French Polynesia, Nauru, New Caledonia, New Zealand, Papua New Guinea, Samoa (American), Samoa (Western), Solomon Island and Vanuatu. In terms of traffic volume, this region is dominated by New Zealand (NZ) and Papua New Guinea (PNG) which accounted for 19% of total telephone traffic in 1982.

(ii) Long haul or Rest of the World: the eight countries that, along with NZ and PNG comprise the top ten customers of OTC, are Great Britain, U.S.A., Canada, Italy, Greece, West Germany, Japan and Hong Kong.

Tariffs: There are four basic tariff instruments: the ISD price T_1, the operator-connected tariff T_2, the person-to-person surcharge S_1 and the collect surcharge S_2. (This distinction between S_1 and S_2 has only been a recent innovation. Historically they have been the same.)

All operator-connected calls have a three minute minimum and are then charged in one minute intervals or parts thereof. For ISD there is no minimum duration and you are charged only for the time used.

Current tariffs are

	T_1	T_2	S_1	S_2
Long Haul	$1.80	2.10	6	7
Short Haul	$1.30	1.50	6	7

Table 2: Values of Design Variables

	ISD CALLS			OPERATOR-ASSISTED CALL		AVERAGE ANNUAL BILL	
Intercept	Peak Price	Off-Peak Price	(Ratio)	Peak Price	Off-Peak Price	If ISD	If Operator-Assisted
1.10	1.80	1.80	1.0	2.10	2.10	13.86	16.17
	1.50	1.50	1.0	1.75	1.75	41.58	48.51
	1.20	1.20	1.0	1.40	1.40	76.23	88.94
	0.90	0.90	1.0	1.05	1.05	223.15	260.34
	1.80	1.50	1.2	2.10	1.75		
	1.50	1.25	1.2	1.75	1.46		
	1.20	1.00	1.2	1.40	1.17		
	0.90	0.75	1.2	1.05	0.88		
	1.80	1.20	1.5	2.10	1.40		
	1.50	1.00	1.5	1.75	1.17		
	1.20	0.80	1.5	1.40	0.93		
	0.90	0.60	1.5	1.05	0.70		
	1.80	0.90	2.0	2.10	1.05		
	1.50	0.75	2.0	1.75	0.88		
	1.20	0.60	2.0	1.40	0.70		
	0.90	0.45	2.0	1.05	0.53		

Table 3a: Optimal Relative Allocation for the CES/translog Model

p_1/p_2 \ z	13.86	41.58	76.23	223.15
1.0	.36	0	0	.19
1.2	0	0	0	0
1.5	0	0	0	0
2.0	.30	0	0	.13

Note: Value of objective function = 93.78

Table 3b: $[\sum_{h=1}^{16} (n_h/n)\ Z_h Z_h']^{-1}$ for the CES/translog model

Regressor	intercept	$\log(p_1/p_2)$	$\log(z)$	$\log(z)*\log(p_1/p_2)$
intercept	15.36	-22.21	-3.76	5.45
$\log(p_1/p_2)$		72.28	5.45	-18.01
$\log(z)$			1.05	-1.52
$\log(z)*\log(p_1/p_2)$				5.09

Table 4a: Optimal Relative Allocation for the CES/translog Model (Constrained)

z p_1/p_2	13.86	41.58	76.23	223.15
1.0	.135	.069	.063	.110
1.2	.019	.060	.062	.036
1.5	0	.056	.062	.013
2.0	.094	.065	.063	.091

Note: Value of objective function = 208.42

Table 4b: $[\sum_{h=1}^{16} (n_h/n)\ Z_h Z_h']^{-1}$ for the CES/translog model

(Constrained)

Regressor	intercept	$\log(p_1/p_2)$	$\log(z)$	$\log(z)*\log(p_1/p_2)$
intercept	30.66	-47.61	-7.19	11.17
$\log(p_1/p_2)$		166.20	11.16	-38.83
$\log(z)$			1.81	-2.83
$\log(z)*\log(p_1/p_2)$				9.75

TELECOMMUNICATIONS DEMAND MODELLING
An Integrated View
A. de Fontenay, M.H. Shugard, D.S. Sibley (Editors)

A GENERAL THEORY OF POINT-TO-POINT LONG DISTANCE DEMAND

ALEXANDER C. LARSON
Southwestern Bell Telephone Company

DALE E. LEHMAN
Fort Lewis College and Bell Communications Research, Inc.

DENNIS L. WEISMAN
Southwestern Bell Telephone Company

The authors wish to express their sincere gratitude to Lester Taylor for helpful comments on an earlier draft of this paper. The authors are constrained to blaming each other for any remaining errors or omissions.

1.0 INTRODUCTION

One of the intriguing aspects of current telecommunications demand theory is that it is nearly identical to the theory of demand for other types of commodities or services. This is curious because telecommunications exhibits a number of unique characteristics that serve to differentiate it from all other commodities. First, telecommunications consumption depends upon the interaction of at least two economic agents who jointly consume the commodity. Second, a telephone call is a shared good, yet the originating caller is billed the entire price of the call. This is unconventional because both economic agents presumably derive utility from the message. Casual theorizing suggests the justification for this rests upon the differential uncertainty between the originating end vis-á-vis the terminating end of a message. Economic agents may not always be willing to pay a charge for a call that is, at the time of economic decision-making, of uncertain origin. This is an important point because it suggests some type of implicit repayment plan between end users who originate messages and end users who derive positive utility from receiving those messages. This observation will serve as a foundation for the development of the formal theory of point-to-point telephone demand developed in this paper.

The subject of point-to-point toll demand has been studied by Pacey (1983)[1], Pacey and Berg (1982)[2], Larsen and McCleary (1970)[3] and Deschamps (1974).[4] The Pacey papers are noteworthy because of the standard functional forms employed. The econometric analysis provides estimation results for 14 selected city pairs, point-to-point, in Mountain Bell territory. Two equations are estimated: a message equation and an average-length-of-conversation equation. The regressors employed include price, income, market size and an indicator variable to distinguish customer type (business vs. residence). The fixed-effects least squares with dummy variables (LSDV) method of pooling was used to estimate parameters using semi-annual data.

The critical observation to make regarding previous research on point-to-point telecommunications demand is that the empirical analyses used so-called "standard" consumer theory. Such theory (for example, Varian, 1984)[5] hypothesizes that the demand for a commodity is a function of the price of that commodity, the price of all other commodities and income. Two questions become salient: (1) given the unique characteristics of the telecommunications commodity, what is the proper theoretical specification of the demand relation; and (2) are there significant differences, theoretical and empirical, between the proper specification of the demand relation and previous research? To these questions we now turn our attention.

1.1 Issues In Point-to-Point Analysis

Though data limitations have hampered efforts to conduct research on point-to-point telecommunications demand in the past, the post-divestiture environment may generate a significant amount of interest in such studies. For example, competition in both interLATA and intraLATA toll markets may foster interest in estimating route-specific price elasticities because they are necessary to derive "optimal" pricing rules. The concern of local telephone companies regarding the threat of bypass is certainly dependent upon the effective price elasticities for toll services. Also, the decision by the OCCs and ATTCOM on whether to enter the intraLATA toll markets will surely depend on the effective price elasticities of demand in these markets.

On the local service side, the issues are equally compelling, as measured service is simply short haul toll. Hence, knowledge derived from observing toll market reaction to rate changes may yield an analytic framework for designing local usage tariffs. In this sense, the point-to-point demand theory discussed in this paper could generalize to issues such as the offering of optional measured service packages. Such issues are critical and will become even more so as the telecommunications industry completes the transition from a monopolistic to a competitive market structure.

1.2 Conceptual Motivation for Point-to-Point Analysis

To gain insight into the determinants of point-to-point telecommunications demand, it is worthwhile to examine some of the functional relationships involved. Suppose that point-to-point telecommunications traffic can be disaggregated into two distinct elements: autonomous traffic and induced traffic. Autonomous traffic is assumed to be independent of the level of traffic between two designated points. Conversely, induced traffic is assumed to bear some mathematical relationship to the volume of traffic between the two points. Two extreme examples of this are as follows. The hypothesis of perfectly reciprocal calling patterns implies that traffic patterns are stabiized in both directions as implicit contracts are observed between economic agents. These agents return messages or minutes in direct proportion to the quantity received. An alternative hypothesis is that of so-called "information content." In this instance, there is a requirement to transfer a given level of information between two economic agents. Hence, an incoming message may alleviate the need for a message to be originated. Once the information constraint is satisfied, the need for further telecommunications between the two points is obviated. These hypotheses imply opposite responses to incoming calls. It is only possible to measure the dominant traffic pattern empirically, however.

2.0 FORMAL THEORY

Previous models of telecommunications demand are based upon utility maximization. The ad hoc formulation normally used is somewhat inconsistent with the heuristic discussions which accompany these models. Consider the following excerpts from Taylor (1980, pp. 29, 53):

> "To the extent that the information conveyed by an incoming calls helps to reduce overall risk, then answering the phone is beneficial ... if a portion of A's calls to B consist of information that B is to convey to C, who does not have a telephone, C's getting a telephone would shorten A's conversation with B, since A could now talk to C directly ... suppose that a person is weighing making three-minute calls to a relative in the middle and at the end of a month or a six-minute call at the end of the month ... the former would be more expensive, but could be of greater value because the information conveyed is more current."

Information yields utility to individuals, not telephone calls per se. Some calls may directly yield utility apart from producing information, and the standard demand analysis is sufficient for such calls. We will confine the discussion to calls which are inputs to the production of information. Utilizing a household production approach (see Becker, 1973)[6] to demand theory, we develop a model more consistent with the joint consumptive nature of telecommunications services.

Consider an economic agent with access to the telephone network. This agent is assumed to derive utility from information and a composite good. Information is produced via a home production function in which incoming and outgoing long distance calls (MTS) are productive inputs. This agent is assumed to choose MTS along with other goods to maximize utility. For simplicity, assume there is only one long distance route, with locations a and b as endpoints. The agent is assumed located at point a. The agent's optimization problem is:

$$\max_{[X^b, Q_{ba}]} U(X, I) \tag{1}$$

$$\text{subject to } I = f(Q_{ab}, Q_{ba})$$
$$\text{and } pX + qQ_{ab} = M.$$

Variables are defined as follows:

X: a composite good

Q_{ab}: MTS from point a to point b

Q_{ba}: incoming MTS from point b to point a

I: information produced as a function of Q_{ab} and Q_{ba}

M: income

p: price of composite good

q: price of MTS

The Lagrangean function for (1) is given by

$$L = U[X, f(Q_{ab}, Q_{ba})] - \lambda(M - pX - qQ_{ab}), \tag{2}$$

where λ is the Lagrangean multiplier.

The first order conditions[1] for (1) are:

$$U_X + \lambda p = O \tag{3}$$

$$U_I f_{Q_{ab}} + \lambda q = O, \tag{4}$$

$$pX + qQ_{ab} = M, \tag{5}$$

where subscripts denote partial derivatives.

These conditions imply that

$$\frac{U_X}{U_I f_{Q_{ab}}} = \frac{p}{q} \tag{6}$$

or, the marginal rate of substitution between X and Q_{ab} equals the price ratio. From (6), the general form of the demand equation for point-to-point MTS is of the form:

$$Q^{*}_{ab} = W(X, p, q\ M, Q_{ba}) \tag{7}$$

The presence of Q_{ba} distinguishes this model from others in the literature, and provides the basis for the empirical work in section 3.0.

2.1 Equilibrium

Equation (7) describes long distance demand originating at point a, as a function of incoming MTS from point b. MTS originating at point b and terminating at point a is determined in a similar manner. The agent at b faces the optimization problem:

$$\max_{[X^b, Q_{ba}]} U^b(X^b, I^b) \tag{8}$$

$$\text{subject to } I^b = g(Q_{ab}, Q_{ba})$$
$$\text{and } pX^b + qQ_{ba} = M^b,$$

where g is agent b's production function for information. The maximization in (8) yields the long-distance demand function

$$Q^{*}_{ba} = Z(X^b, p, q, M^b, Q_{ab}). \tag{9}$$

Equation (7) describes agent a's demand as a function of agent b's calls, while equation (9) describes agent b's calls as a function of agent a's calls. A Nash equilibrium is attained by the simultaneous solution of (7) and (9). Thus, estimation of long distance point-to-point demand requires simultaneous estimation of these two demand equations, with explicit accounting for the influence of return traffic on originating MTS.

2.2 Two Extreme Cases

There are two polar hypotheses regarding the functional relationship between originating and terminating MTS. This is to say,

$$Q_{ab} = h(Q_{ba}) \tag{10}$$

where the mathematical properties of function h depend upon how information is produced.[2]

The first hypothesis is that of reciprocity. Under the extreme form of reciprocity, terminating MTS must be returned in order to be productive. This describes a fixed coefficients production function for information with no substitutability between originating and terminating MTS, i.e.

$$I = f(\min(Q_{ab}, Q_{ba})) \text{ with } f' > 0 \text{ and } f'' < 0. \tag{11}$$

The customers optimization problem under reciprocity is represented in Figure 1.

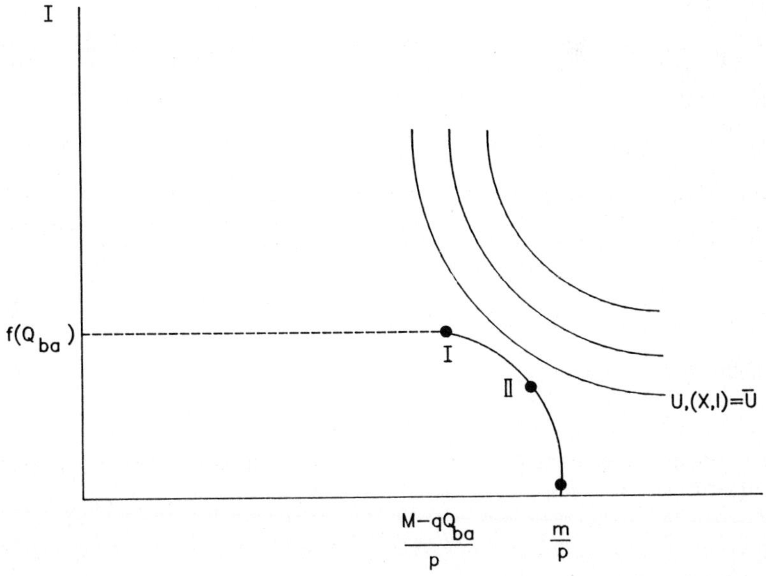

Figure 1: CONSUMER OPTIMIZATION UNDER RECIPROCITY

The indifference curves are drawn over information, I, and all other goods, X. The budget line illustrating combinations of X and Q_{ab} that can be purchased with income M is linear: the budget constraint drawn in Figure 1 is not. The latter shows the attainable set of (X,I) bundles, so the production function for information is embedded in this budget constraint. The concavity of the budget constraint results from the assumption of diminishing marginal returns in the production of information. Under reciprocity, however, purchases of MTS cease to be productive once outgoing MTS exceeds terminating MTS. When $Q_{ab} = Q_{ba}$, income of $M - q\,Q_{ba}$ remains for the purchase of $\frac{M - q\,Q_{ba}}{p}$ units of good X. So, the solid portion of the budget constraint applies for $Q_{ab} < Q_{ba}$, so that $Q^{*}_{ab} = \min(Q_{ab}, Q_{ba})$ and $I = f(Q_{ab})$. Point a results from setting $Q^{*}_{ab} = Q_{ba}$ with purchases of X equal to $(M - qQ_{ba})/p$. Further consumption of Q_{ab} would produce no additional information (accounting for the flat portion of the budget constraint), since $I = f(Q_{ba})$ for $Q_{ab} > Q_{ba}$. From Figure 1, it is clear that there are two subcases under reciprocity, depending upon whether Q^{*}_{ab} is greater or less than Q_{ba}:

(I) $Q^{*}_{ab} = Q_{ba}$, which implies that $\frac{\partial Q^{*}_{ab}}{\partial Q_{ba}} = 1$ (except for the marginal case where the individual moves from case I to case II) and

(II) $Q^{*}_{ab} < Q_{ba}$. Then $\frac{\partial Q^{*}_{ab}}{\partial Q_{ba}} = 0$. This is the least interesting case and will not be discussed here.

The more general form of reciprocity yields $h'(Q_{ba}) > 0$ in (10) with $h'(Q_{ba}) = 1$ for case I above.

The other polar extreme is the <u>information content hypothesis.</u> Under this hypothesis, originating and terminating MTS are perfect substitutes, so that $I = f(Q_{ab} + Q_{ba})$. In general, the more substitutable are Q_{ab} and Q_{ba} in the production of information, the more likely that Q_{ab} will vary inversely, so that $h'(Q_{ba}) < 0$ in (10). Since calls are made to provide information content, more incoming MTS allows for less outgoing MTS and more consumption of other goods.

MTS calling patterns probably support both the reciprocal and information content hypotheses.

3.0 MODEL SPECIFICAION

This paper makes the case that a proper specification of point-to-point MTS demand equations should include reverse traffic as an included endogenous variable in each respective structural equation. Two polar hypotheses regarding MTS calling patterns were discussed, those of reciprocity and information content. Under the reciprocity hypothesis, the coefficient on the reverse traffic variable should be positive. Under the information content hypothesis, the coefficient on the reverse traffic variable should be negative.

3.1 On Market Size

In examining the demand for point-to-point long distance, the concept of market size is extremely important. This is so because the larger the network, the larger the aggregate market community of interest and hence, the more calls we would expect. It is also true because there are consumption externalities involved in the usage of point-to-point long distance.

Previous models of point-to-point long distance have used various measures of market size. A measure of market size should measure the number of subscribers who comprise the meaningful community of interest, i.e., those who make and/or receive calls between the two points defining a route. Pacey (1983) used population as this market size variable. This type of measure is quite common and is logical, since the greater the population in each point, the more likely is the event of point-to-point communication over the route. We have chosen to augment the Pacey market size variable by using the product of the populations at each point. This variable measures the number of likely connections there are between the points of a given route. Regardless of which variable is used, the order of magnitude of the coefficient on market size is an empirical question.

The statistics in Table 1 suggest that conventional measures of market size at least have the potential to be of little value in capturing the true community of interest for specific routes. Of note here is the fact that the traffic ratios lie in the proximity of .90 for the majority of the nine routes while the population ratios range from a low of 1.09 to a high of 36.96. This suggests a core community of interest in each point that is actually a very small subset of the population. Thus, market size variables should be used with care and the assumptions necessary for their use should be known and acknowledged by the econometrician.

Statistical analysis and matching of the calling patterns of calls from City A to City B at the subscriber level show that fairly high proportions of such message volumes terminate with subscribers at point B who in turn placed calls to subscribers in City A (see Table 2). Thus, the phenomenon of call reciprocity takes place for a large percentage of route-specific calls. The theory exposed in Section 2 dictates that return calling patterns should be included in demand models. Table 2 shows that these return calling patterns are quite pervasive.

TABLE 1 - INTRALATA MTS POINT-TO-POINT DATA COMPUTED FROM 1983: III DATA		
Route No.	Traffic Ratio*	Population Ratio**
1	0.89	6.28
2	0.89	7.47
3	0.90	4.78
4	0.90	7.54
5	0.89	12.45
6	0.89	36.96
7	1.00	1.36
8	0.83	24.96
9	0.87	20.59
Sample Mean	.90	13.60
Sample S.D.	.045	11.62
Sample CV	.050	.854

* Ratio of City A originating DDD MTS minutes traffic to City B DDD MTS minutes originating traffic. Source: Centralized Message Data System (CMDS) 5% MTS traffic sample

** Ratio of City A population to City B population.

TABLE 2 - ANALYSIS OF CALLBACK PATTERNS
DDD INTRALATA MTS CALLS
CALLBACK RATIOS*

Route	Direction	Bus.-Orig.	Res.-Orig.	All
1	A to B	.38	.54	.49
1	B to A	.33	.43	.41
2	A to B	.35	.54	.46
2	B to A	.31	.39	.36
3	A to B	.34	.44	.39
3	B to A	.27	.35	.31
4	A to B	.31	.46	.38
4	B to A	.27	.35	.31
5	A to B	.39	.46	.43
5	B to A	.38	.43	.42
6	A to B	.39	.56	.50
6	B to A	.34	.43	.41
7	A to B	.42	.56	.51
7	B to A	.35	.45	.43
8	A to B	.27	.46	.38
8	B to A	.28	.37	.34
9	A to B	.30	.49	.42
9	B to A	.27	.38	.35

* A callback ratio is defined as the proportion of MTS minutes from A to B (B to A) placed to subscribers who placed at least one call from B to A (A to B).

Source: June 1985 Customer Record Information System (CRIS) data

4.0 ECONOMETRIC ESTIMATION

The parameters of the two-equation system shown in Section 3.1 were estimated with two-stage least squares (2SLS) using pooled cross-section and time series data from 9 individual city pairs.[3] The equations are as follows:

$$\ln Q_{12t} = \Sigma \kappa_i D_{it} + \sum_{i=0}^{k} \pi_i \ln P_{t-i} + \sum_{i=0}^{m} \tau_i \ln Y_{1\,t-i} + \gamma \ln POP_t + \beta \ln Q_{21t} + U_{it}\,, \tag{12}$$

$$\ln Q_{21t} = \sum W_i D_{it} + \sum_{i=0}^{n} \lambda_i \ln P_{t-i} + \sum_{i=0}^{q} \xi_i \ln Y_{2\,t-i} + \theta \ln POP_t + \delta \ln Q_{12t} + U_{2t}. \tag{13}$$

Variables are defined as follows:

Q_{12t} : Total minutes of intraLATA long distance usage billed to Bell residence subscribers. This usage is for direct-distance-dialed (DDD) calls only, aggregated over all rate periods, that originate from point 1 and terminate in point 2 at time t. Variable Q_{21t} is defined similarly and represents reverse traffic that originates from point 2 and terminates at point 1 at time t. Source: CMDS

P_t : Fixed-weight Laspeyres price index for the type of long distance phone call described above. Expressed as average revenue per minute and deflated by state CPI or SMSA CPI, where appropriate. Fixed quantity base is October 1981. CPI data used to express price and income variables in real terms are from the Bureau of Labor Statistics.

Y_{it} : Real per capita personal income for point of origination i, (i=1,2). Derived from county-level wages and salaries disbursements data. Source: Bureau of Labor Statistics

POP_t : Market size variable computed by multiplying the population of City A by the population of City B. Source: Bureau of Economic Analysis

D_{it} : Route-specific intercept term, i=1,2,...,9 .

Structural equations are in logarithmic form and the price and income terms are dynamic.

The nature of the error terms of each structural model is of particular importance if the estimators are to be efficient. The following error structure is assumed.

Let U_{1it} be the nm X 1 vector of disturbances from the first structural equation for cross-section i and time period t, (i =1,2,....,m and t =1,2,...,n).

We assume that:

$$E\,(U_{1it}\,U'_{1jt}) = \begin{cases} \sigma_i^2 W_i,\ i=j \\ \sigma_{ij}\,\mathrm{I},\ i \neq j, \end{cases}$$

where

$$W_i = \begin{bmatrix} 1 & \rho_i & \rho_1^2 & \cdots & \rho_i^{n-1} \\ \rho_i & 1 & & & . \\ . & & . & & . \\ . & & & . & . \\ . & & & . & . \\ . & & & & \rho_i \\ \rho_i^{n-1} & . & . & . \ \rho_i & 1 \end{bmatrix}$$

Thus, we assume that disturbances are serially correlated with an AR(1) process in a given cross-section. We also assume that disturbances are cross-sectionally heteroskedastic and contemporaneously correlated across cross-sections. Assumptions on disturbances in the second structural equation are defined analogously. In addition, one final assumption must be made regarding disturbances, i.e.,

$$E(U_{1it}U'_{2it}) = O.$$

Thus, it is assumed that disturbances across structural equations are not correlated.

Derivation of 2SLS estimates from the pooled time series and cross-section sample was not computationally difficult. The standard fixed effects least squares with dummy variables (LSDV) model was used to estimate reduced forms for each equation. Fitted values were then computed for each vector of endogenous variables. These fitted values were then used as regressors in the second stage of estimation. Direct 2SLS estimates were obtained by estimating each structural equation, using the fitted values for the included endogenous variables. These estimates are posted in Section 5 in Tables 3 and 4.

Dynamic effects of price and income were estimated using Almon polynomially-distributed lags. In each case, first degree polynomials were used with the furthermost lag coefficients constrained to equal zero.

Several tests were performed to validate the statistical properties of the model. The structural equations were tested for autocorrelation using a methodology developed by Harvey and Phillips (1981). The presence of cross-sectional heteroskedasticity was tested using a likelihood ratio test described in Fomby, Hill and Johnson (1984).[7] This test did not require full maximum likelihood estimates of parameters, though at the expense of some power. Normality of disturbances from the structural equations was tested using the Shapiro-Wilk statistic. The validity of the linear restrictions imposed by pooling the data (given the double-logarithmic functional form) was tested using the weaker mean square error criteria suggested by Wallace (1972).[8] In addition, the double-logarithmic functional form was chosen using the Box-Cox transformation and simple iterative GLS. Four functional forms were evaluated (linear, semi-logarithmic, inverse semi-logarithmic and double-logarithmic) using the appropriate likelihood function.[4]

5.0 EMPIRICAL RESULTS

Empirical results for both structural equations are shown in Tables 3 and 4. The appropriate interpretations are shown in Table 5. Empirical results for both equations estimated with no exogenous variables included in the model are shown in Table 6. Table 7 yields a comparison of long-run price and income elasticities as estimated from a simultaneous system and from a system of independent equations.

Results in Table 5 show price elasticities that are quite a bit higher than those shown in many other empirical studies of short-haul toll. Elasticities, while still less than unity, are still far higher than would be suggested by conventional econometric models that do not include reverse MTS traffic flows as a regressor. Also of interest are the long run income elasticitics for each structural model. These results are similar to those of Pacey (1983) but different from those of Griffin (1982). They indicate, quite plausibly, that point-to-point residence DDD long distance is a normal good.

Another notable result is the fact that the coefficients on reverse traffic are positive and significant, providing support for dominance of the reciprocity hypothesis. This suggests that the origination of MTS usage induces additional MTS usage. These coefficients are fairly close for each structural equation. The fact that they are positive is important to a proper analysis of consumer welfare. It suggests that price changes affect traffic flows in each direction and the reverse traffic flows in a given direction affect MTS volumes further still. This flow - through effect is reflected in reported elasticities in the left-hand column of Table 5.

Estimation of the structural equation for City A to City B showed that the conventional market size variable had both a small and an insignificant effect on minutes volumes. While such a result is at first puzzling, it is feasible. If, for this direction of MTS traffic, the growth in market size caused increased volumes of traffic from City B to City A, then the reverse traffic term may be all that is necesary, along with price and income, to explain traffic from City A to City B.

Because this simultaneous system differs markedly from conventional models, a comparison of empirical results derived from a model which does not include return traffic as an endogenous regressor may be useful. Table 7 shows this comparison.

The results summarized in Table 7 do not show marked differences in price and income elasticities produced by each set of models. The price elasticities derived from the simultaneous equations system are similar to those produced by the conventional model specification but they still yield strong implications for the proper economic analysis of proposed pricing policies. The structural equations seem to indicate that price effects alone, with no flow-through effects incorporated, are actually quite inelastic. This can be seen by examining the simple price elasticities suggested by the results in Table 3 and Table 4, -.18 and -.26, respectively. The full multiplier effect, of course, occurs when price changes affect unidirectional MTS volumes, which then complete the full effect via reciprocity. This suggests that when conventional model specifications (which do not include return traffic) are used, the price coefficient is biased because it includes both price effects and resulting flow-through effects. All other coefficients would be biased as well.

TABLE 3 - EMPIRICAL RESULTS

CITY A TO CITY B

	ln P	ln Y	ln Q_{21}
	$\sum_i \pi_i = -.18$	$\sum_i \tau_i = .27$	.75
	(3.3)	(3.3)	(15.6)
where:			
	$\pi_0 = -.07$	$\tau_0 = .08$	
	$\pi_1 = -.05$	$\tau_1 = .06$	
	$\pi_2 = -.04$	$\tau_2 = .05$	
	$\pi_3 = -.02$	$\tau_3 = .04$	
		$\tau_4 = .03$	
		$\tau_5 = .01$	
	n = 243		D.W. = 1.90

TABLE 4 - EMPIRICAL RESULTS
CITY B TO CITY A

ln P	ln Y	ln POP	ln Q_{12}
$\sum_i \lambda_i = -.26$	$\sum_i \xi_i = .23$	.30	.67
(3.5)	(2.0)	(2.4)	(7.2)

where:

$\lambda_0 = -.11$	$\xi_0 = .07$
$\lambda_1 = -.08$	$\xi_1 = .06$
$\lambda_2 = -.05$	$\xi_2 = .04$
$\lambda_3 = -.03$	$\xi_3 = .03$
	$\xi_4 = .02$
	$\xi_5 = .01$

n = 243 D.W. = 1.78

All summary statistics refer to the transformed model. Student's t-statistics are given in parentheses. Models were corrected for autocorrelation, heteroskedasticity and mutual correlation.

TABLE 5 - SUMMARY OF LONG-RUN PRICE AND INCOME ELASTICITIES*

	PRICE	INCOME
City A to City B	-.75	.54
City B to City A	-.76	.46

* Reported elasticities are for price (income) effects in conjunction with the resulting flow-through effect on reverse traffic.

TABLE 6 - EMPIRICAL RESULTS
NO INCLUDED ENDOGENOUS VARIABLES
IN MODEL SPECIFICATION

CITY A TO CITY B

ln P	ln Y	ln POP
-.82	.40	.66
(26.2)	(4.7)	(9.2)
n = 243	$\bar{R}^2$ = .99	D.W. = 2.12

CITY B TO CITY A

ln P	ln Y	ln POP
-.73	.62	.68
(17.4)	(4.0)	(4.9)
n = 243	$\bar{R}^2$ = .99	D.W. = 1.93

All summary statistics refer to the transformed model. Student's t-statistics are given in parentheses. Models were corrected for autocorrelation, heteroskedasticity and mutual correlation.

NOTE: Only long-run coefficients are reported in this table. All price and income variables have the same dynamic specification as listed in Tables 3 and 4.

TABLE 7 - SUMMARY OF LONG-RUN PRICE AND INCOME ELASTICITIES*

CITY A TO CITY B

	MODEL SPECIFICATION	
	Incl. Return Traffic	**Excl. Return Traffic**
Price	-.75	-.82
Income	.54	.40

CITY B TO CITY A

	MODEL SPECIFICATION	
	Incl. Return Traffic	**Excl. Return Traffic**
Price	-.76	-.73
Income	.46	.62

* For model specifications which include reverse traffic, reported elasticities are for price (income) effects in conjunction with the resulting flow-through effect on reverse traffic.

6.0 WELFARE ANALYSIS

The theoretical analysis predicts that changes in Q_{ba} shift the demand curve for long-distance. Consider the demand curves shown in Figure 2.

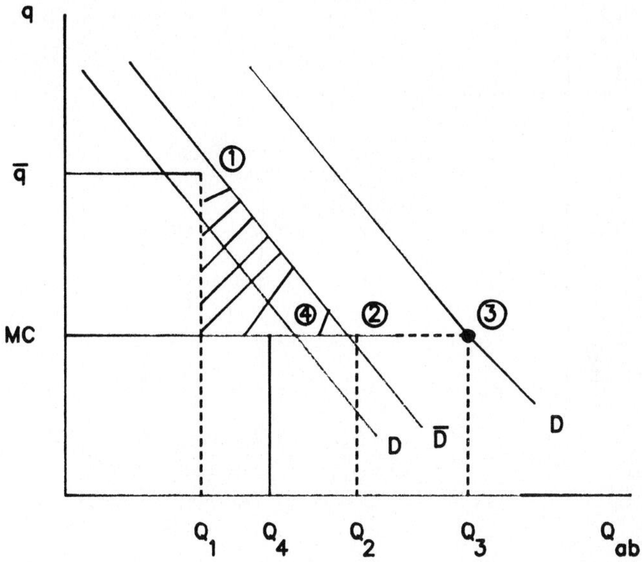

FIGURE 2: THE EFFECT OF CHANGES IN Q_{ba} ON Q_{ab}

$\bar{D}$ is the demand curve (7) with $Q_{ba} = \bar{Q}_{ba}$. The current long-distance regulated price, $\bar{q}$, yields point 1 on $\bar{D}$. Consider lowering price to the marginal cost, indicated by MC. If $Q_{ba} = \bar{Q}_{ba}$, then quantity demanded will be Q_2. But, under reciprocity (i), when q decreases $\bar{Q}_{ba}$ increases (say to $\hat{Q}_{ba}$). If reciprocity holds, then $\bar{D}$ will increase to $\hat{D}$, where $Q_{ba} = \hat{Q}_{ba} > \bar{Q}_{ba}$, and the quantity demanded will be at $Q_3 > Q_2$. Under the information content hypothesis (ii), $\bar{D}$ will decrease to D, and the quantity demanded will be at $Q_4 < Q_2$ (it is possible that $Q_4 < Q_1$). Depending upon which hypothesis holds, points 1 and 3 or 1 and 4 can be thought of as lying on an effective demand curve which illustrates the direct effect of a change in price on quantity, as well as the indirect effect of price on incoming calls, and the resulting impact on quantity demanded. The welfare gains from marginal cost pricing equals the shaded triangle, but this should be measured along D or $\hat{D}$, rather than $\bar{D}$, since Q_{ba} will change. This area will yield the direct effect of the price change on welfare due to a price decrease.[5] The total welfare change will consist of the increased utility due to the lower price and the change in utility due to the change in incoming MTS.

A continuous version of this is shown in Figure 3, where for each $q_i < \bar{q}$, account is made of the change in Q_{ba} and its effect on Q_{ab} as well as the direct effect of q_i on Q_{ab}.

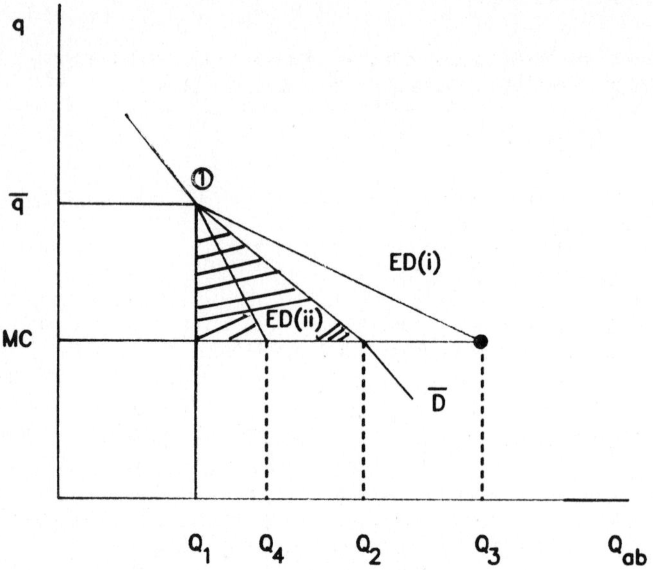

FIGURE 3: EFFECTIVE DEMAND CURVES

Since small (large) price decreases cause small (large) changes in Q_{ba}, the effective demand curve will pivot from point 1 on D. ED(i) and ED(ii) represent the effective demand curves corresponding to hypotheses I and II, respectively.

Our story can be summarized by paraphrasing Taylor's (1980, pg. 159) conclusion regarding the indirect effect of price on usage via induced effects on access decisions:

"The moral of the story is as follows: If the focus is entirely on estimating the effects on usage of changes in the price of usage, an access/use framework is not critical - so long, that is, that usage is assured to be conditional on the number of main stations - since any impact of the price change on the number of main stations is probably small enough that it can be ignored. However, if the focus is on estimating the impact on usage of changes in income, an access/use framework is critical, because to ignore the indirect effect on usage that arised through the adjustment in main stations is likely to lead, in most cases, to a serious underestimate of the total effect of a change in income."

In our case, price changes will yield significant effects upon both initiated and incoming usage, and the latter significantly affects the former. The empirical results of Section 5.0 support this. Failure to account for the direct effect of price on usage as well as the indirect effect of price on usage through increased return traffic will lead to a serious bias in the effect of a change in price on usage.

7.0 FUTURE DIRECTIONS AND CONCLUSION

The theory of point-to-point toll demand should prove to be a fruitful area of research. The advent of competition in both interexchange and intraexchange telecommunications markets will require more disaggregated types of demand analysis techniques as competitors attempt to establish optimal pricing policies.

This paper has argued that previous research on point-to-point toll demand may not have analyzed the problem correctly. The omission of a reverse traffic regressor in the estimated demand functions may have led to a bias in the resultant price elasticities. Should further empirical research confirm this hypothesis, it could have major policy implications ranging from optimal pricing of measured service to intraLATA toll competition and to bypass and beyond. In addition, the inclusion of a reverse traffic regressor reveals possible reciprocal calling patterns. This may suggest that a prime determinant of the number of originating calls is not only price, but the number of calls received at some point in the recent past. This reciprocity hypothesis confirms earlier intuition that "calls propagate calls."

There is considerable future research to be done on point-to-point analysis before any firm policy conclusions can be drawn. Estimation methods using subscriber-specific survey data and the effects of competition need further investigation. And of course, more research on the appropriate market size variable, both theoretical and empirical, would yield extremely valuable contributions to demand analysis of telecommunications services in general. Until such research is conducted, the policy statements made in this paper should be considered the basis of an evolving body of knowledge concerning the demand for point-to-point telecommunications.

FOOTNOTES

1. The sign of $\frac{\partial Q_b^*}{\partial Q_{ba}}$ can be derived by total differentiation of the first order conditions (3), (4) and (5) with respect to Q_{ba} , and use of the implicit function theorem and Cramer's Rule, to yield:

$$\text{sign}\ \frac{\partial Q_{ab}^*}{\partial Q_{ba}} = \text{sign}\left\{ -qU_{XI}\, f_{Q_{ba}} + \eta U_I f_{Q_{ab}} Q_{ba}\ [1 - E(I)]\right\}$$

where η is the elasticity of substitution between Q_{ab} and Q_{ba} in the production of information, and $E(I)$ is the elasticity of marginal utility of information. Thus, the properties of function h depend upon the technology of information production and upon the form of individual preferences.

2. See footnote 1 and let η get large.

3. High density IntraLATA MTS routes were chosen for initial inclusion in the study based on relatively high traffic volumes. Those included in the pooled sample were chosen so as to produce homogeneous cross-sections which would pass the tests of linear restrictions imposed by the LSDV model. All routes consist of a large metropolitan area and a relatively small suburb or town. The phrase "City A" always refers to the metropolitan point of the route, e.g., in Table 2 the route direction "A to B" always means traffic volumes are from a large metropolitan area to a small population center.

4. The authors acknowledge that such a methodology is useful but not technically correct. While reduced forms have only exogenous variables as regressors, making this procedure viable, the strutural equations have one included endogenous variable each. This suggests that limited information maximum likelihood (LIML) would be required to test structural equations simultaneously for functional form and violations of classical assumptions. The computational difficulty and high cost of such a procedure would soon outweigh its benefits.

5. Subject to the Willig (1976) conditions insuring that the compensated and ordinary demand curves are not significantly different over this range of prices.

REFERENCES

[1] Pacey, Patricia L., "Long Distance Demand: A Point-to-Point Model," Southern Economic Journal , April 1983, pp. 1094-1107.

[2] Berg, Sanford V., and Patricia L. Pacey, "Impact of Deregulation on Point-to-Point Demand in the U.S.A.", Telecommunications Policy Dec. 1982.

[3] Larsen, W.A. and S. J. McCleary, "Exploratory Attempts to Model Point-to-Point Cross-Sectional Interstate Telephone Demand." Unpublished Bell Laboratories Memorandum, July 1970.

[4] Deschamps, P.J., "The Demand for Telephone Calls in Belgium, 1961-1969," paper presented at the Birmingham International Conference in Telecommunications Economics, Birmingham, England, May 1974.

[5] Varian, Hal R., Microeconomic Analysis , New York: W. W. Morton S. Company, 1984

[6] Taylor, L.D., Telephone Demand: A Survey and Critique , Cambridge, Mass.: Ballinger Publishing Co. 1980

[7] Becker, G.S., (1973) "On the New Theory of Consumer Behavior." Swedish Journal of Economics , Volume 75, pp. 378-395.

[8] Fomby, Thomas B., R. Carter Hill and Stanley R. Johnson, Advanced Econometric Methods , New York: Springer-Verlag, 1984, pp. 188-189

[9] Wallace, T.D. (1972) "Weaker Criteria and Tests for Linear Restrictions in Regression," Econometrica , 40, pp. 689-698.

[10] Griffin, James M., "The Welfare Implications of Externalities and Price Elasticities for Telecommunications Pricing," Review of Economics and Statistics , February 1982, pp. 59-66.

[11] Willig, "Consumer's Surplus Without Apology," American Economic Review , 66, 4, pp. 589-597, 1976.

TELECOMMUNICATIONS DEMAND MODELLING
An Integrated View
A. de Fontenay, M.H. Shugard, D.S. Sibley (Editors)

Relation of conversation time to message frequency in residential usage

J. G. Veitch

AT&T Bell Laboratories
Murray Hill, NJ

ABSTRACT

This paper models the relation between two common demand quantities in residential telephone usage, *message frequency,* and total time of calling (called *total conversation length* or *total minutes* in this paper). This relation is modeled at the individual customer level, so the results apply whenever we have customer level data or a customer level model relating to these demand quantities. For example, given a change in the rate structure, one may have a model for change in message frequency, which one wishes to use to predict change in total minutes. A second application arises in simulation. Given a marginal distribution of message frequency across a population, one can apply the model to simulate the joint distribution of message frequency and total minutes, and find a marginal distribution of total minutes. This paper does not attempt to model either the marginal distribution of message frequency alone, or the marginal distribution of total minutes alone.

We model the relation between total minutes and message frequency as a conditional distribution of total minutes given message frequency. More precisely, suppose we observe total conversation length and number of messages made for some sample of residential customers in a fixed time period. Using this data, we estimate a customer specific distribution of total minutes conditional on the message frequency of this specific customer. The conditional distribution is given by the sum of independent log-normals parametrized by a customer specific average and variance, where each term in the sum is a *holding time* (i.e. the message length or call length). These parameters are estimated by a Bayes procedure based on a parametric approximation to the empirical distribution of expected holding time per customer (over the population) and another parametric approximation to the distribution of holding time per message. Some simple formulas for finding these parameters in a specific population are given.

This procedure is tested against several other possible procedures and does the best in minimizing the absolute error of prediction, especially in the higher usage segments. Of the other procedures that empirical Bayes is tested against, the next best is to estimate a customer's total minutes by multiplying his message frequency by a holding time calculated as the mean of all individual observed average holding times. The worst of these procedures is to estimate a customer's total minutes from his message frequency by multiplying this message frequency by his average holding time observed in some other time period.

1. Introduction

Different types of long distance usage, such as operator handled, direct distance dialing (DDD) during day rates, etc., in the residential sector are often summarized by one of two different measures: by the number of messages made (or message frequency), or by total conversation time. It is clear that these two measures are related; for example, the more messages a user initiates, the greater his total time is likely to be. Sometimes, even though one wishes to be know both quantities, it is simpler to only have to deal with one of these demand quantities.

For example, one might wish to model the long distance minute and message usage for some telecommunications scenario (like a tariff change). For costing purposes one needs to know both number of calls and total time, as there is an initial setup cost when a call is initiated and there is a variable cost dependent on the *holding time* or *call length*. Hence in this case one would like to predict change in terms of minutes given some predicted change in terms of message frequency alone, rather than building two models, one for changes in message frequency, and one for changes in total minutes.

In this paper we model the bivariate relation between total minutes and message frequency by finding the conditional distribution of total minutes given message frequency. We shall assume during the sequel we observe usage over a unit time period, so message frequency is just the number of messages. We observe only night-weekend DDD residential service, but there is no reason to believe the methods developed in this paper will not apply to other types of service. We derive this model by considering individual customer level data. To find the full bivariate distribution, one must know either the marginal distribution for message frequency or the marginal distribution for total minutes. We make no attempt to model the marginal distributions (i.e. the distribution of total minutes alone or the distribution of message frequency alone). Such work on modeling the marginal distributions can be found in, for example, Pavarini [1] or Kearns [2].

Bosch [3], has attacked a similar problem to the one we are interested in here. Bosch models the bivariate relation by assuming that a power transform (in the case of business calling) or a log transform (in the case of residence calling) of both variables, mean holding time and message frequency, are bivariate normal, with correlation coefficient to be estimated from data. Bosch also requires that the data be at the individual customer level. Bosch's approach has the advantage that it is as easy to find conditional distributions for message frequency given total minutes as it is to find the conditional distribution of total minutes given message frequency. This is not true for our approach. Bosch's approach has a disadvantage in that the assumption of bivariate normality may not always hold. It certainly fails for the residential data we have (it is clear that the points in the scatter plot in Figure 1, where we plot mean holding time versus message frequency on a log-log scale, are far from bivariate normal).

We begin by describing the dataset in Secton II. We then give short summary of the empirical Bayes method of estimation in Section III. In Section IV we empirically justify a model for total time given n messages from a particular customer as the sum of n independent identically distributed log-normals. We also empirically justify that expected holding times for our set of customers are approximately log-normal and are independent of message frequency (or number of messages, as we assume unit time). In Section V we discuss some shortcomings of this model.

In Section VI, we use this model to find a customer specific conditional distribution for total conversation time given message frequency and given observed data. We develop Bayes estimates for customer specific expected holding times. These estimates parametrize the log-normals comprising the sum of individual holding times. The Bayes estimate is based on the heuristic distribution for expected holding times across all customers and the heuristic distribution for holding time per message found in section IV. In Section VII, we implement the Bayes procedure for data in one time period to estimate customer specific expected call length and use this to predict total minutes given the message frequency in a second, separate period. We then test the Bayes procedure by comparing the predicted total minutes for the second period against the actual, observed total minutes in this period. The errors seen were compared against several other methods for predicting total minutes and we found that the error from the Bayes procedure was uniformly smallest. Of the other procedures that the Bayes procedure is tested against, the next best is to estimate total minutes by multiplying the observed number of messages by the population wide mean holding time. The worst of these procedures is to estimate minutes from messages by multiplying each individual's observed number of messages in the second time period by his observed average holding time in the first period.

2. The dataset

This dataset consists of 3490 residential customers, close to a census from one exchange in Los Angeles, California. Each customer has monthly aggregate interstate usage statistics for at least 9 out of 11 months, February to December 1983. Since customers who stop service may have their number reassigned to a new customer after a few months, and the data was identified by phone number, phone numbers with 3 or more months of missing data were not used. The apparent number of residential customers showed month to month fluctuations from a low of around 3500 to a high of 5000, but most months had around 4200 or so. This "missing data" may bias the results to an unknown extent; however it is reasonable to believe that the results for the 3490 are fairly representative.

The monthly statistics included total mileage called, number of messages, and total time called, broken down by rate period and type of service (this last to a lesser extent). All this data is only for interstate toll calling. We shall only look at night-weekend DDD usage in detail; henceforth, unless otherwise explicitly stated, assume "usage" refers to usage in this category only. Before looking at the relation between total (long-term) usage at the individual level and average length of message, we briefly discuss empirical Bayes techniques. For a more general presentation of these techniques, see Ledermann, Chapter 15 [4].

3. Empirical Bayes techniques

We motivate the use of empirical Bayes techniques in the context of our setting. Say we observe an individual (denoted by i), who makes one message of length 100 minutes, and say that we estimate in some second period that this individual is likely to make ten messages. How should we estimate total time usage for this individual in the second period? Prior belief about typical holding times for any customer suggest estimating total minutes as ten messages times 100 minutes per message, or 1000 minutes may be unreasonably large, even though 100 minutes in a given message may not be unusual. The Bayes procedure gives us a way to combine our prior belief about typical holding times with this particular observation of an average holding time of 100 minutes per message.

Our basic approach is to model customer i's holding time for his jth message as a random variable (say $L_{i,j}$). We shall assume $L_{i,j}$ are independent, identically distributed random variables specific to customer i (the evidence for this assumption is discussed in the next section). Note that this implies that customer i has some expected holding time, namely $E[L_{i,1}]$. This expected holding time is unobservable and must be estimated. His total time for the second period is $\sum_{j=1}^{10} L_{i,j}$, i.e. the sum of the ten holding times for each message. Hence we might estimate customer i's total minutes in this second period by the number of messages multiplied by his estimated expected holding time. Conversely, note that if we had an estimate for customer i's total minutes in the second period and wanted to estimate number of messages, it would be reasonable to divide the time by estimated expected holding time. So we now need a way to estimate this customer's expected holding time.

Assume now that we know the distribution of expected holding times among the population of all individuals. If we make a random choice of individual from the population, we generate an associated random variable (r.v.), namely the expected holding time for the chosen individual. Call this r.v. Λ, with probability density $g(y)$. This r.v. can be thought of as representing our prior beliefs about what a customer's expected holding time might be in the absence of any other information about this customer. Then if the density of observed average holding time based on n messages, (denoted by $\bar{L}$), given $\Lambda = y$ is $f(x \mid y)$, we can represent the distribution of Λ given an observed $\bar{L}$ (denoted as $\Lambda \mid \bar{L}$) by

$$h(y\,|\,x) = \frac{f(x\,|\,y).g(y)}{\int f(x\,|\,y).g(y)\mathrm{d}y} \tag{1}$$

This new r.v., $\Lambda|\bar{L}$ implicitly depends on n, and we shall just suppress the n. $\Lambda|\bar{L}$ represents the way observing an average holding time based on n messages modifies our belief about what this customer's expected holding time might be, so we might estimate our customer's expected holding time by the conditional mean

$$E(\Lambda\,|\,\bar{L}) = \int y.h(y\,|\,\bar{L})\mathrm{d}y \tag{2}$$

$E[\Lambda|\bar{L}]$ will be closer to the population average than $\bar{L}$ alone (i.e. the observed average holding time). Hence in the example above, if E(Λ) is smaller than 100 minutes, one would estimate total time usage as something smaller than 1000 minutes. This estimation technique is an example of Bayes estimation. We shall use exploratory data analysis to derive appropriate forms for f, g, and h, so the technique is known as empirical Bayes.

Suppose we represent our observed data on each individual by a r.v. X, (corresponding to $\bar{L}$ in our example above), and our prior belief about the population by the r.v. Y (corresponding to Λ in our example above). If $X|Y$ is normal, mean Y, and variance independent of Y, say $\sigma^2(X)$, and Y is also normal, variance $\sigma^2(Y)$, then $Y\,|X$ is normal and (2) reduces to the weighted average:

$$E[Y|X] = X\left[\frac{\sigma^2(Y)}{\sigma^2(X)+\sigma^2(Y)}\right] + E[Y]\left[\frac{\sigma^2(X)}{\sigma^2(X)+\sigma^2(Y)}\right] \tag{3}$$

However, in our case we need to transform both our data and our "prior belief": log(Λ) is approximately normal, and the observed average of the log individual holding times is also approximately normal. Hence, to use (3), we need to log transform the original variables and then, in order to understand the results we get, we need to relate our results on the logged variables back to results on the original variables. We do this in Section VI.

4. Justification of the model

Figure 1 is a log-log plot of average call length for each customer versus his total messages in Feb.-Dec. 1983. This plot suggests that overall average call length is essentially unaffected by number of messages and that the variance of observed average call length across individuals is a decreasing function of the total number of messages made. This suggests a simple model hypothesizing independence of holding time and number of messages.

Let Λ_i be individual i's expected holding time. Again, if we pick an individual at random from the population, then his expected holding time is a random variable (r.v.), denoted by Λ. Suppose that we model the length of the jth phone call of individual i by

$$L_{i,j} = (1+\varepsilon_j).\Lambda_i \tag{4}$$

where the ε_j are independent and identically distributed (i.i.d.) mean zero random variables, and $(1+\varepsilon_j)$ is assumed positive. We shall generically denote the distribution of message length for individual i by the random variable (r.v.) L_i, and we shall denote the result of randomly choosing a customer and selecting a particular call length by the r.v. L. Note we are assuming that message lengths for individual i (the r.v.'s L_i) are i.i.d. and have variance independent of i. Then, if we consider the population of customer expectations, Λ_i, $i = 1, \cdots, N$ for the N customers, we should firstly see that conditioning on n messages, and calculating expectations across both ε_j and the customer population:

$$E(\bar{L} \mid n \text{ messages}) = \lambda \tag{5}$$

where λ is the unconditional population expectation, E[Λ], and $\bar{L}$ is the average length of n independent calls. After some straightforward calculation, we should secondly see:

$$\text{var}(\bar{L} \mid n \text{ messages}) = \sigma_\Lambda^2 + E(\Lambda^2).\ \frac{\sigma_\varepsilon^2}{n} \tag{6}$$

where $\sigma_\Lambda^2 = \text{var}(\Lambda)$ and $\sigma_\varepsilon^2 = \text{var}(\varepsilon_j)$. Hence conditional variance is hyperbolic in n.

Figure 1

Mean call length plotted by frequency of calls in the Control exchange 213466

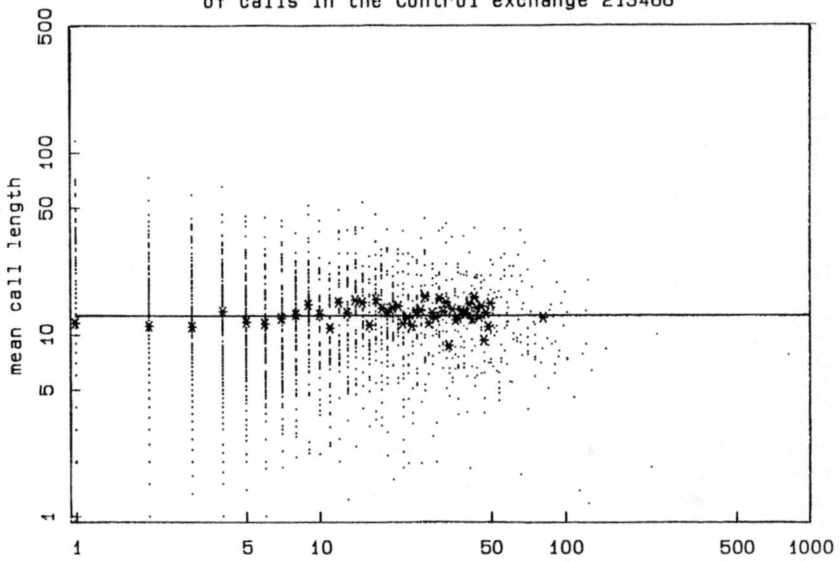

Total messages in 1983
Added line is the overall mean of all messages

These predictions look like good approximations to the observed data. Figure 1 plots observed average call length versus number of messages made for the 3490 individuals in the dataset. The stars are the observed conditional mean times for total usage from 1 to 50 messages. Above 50 messages it is difficult to calculate stable estimates for the conditional means because for any particular number of messages there are relatively few or no individuals making that many calls. We just aggregated all such individuals together; their mean number of messages was about 82, and their average aggregate message length is plotted at 82 messages. The horizontal line is the global mean (12.40 mins), which should approximate λ very closely with this much data. This plot of conditional means of average call length is repeated in Figure 2, where each conditional mean has a 99% Confidence Interval (C.I.) surrounding it. We used the variance estimate from (7) below, rather than individual variance estimates for each call total; using the individual variance estimates would tend to broaden most of the C.I.'s because many of the groups are small. Notice that the global mean lies within every such 99% C.I.

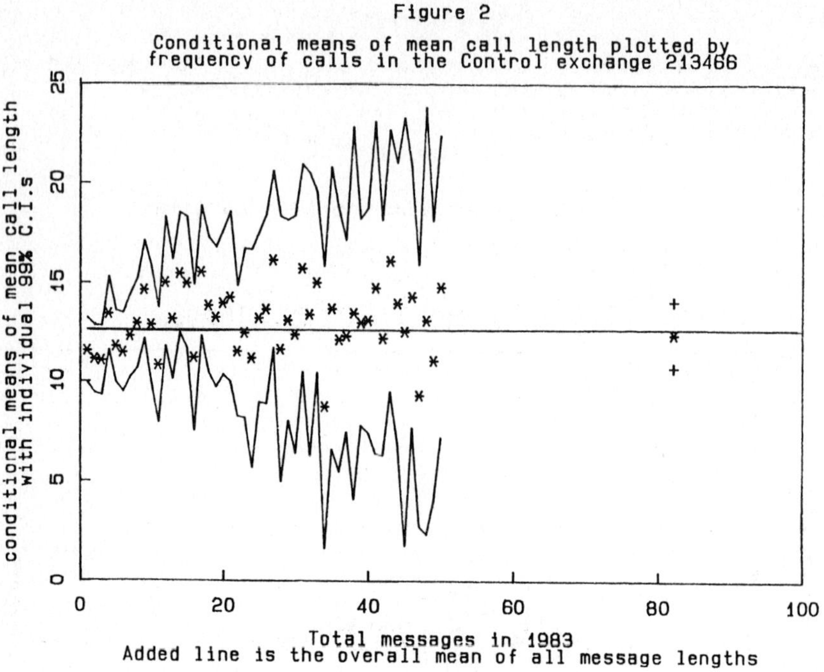
Figure 2
Conditional means of mean call length plotted by frequency of calls in the Control exchange 213466
conditional means of mean call length with individual 99% C.I.s
0
5
10
15
20
25
0
20
40
60
80
100
Total messages in 1983
Added line is the overall mean of all message lengths

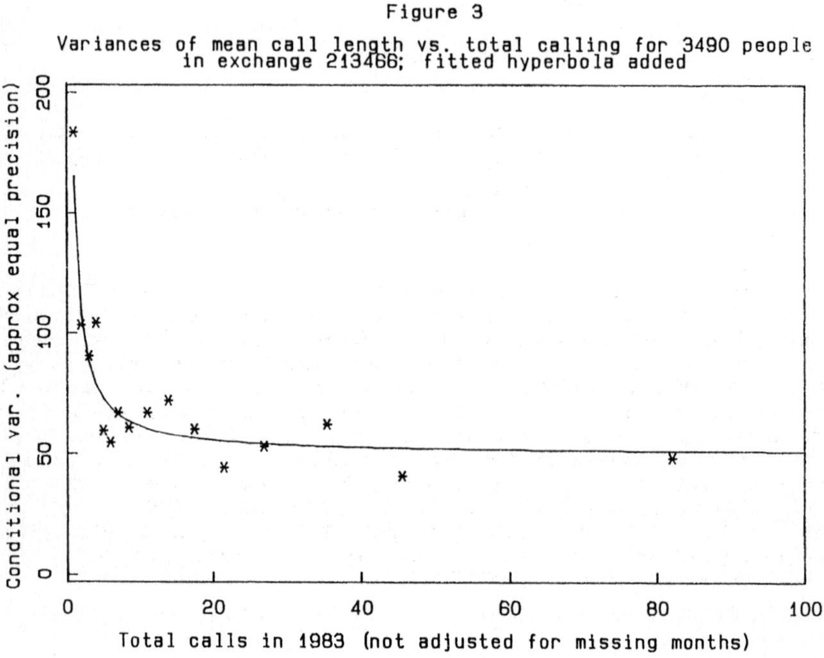
Figure 3
Variances of mean call length vs. total calling for 3490 people in exchange 213466; fitted hyperbola added
Conditional var. (approx equal precision)
0
50
100
150
200
0
20
40
60
80
100
Total calls in 1983 (not adjusted for missing months)

Figure 3 is a plot of observed (conditional) variance of observed average holding times given number of messages. Neighboring values of total messages made were grouped so that approximately the same number of individuals (around 100, except for the first few message totals, where there are more) are in each group. The midpoints of each group together with the group's average call length variance is plotted. The approximately equal group sizes mean each point (other than the first few) has roughly the same precision. A fitted hyperbola appears; the formula is (cf. (6))

$$y = a + \frac{b}{x} \qquad (7)$$

where $a = \sigma_\Lambda^2$ is 50 and $b = E(\Lambda^2).\sigma_\varepsilon^2$ is 115. Hence we estimate σ_ε^2 by $b/E(\Lambda^2)$. As $E(\Lambda^2) = (\lambda^2 + \sigma_\Lambda^2)$, and since we can estimate λ^2 as 12.4^2 and σ_Λ^2 as 50, we can estimate σ_ε^2 as about $115/(154+50) \approx 0.56$. This estimates individual i's call to call variability relative to his expected holding time Λ_i. Note our model assumes this is the same for every individual.

The multiplicative model (4) was initially suggested by observing that the individual call lengths are very skew to the right; this suggested a log transformation might be appropriate. In fact, if we do a QQ plot of the log of single messages against the Normal, this is approximately linear (see Fig. 4), so we shall assume the errors in model (4) are approximately log normal.

Figure 4

QQ plot against normal of lengths of single calls

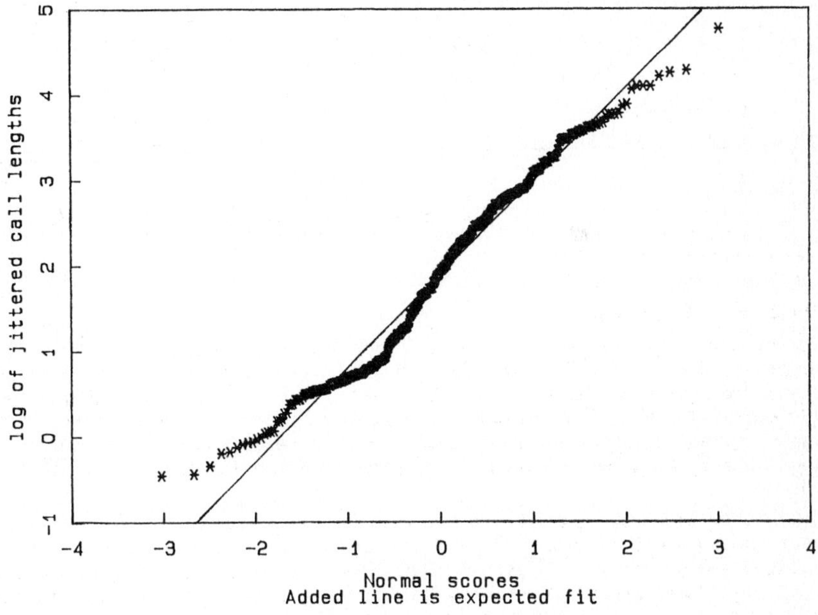

It remains to estimate the population distribution of Λ. We did this by simply considering only individuals making greater than 20 messages - applying (7) implies the variance contribution due to call to call variation should be less than 10%. A histogram of the logs of average call length for 456 individuals making greater than 20 messages appears in Fig. 5. This histogram is clearly

skewed to the left, but on doing a similar QQ plot (not shown) to that done for the individual call lengths, we see that this distribution is also reasonably close to normal; therefore we shall assume the population of Λ_i's is log normal too. This agrees with Pavarini's findings [1].

Figure 5

Histogram of the log of mean call lengths for individuals making greater than 20 calls in 1983

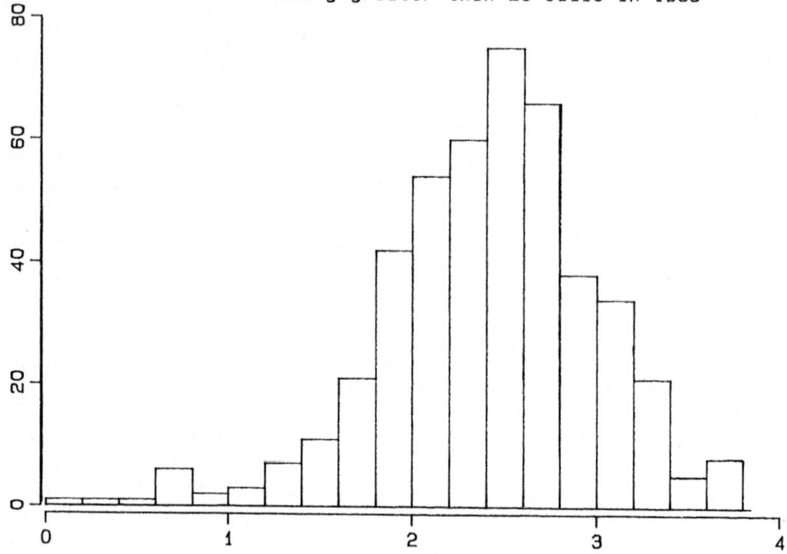

5. Problems with the model

There are at least three drawbacks to our model worth discussing. One is that expected holding time does not depend on number of messages. Bosch [3], gives data showing that this may not always be true, especially in business. For residence data, Bosch calculates the correlation between $\log(L)$ and log of number of messages and finds a negative correlation in some exchanges (i.e. the mean of expected holding times decreases as number of messages increase). This would be the case in our data if the points in Figure 1 formed a pattern which moved down as we moved along the X-axis.

His assumption of bivariate normality implies that the conditional variance of $\log(\bar{L})$ given number of messages is n should be constant. Figure 3 shows this is far from true for our data. However, his method suggests one way we might incorporate such data into our model. Using the correlation to find $E[\log(\Lambda)\,|\,\log(n \text{ messages})]$ is equivalent to using linear regression for data where the conditional variance is constant. By analogy, we might use weighted regression, where we weight each observation by the inverse of its conditional variance. A data analytic method to handle this could fit an ordinary regression, and then estimate the conditional variance of the residuals using similar methods to the way we produced formula (7). Using these preliminary conditional variances we fit a weighted regression, and iterate until there is no more change. A second possibility would be to fit a robust regression directly.

A second drawback arises if we are interested in simulating the full bivariate distribution knowing distributions of expected holding times and a marginal distribution of total minutes, not of message frequency. In principle, the full bivariate distribution can be determined from this information, but in practice it will be difficult to get. This difficulty arises only in the simulation case, not the case where one wishes to predict message frequency given total minutes and one has prior data which one can use to estimate expected holding time for each individual.

A third drawback arises from the assumption that expected holding time for each individual never changes. One can imagine scenarios where tariff changes might change the holding time substantially. Whether or not the model can be adapted to handle this case will probably depend on how well we can guess at how the tariff change will affect each individual's holding time. However, holding time tends to be much more constant under tariff changes than message frequency.

6. Estimation of individual expected holding time

If someone makes no calls at all, estimate his expected holding time as λ. How much we are likely to be off is measured by the square root of the variance across the population, (denote this variance by σ_Λ^2). If individual i makes n_i messages, of average length $\bar{L}_i$, we shall estimate his long-term average call length, Λ_i, by a Bayes procedure which we now derive.

First note that (4) implies $\log(L)$ is the sum of two independent r.v.'s:

$$\log(L) = \log(1+\varepsilon) + \log(\Lambda) \tag{8}$$

We assume that the average of an individual's logged call lengths, $\overline{\log(L)}$ given $\log(\Lambda)$ is normal. This amounts to assuming that the error term $\log(1+\varepsilon)$ is normal. That this is consistent with the evidence given in the previous section comes from noting that $\log(\Lambda)$ and $\log(L)$ are both normal; even if it is not quite true for a small n_i, as n_i increases the Central Limit Theorem will make it reasonable.

Our aim is to apply a version of equation (3). Hence we have that $X = \overline{\log(L)}$ given $Y = \log(\Lambda)+E[\log(1+\varepsilon)]$ is normal with mean Y and variance $\sigma^2_{\log(1+\varepsilon)}$. By the previous section, Y is assumed normal, mean $E[\log(\Lambda)+E[\log(1+\varepsilon)]$, and variance given by the variance of $\log(\Lambda)$, say $\sigma^2_{\log\Lambda}$.

Applying (3), we display the Bayes estimate for $\log(\Lambda_i)$, given individual i makes n_i messages, as given by the weighted average:

$$E[\log(\Lambda_i) \mid \overline{\log L_i}] = \overline{\log(L_i)}.p + (E[\log(\Lambda)]+E[\log(1+\varepsilon)])(1-p) \tag{9}$$

where p is given by $\left[\dfrac{\sigma^2_{\log\Lambda}}{\sigma^2_{\log\Lambda} + \sigma^2_{\log(1+\varepsilon)}/n_i}\right]$.

Hence, to apply (9), we must estimate the mean and variance of $\log(\Lambda)$ and also the mean and variance of $\log(1+\varepsilon)$. One approach to estimating these quantities is to use the estimates for $E[\Lambda]$, σ_Λ^2 and σ_ε^2 developed using (6), and then use relations between moments of normal and log-normal r.v.'s. These relations are derived in Appendix A, since we shall be using them a little later.

A safer method is to directly estimate these quantities from the data somehow. For example, to estimate $\sigma^2_{\log\Lambda}$ we directly find the variance of the log of observed average holding times for a group with high message frequency, even though this procedure will tend to overestimate (it will include the remaining variation in the observed average call lengths). Using the group included in Fig. 5, this method gives the value .33, versus the value .29 got by using the log-normal assumptions; i.e. got by using the formula (cf. Equation (A6) in the Appendix):

$$\sigma^2_{\log\Lambda} = \log(\sigma_\Lambda^2/\lambda^2 + 1) \tag{10}$$

(the two values are not significantly different, but we shall use the first, .33).

Similarly, if one wishes to avoid log-normal assumptions one can directly estimate E[log(Λ)] using the group in Fig. 5. This estimate is 2.41, otherwise one is forced to use an estimate using log-normal assumptions (cf. relation (A5) in the Appendix):

$$E\,[\log(\Lambda)] = \log\left[\frac{\lambda^2}{(\sigma_\Lambda^2+\lambda^2)^{1/2}}\right] \tag{11}$$

which in this case is 2.37 (again, not a significant difference). We prefer the first estimate.

We cannot easily use the data to directly estimate the mean and variance of log(1+ε), so here we use the log-normal assumption. Equation (A5) in the Appendix gives the estimate for the mean of log(1+ε) as:

$$E\,[\log(1+\varepsilon)] = -\frac{1}{2}.\log(\sigma_\varepsilon^2+1) \tag{12}$$

recalling that the log-normal r.v. (1+ε) has expectation one. This is estimated by value found using (7), i.e. −.5log(.56+1) or about -0.22. Similarly, using (A6) gives the variance for $\sigma^2_{\log(1+\varepsilon)}$ as $\log(1+\sigma_\varepsilon^2)$, estimated by log(1+.56), or 0.44.

Lastly, our data is set up so that we do not have the average of the log call lengths, $\overline{\log L_i}$, for any individual i having n_i greater than one. A simple Taylor series expansion suggests using

$$\overline{\log(L_i)} \approx \log(\bar{L}_i) - \frac{n_i-1}{2.n_i}\sigma_\varepsilon^2 \tag{13}$$

For a derivation see relations (A6)-(A10) in the Appendix.

Denote the conditional mean in equation (9) by W_i. It can be shown (e.g. see [4]) that the conditional distribution of $\log(\Lambda_i)$ given $\overline{\log(L_i)}$ is normal with estimated mean given by W_i and estimated variance, denoted by σ_W^2, given by:

$$\sigma_W^2 = \frac{(\sigma^2_{\log(1+\varepsilon)}/n_i).\sigma^2_{\log\Lambda}}{\sigma^2_{\log\Lambda}+\sigma^2_{\log(1+\varepsilon)}/n_i} \tag{14}$$

Hence we find the empirical Bayes estimate for individual i's expected holding time, Λ_i, by using

$$E\,[\Lambda_i\,|\,\bar{L}\,] = \exp(W_i + \frac{1}{2}\sigma_W^2) \tag{15}$$

with the error in our estimate having a variance given by the empirical Bayes estimate of variance:

$$\hat{\sigma}_i^2 = \exp(2.W_i + \sigma_W^2)\left[\exp(\sigma_W^2) - 1\right] \tag{16}$$

These last two equations follow directly from relations (A3) and (A4) in the Appendix.

Note that knowledge of the time series nature of usage is unnecessary for making these Bayes estimates; the input data for this model depends only on having knowledge of total time and number of messages made for individual customers aggregated over some block of time.

7. Empirical test of the method

The Bayes procedure developed in the preceding section was tested by comparing against several other techniques. We took three months (Feb., March, and April) of night-weekend minutes and night-weekend messages, aggregated over the three months, for each individual. We used this data to predict total night-weekend minutes for each customer given total night-weekend messages for each customer in the period October, November and December. As a measure of prediction accuracy we used the mean of the absolute error of prediction across all customers, i.e., if there are

N customers and we observe Y_i minutes for customer i, but predict $\hat{Y}_i(P)$ using some procedure (denoted by P), the mean absolute error of prediction for procedure P, denoted mae(P), is:

$$\text{mae}(P) = \frac{1}{N}\sum_{i=1}^{N} |Y_i - \hat{Y}_i(P)| \tag{17}$$

We used the mean absolute error instead of the more classical mean square error because it is more robust; however we checked that the results of this section are similar for mean square error.

We shall denote the predictions got from the Bayes method by $\hat{Y}_i(B)$; here is the exact procedure:

(B) Predict minutes by taking the number of messages in the Oct-Dec period and multiplying this by the Bayes estimate for average call length for the individual in the Feb-April period. If there are no messages in this early period, use the global mean call length. Call this procedure (B).

The alternative procedures we used are as follows:

(*) Predict minutes by taking messages in the Oct-Dec period and multiplying this by the global mean call length found in the previous section (12.4 minutes). Call this procedure (*). This will be our baseline procedure for comparing other methods against.

(M) Predict minutes by taking messages in the Oct-Dec period and multiplying this by the observed mean call length for the individual in the Feb-April period. If there are no messages in this early period, use the global mean call length. Call this procedure (M).

(R) Predict minutes by taking messages in the Oct-Dec period and multiplying this by an estimate of average call length got by regressing Oct-Dec average length of call on Feb-April average length call. This regression was actually done in square roots so as to stabilize the conditional variances. Again, if there are no messages in this early period, use the global mean call length to estimate the individual's average call length. Call this procedure (R). We include this procedure as a test of how well one might expect to do if one incorporates information from the prediction period. Under the circumstances we are doing the prediction this data would normally be unavailable, and hence this procedure would be unusable. On inspection of Table I (explained below) one sees that the Bayes procedure does marginally better.

We split up the population into seven segments, based on total night-weekend minutes in the Feb-Apr period, and found the mean absolute error in each segment for each method. These are presented in Table I. A comparison of the methods is given in Fig. 6, where the ratio of the mean absolute error for each method to the mean absolute error for method (*) is plotted against the observed mean total minutes in the Oct-Dec period for each segment. Hence the straight line at height 1 represents method (*). The regression method is omitted as it will be usually unusable. It is clear the Bayes predictions are easily the best in the high use segments; but elsewhere, the Bayes method is only a fairly small improvement over method (*). The straight use of previous averages to predict future averages (method (M)) is the worst predictor, except in the very high usage segment. We also found the global mean of the predicted minutes for each method and these are presented in Table II. It is worth noting that the Bayesian prediction seems to consistently under-predict, in aggregate, but seems to be doing a little better than the baseline method (*) in the heavier usage segments. This does generalize to other data, but we shall not pursue this problem here.

The practical consequence of this is that if you can't use individual level Bayes prediction for finding minutes from number of messages made, the next best thing is to just multiply the observed number of messages by some global mean length of call. In this example, the worst of these procedures is attempting to estimate minutes from messages by using the unadjusted individual average call length.

Figure 6
Mean absolute prediction error relative to (*)

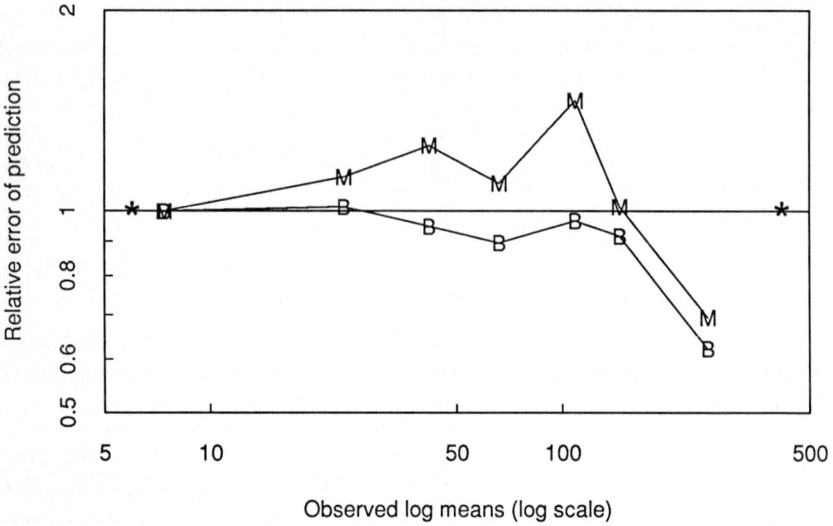

Table I

Mean absolute error of prediction for 3 month Oct-Dec total minutes					
Segment (mins/month in Feb-Apr 1983)	m.a.e. Bayes (B)	m.a.e. base (*)	m.a.e. Mean (M)	m.a.e. Regression (R)	Number in segment
0	4.6	4.6	4.6	4.6	1883
0-10	13.7	13.5	15.2	13.0	736
10-20	18.3	19.3	24.2	19.0	312
20-40	29.2	32.6	35.9	31.1	284
40-60	40.1	41.5	60.8	42.4	115
60-100	63.1	68.8	69.8	68.1	95
100+	71.4	114.8	79.7	92.1	65
all	13.8	15.1	16.2	14.4	3490

Table II

Comparison of segment means for predicted minutes by method (three month totals)						
Segment (mins/month in Feb-Apr 1983)	Mean Bayes (B)	Mean base (*)	Mean Mean (M)	Mean Regression (R)	Mean Observed	Number
0	7.9	7.9	7.9	7.9	7.4	1883
0-10	17.1	28.1	15.2	26.3	23.8	736
10-20	34.5	41.9	45.8	47.1	41.7	312
20-40	59.0	68.1	79.1	78.2	66.0	284
40-60	93.8	101.6	136.9	121.2	108.7	115
60-100	125.2	124.3	168.3	150.7	144.9	95
100+	218.3	202.0	294.6	250.5	256.0	65
all	26.3	30.0	32.5	33.1	30.4	3490

Appendix A

In this Appendix we derive the relationships used previously between normal and log-normal variables. The moment generating function for a normal r.v., Z, mean $E[\,Z\,]$, and variance var(Z) is just

$$E[\,e^{tZ}\,] = \exp\left[E[\,Z\,].t + \frac{1}{2}\mathrm{var}(Z).t^2\right] \tag{A1}$$

Then $U = e^Z$ is log-normal, so

$$E[\,U\,] = E[\,e^Z\,] = \exp\left[E[\,Z\,] + \frac{1}{2}\mathrm{var}(Z)\right] \tag{A2}$$

and

$$E[\,U^2\,] = E[\,e^{2Z}\,] = \exp\left[2E[\,Z\,] + 2\mathrm{var}(Z)\right] \tag{A3}$$

and so the variance of U, denoted by var(U), is

$$E[\,U^2\,] - E[\,U\,]^2 = \exp\left[2E[\,Z\,] + \mathrm{var}(Z)\right].\left[e^{\mathrm{var}(Z)} - 1\right] \tag{A4}$$

From Equations (A2) and (A4), knowing E[U] and var(U), we can solve for E[Z] and var(Z) giving:

$$E[\,Z\,] = \log\left[\frac{E[\,U\,]^2}{(E[\,U\,]^2 + \mathrm{var}(U))^{1/2}}\right] \tag{A5}$$

$$\mathrm{var}(Z) = \log\left[1 + \frac{\mathrm{var}(U)}{E[\,U\,]^2}\right] \tag{A6}$$

A justification of Equation (13) follows: Let $X_1, \cdots, X_n$ be i.i.d. random variables, and let $\bar{X}$ be their average. The X_j's play the role of each call length (i.e. the r.v. X is equivalent to the r.v. L_i

for individual i). Then if r_j is the residual $\bar{X} - X_j$,

$$\frac{1}{n}\sum\log(X_j) = \frac{1}{n}\sum\log(\bar{X} - r_j)$$

$$= \log(\bar{X}) + \frac{1}{n}\sum\log(1 - r_j/\bar{X})$$

$$\approx \log(\bar{X}) + \frac{1}{n}\sum\left[-r_j/\bar{X} - \frac{1}{2}(r_j/\bar{X})^2\right]$$

(by a Taylor series expansion)

$$= \log(\bar{X}) - \frac{1}{n}\sum(r_j/\bar{X})^2$$

(since the residuals sum to zero)

$$\approx \log(\bar{X}) - \frac{n-1}{n} var(X)/\bar{X}^2 \tag{A7}$$

(approximating the expectation of the mean sum of squares by its expectation)
So if X is given by a multiplicative model such as:

$$X_j = (1 + \varepsilon_j).K \tag{A8}$$

where ε_j are i.i.d., mean zero r.v.'s, then

$$\frac{\mathrm{var}(X)}{\bar{X}^2} \approx \frac{\mathrm{var}(\varepsilon).K^2}{K^2} = \mathrm{var}(\varepsilon) \tag{A9}$$

(approximating K by $\bar{X}$). Note K is a constant since we are working with one fixed individual i only. Putting (A7) and (A9) together we get

$$\frac{1}{n}\sum\log(X_j) \approx \log(\bar{X}) - \frac{n-1}{n}\mathrm{var}(\varepsilon) \tag{A10}$$

This is equivalent to equation (13).

References

[1] Pavarini, C., "Residence and Business Flat Rate Telephone Usage", Bell Laboratories MF-76-3441-9, 7 July 1976.

[2] Kearns, T.J., "Properties of Subscriber Distributions of Local Telephone Usage", Bell Laboratories MF-77-3441-5, 1 April 1977.

[3] Bosch, H.B., "Revenue Calculation in the Presence of Correlated Calling Rates and Call Durations", Bell Laboratories MF-80-9541-01, 21 February 1980.

[4] Ledermann (1984), *Handbook of Applicable Mathematics,* **IV:** Statistics, Part B, Wiley

TELECOMMUNICATIONS DEMAND MODELLING
An Integrated View
A. de Fontenay, M.H. Shugard, D.S. Sibley (Editors)

GENERALIZED GAMMA FAMILY REGRESSION MODELS FOR LONG DISTANCE TELEPHONE CALL DURATIONS

Trudy Ann CAMERON and Kenneth J. WHITE

Department of Economics, University of California, 405 Hilgard Avenue, Los Angeles, CA, USA 90024, and Department of Economics, University of British Columbia, 1873 East Mall, Vancouver, BC, Canada V6T 1Y2

In regression applications where the dependent variable cannot logically take on negative values, the Generalized Gamma (GG) distribution can be adopted as a flexible alternative to the usual Normal conditional distribution. This paper offers a comprehensive assessment of the impact of different distributional assumptions (the GG and its many special cases) upon regression parameter estimates in both linear and log-linear specifications for a model of long-distance telephone call durations.

1. INTRODUCTION

In regression applications where the dependent variable cannot logically take on negative values, the usual normal (N) conditional probability density function is often inappropriate. In these circumstances, the Generalized Gamma (GG) distribution can be adopted as a flexible distribution for regression error terms. This distribution contains the simple gamma (G), Weibull (W), Exponential (E), Lognormal (LN), and other distributions as special cases. Discrimination among these special cases is achieved by examination of the appropriate Wald, Lagrange Multiplier, or Likelihood Ratio test statistics. McDonald [16] has used the simple GG distribution to estimate the parameters of an income distribution and McDonald and Newey [17] consider regression models. In this paper, we utilize the GG family in regression models for the durations of long distance telephone calls. The existing literature observes that call durations vary with a number of characteristics of the call, such as marginal price per minute, distance, time-of-day, type and origin of call. Previous regression models of call duration have assumed normal errors (usually on aggregated data); Weibull and Gamma distributions have been used, but only in a non-regression context. This paper therefore represents a synthesis of several previous avenues of research pertaining to telephone call durations. However, we should stress that we do not pretend to offer a comprehensive model of long-distance demand. Our data base for this study is limited to telephone company billing records, so we cannot implement a full behavioral model. Since we are unable, therefore, to estimate a specification which is consistent with an idealized economic model of long-distance demand, readers should be aware of the potential limitations of the tentative policy conclusions drawn here. Nevertheless, our applications illustrate an innovative methodology for a detailed assessment of the duration component of this demand.

In Section 2, a brief description of related research on the demand for telephone services is provided. Section 3 focuses on call duration as one important aspect of telephone demand and develops a regression model with generalized gamma errors for analyzing the determinants of call duration. The theoretical regression models described in Section 3 are rendered operational in Section 4 with the selection of specific explanatory variables. The estimation results are discussed and compared to the findings of previous analyses. The important methodological issue of discrimination among various models in the generalized gamma family is addressed in Section 5. Finally, Section 6 describes some policy implications of the estimated equations.

2. RELATED RESEARCH ON TELEPHONE DEMAND

Park, Wetzel, and Mitchell [19] use aggregated data on local telephone calls with the objective of measuring prices elasticities both for the total number of calls and total minutes of conversation. Pacey [18] concentrates on the estimation of "point-to-point" price elasticities for intercity long distance services, concluding that "more disaggregated data need to be made available in order to estimate the mean duration of a call." (We allow mean duration in a disaggregated model to be a function of a whole range of explanatory variables, while still constraining the distribution of durations to be strictly non-negative.) Rea and Lage [21] utilize time-series/cross-sectional data on calls to 37 foreign countries, stating that "...since demand for telegraph and telephone services arises from both household and business sectors, it would be desirable to estimate disaggregated functions. However, the data are not available, and the assumption must be made in estimating the demand equations that meaningful aggregate relationships exist." (We estimate a disaggregated model of the durations of overseas calls, as well as for calls with destinations in Canada and the U.S.)

Gale [7,8] made several studies of call duration and reported a number of basic results. Among these were that (1) the mean duration of a toll call is longer, the more distant the call. (2) Evening calls are longer than daytime calls. (However, since night rates are lower than daytime rates, the difference incorporates an unknown price effect.) (3) Person-to-person calls are longer than station-to-station calls, and (4) collect calls and calls billed to third parties are longer than "paid" calls. However, these are only pairwise relationships; our regression models distinguish these effects, *ceteris paribus*.

The simple, unconditional distribution of telephone call durations has been explored by Wong [26] using the Weibull distribution, by Pavarini [20] using the Powernormal distribution, and by Curien [4] using the exponential and Erlang distributions. De Fontenay, Gorham, Manning and Lee [5] estimate separate Weibull distributions for the average lengths of calls with destinations in different mileage bands to obtain mean duration values used in a subsequent ordinary least squares (OLS) regression to derive price and income elasticities of demand. Recently, Veitch [24] has explored the conditional distribution of total minutes given message frequency. He assumes log-normality.

In this study, we attempt to consolidate the methodology used in the previous studies to show how the determinants of duration can be explored using maximum likelihood estimation of a regression model with non-normal errors. We examine a sample of completely disaggregated data for long-distance phone calls over a twenty-four hour period for the Canadian province of British Columbia. As deregulation of long-distance telephone markets progresses, further work on the demand for such services is warranted. Our data permit separate analyses of particular types of calls. For example, we have chosen (a) residential calls to Canada and United States destinations and (b) both business and residential calls to overseas destinations.

3. DISTRIBUTIONAL ASSUMPTIONS FOR CALL DURATION

In the linear regression model with normal errors, a non-zero probability is associated with negative values of the dependent variable, even though the regression line might be strictly positive, thus allowing prediction intervals which include values that may be theoretically (and empirically) impossible in some contexts. Lawless [13] provides a comprehensive analysis of "lifetime" models for product testing (often called "accelerated failure time" (AFT) models) which fall into this category. More recently, Heckman and Singer [10] have explored the estimation of these models in studies of unemployment durations where the data are censored.[1]

In contrast to unemployment duration data, the telephone call durations modeled here are very short relative to the sample period. Since the data consist of all calls initiated during a twenty-four hour sample period, and complete durations are recorded, censoring is not a problem. Furthermore, since individual call durations are so very short, *ceteris paribus* may more readily be assumed to hold during each call. Most importantly, the rates applicable to each call are determined solely by the time of initiation of the call, even if the call itself spans two rate periods.

3.1. Linear Models

We focus first on the class of "linear" regression models where duration (t) for a particular call is assumed to be a linear function of a set of exogenous variables (x) so that $E(t|x) = x'\beta$. While this allows for negative fitted values of the dependent variable, it does permit a simple interpretation of the coefficients, which are analogous to those in linear OLS regressions.

The simple GG probability density function (see Johnson and Kotz [12,p. 197] for a single observation (t) on the random variable T is given by:

$$(1) \qquad f(t) = \{ct^{ck-1}/[b^{ck}\Gamma(k)]\} \exp[(-t/b)^{c}], \quad t \geq 0$$

where b is a "scale" parameter, c and k are "shape" parameters and Γ is the mathematical Gamma function. The simple Gamma (G) distribution is obtained when c = 1, while the Weibull (W) imposes k = 1. For the Exponential (E) model, c = k = 1, while as k approaches ∞, the lognormal (LN) distribution results.[2]

The mean of the GG distribution is:

$$(2) \qquad E(t) = b\Gamma(k + 1/c) \,/\, \Gamma(k)$$

If we wish to make this density conditional upon a (p x 1) vector of explanatory variables, x, we can adopt the usual assumption that the scale parameter varies with x while the shape parameters are constant.[3] Thus for a "linear" regression model, $t = x'\beta + \epsilon$, with t distributed GG:

(3) $$E(t|x) = b(x)\Gamma(k + 1/c) / \Gamma(k) = x'\beta.$$

This means that we must substitute b(x) for b wherever the latter appears in the density function (1).[4] We find that:

(4) $$b(x) = x'\beta\ \Gamma(k) / \Gamma(k + 1/c) = x'\beta/F$$

where $F = \Gamma(k + 1/c) / \Gamma(k)$. The conditional density function for t is now:

(5) $$f(t|x) = \{ct^{ck-1} /[(x'\beta/F)^{ck}\ \Gamma(k)]\} \exp[-(tF/x'\beta)^c],\ t \geq 0.$$

The joint density of n independent observations on the variable $T = (t_1, t_2, \ldots, t_n)$ yields the log-likelihood function:

(6) $$L = n \log c - n \log \Gamma(k) + nck \log F + ck\ \Sigma \log t_i^* - \Sigma \log t_i - \Sigma (t_i^*F)^c,$$

where all sums are over i from 1 to n, and $t_i^* = t_i/x_i'\beta$. A detailed Appendix (available from the authors) gives the first and second derivatives for this likelihood function, $L^{(1)}$ and $L^{(2)}$.

Given the maximum likelihood estimates of the parameters (β, k, c), the negative of the inverse of the matrix of second derivatives can be used to estimate the asymptotic covariance matrix of parameter estimates, V. These estimates can be used for Wald tests involving hypotheses about the coefficients. In particular, specific tests concerning the values of the shape parameters can be used to aid in model selection.

The necessary equations for maximum likelihood estimation of the E, G, and W models can easily be derived by substitution of the appropriate constraints. The maximum likelihood estimation of a model with LN errors is obtained by maximizing the LN likelihood function also given in the separate Appendix.

3.2. Log-linear Models

Accelerated failure time models such as those discussed in Lawless [13], generally assume that y = log t and $E(y|x) = x'\beta$. These models have the advantage of forcing the fitted value of t to be positive, regardless of the sign of the inner product, $x'\beta$. However, this procedure makes the model log-linear rather than linear. When we have no *a priori* reason to prefer either a linear or a log-linear specification, we might wish to explore both functional relationships between t and the explanatory variables. For the log-linear generalized gamma model, the appropriate conditional density function is:

(7) $$f(t|x) = \{ct^{ck-1}/[(b(x))^{ck}\ \Gamma(k)]\} \exp[(-t/b(x))^c].$$

It is convenient to redefine the parameters as $b(x) = \exp(x'\beta)$ and $\sigma = 1/c$. If we then let Y = log T, the conditional density for Y becomes:

(8) $$f(y|x) = \{1/[\sigma\Gamma(k)]\} \exp[(wk) - \exp(w)],$$

where we simplify by letting $w_i = (y_i - x_i'\beta)/\sigma$. The log-likelihood function can be written as:

$$L = -n \log \sigma - n \log \Gamma(k) + k \Sigma w_i - \Sigma \exp(w_i). \quad (9)$$

Once again, derivatives for this function are provided in the separate Appendix. The formulas for the E, G, and W models are again easily derived. Note that the log-linear LN model is simply ordinary least squares (OLS) with y = log(t) as the dependent variable.

4. ESTIMATION AND RESULTS

Data were obtained on a stratified sample of approximately 65000 long distance telephone calls originating in the Canadian province of British Columbia on July 13, 1983. From this sample, a subsample of 21738 residential calls to Canada and the U.S. (excluding Alaska) were found. A second subsample consists of both business and residential calls to all overseas destinations, yielding a total of 4934 calls. Table 1 summarizes the sample.[5,6] (The separate Appendix provides a more complete description of the data.)

Table 1

MEANS (STANDARD DEVIATIONS) OF DATA: LONG DISTANCE CALLS

Variable	Description	Canada and U.S. Destinations (n = 21738)	Overseas Destinations (n = 4934)
DUR	duration in minutes	6.566 (7.939)	7.738 (7.512)
LOG(DUR)	log of duration	1.368 (1.007)	1.635 (0.948)
RATE	marginal rate ($)	0.374 (0.188)	2.111 (0.468)
LOG(DIST)	log of distance (mi)	4.515 (1.424)	8.567 (0.193)
EVE	=1 if evening call	0.462	-
NIGHT	=1 if night call	0.083	0.248
BUS	=1 if business call	-	0.303
COLL	=1 if collect call	0.065	-
CARD	=1 if credit* call	0.056	0.036
PERS	=1 if person** call	0.008	0.039
NE	=1 if originated in NE	0.379	0.109

* credit card call
** person-to-person call

4.1. Canada and U.S. Destinations

4.1.1. Linear Models

The dependent variable in all cases is the duration of each long distance telephone call (in minutes). For each alternative model, the estimated coefficients, shape parameters, and regression statistics for this sample appear in Table 2.A.[7] The column headed N gives the results for a linear normal model estimated by OLS. Among the linear specifications, the GG

Table 2 - MAXIMUM LIKELIHOOD ESTIMATES*, CANADA AND U.S. DESTINATIONS

	A. LINEAR REGRESSION MODELS (Dep. Var. = DUR)						B. LOG-LINEAR MODELS (Dep. Var. = log(DUR))				
	N	E	G	W	GG	LN	E	G	W	GG	LN
RATE	-1.626 (.7135)	-3.770 (.7581)	-3.770 (.6774)	-3.677 (.7117)	-3.423 (.7124)	-3.422 (.7122)	-.0010 (.1051)	-.0010 (.0939)	.0046 (.0988)	.0279 (.0894)	.0280 (.0894)
LOG(DIST)	1.238 (.0885)	1.428 (.1001)	1.428 (.0894)	1.418 (.0939)	1.331 (.0936)	1.330 (.0936)	.1645 (.0132)	.1645 (.0118)	.1633 (.0124)	.1482 (.0111)	.1482 (.0111)
EVE	2.716 (.1578)	2.057 (.1459)	2.057 (.1304)	2.089 (.1372)	1.815 (.1347)	1.815 (.1247)	.4773 (.0223)	.4773 (.0199)	.4787 (.0209)	.4318 (.0198)	.4318 (.0198)
NIGHT	2.867 (.3043)	1.947 (.3211)	1.947 (.2869)	2.021 (.3026)	1.370 (.2852)	1.370 (.2851)	.5026 (.0432)	.5026 (.0385)	.5082 (.0405)	.4172 (.0381)	.4171 (.0381)
COLL	.4210 (.2120)	.4956 (.1832)	.4956 (.1637)	.4725 (.1715)	1.031 (.1899)	1.031 (1.899)	.0955 (.0279)	.0955 (.0249)	.0914 (.0261)	.1740 (.0266)	.1740 (.0266)
CARD	-.9992 (.2267)	-.6006 (.1351)	-.6006 (.1207)	-.6274 (.1261)	-.3008 (.1432)	-.3007 (.1432)	-.1399 (.0298)	-.1398 (.0266)	-.1452 (.0279)	-.0709 (.0284)	-.0709 (.0284)
PERS	-.4133 (.5792)	-.1044 (.4668)	-.1045 (.4171)	-.1547 (.4337)	.4776 (.5087)	.4783 (.5087)	-.0600 (.0765)	-.0600 (.0683)	-.0674 (.0716)	.0284 (.0726)	.0284 (.0726)
NE	-.3780 (.1064)	-.4520 (.0737)	-.4520 (.6586)	-.4496 (.0692)	-.4444 (.0728)	-.4445 (.0728)	-.0789 (.0140)	-.0789 (.0125)	-.0780 (.0131)	-.0731 (.0133)	-.0731 (.0133)
intercept	.2644 (.1821)	.5508 (.1396)	.5508 (.1247)	.5694 (.1311)	.9231 (.1362)	.9237 (.1362)	.8443 (.0247)	.6172 (.0237)	.8760 (.0233)	-26029 (8072.6)	.4747 (.0228)
k	-	1	1.252 (.0108)	1	41259409 (419.69)	-	1	1.255 (.0108)	1	3320932. (1816400)	-
c	-	1	1	1.068 (.0052)	.00016348 (.00000078)	-	1	1	.9359 (.0046)	1733.5 (474.48)	-
σ	-	-	-	-	-	.9524 (.0046)	-	-	-	-	.9511
L	-80661.4	-61253.3	-60931.8	-61166.5	-59525.8	-59525.9	-61232.0	-60905.1	-61144.3	-59492.3	-59492.0

* asymptotic standard error estimates in parentheses;
k = gamma shape, c = Weibull shape, σ = lognormal shape parameter.

model achieved the highest value of the log-likelihood function as expected. It is seen that the linear LN estimates are extremely close to those of the GG model. This should be the case since the estimated value of k in the GG model was very large. Estimation of the GG model is quite difficult in this range due to a relatively flat likelihood surface over very large variations in k. Neither quasi-Newton algorithms nor the Goldfeld-Quandt [9] Quadratic Hill-Climbing (GRADX) algorithm worked well in this range.

The fitted coefficients of the "linear" model suggest that a ten cent increase in marginal rates should decrease call duration by .34 minutes or about 20 seconds. A one percent increase in the distance between call origin and call destination was found to increase expected duration by 1.3 minutes. This supports previous findings summarized by the phrase "the longer the haul, the longer the call" (see Taylor [23]). Duration is often expected to increase with distance because the frequency of calls declines.

The dummy variables for the time-of-day rate periods indicate that evening calls are about 1.8 minutes longer, and night calls are about 1.4 minutes longer, on average, than day calls. This time-of-day effect is distinct from the influence of the lower marginal rates charged during these off-peak periods. Person-to-person calls are a half a minute longer than station calls but the difference is not statistically significant. Collect calls are longer than paid calls by about one minute and statistically significant. This may be explained by the fact that the initiator of the call might not be paying for the call. Credit card calls, on the other hand, are about 20 seconds shorter, on average. This is difficult to explain, but may reflect the fact that the caller is often using someone else's residence telephone. Courtesy may require that such calls be kept "short."

The coefficient on the NE (northeast) dummy variable indicates that calls originating in this generally less-populated area of British Columbia tend to be about a half-minute shorter than those originating in other regions. The greater geographical dispersion of the population in this region may mean that a greater proportion of calls made by these subscribers must be long distance calls. Hence we may be observing a substitution effect between frequency of calls (or number of different destinations called) and duration of each individual call, subject to the subscriber's overall budget constraint.

It is important to note that while the estimated coefficients of the GG and LN models are very similar, they are sometimes substantially different from those of the E, G, and W models. The simple OLS coefficients are markedly different from any of the other models and the calculated OLS log likelihood function is roughly 20000 lower than for the other models.

4.1.2. Log-linear Models

Table 2.B gives the results for the Canada/U.S. sample when the same family of conditional distributions is used for duration, but it is assumed instead that the *log* of duration is linearly related to the explanatory variables. Optimization proceeds using the associated conditional distributions for the logarithmically transformed variable. Point estimates of the slope parameters thus represent the expected *percentage* change in duration as a result of a one-unit change in each explanatory variable.

Again, the E, G, and W models yield very similar parameter estimates, but the LN and GG specifications result in a considerably higher maximized value of the likelihood function and (for some variables) different parameter estimates. All computed log-likelihood functions in the log-linear models were transformed to be made comparable to those in linear models by using the appropriate Jacobian transformation as discussed in Box and Cox [1964]. Although the linear and log-linear models are not nested, the calculated likelihood functions of the log-linear models attained higher values suggesting that these models are preferred for these data. Unfortunately, global convergence of the GG model was not obtained despite extensive computational search. This was indicated by the fact that the calculated likelihood function in the LN model (which is easily computed by ordinary least squares) actually exceeded the calculated likelihood function of the GG model by a small number. The GG estimates reported in Table 2.B (and later in Table 4) use the best GG estimates for the log-linear model that could be obtained.

4.2. Overseas Destinations

4.2.1. Linear Models

Table 3.A exhibits the estimated results for the subsample of overseas calls. The N column again gives "naive" OLS parameter estimates. The linear GG regression model achieved a substantially higher value for the log-likelihood function than all the other linear models. In addition, the linear models appear to be marginally better than the log-linear models (they yield a higher maximized value for the log-likelihood function), in contrast to our results for the Canada/U.S. sample. The estimated coefficients imply that a ten cent increase in the marginal rate would be expected to decrease average duration by only .08 minutes, in contrast to the .34 minutes observed for the Canada/U.S. sample. Distance, however, has no significant impact on call duration, in sharp contrast to our results for the previous sample. The coefficients on the dummy variable for the off-peak nighttime period suggest that such calls are marginally longer by .73 minutes, an amount which is statistically significantly different from zero. Business calls are shorter by about a half a minute and also statistically significant. Credit card calls and calls billed to a third number are longer by about 1.4 minutes, also a significant difference. The largest increase in duration is observed for person-to-person calls, which are longer by almost three and a half minutes and statistically significant. Finally, in contrast to the results for the Canada/U.S. sample, we find that for the overseas calls the region of origin (NE) has no significant influence on call duration.

4.2.2. Log-linear Models

In Table 3.B the estimated parameters are displayed for the log-linear model using the overseas data. Once again, the estimated parameters for the GG and LN models are closer to each other than they are to the estimates for the other special cases of the GG model. None of the estimated slope parameters are more than just marginally significantly different from zero. It seems that the explanatory variables available on billing records are not particularly reliable predictors of the expected duration of an overseas call.

Table 3 - MAXIMUM LIKELIHOOD ESTIMATES*, OVERSEAS DESTINATIONS

	A. LINEAR REGRESSION MODELS (Dep. Var. = DUR)						B. LOG-LINEAR MODELS (Dep. Var. = LOG(DUR))				
	N	E	G	W	GG	LN	E	G	W	GG	LN
RATE	-.0999	-.1845	-.1856	-.0697	-.8250	-1.013	-.0190	-.0189	-.0054	-.0973	-.1184
	(.4431)	(.4680)	(.3998)	(.4149)	(.4097)	(.4162)	(.0611)	(.0522)	(.0534)	(.0556)	(.0557)
LOG(DIST)	-.7857	-.6523	-.6508	-.7841	-.1674	-.1368	-.0973	-.0974	-.1116	-.0466	-.0453
	(.7591)	(.7742)	(.6614)	(.6829)	(.6888)	(.7036)	(.1034)	(.0883)	(.0902)	(.0948)	(.0954)
NIGHT	1.012	.9656	.9651	1.009	.7343	.6730	.1225	.1226	.1272	.0907	.0794
	(.3896)	(.4197)	(.3586)	(.3681)	(.3869)	(.3998)	(.0529)	(.0452)	(.0461)	(.0486)	(.0489)
BUS	-.1490	-.1925	-.1929	-.1045	-.5642	-.6425	-.0225	-.0225	-.0110	-.0699	-.0780
	(.2396)	(.2429)	(.2075)	(.2141)	(.2206)	(.2267)	(.0321)	(.0275)	(.0280)	(.0298)	(.0301)
CARD	1.315	1.445	1.445	1.440	1.424	1.419	.1655	.1655	.1649	.1609	.1572
	(.5816)	(.6997)	(.5978)	(.6123)	(.6490)	(.6709)	(.0778)	(.0665)	(.0677)	(.0722)	(.0731)
PERS	1.890	2.017	2.017	1.850	3.394	4.048	.2334	.2335	.2140	.3726	.4245
	(.5621)	(.7007)	(.5987)	(.6032)	(.7437)	(.7966)	(.0752)	(.0642)	(.0654)	(.0705)	(.0706)
NE	-.0409	-.0253	-.0252	-.0631	.2306	.2924	-.0040	-.0040	-.0090	-.0245	-.0344
	(.3447)	(.3569)	(.3049)	(.3104)	(.3435)	(.3589)	(.0461)	(.0394)	(.0401)	(.0429)	(.0433)
intercept	14.36	13.41	13.40	14.30	10.79	11.14	2.878	2.564	3.022	-10.80	2.251
	(5.934)	(5.982)	(5.111)	(5.269)	(5.365)	(5.492)	(.8040)	(.6872)	(.7011)	(4.234)	(.7454)
k	-	1	1.370	1	21.66	-	1	1.370	1	21.27	-
			(.0249)		(8.338)			(.0249)		(8.074)	
c	-	1	1	1.150	.2311	-	1	1	.8696	4.289	-
				(.0121)	(.0451)				(.0092)	(.8257)	
σ	-	-	-	-	-	.9394	-	-	-	-	.9404
						(.0095)					
L	-16929.4	-15011.0	-14873.6	-14930.1	-14746.4	-14759.4	-15011.3	-14874.1	-14930.4	-14747.5	-14760.7

* asymptotic standard error estimates in parentheses;
k = gamma shape, c = Weibull shape, σ = lognormal shape parameter.

It is important to draw attention to the possible consequences of using parameter estimates from a regression model with misspecified errors. Tables 2 and 3 show that, relative to the GG estimates, the implied influence of marginal rates is sometimes substantially distorted by the E, G, or W shape parameter restrictions (and especially by the Normal assumption in the linear model). In the log-linear models, the effect of rate becomes insignificant in these special cases. Thus, not only an inappropriate functional form, but also the choice of an inappropriate error distribution could have significant implications for rate policy decisions. Recall that previous studies of duration have been inclined to use the W or the E distribution. For the overseas sample, both the linear and the log-linear E, G, and W models imply that business calls are not significantly shorter. The GG and LN specifications suggest that they are.

Section 2 described a number of results from previous studies. The positive effect of distance on duration is strongly supported by our models for Canada/U.S. data, but we do not find the same result for the overseas calls. Perhaps the incremental effect of distance diminishes with distance, and the overseas calls are at distances beyond where this effect vanishes. For previous studies, the result that night calls are longer than day calls was derived without controlling for differing marginal rates. With regression, we have been able to isolate the *separate* effect of time-of-day on duration. We find little support in the Canada/U.S. sample for the result that person-to-person calls are longer than station-to-station calls, but this is strongly corroborated by the overseas destination sample. Collect calls and calls billed to third parties have previously been found to be longer than sent-paid calls. We find in the Canada/U.S. sample that collect calls are longer, but third party (credit card) calls are shorter. There are no collect calls in the overseas sample, but third-party calls are (somewhat surprisingly) significantly longer by more than a minute.

5. MODEL DISCRIMINATION

Why might it be important to be able to distinguish the precise shape of the conditional distribution of durations for calls with a given set of characteristics? Two potential concerns can be considered. In Figures 1 and 2 we present some diagrams to illustrate the differences in the fitted models. Since the shape of the conditional distribution for t depends upon the x vector, Figure 1 depicts the fitted conditional distributions for t, in the linear models, for a "representative" category of telephone calls from Vancouver, British Columbia, to Toronto, Ontario (RATE = \$1.05, DIST = 2098 miles, all dummy variables = 0). In Figure 2, the domain of the conditional density functions for the log-linear models is $y = \log t$. Figure 2 compares the transformed densities for the GG model with those for each of its special cases, along with the symmetric normal density appropriate to the log-linear normal model. (It is interesting to observe that the transformed densities in the log-linear models appear much more alike than the densities of the linear model.)

As a first reason for concern about distributional shape, note that for the linear models in Figure 1, the fitted proportion of calls one minute or less in duration differs dramatically between the GG model and the W model. The latter suggests almost twice as many calls under one minute than does the former. Several smaller long-distance companies in the U.S. are experimenting with billing algorithms which round call durations up to

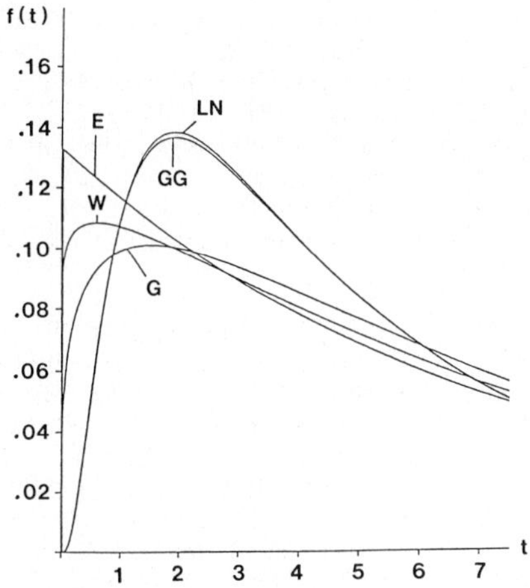

FIGURE 1

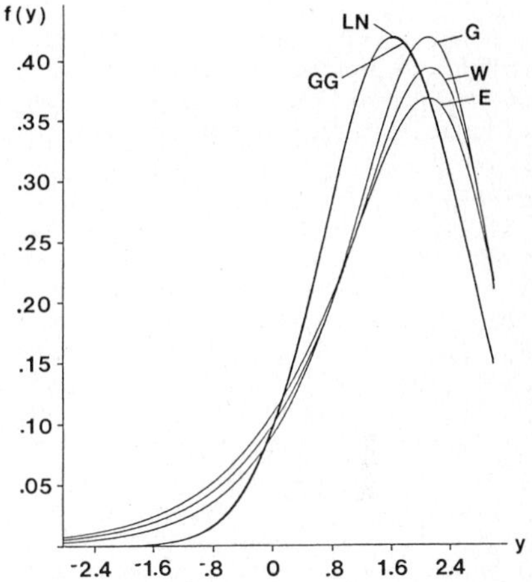

FIGURE 2

the nearest tenth of a minute, rather than to the nearest whole minute. If one could assume that the distribution of durations would be more-or-less unaffected by this refinement, the two different distributions would predict different consequences for company revenues.

A second circumstance under which the precise distributional shape would be important occurs when the telephone company offers discounts on extra minutes for calls longer than a given duration. The "thickness" of the distribution's tail would suggest the proportion of customers in this category which would be likely to respond. On the other hand, it is conceivable that consumer complaints about congestion might lead a company to attempt to influence the duration distribution relatively in favor of shorter calls, in order to increase the number of messages which can be carried by a system during peak periods. A surcharge might be considered for each minute of duration beyond some threshold level. Different distributional assumptions will again yield different predictions about the proportion of consumers likely to be affected.

While the fitted densities appear to differ considerably across models, it is important to assess whether the alternative formulations yield results which are *statistically significantly* different. We must appeal to the null hypotheses of model equivalence. Lee [14] proposes Lagrange Multiplier (LM) tests for specific accelerated failure time models. This procedure can be used here to test estimates of a restricted model (for example: linear E) against a more-general model (for example: linear W). Alternative tests are the Wald and the Likelihood Ratio (LR) tests. It is computationally convenient to employ the Hessian formulas from our separate Appendix in computing the Wald and LM tests. Following Amemiya [1, p. 350], the appropriate test statistics are:

(10) $$LR = 2\ [\ L(\gamma) - L(\gamma^*)\ ]$$

(11) $$Wald = (\gamma - \gamma^*)'\ [V(\gamma)]^{-1}\ (\gamma - \gamma^*)$$

(12) $$LM = [L^{(1)}(\gamma^*)]'\ \{L^{(2)}(\gamma^*)\}^{-1}\ [L^{(1)}(\gamma^*)]$$

where γ is the vector of unrestricted estimated parameters and γ^* is the vector of parameters subject to a set of distribution restrictions. In our case, γ includes the parameters (β,k,c) and $V(\gamma)$ is the estimated covariance matrix of parameters.

Comparison of the maximized value of the log-likelihood function under alternative distributional assumptions is a simple method for choosing among the GG, G, W, LN, and E models. Hypotheses concerning the adequacy of special cases such as the G, W, or LN models (i.e. the validity of shape parameter restrictions) can be formally tested using the appropriate LR, LM, or Wald tests against the more-general GG distribution. Similarly, these formal tests can be used to compare the E model against either the G or W models. Tables 4 and 5 give the values of the test statistics for each pairwise comparison between nested models (within the linear and log-linear families), for the Canada/U.S. and overseas samples respectively. As we might expect in very large samples, the null hypotheses of model equivalence are soundly rejected in all cases. Although the computed $\chi 2$ statistics for the three tests are often quite different numerically, there is apparently no conflict in the test results. In all cases, the tail probability values associated with the observed values of the test statistics are extremely small.

Nevertheless, it is interesting to note the relative magnitudes of these test statistics. The Berndt-Savin [2] inequality (Wald $\geq$ LR $\geq$ LM) for linear normal models is not expected to hold here and it does not. We find the LM and Wald statistics to be quite close in comparing E to either G or W models, while the corresponding LR statistic is higher. Tests against the GG model show much less agreement. In this case the LR or LM statistic is always the largest. Magee [15] shows that conflict among the test statistics can sometimes be explained by examination of the third and fourth derivatives of the log-likelihood function. His results indicate that when the third and fourth derivatives are negative, the LR statistic will be the smallest. In cases where the third derivative is not negative, the LR statistic will be the largest. As an illustration of Magee's results in the present case, we can easily compute the third and fourth derivatives of the likelihood function for the G distribution and examine the ranking of the test statistics of the E distribution against the G distribution in the linear model. The third and fourth derivatives of the linear G log-likelihood function with respect to shape parameter k are:

(13) $$L^{(3)}(k) = n[-k^{-1} - \Psi^{(2)}(k)],$$

(14) $$L^{(4)}(k) = n[2k^{-3} - \Psi^{(3)}(k)],$$

where $\Psi^{(i)}$ is the ith derivative of the Ψ function, which is the derivative of the logarithm of the Γ function. Tables 4 and 5 show that the LR statistics in this test always exceed the corresponding LM and Wald statistics. For the linear G model, evaluation of the third and fourth derivatives at the maximum likelihood estimates of k resulted in $L^{(3)}$ =

Table 4

SPECIFICATION TEST* STATISTICS: CANADA AND U.S. DESTINATIONS

	LR	LM	Wald
linear models:			
E:G	643.04	548.53	547.95
E:W	173.58	169.05	169.18
G:GG	2811.52	4591.80	1625357600000.**
W:GG	3281.32	771.29	9664659500.**
LN:GG	0.14	-	-
log-linear models:			
E:G	653.84	926.19	556.38
E:W	175.36	197.0	195.47
G:GG	2825.62***	7822.0	13.332
W:GG	3304.10***	780.37	3.3425

* 5% critical value for $\chi^2(1)$ is 3.84
** reflects extreme k and c estimates in Table 2.A
*** using highest value of likelihood achieved for GG

Table 5

SPECIFICATION TEST* STATISTICS: OVERSEAS DESTINATIONS

	LR	LM	Wald
linear models:			
E:G	274.8	220.54	220.15
E:W	161.8	152.07	152.56
G:GG	254.4	349.34	290.50
W:GG	367.4	112.26	6.1381
LN:GG	26.0	-	-
log-linear models:			
E:G	274.54	465.16	219.94
E:W	161.78	209.16	201.66
G:GG	253.14	104.66	15.865
W:GG	365.90	112.04	6.304
LN:GG	26.44	-	-

* 5% critical value for $\chi^2(1)$ is 3.84

14874 and $L^{(4)}$ = -37962 for the Canada/U.S. sample. Under these circumstances, Magee reports that we should obtain LR $\geq$ max(W,LM) which is exactly what happens. While computation of the third and fourth derivatives is often quite difficult, we have shown that in a few cases, it is relatively simple and can be used to explain conflicts among the test statistics. Examination of Table 4 and 5 shows that the LR statistic was the largest in most tests except for some the the GG tests.

Model discrimination among the G, W, and LN models is more difficult since these are non-nested. However, the distribution of the likelihood ratio test statistics for some non-nested tests have been tabulated. For example, Dumonceaux and Antle [6] provide small-sample tables (derived by Monte Carlo methods) for the log-linear W and LN models. Unfortunately, these tables are not sufficiently detailed for use here.

6. POLICY IMPLICATIONS

One outcome of deregulation in long-distance markets has been decreases in marginal rates. This is likely to increase both the number of calls and the mean duration of calls. The fitted models obtained above can be employed to shed some light on current rate-setting issues facing telephone utilities. Although the model is not designed to predict the number of new calls induced by a change in rates, one can predict the influence of hypothetical long-distance rate reductions upon the durations of existing calls.

The estimated coefficients reported in Table 2 and 3 can be used with the summary statistics reported in Table 1 to generate estimates of the "rate elasticity of duration" at the mean values of the observed data. In the Canada/U.S. sample, these estimated rate elasticities are nearly zero for the log-linear regressions. Table 6 shows the implied rate elasticities

Table 6

RATE ELASTICITIES OF DURATION (AT SAMPLE MEANS)

Distr.:	N	E	G	W	GG	LN
Canada and U.S. Destinations:						
linear:	.093	.215	.215	.210	.195	.195
log-linear:	-	.0004	.0004	.0017	.0105	.0105
Overseas Destinations:						
linear:	.027	.051	.051	.019	.225	.277
log-linear:	-	.040	.040	.011	.205	.250

of duration for each sample, each distributional assumption, and each functional form of the regression relationship. Duration is consistently rate-inelastic, but considerable variation can be observed across models. In particular, for the Canada and U.S. destinations, the point estimates are comparable across the different distributional assumptions, but differ drastically between the linear and log-linear functional forms. In contrast, for the overseas destinations, the estimated rate elasticities differ quite markedly over the range of distributional assumptions, but within a particular distributional assumption, are comparable across functional forms. Clearly (depending on the precise data being utilized), the policy implications of the fitted model can depend on the choice of distribution and/or the choice of functional form.

As a numerical illustration, assume that these estimated point elasticities can be extrapolated across a relatively large, finite rate change. Then, for the Canada/U.S. sample, a 25% reduction in marginal rates as a result of deregulation (reductions of this magnitude are likely) could increase durations at most by 5% according to these results. The expected 20-25% loss in revenue on existing calls would have to offset by a moderate increase in the number of calls or increased revenue from other services (which might be obtained by increased local service charges). As new services such as MCI and Sprint in the U.S. and perhaps CNCP in Canada offer these rate reductions, we are likely to observe substantial increases in their business at the expense of traditional companies such as AT&T and Bell Canada. However, our results indicate that total industry revenue will decline unless it is offset from other sources. In Canada, the Canadian Radio-Television and Telecommunications Commission (CRTC) has recently recognized that increases in local rates may not be socially desirable (see Surtees [22]) and surprised the nation by rejecting the CNCP bid to enter the long-distance market despite the promise of lower rates. However, the decision is expected to be temporary as CNCP is expected to submit a revised proposal. The CRTC did recognize that competition in the long distance market would be desirable but is concerned about the implications for the universality and price of local service. This controversy is likely to be settled only by a Federal Cabinet decision (currently governed by the Conservative Party, which is sympathetic towards deregulation).

Although deregulation has had little impact on overseas rates our overseas results indicate similar rate elasticities of duration to those in the Canada/U.S. sample.

7. CONCLUSIONS

This paper has demonstrated that in regression applications where the dependent variable is strictly non-negative the usual normal conditional probability density function may be inappropriate. Whereas a log-linear regression model with normal errors may sometimes suffice, we have experimented with a wide range of linear and log-linear models with error distributions in the GG family of distributions.

In our application of these alternative models to long-distance telephone call durations, we have utilized two samples, two assumptions about functional form, and five major assumptions about error distributions (the GG and its four special cases). We have found that different error assumptions can result in considerably different point estimates for the regression coefficients. We have quite clearly refuted the adequacy (for these applications) of the Exponential, Weibull, or Gamma distributions for characterizing durations, despite their popularity in the literature. Lagrange Multiplier, Wald, and Likelihood Ratio tests have been employed to distinguish between nested pairs of models. However, the two classes of models, linear and log-linear, are not formally comparable by these methods because they are non-nested.

Estimation of the GG model reveals that the simple log-linear model with normal errors seems to provide a good fit for our sample of calls to Canada and U.S. destinations. In contrast, the linear model with GG errors fits better for our sample of overseas calls. The regression technique improves upon previous pairwise correlation studies between duration and other variables; it also formalizes the fitting of non-normal distributions for different categories of calls. Qualitatively, our results regarding the determinants of duration are generally consistent with previous findings, but the regression model represents a more systematic mode of analysis and should therefore yield more reliable quantitative estimates. The estimated rate elasticities indicate that rate reductions are likely to result in only a small increase in call duration. As a result, the revenue loss on existing calls must be offset by a substantially increased number of calls or other services or higher charges for current services such as local calls.

All of the quantitative results presented here are derived strictly for existing calls. The formulation of the model precludes any consideration of the relationship between call durations and call frequencies. This work must therefore be incorporated into a more-complete model of long-distance telephone demand which takes account of both duration and frequency.

ACKNOWLEDGEMENTS

The authors wish to thank D. Gorham, S. Haun, D. Smeaton, T. Wales, and three anonymous referees for helpful comments and assistance.

NOTES

[1] Estimation becomes complicated when typical durations are long (both relative to the sample period and in real time), and the sample will often contain a substantial proportion of incomplete spells. In these contexts, the hazard function is a more useful statistical concept upon which to base an econometric analysis.

[2] Technically, as k approaches ∞, this GG distribution corresponds to a lognormal distribution with parameters μ and σ. The relationships between the parameters are $b^c = \sigma^2 c^2$ and $k = (c\mu+1)b^c$. (See McDonald [16].)

[3] For example, the mean of the simple G distribution is the product: bc. The lease restrictive assumption would allow both $b = b(x)$ and $c = c(x)$, but then the fitted regression line would be quadratic in x: $b(x)c(x)$. Holding the shape parameters constant is an assumption no stricter than that of homoscedasticity in the normal regression model.

[4] Straightforward substitution of just $x'\beta$ for $b(x)$ would mean that the fitted value of t would be a non-linear function of the estimated parameters. This complicates the process of inference.

[5] For this study, we have access only to one day's data on individual calls. This precludes any simple method of including household-specific

variables among the regressors. Unfortunately, we must also overlook any potential endogeneity of the explanatory variables.

6 The RATE variable was constructed with considerable care by identifying the precise type (and point-to-point distance) of each call and consulting telephone company billing schedules to determine the price of one extra minute, given the actual duration of the call.

7 The first and second derivatives derived in the separate Appendix have been incorporated in the quasi-Newton algorithms utilized by the MLE command in Version 5 of White's [25] SHAZAM Econometrics Computer Program. Convergence was generally attained rapidly except in the GG case.

REFERENCES

[1] Amemiya, T., "Nonlinear Regression Models," Ch. 6 in Griliches and Intriligator, *Handbook of Econometrics*, North-Holland, 1983.

[2] Berndt, E. and N.E. Savin, "Conflict Among Criteria for Testing Hypotheses in the Multivariate Linear Regression Model," *Econometrica*, 45 (1977), 1263-78.

[3] Box, G., and D. Cox, "An Analysis of Transformations," *Journal of the Royal Statistical Society, Series B*, (1964), 211-52.

[4] Curien, N., "Modelisation de l'effet des tarifs sur la consommation et le trafic; Application a l'etude de la taxation des communications locales a la duree," Note DGT, 1981.

[5] de Fontenay, Alain, Debra Gorham, J.T. Marshall Lee, and George Manning, "Stochastic Demand for a Continuum of Goods and Services: The Demand for Long Distance Telephone Services," Paper presented to the Transportation and Public Utilities Group Session of the American Economic Association Meetings, New York, NY: December 29, 1981.

[6] Dumonceaux, R., and C. Antle, "Discrimination Between the Lognormal and Weibull Distributions," *Technometrics*, 15 (1973), 923-26.

[7] Gale, W.A., "Duration of Interstate Calls, March 1969," unpublished Bell Laboratories Memorandum, December 1971.

[8] Gale, W.A., "Elasticity of Duration for Intrastate Calls," unpublished Bell Laboratories Memorandum, October 1974.

[9] Goldfeld, S., and R. Quandt, "Maximization by Quadratic Hill-Climbing," *Econometrica*, 34 (1966), 541-51.

[10] Heckman, James, J,. and Burton Singer, "Econometric Duration Analysis," *Journal of Econometrics*, 24 (1984), 63-132.

[11] Jensik, John M., "Dynamics of Consumer Usage," in J.A. Baude et al. (eds.) *Perspectives on Local Measured Service*, Kansas City, MO: Telecommunications Industry Workshop Organizing Committee, 1979.

[12] Johnson, N.L., and S. Kotz, *Continuous Univariate Distributions (1)*, New York, NY: John Wiley and Sons, 1970.

[13] Lawless, J.F., *Statistical Models and Methods for Lifetime Data*, New York, NY: John Wiley and Sons, 1982.

[14] Lee, Lung-fei, "Maximum Likelihood Estimation and a Specification Test for Non-normal Distributional Assumptions for the Accelerated Failure Time Models," *Journal of Econometrics*, 24 (1984), 159-79.

[15] Magee, L., "Sufficient Conditions for Inequalities for LR, W, and LM Tests from Taylor Series Expansions," McMaster University, 1985.

[16] McDonald, James B., "Some Generalized Functions for the Size Distribution of Income," *Econometrica*, 52 (1984), 647-63.

[17] McDonald, James B., and Whitney Newey, "A Generalized Stochastic Specification in Econometric Models," Paper presented at the 1984 North American Summer Meeting of the Econometric Society, Stanford, CA, June 29.

[18] Pacey, Patricia L., "Long Distance Demand: A Point-to-Point Model," *Southern Economic Journal*, 49 (1983), 1094-107.

[19] Park, Rolla Edward, Bruce M. Wetzel, and Bridger M. Mitchell, "Price Elasticities for Local Telephone Calls," *Econometrica*, 51 (1983), 1699-730.

[20] Pavarini, Carl, "The Effect of Flat-to-Measured Rate Conversions on Local Telephone Usage," in J. Wenders (ed.) *Pricing in Regulated Industries: Theory and Application*, Denver, CO: Mountain States Telephone and Telegraph Co., 1979.

[21] Rea, John D. and Lage, Gerald M., "Estimates of Demand Elasticities for International Telecommunications Services," *Journal of Industrial Economics*, 26 (1978), 363-81.

[22] Surtees, Lawrence, "CNCP's Telephone Bid Rejected," *Globe and Mail*, August 30, 1985, 1-2.

[23] Taylor, Lester D., *Telecommunications Demand: A Survey and Critique*, Cambridge, MA: Ballinger, 1980.

[24] Veitch, J.G., "Relation of Conversation Time to Message Frequency in Residential Usage," AT&T Bell Laboratories, Murray Hill, New Jersey, 1985.

[25] White, Kenneth J., "A General Computer Program for Econometric Models- SHAZAM" *Econometrica*, 46 (1978), 239-40.

[26] Wong, T.F., "Identifying Tariff-Induced Shifts in the Subscriber Distribution of Local Telephone Usage," Bell Laboratories, 1981.

SECTION III: USER MARKET STRUCTURE

III.1. Access and Market Structure

TELECOMMUNICATIONS DEMAND MODELLING
An Integrated View
A. de Fontenay, M.H. Shugard, D.S. Sibley (Editors)

COMPETITION WITH LOCK-IN

Joseph FARRELL

GTE Laboratories Incorporated, and
University of California, Berkeley
Berkeley, CA 94720 USA.

1. INTRODUCTION

Once a buyer begins to buy from a particular seller, he may become locked in: competing goods that were good substitutes before are now less good substitutes. There is less competition ex-post than there was ex-ante.

Examples of this are common. A customer who chooses a particular long-distance carrier in an equal-access exchange may face explicit charges as well as nonpecuniary costs in changing to another carrier. A large user who has hard-wired bypass of a local loop to a long-distance carrier will face costs of changing to another supplier. A firm that locates a manufacturing plant near a major supplier gives that supplier some power to raise price without inducing substitution, even though it had no advantage ex ante. A person who chooses a doctor will normally be somewhat reluctant to change, even if there is no evidence that the doctor is better than another. A buyer of cars, cameras or computers may be obliged to buy upgrades, spare parts or accessories from the maker of the original equipment. This phenomenon of "lock-in," "switching costs," or "inertia" has attracted some attention recently in the economics literature: see for instance Farrell (1986), Farrell and Gallini (1986), Farrell and Shapiro (1986), Green and Scotchmer (1986), Klemperer (1986), Scotchmer (1986), Summers (1985), Sutton (1980), and von Weizsacker (1984). Generally, these treatments have focused on the effects of lock-in on firms' pricing policies. Klemperer and Summers emphasise artificially created switching costs (for instance, airlines' "frequent flyer" discounts). Porter (1985) makes "switching costs" an important part of his analysis of competitive advantage. For a discussion of customers' "conversion costs" in the computer industry, see Fisher, McGowan and Greenwood (1983).

These examples involve the formation of relationship-specific capital, using the term "capital" in its most general meaning of an asset that lasts in time. By saying that an asset is relationship-specific, we mean that its best use outside the relationship is strictly less valuable than its use within. Besides these examples of rational lock-in effects, we also observe brand loyalty, especially in consumer purchases, even when there is no apparent specific capital -- in other words, when there is no "objective" reason for inertia. From the seller's point of view it may matter little what is the source of the inertia.

Lock-in is important not only in markets with posted prices and many buyers, but also, for instance, in the procurement problem faced by the Defense Department, a city contracting for a cable-TV franchise, or any buyer of custom-designed goods. Once the initial contract is awarded to one supplier, that supplier may have considerable ex-post monopoly power, even though before the contract was awarded there were many equally qualified sellers clamoring to be selected.

Lock-in is important also for regulation. Once a provider of some service has become entrenched, competitive pressure may no longer do an adequate job of disciplining price, service quality, and so on -- even if ex ante there were many bidders for the "franchise". This is one of the problems with the "competitive franchising" alternative to administrative regulation suggested by Demsetz (1968). Demsetz proposed that sellers "bid" on "the price" at which they will serve demand. In a simple model, this effectively makes sellers reveal the true level of average costs, and promise to service demand at that price -- a result that would probably outperform practical administrative regulation. But since prices will have to change over time in response to cost changes, such a bid would have to be a complicated function of observable aspects of costs and demand data, if there is to be any chance of achieving efficiency. Such complex long-term contracts are notoriously hard to write and to enforce, and it might be that contract enforcement would come to much the same thing as administrative regulation. On the other hand, lock-in means that it would be hard to have re-franchising too often. Cable TV regulation (by cities) has encountered precisely these problems.

Three important problems arise in a market with lock-in that do not arise without. First and most obviously, the ex-post monopoly power may be exploited by the seller. In other words, a seller who has acquired some "locked-in" customers may raise prices (see Klemperer 1986, Farrell and Gallini, 1986), lower service quality (see Shepard 1986), cut back on research or other expenditures that make the product attractive, or the like. Second, even if the price-gouging problem were solved, there is another supply problem: if sellers may go bankrupt (or leave the market for other reasons), buyers will have to try to predict the likelihood of that, and choose their supplier with that fear in mind, in a way that does not apply in a standard market in which there are no costs to leaving a sinking ship. Third, if different sellers may be more or less successful in tracking technological progress, then buyers will be concerned to predict whose products or services will be best in the future, not only whose are best now. In particular, if there are network externalities in consumption, then there is an advantage to buying from the seller who will have greatest market share in the future, even if his share now is low (see Katz and Shapiro (1986a), Farrell and Saloner (1986)).

In this paper, we discuss some economic problems generated by lock-in. In Section 2, we summarise existing work on price effects. In Section 3, we discuss the problems of bankruptcy and technological progress with lock-in, with particular reference to the microprocessor industry. Section 4 describes some active strategies with which buyers can sometimes mitigate the problems of lock-in. Section 5 concludes.

2. PRICE COMPETITION

Lock-in has two competing effects on price. On the one hand, buyers who are locked in can be exploited by their supplier, if no contract prevents this. (We discuss contracts below.) This exploitation can potentially far exceed the simple degree of lock-in (switching costs), for each seller can exploit his locked-in buyers by charging a price a little higher than do his competitors -- who are themselves doing likewise. As Klemperer (1986) has emphasised, this can lead to monopoly pricing, in much the same manner as in Diamond's (1971) search model. (Summers (1985) and Green and Scotchmer (1986) have related results.)

Clearly, locked-in customers are profitable. But this fact itself creates competition -- competition to capture buyers! When market share is valuable, as when buyers are locked in, a seller's marginal-revenue curve is shifted upwards, so that competition becomes fiercer. Thus, if new buyers can be effectively separated (charged different prices) from old, as in Klemperer's two-period models, we find ex-ante competition followed by ex-post monopoly. This competition may lower profits so much that firms prefer to reduce switching costs by making their products compatible -- see for instance Klemperer (1986) or Katz and Shapiro (1986b).

When new and old buyers must be charged the same price, however, then there is no clearcut "ex-ante" and "ex-post". Each seller must compromise between his desire to exploit his locked-in buyers and his wish to attract new buyers. The importance of going after new buyers depends on how profitable they will be -- how much they will be exploited -- in the future. At the same time, the extent to which he wishes to exploit the old buyers depends on the relative importance of attracting new buyers. This problem, therefore, cannot be properly tackled in a two-period model, but demands a many-period treatment. Unfortunately, such a treatment has (so far) proven mathematically intractable. Von Weizsacker (1984) and Green and Scotchmer (1986) simplified the problem by using solution concepts that ignore some part of the strategic intertemporal interaction between sellers; they effectively assumed away competitors' price reactions to a seller's change in price. Farrell and Shapiro (1986) solved for perfect equilibrium (thus taking account of such reactions), but were able to do so only by drastically simplifying the structure of demand.

One conclusion of these models, emphasised especially by Farrell (1986) and Farrell and Shapiro (1986), is that firms with many locked-in buyers will be relatively less willing to cut prices so as to attract new buyers: their marginal-revenue curves are always lower than those of less well-endowed rivals, because any price cut must be given to locked-in buyers as well as to new buyers. This is a "fat-cat" result, in the sense of Fudenberg and Tirole (1984): the large firm is too "fat" to compete effectively for the new buyers. An interesting consequence of this is that buyers may not always wish to patronise the cheapest firm, even if all products are identical and if they are not yet locked-in. The reason is that, if all new buyers go to the cheapest firm, it may well become a large firm as a result, and will therefore be interested more in exploiting its locked-in buyers than in competing for new; thus its price is likely to be high. Therefore, buying from the cheap and much-patronised firm now may lock a buyer in to what will become an expensive firm soon. It may be wiser to "flee the crowd" and buy from a smaller, if slightly more expensive, seller. Whether we see such behavior in practice, however, is questionable.

3. OTHER PROBLEMS

Price gouging is by no means the only problem for buyers in a market with lock-in. Interruption of supply, or technological backwardness on the part of a supplier, may be equally or more damaging, and may be much harder to control contractually. What can we say about these problems?

In the microelectronics industry, products such as personal computers are designed around a microchip that incorporates an "architecture" that is often proprietary to the microchip supplier. To change to a new architecture involves extensive redesign of the entire product, and this is a very substantial switching cost indeed. Because both buyer and seller of the chip are firms, and because this switching cost is large, it might seem appropriate to solve the problem of lock-in by vertical integration or by detailed contracting. Surprisingly, vertical integration is not widespread in the United States microelectronics industry. There is much more vertical integration in the Japanese industry. See Ferguson (1985) for an extended discussion of this, its historical causes, and its possible implications. Shepard (1986) also reports that, while long-term price contracts are common in this industry, they seldom specify such other important features of performance as delivery times.

These architectures are not the same as the chips. Rather, as technical progress winds its rapid way along, the chips are updated (typically every two or three years) by each chip manufacturer within its own architecture. For instance, at the time of writing, Intel has recently introduced the 80386 chip, which is compatible with but an advance on its 80286 and 8086 chips. Therefore, in choosing a chip manufacturer, one is choosing to trust a firm to keep up with (or preferably lead) technical change. Moreover, if a chip maker goes bankrupt then it is by no means guaranteed that someone else will take over and develop that architecture. So the choice of an architecture also involves trusting a firm to stay in business.

These problems may be more important than even quite large price differences. For instance, if performance in the industry as a whole is improving at 20% per year, and if an unwise choice of architecture means that the improvement in the product one is locked into is only half that, then performance will be about 30% behind the industry after only four years. This could well prove a fatal problem for the product. Moreover, if the architecture is not developing quickly enough, then it will be attracting few if any new buyers, and this could lead either to bankruptcy of the supplier or to incentives to raise prices: given that the seller is attracting few if any new buyers in any case, he may be tempted to raise price and exploit his locked-in buyers.

How does a buyer choose a seller that he can trust to keep up with the industry's progress? Reputation may come in here, giving an advantage to those firms that have been long established (though not of course to those that have performed poorly in the past). Size (in the market) is another major advantage, for three reasons. First, size will support generous research and development budgets. Second, size makes bankruptcy relatively unlikely. Third, in the event of bankruptcy, it is likely that size will make others pick up the architecture. Moreover, there may be standardization advantages to "going with the crowd".

These advantages of size have a "positive-feedback" effect on competition for market share. At any stage, the largest seller has an advantage due to size. This effect is analytically akin to the presence of economies of scale, learning by doing, or network externalities. Especially if the effects are dynamic (if scale in one period gives an advantage in subsequent periods, as well as contemporaneously), entry may be made relatively difficult, and early leaders may achieve a lasting benefit. Of course, if the industry is competitive from the beginning, that merely shifts the locus of competition to the early stages, in much the same way as lock-in itself does so with competition for market share.

To the extent that buyers look for signs of continuity of supply, there are advantages to being perceived as relatively unlikely to suffer strikes, go bankrupt, or become capacity-constrained. Again, all these things are important contemporaneously in any market; but when lock-in is important, buyers' predictions of future values become essential also.

4. STRATEGIES FOR STRATEGICALLY ACTIVE BUYERS

In the models discussed above, buyers have no strategic power. That is, each new buyer selects the seller that offers him the best available deal, but buyers cannot affect the set of options offered. This is often a reasonable assumption, for instance in most consumer markets, but there are many important markets in which it is not the case. Examples include government procurement, regulation, and the microelectronics market discussed above. More generally, it includes cases of bilateral monopoly power, such as the case where a major user of fuel is considering a choice of supplier. See Joskow (1985). But market power is not necessary on either side for there to be active negotiation of contract terms. Rather, what is required is presumably some combination of absolute size of transaction and relative importance for both parties; and although size tends loosely to go with market power, there is no necessary connection. In this section, we ask what strategies a buyer might usefully follow in order to mitigate the problems discussed above.

We focus on three main strategies: long-term contracts, vertical integration, and second-sourcing.

4.1 Long-Term Contracts

Perhaps the most obvious protection against the kind of opportunism that lock-in may produce is to sign long-term contracts. When buyers are strategically active, such contracts are sometimes used. For instance, Joskow (1985) describes long-term contracts between power producers and their suppliers of coal. Such contracts are obviously useful weapons against price gouging. However, when there is uncertainty about future costs, demand, value for the product, etcetera, and when important features of behavior (intensity of research and development effort, good-faith efforts to reduce delivery lags, etc.) are difficult to observe or to contract on, long term contracts are far from a full solution. Shepard (1986) reports, for instance, that although prices are often determined by such contracts in the microelectronics industry, delivery times seldom are.

An ideal long-term contract is equivalent to an ideal scheme of regulation. In either case, the goal is to formulate rules under which efficient decisions result from a process in which the interests of the better-informed party do not coincide with those of the other party. In the economics literature, this is studied under the heading of the "principal-agent problem". The basic lesson is clear enough: the informational asymmetry causes problems. The problems of regulation are not simply a matter of government intervention in the market: rather, the private market also suffers from them in a closely related form. For example, an ex-ante efficient long-term contract will normally offer some protection to the seller against cost increases -- it will not specify prices independently of costs. To that extent, the incentives to keep costs down are diluted. Likewise, if the seller is also selling to other buyers, and there are joint costs, then the problem of how to allocate the joint costs between the contractual output and other output is the same problem in essence as the regulatory problem of how to allocate joint costs between regulated and unregulated activities.

4.2 Vertical Integration

If a contract does not work, one alternative is vertical integration: put a top executive in charge of both buyer and seller, charge her to maximize joint profits, and the problems go away. So at least it would seem in theory. What then prevents firms from integrating almost universally? -- after all, any intelligent industrial organization economist can think of several kinds of problem that might arise between two firms in any particular industry. Clearly, our "theory" of vertical integration has its problems: there are costs to integration. Some can be seen in the context of lock-in.

For instance, if there are economies of scale in the upstream industry, or benefits from making a product a de facto industry standard, and if downstream firms are unwilling to buy from a vertical partner of a downstream rival, then vertical integration may involve the sacrifice of these economies of scale.

A more systematic flaw in vertical integration is that the "upstream" division may come to depend on its internal customer for viability, creating political pressure on the downstream user to buy internally. While the problems of price gouging presumably can be prevented by our senior executive, other problems such as "laziness" are harder to police.

4.3 Second Sourcing

One way to achieve ex-post competition is to insist on product standardization or compatibility among suppliers, so that switching costs become very much smaller (perhaps negligible). Then the seller individually has no power to behave opportunistically. This strategy has been followed by the Department of Defense, and is often used in industry also. For example, Intel has made second-sourcing arrangements with various competitors, and Xerox has openly licenced its Ethernet local-area-network technology. In AM stereo, Motorola has also followed a low-price licensing strategy. Evidently the strategy is attractive to buyers; it is perhaps less obvious that it may be attractive to sellers also, but the seller must attract new buyers at some stage, and must either convince them that he will not exploit them once locked in, or else accept a lower level of demand than he could have had. Farrell and Gallini (1986) analyze the incentives for a monopoly innovator of a product with lock-in voluntarily to allow entry (or to licence its product free of royalties) after an initial period, and show that, if the lock-in is severe enough, it often pays to do so, in order to attract new buyers. On this strategy, see also Shepard (1986).

What are the social costs of second-sourcing? We identify two here: one related to costs, and the other to the process of standardization that is often accelerated by second-sourcing agreements.

In the absence of sunk costs of production, compulsory licensing of product design is sufficient to produce ex-post competition. But when there are sunk costs, for instance if there is a learning curve, then it is necessary in general to do more: the knowledge in principle of how to produce a product may not be enough to make one an active competitor. Thus, free entry and public knowledge of the technology may not suffice to deregulate local telephone service without generating monopoly problems. Maintaining an active competitor may be costly, since it involves splitting the orders so as not to let one seller go too much further down the learning curve than the other.

From a policy point of view, there are pitfalls in imposing or accelerating standardization. It is often difficult to change a standard once it is in place (see David, 1985; Farrell and Saloner, 1985, 1986), and the buyer concerned with mitigating lock-in problems may be more interested in the short-run competitive effects than in the implications for long-term social benefits from the standard that gets adopted. This raises the possibility that second-sourcing, in encouraging early standardization, may sometimes be harmful.

5. CONCLUSION

We have identified an important problem in the theory of competition that has until recently received little attention from economists. Whether rationally or not, buyers are often "loyal" to suppliers. As a result, no static formulation of the degree of competition is adequate: there may be intense competition to "capture" new buyers, while at the same time monopolistic practices may prevail in the price and other treatment of "old" customers.

When buyers have strategic power, they can mitigate the effects of switching costs. We have briefly discussed three potentially useful strategies: long-term contracts, vertical integration, and second-sourcing requirements. In general, none of these strategies is ideal, however, so we can expect to see (as we do see) problems of lock-in persisting despite buyers' strategic actions. Furthermore, buyers' attempts to predict features of sellers' behavior and performance in the future may lead to biases towards (for example) large sellers, which may affect the efficiency of competition. If buyers attempt to influence the course of events through long-term contracts or through second-sourcing requirements, these actions may themselves have efficiency effects comparable to those of regulation.

FOOTNOTES

1. I am indebted to Charles Ferguson for conversations on the microelectronics industry. Any misunderstandings are mine.

2. Zilog's Z8000 chip, for instance, was initially very attractive, but Zilog failed to update the chip for many years, with the result that it is now nearly obsolete (Ferguson, 1985, p. 47).

3. For instance, MOS Technologies, the supplier of the 8-bit 6502 chip used in the Apple IIe, went bankrupt. Although there is now a successor corporation (Western Design Center) that has recently announced a successor chip within the MOS architecture, there has been much more of a lag (ten years) in updating the MOS chip than would have been likely had MOS not gone bankrupt. (Ferguson, 1985, p. 47.) It is difficult to take over the architecture of a bankrupt concern (even if it is very valuable) because much of the essential knowledge is in people's heads. Thus it is not simply an asset that will be transferred on bankruptcy.

REFERENCES

P. David, "The Economics of QWERTY," American Economic Review, 75, May 1985, 332-336.

H. Demsetz, "Why Regulate Utilities?" Journal of Law and Economics, 11, 1968.

P. Diamond, "A Model of Price Adjustment," Journal of Economic Theory, 3, 1971.

J. Farrell, "A Note on Inertia in Market Share," Economics Letters, 21, (1986), 73-75.

---- and N. Gallini, "Second-Sourcing as a Commitment," mimeo, Berkeley, (1986).

---- and G. Saloner, "Standardization, Compatibility and Innovation," Rand Journal of Economics, 16, (1985a), 70-83.

---- and ----, "Compatibility and Installed Base: Innovation, Product Preannouncements, and Predation," American Economic Review, 76, December 1986, forthcoming.

---- and ----, "Economic Issues in Standardization," in Telecommunications and Equity, proceedings of the 1985 Telecommunications Policy Research Conference, J. Miller, editor, North-Holland 1986.

---- and ----, "Competition, Compatibility, and Standards," in Proceedings of INSEAD conference on Standardization, H.L. Gabel, editor, North-Holland, to appear.

---- and C. Shapiro, "Dynamic Competition with Lock-In," mimeo, Berkeley and Princeton.

C. Ferguson, "American Microelectronics in Decline: Evaluation, Analysis and Alternatives," mimeo, MIT, 1985 (draft).

F. Fisher, J. McGowan, and J. Greenwood, Folded, Spindled and Mutilated, MIT Press, 1983.

D. Fudenberg and J. Tirole, "Learning by Doing and Market Performance," Bell Journal of Economics, 14, Autumn 1983.

---- and ----, "The Fat-cat Effect, the Puppy-Dog Ploy, and the Lean and Hungry Look," American Economic Review, 74, (1984).

J. Green and S. Scotchmer, "Bertrand Competition with a Distribution of Switch Costs and Reservation Prices," mimeo, Harvard, 1986.

W. Hanson, "Bandwagons and Orphans: Dynamic Pricing of Competing Technological Systems subject to Decreasing Costs," Stanford thesis 1985, mimeo University of Chicago.

P. Joskow, "Vertical Integration and Long-Term Contracts," MIT mimeo, 1984.

M. Katz and C. Shapiro, "Technology Adoption in the Presence of Network Externalities," Journal of Political Economy, 94 (1986a), 822-841.

---- and ----, "Product Compatibility Choice in a Market with Technical Progress," Oxford Economic Papers, 38, (1986b), forthcoming.

P. Klemperer, "Markets with Consumer Switching Costs," Stanford GSB thesis, 1986.

M. Porter, Competitive Advantage, The Free Press (Macmillan), 1985.

S. Scotchmer, "Market Share Inertia with More than Two Firms: An Existence Problem," Economics Letters, 21, (1986), 77-79, and "Erratum," Economics Letters, forthcoming.

A. Shepard, "Licensing to Enhance Demand for New Technologies," Yale mimeo, 1986.

L. Summers, "Frequent Flyer Programs and Other Loyalty-Inducing Economic Arrangements," mimeo, Harvard, 1985.

J. Sutton, "A Model of Stochastic Equilibrium in a Quasi-Competitive Industry," Review of Economic Studies, 47, (1980), 705-722.

C. von Weizsacker, "The Costs of Substitution," Econometrica, 52, September 1984, 1085-1116.

O. Williamson, Markets and Hierarchies, The Free Press, 1975.

TELECOMMUNICATIONS DEMAND MODELLING
An Integrated View
A. de Fontenay, M.H. Shugard, D.S. Sibley (Editors)

ALGORITHMIC ANALYSES OF OLIGOPOLISTIC COMPETITION IN SPACE

Balder VON HOHENBALKEN and Douglas S. WEST

Department of Economics, University of Alberta
Edmonton, Alberta, Canada T6G 2H4

1. INTRODUCTION

Until quite recently, the telephone industry in the U.S. was virtually monopolized by one firm, AT&T. Intercity long distance service was supplied by AT&T Long Lines, while local service to much of the country's population was provided by local Bell System Operating Companies. (For a comprehensive discussion of the development of AT&T and the telecommunications industry, see Brock [1].) Cracks began to appear in the monopoly in the 1960's and 1970's as new firms successfully challenged AT&T's hold over intercity telecommunications services. Then, in 1982, it was announced that the Justice Department and AT&T had reached an agreement that would settle the antitrust complaint filed by the Justice Department in 1974 against AT&T and its subsidiary Western Electric. The relief prescribed in the Modified Final Judgment (MFJ) included (1) a requirement that AT&T divest itself of its local operating companies, (2) that these companies provide equal access to all interexchange carriers, and (3) permission for AT&T to retain Long Lines, Western Electric, and Bell Labs. (For a more detailed review of the MFJ, see Crandall and Owen [2].)

The breakup of AT&T has important implications for the way in which economists will theoretically model and empirically analyze the telecommunications industry. For example, in the past, studies of telecommunications demand did not need to concern themselves with differences in demand faced by the firm and the industry. Instead, economists could focus on the demands for different telephone services, such as demands for access, local calls, toll calls, and call duration, the objective apparently being to estimate various price and demand elasticities (see Taylor [5]). With the recent entry of new firms into the intercity telecommunications industry, a new dimension to the study of telecommunications demand has been revealed: the demand for access to a given firm's intercity telecommunications network. Given that there is likely to be a small number of firms competing in the industry, the tools of oligopoly theory will need to be used to study this problem. The question then arises as to which tools and techniques of analysis should be used.

In part the answer will depend on whether long-distance carriers interconnect with local exchange carriers or whether they bypass the local networks with direct connections between themselves and their customers. If, as Jackson and Rohlfs [3] predict, bypass becomes quite prevalent over the next 10 years, and if firms charge prices for access that are a function of the distance between the consumer and the firm, then competition among long-distance carriers will be localized in space and the techniques of spatial competition analysis can be brought into play. These techniques have served us well in our studies of oligopolistic competition in the supermarket industry, an industry where location is obviously an important decision variable. Since the economics of bypass may make intraurban location an important decision for long-distance firms, spatial competition analysis may also have a useful role to play in the theoretical modeling and empirical investigation of this industry. It is the purpose of our paper to outline why locational choice might need to be considered in the analysis of oligopolistic competition among telecommunications firms, and to explain how the techniques of spatial analysis that we have developed might be used to model the intraurban spatial demands faced by these firms.

In the next section, we discuss in more detail why the intraurban location of a long-distance firm vis-a-vis its rivals might have an impact on the demand it faces. In Section 3, our algorithmic procedure for calculating market areas of establishments when distances are Euclidean is presented. Section 4 discusses several ways in which city limits and other topographical features of a city can be modeled, while Section 5 describes our algorithm for finding market areas when distances are Manhattan. Sections 6 and 7 then consider how our algorithms can be extended to find market area populations and the market areas that result when firms charge different prices. Section 8 contains a summary and some concluding remarks.

2. SPATIAL COMPETITION AMONG TELECOMMUNICATIONS FIRMS

The intraurban demand faced by a given long-distance firm does not seem at present to depend on where the firm's central switching station is located within the city. This is because of the way in which most consumers obtain access to a firm's long-distance network and the pricing system which has been used. Firms offering switched long-distance service interconnect with local exchange facilities. That is, a typical residential customer's long-distance call will proceed along local phone lines before it is switched to a firm's long-distance network. The interconnection charges which AT&T's competitors (sometimes known as Other Common Carriers or OCCs) agreed to pay to the local telephone companies were set out in an agreement (reached in December 1978) on exchange network facilities for interstate access, now known as the ENFIA tariff. This tariff included a fixed charge per month per access line plus a charge per minute based on an allocation of local exchange costs to interstate toll use. (See Brock [1, p. 228] for details.) Neither part of the tariff depends on where within a city a long-distance call originates, and the prices charged individual OCC subscribers would presumably not depend on it either. In 1983, the FCC announced a long-term solution to access charges in which telephone subscribers and interstate carriers would both pay explicit prices for access to local exchange facilities. According to Wiley [13, p. 41], telephone subscribers were to pay for the "interstate costs associated with the copper wire which runs from homes and offices, over telephone poles or through underground conduits, to the telephone company central office. Interstate carriers will pay for the lines and switches necessary to connect their interstate networks to the local telephone company". Each subscriber was to pay a flat monthly fee regardless of whether any long-distance calls were made. Under this pricing system, the consumer's demand for the services of a given carrier again would not depend on his location vis-a-vis the firm.

In general, as long as switched long-distance service involves interconnection with local phone networks and the price of access is not distance-related, the demand facing a given long-distance firm will not depend on the distance between the firm and the potential consumers of its services. However, telecommunications firms in the future might seek to bypass the local network entirely for some portion of their long-distance business. In particular, as Crandall and Owen [2, p. 68] point out, long-distance carriers and their high volume customers might find bypass attractive if the cost of bypass, accomplished through a waveguide system (e.g., coaxial or fiber optic cables) or a microwave system, is less than local exchange access fees.[1] Bypass might also be desirable if the local operating companies do not provide the specialized services demanded by certain large users of telecommunications services. If bypass of the local network is primarily achieved by waveguide or microwave systems (as seems likely given the cost of earth stations and the frequency congestion in densely populated urban areas that a satellite-based system would entail), then the firm's intraurban location decision may have an impact on the demand which it faces.

To illustrate this point, consider the following example: assume that there are two long-distance firms occupying different locations in a city. Both firms have identical long-distance networks, toll charges, and quality of service. Suppose that both firms bypass the local network and that they charge monthly access fees which are a function of the distance between their customers and their switching stations. In this case, consumers will patronize the closest firm since the nearest firm will offer the lowest delivered price for long-distance services. Alternatively, suppose the firms wish to charge a uniform access fee (which need not be the same for both firms) to all of their customers, and the access fee chosen must be sufficient to cover the direct cost of bypassing the local network. In this case, the location of the firm vis-a-vis its rival will affect the demand it faces because the demand facing the firm depends on which set of consumers it can serve at an access fee no greater than its rival. (The marginal consumer will be the one for which the marginal cost of serving him raises the uniform access fee above that of the rival firm.)

Thus the way in which access fees are set will determine whether the firm's location affects the demand it faces. If there is a spatial dimension to the intraurban demand for long-distance telecommunications services, then the tools of spatial competition theory can be used to help model the firm's optimal location and price decisions. The model builders will need to consider the spatial distribution of consumers and their demands, market growth, the nature of capital (e.g., the extent to which it is mobile) and the costs of producing long-distance telecommunications services (e.g., extent to which there are scale economies and that capital costs are sunk costs), rival firm beliefs and entry deterrence strategies. If we have such a model, we might wish to test its predictions, and this would probably entail examining the timing of entry, the neighbor relations among firms, the market areas of firms as well as the number of people served within each market area. We might also wish to study how market areas and their populations are affected by firms setting different prices. It is for this type of applied research that our techniques of spatial analysis are especially suited. The remainder of this paper will present a survey of the algorithms used by us to test for spatial predation (Von Hohenbalken and West [10, 11]), a hierarchy of shopping centers (West, Von Hohenbalken, and Kroner [12]), and differences in Euclidean and Manhattan market areas (Von Hohenbalken and West [8,9]).

3. MARKET AREAS BASED ON EUCLIDEAN DISTANCES

We begin our discussion by reviewing our most basic procedure for calculating market areas and their adjacencies. The following assumptions are made: there are several firms serving the market and they are differentiated from each other only on the basis of the location of their urban establishments (which might be stores, plants or switching stations). Delivered prices (e.g. access fees) are an increasing function of Euclidean distance and all firms charge the same mill price. (The implications of relaxing this assumption are discussed later in the paper.) Consumers are cost-minimizers and this, in conjunction with our equal-price assumption, implies that consumers will patronize the nearest establishment.

With these assumptions the calculation of all geographic market areas of urban establishments is equivalent to finding the Euclidean Voronoi diagram of a point set in the plane. Figure 1 illustrates one such Voronoi diagram; it shows the market areas of establishments owned by three firms, P, V, and C, in north and south Edmonton. A Voronoi diagram (so called for historical reasons) partitions the plane into individual nearest-point sets; each such Voronoi set is associated with a point s. From a Voronoi diagram one may glean the size and the adjacencies of the Voronoi sets M, and these attributes can be interpreted as the market size and the neighbor relations of a given set of establishments. Under the Euclidean metric every M is a convex polygon; if M is bounded and its vertices $v^1,\ldots,v^m$ are given in counterclockwise cyclical order, the content of M (the size of the market area) is easily computed[2] by

$$\text{area of } M = 1/2\{\det[v^1,v^2]+\ldots+\det[v^m,v^1]\}.$$

We now introduce our algorithm to calculate the ordered vertex set of any bounded Voronoi set of the origin (one can do this without loss of generality because the whole given point set - the locations of the establishments - can be translated such that any location comes to lie in the origin). To fix ideas consider the origin and one other point, say s; brief reflection and possibly a simple diagram reveals that the part of the plane that is closer to the origin than to s is a halfplane which is bounded by the perpendicular bisector of the segment between 0 and s; the equation of this bisector is $s(x-s)/2=0$. The same will be true for other points, and thus the <u>external representation</u> (by halfplanes) of the Voronoi set M of the origin with respect to points s^i, $i=1,\ldots,n$, is the intersection of n such halfplanes, given by

$$s^i(x-s^i)/2 \leq 0, \text{ all } i. \tag{1}$$

The set of inequalities (1) is useful to check whether some point z does lie in M or not, but information about neighbors, vertices and size of M is only implicitly present. To gain direct access to these statistics one has to convert to the <u>internal representation</u> of M, which involves finding its vertex set. The following algorithm reveals the vertices of M in counterclockwise order; (we assume that M is bounded, possibly by adding constraints).

<u>Euclidean algorithm:</u>

<u>initial step:</u> Let s^1 be the point closest to 0; line 1 (its associated perpendicular bisector) then furnishes the first segment of the boundary of M. Set t=1 and go to main step.

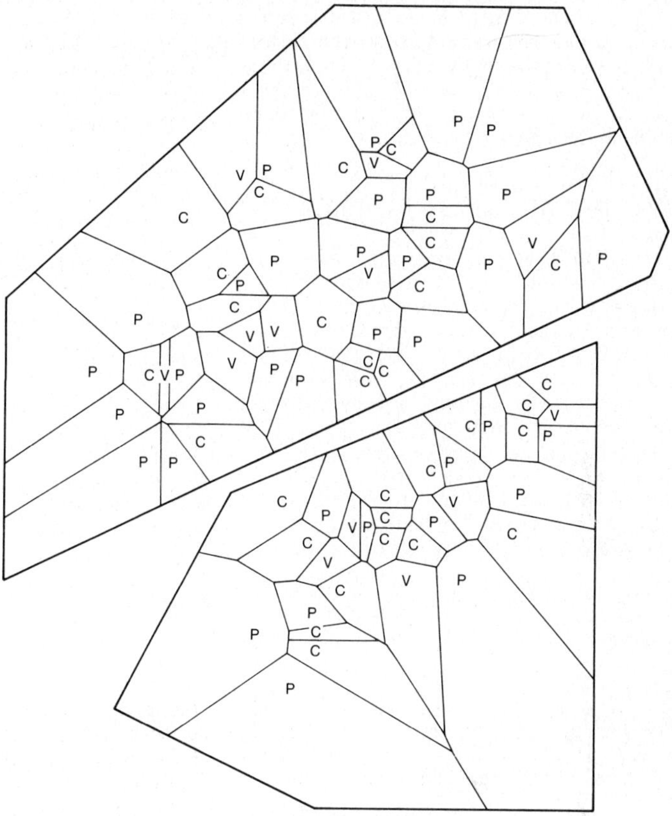

FIGURE 1
Euclidean market areas, convex boundary (Edmonton 1973).

<u>main step:</u> given line t, take as line t+1 that bisector
(a) whose angle with line t is positive and less than π
(b) whose intersection point with line t satisfies all inequalities (1);
terminate if line t+1 is the initial line 1; otherwise set t=t+1 and repeat main step.

<u>Comments:</u> Line 1 in the initial step is certain to be part of the boundary of M because if it were not another line would have to lie between line 1 and the origin, contradicting nearestness.

Criterion (a) in the main step enforces strict counterclockwise progression around the boundary, while (b) admits only the one proper successor line.

The procedure will terminate because every iteration disposes of another point out of a finite number.

The algorithm is greatly accelerated by first ordering the points s^i according to their distance from 0, and then examining them in this order in every iteration. Apart from the ordering operation the average running time is then independent of n, the number of given points.

The set of neighbors is obtained by merely recording the indices of the consecutive boundary pieces.

4. CITY BOUNDARIES AND TOPOGRAPHICAL FEATURES

While the interior Voronoi sets are clearly bounded, the Voronoi diagram as a whole is not. To obtain economically meaningful market areas of establishments on the periphery of a city, city limits need to be imposed. In addition, certain types of firms (e.g. retail) might have their market areas constrained by such geographic features as a river or limited access highway that divides a city, and we would therefore wish to impose these constraints on our Voronoi set calculations. We might also wish to calculate market areas that only contain certain types of consumers, for example, households; we would then wish to carve away those portions of the market areas (e.g. parks) that do not contain the consumers of interest. We shall discuss three different approaches to impose topographical features onto a given Voronoi diagram; all have been tried in our earlier work (Von Hohenbalken and West [9, 10, 11]).

The first method, a version of what we termed "fence post approach", allows the imposition of a piecewise linear boundary onto a city that has a basically convex shape. It simply adds some inequalities to those that represent the perpendicular bisectors; these additional inequalities correspond to the edges of the "city limits" to be imposed. The city limits are laid out easiest by placing "fence posts" (=vertices) around the city such that they define a convex polygon, and then deriving the equations of the straight lines through all pairs of mutually adjacent vertices (see Figure 1).

A more sophisticated version of the "fence post" approach that allows the modeling of nonconvex boundary features relies on the Boolean set operations (complementation, intersection, union and differences). We shall first discuss the relations between these operations in the context of polygonal boundaries, and then give a brief overview of the algorithm.

Following a convention deeply rooted in the theory of determinants we represent a polygon A by its counterclockwise ordered vertex set $\{v^1,\dots,v^m\}$; loosely, we shall call this vertex set the ccw boundary of A. A polygon with a ccw boundary has area that is <u>positive</u> when obtained by Green's formula $1/2\{\det[v^1,v^2]+\dots+\det[v^m,v^1]\}$. If the order of the vertex set is reversed, it represents the <u>complement</u> of A; ~A has thus a clockwise (cw) boundary, and its A-shaped hole has negative area.

Consider next two bounded polygons A and B, given by their ccw boundaries. If one has a method (as we do) to find the ccw boundary of their intersection $A \cap B$, then the other Boolean operations can also be carried out easily.

Because the set difference $A \setminus B$ is equivalent to $A \cap {\sim}B$, a ccw boundary for $A \setminus B$ will be obtained if the ccw boundary of A and the <u>cw</u> boundary of ~B are used as arguments in the intersection algorithm. B\A is obtained symmetrically.

The union A∪B is equivalent to ~(~A∩~B) by de Morgan's law, and therefore a ccw boundary for A∪B can be found by using the intersection method with the cw boundaries for ~A and ~B, and then reversing the resulting vertex set.

We actually have devised a boundary intersection method alluded to above. The algorithm takes as inputs the ordered (ccw or cw) vertex sets of two bounded, not necessarily convex polygons that are simply connected (i.e., the boundaries must be simple closed curves). The output consists of one or several labeled vertex sets (with either one or both cyclical orientations) that represent the possibly multiply connected or even disconnected set that results.[3]

The operation of the algorithm may be outlined as follows: start anywhere on the boundary of one of the two sets, moving along in counterclockwise direction. At the first intersection where the boundary of the other set crosses from right to left, switch to the other boundary; proceed as before, but now switch boundaries every time a crossing is encountered; stop when the first switching point reappears. Proofs and examples may be found in Von Hohenbalken [7]. In Figure 2, Edmonton's river and its city limits have been imposed on each Voronoi set in this fashion; (the picture also includes Edmonton's 1981 census tracts - in light lines - whose intersection and use are taken up later).

A third method to impose topographical features on a Voronoi diagram is to augment the originally given point set (of, say, certain establishments) by strategically placed "dummy points" (the procedure had been called "dummy store approach" in our chain store application; see Von Hohenbalken and West [9] and West, Von Hohenbalken, and Kroner [12]). Like other points in a Voronoi diagram the dummy points acquire their own nearest point sets and thereby "push back" or "hem in" the Voronoi sets of the original and relevant points. Naturally the nearest point sets of the dummy points are of no interest in themselves and their vertex sets are not computed. Most gaps, holes and boundaries of Voronoi diagrams can be modeled "softly" by dummy points, as the recalculated market areas of firms P, V and C in Edmonton demonstrate in Figure 3. This approach is undemanding computationally and even permits special interpretations in some applications (see Section 6 below).

5. MARKET AREAS UNDER THE MANHATTAN METRIC

With our Euclidean algorithm and procedures for modeling topographic features of a city, we can find the market areas and their adjacencies of all establishments of a given type in a city. It is assumed when doing these calculations that delivered prices are in fact a function of Euclidean distance. Such an assumption seems appropriate if we wish to find the intraurban market areas of interstate carriers using microwave radio bypass technology. Estimates made by Jackson and Rohlfs [3] of the monthly costs per trunk of a microwave bypass system suggest that such costs (and, by implication, cost recovering access fees) will be an increasing function of the Euclidean distance spanned by the bypass link, but not continuously so. If instead of microwave a fiber optic waveguide bypass technology is used, then the assumption of Euclidean distance may not be

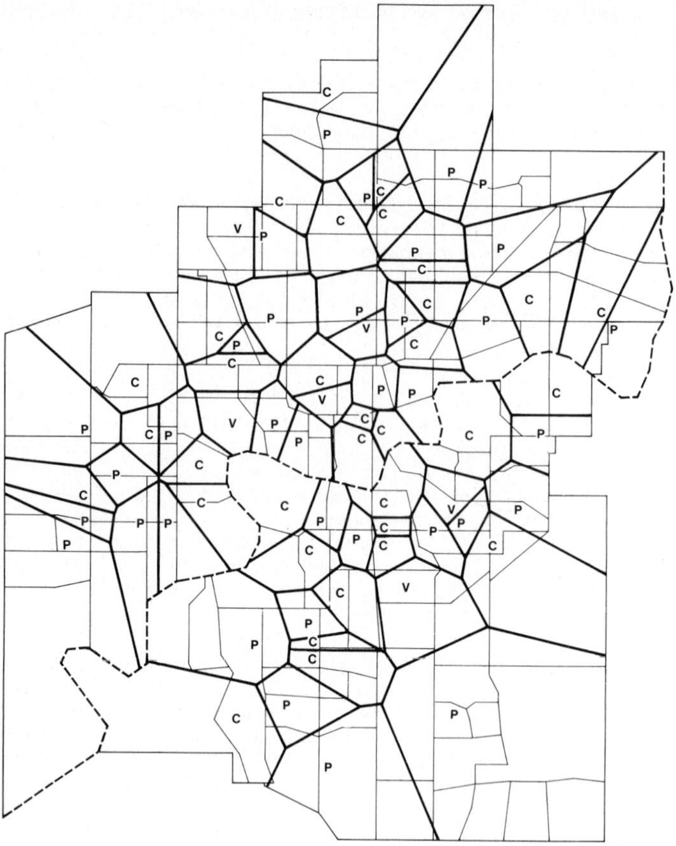

FIGURE 2
Euclidean market areas, census tracts, non-convex city boundary (Edmonton 1982).

appropriate. This is because a fiber optic system would likely be installed along existing right-of-ways either through ducts or over poles, and these would be expected to follow the (usually rectangular) road network. The right-of-way fees would thus be a function of city-block or Manhattan distance, and we would then wish to calculate market areas using the Manhattan metric (on the assumption that interstate carrier access fees are set to recover the costs of bypass). In some applications where the Manhattan metric would seem to be the appropriate one (e.g., calculating the market areas of supermarkets), one might nonetheless be able to achieve a good approximation to the results that would be obtained with the Manhattan metric by using the Euclidean metric. For example, in an earlier paper (Von Hohenbalken and West [9]), it was shown that the Voronoi diagram of P, V and C establishments in Edmonton under the Manhattan metric resembles the Euclidean one to a fair degree. However,

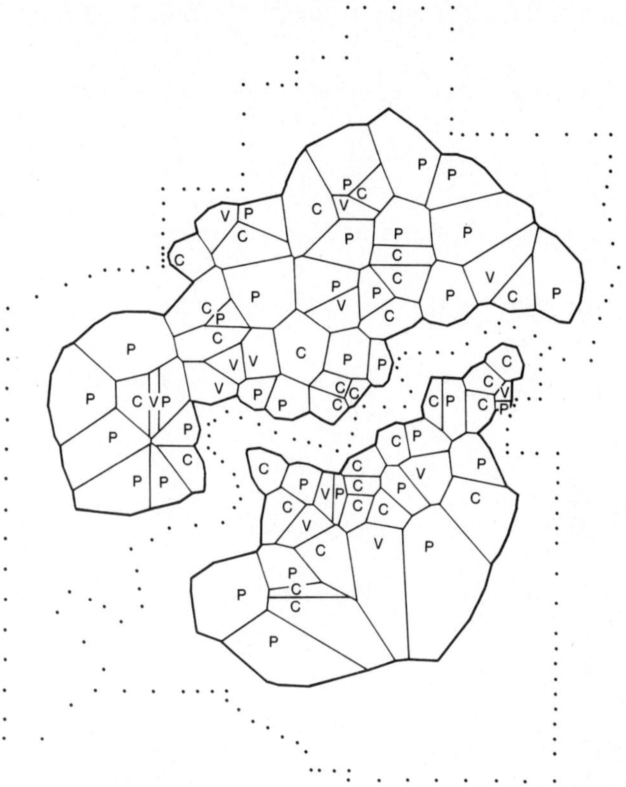

FIGURE 3
Euclidean market areas, dummy point boundary (Edmonton, 1973).

if there are only a few establishments in the city, the approximation may be weaker, and then precise calculations using the Manhattan metric would be desirable.

Under the Manhattan metric, the boundaries between two establishments are not straight lines anymore, but consist of diagonal line segments that continue either vertically or horizontally. The pictures in Figure 4 (which are drawn, w.l.o.g., for a NS-EW coordinate = road system) show the different forms of Manhattan boundaries: (4a) and (4b) depict the four usual cases; the dashed boxes show the locus of all possible shortest paths between points s^1 and s^2 (note that there are many such paths); the boundary itself (the solid line) is the set of all points in the plane which are equally far from s^1 and s^2, and the reason for its "zig-zag" shape is the diamond-shaped "balls" of the Manhattan metric (see Von Hohenbalken and West [9]).

(4c) and (4d) show two singular cases, with s^2 lying precisely north or east (4c) and precisely northeast or southeast (4d) of point s^1. The

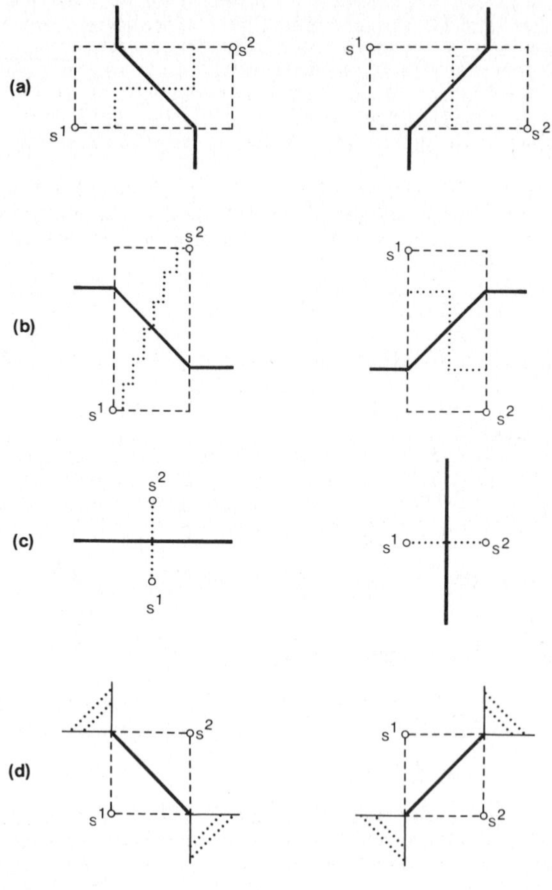

FIGURE 4
All types of Voronoi boundaries under the Manhattan metric.

latter case is remarkable because even outside the dashed box there are many points at a given and equal distance from s^1 and s^2 (the dotted lines are examples). Interpreted this means that all customers living on such "plateaus" would be indifferent between establishments s^1 and s^2, if distance is the criterion; this naturally raises the empirical problem of how to divide plateaus "equitably" between establishments.

Despite the above complications, the Manhattan nearest point sets can still be viewed as polygonal market areas, whose adjacencies and sizes may be interpreted and computed as before. The lack of convexity, however, renders inoperable the usual analysis based on linear inequalities, precisely because a nonconvex set cannot be represented as the intersection of halfplanes; to define a nonconvex polygon, the plain vertex set is equally insufficient; the vertices must be given in cyclical order. The algorithm to compute such ordered vertex sets has been developed in Von Hohenbalken and West [8]. We confine our discussion here to the development of the Manhattan boundary of the origin in the non-negative orthant, because any given point set can be translated such that the point of interest (i.e., the establishment whose market area is to be found) falls in the origin; surrounding points lying in other quadrants can be rotated into the non-negative one by an orthogonal transformation.[4]

We begin by splitting the points in the first quadrant into those with vertical boundary pieces (below the 45° line) and those with horizontal ones (above 45°). Each boundary piece can be represented by a single boundary vertex, namely

$$a^i = (a_1^i, a_2^i) = ((p_1^i - p_2^i)/2,\ p_2^i)$$

if p^i lies below the 45° line, and

$$b^j = (p_1^j,\ (p_2^j - p_1^j)/2)$$

if p^j is above the 45° line. Figure 5 shows that the a's and b's can lie anywhere in the quadrant; it also serves to demonstrate that only those a's and b's participate in the final Voronoi boundary, that are not dominated under the following partial orders:

Definition: A boundary vertex a^i is dominated if there exists a vertex a^k such that

$$a_1^k + a_2^k \leq a_1^i + a_2^i \text{ and } a_1^k \leq a_1^i.$$

A boundary vertex b^j is dominated if there exists a vertex b^ℓ such that

$$b_1^\ell + b_2^\ell \leq b_1^j + b_2^j \text{ and } b_2^\ell \leq b_2^j.$$

Given these criteria one proceeds to eliminate the dominated vertices, which is a nontrivial but well-known exercise. By ordering the remaining a-vertices and b-vertices by their coordinate sums one obtains a counterclockwise ordered set of a-vertices and a clockwise ordered set of b-vertices. All the above can be verified in Figure 5; nondominated vertices are emphasized by circles and ordered; also shown are the two components of the ultimate Voronoi boundary.

Let $\{a^1, a^2, \ldots, a^s\}$ and $\{b^1, b^2, \ldots, b^t\}$ be the ordered sets of nondominated vertices of the two boundary components. To define these components as sequences of segments one fills the interstices in the above sets to obtain

$$A = \{\bar{a}^1, a^1, \bar{a}^2, a^2, \ldots, \bar{a}^s, a^s, \bar{a}^M\} \text{ and}$$

$$B = \{\bar{b}^1, b^1, \bar{b}^2, b^2, \ldots, \bar{b}^t, b^t, \bar{b}^M\}.$$

In the new ordered sets the barred characters mean

$$\bar{a}^1 = (a_1^1 + a_2^1, 0),\ \bar{b}^1 = (0, b_1^1 + b_2^1);$$

for $i,j \neq 1, M$

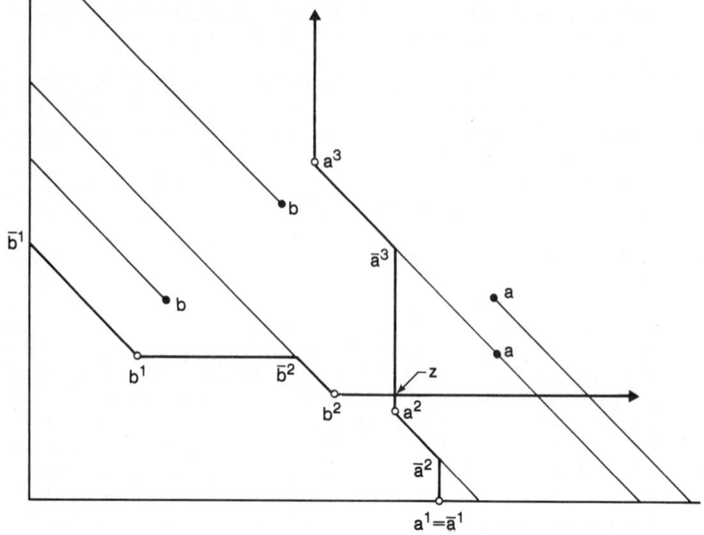

FIGURE 5
"Vertical" and "horizontal" components of Manhattan Voronoi boundary in the first quadrant.

$$\bar{a}^i = (a_1^{i-1},\ a_1^i + a_2^i - a_1^{i-1}),\ \bar{b}^j = (b_1^j + b_2^j - b_2^{j-1},\ b_2^{j-1})$$

and

$$\bar{a}^M = (a_1^s,\ M),\ \bar{b}^M = (M,\ b_2^t);$$

(M is a fixed large number). Roughly speaking the barred points have one coordinate in common with their precursors, and they lie on the same diagonal as their successors. (See Figure 5.) The terminal segments $[a^s, \bar{a}^M]$ and $[b^t, \bar{b}^M]$ are chosen long enough so as to reach well beyond the component emanating from the other axis. This and the fact that both components start on one axis and do not diverge from the other ensure that the two components must cross, thereby closing the quadrant's Voronoi boundary.

Given the ordered sets A and B the final task is actually to accomplish the intersection of the two saw-toothed components. For this purpose we must be able to decide when two segments meet: let

$$[a,c] \equiv \{x \mid x = a + \lambda(c - a),\ 0 \leq \lambda \leq 1\}$$
$$[b,d] \equiv \{x \mid x = b + \mu(d - b),\ 0 \leq \mu \leq 1\}.$$

By equating

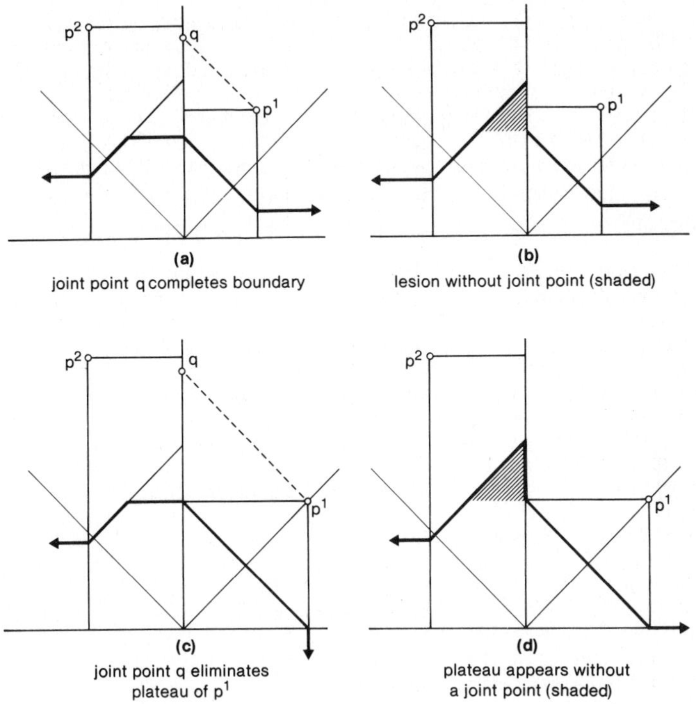

FIGURE 6
The effect of joint points to fuse quadrants and control plateaus.

$$a + \lambda(c - a) = b + \mu(d - b)$$

one has two simultaneous equations (because a, c, b, $d \in R^2$) which can be solved for λ^*, μ^* as long as the two segments are not parallel. If, in addition, both λ^* and μ^* lie in [0,1], the segments intersect, with the intersection point z being

$$a + \lambda^*(c - a) \text{ or } b + \mu^*(d - b).$$

The point common to the curves defined by A and B is found by intersecting the segments in A consecutively with those in B, omitting those in B that are outside the range of A (see Figure 5).

To form the final ordered set that represents the Voronoi boundary of the quadrant one truncates A and B beyond the intersection point z; then one concatenates the remaining part of A with z and the reversed remainder of B, obtaining a set of boundary vertices that is ordered counterclockwise throughout.

Given the Voronoi boundaries for all four quadrants have been found, the question arises how they are fused together. In fact this is being taken care of before one starts by augmenting the point set P with a joint point q on each of the four axes, which ensure that the boundary pieces of adjacent quadrants meet satisfactorily. Figures 6a and 6b show Voronoi boundaries with and without the ancillary piece generated by joint point q; if q is missing a lesion appears. As can be seen from Figures 6a and 6b the joint point q between the first and second quadrant lies on the ordinate and has the same (Manhattan) length as p^1; p^1 itself is the shortest vector of P that lies on or above the center lines of the adjacent quadrants. In other words, the second coordinate of q equals $\min(|p_1^1| + |p_2^1|)$, taken over all vectors in P with polar angles θ, $\pi/4 \leq \theta \leq 3\pi/4$.

It turns out that joint points play a decisive role in the algorithmic control of "thick" boundaries. It will be remembered from Figure 4d that points of P that lie on a 45° line induce the boundary between themselves and the origin to have plateaus with positive content. If 45° points are treated like any other point in P, plateaus will be excluded from the Voronoi set V (see Figures 6c and 6d). If, however, 45° points are not permitted to compete for joint points in the minima above, (i.e., only points with polar angles $\pi/4 < \theta < 3\pi/4$ are admitted), all potential plateaus will be included in V. As Figure 6d shows, the lesion in Figure 6b is then turned into the plateau required in this special case.

After everything is computed and done, it emerges that the Manhattan Voronoi sets have practically the same neighbors as their Euclidean cousins, despite having more than twice as many sides and vertices. The Manhattan Voronoi algorithm can't be programmed as compactly as the Euclidean method, and thus, in APL at least, uses more CPU time. Nonetheless it is the Euclidean computations which are intrinsically more complex because of their greater "simultaneity".

6. MARKET AREA POPULATIONS - VIA CENSUS TRACTS AND VIA MASS POINTS

The market areas calculated using either the Euclidean or Manhattan metrics are subject to different interpretations. First, one might interpret them as catchment areas in the sense that consumers residing within a given establishment's market area will find that establishment the least costly one to patronize. This interpretation is all that is required if one is interested in, say, comparing the territories that different firms can serve at lower costs or delivered prices than their rivals. Second, in some applications, one might like to be able to interpret the calculated market areas as proxies for sales. If consumer demand for the product(s) sold by an appropriately defined set of establishments is relatively inelastic and if consumers are approximately uniformly distributed over the city, then market area size and establishment sales should be highly correlated. Problems arise in interpreting market areas this way if, as is likely, the urban population is not uniformly distributed. There is ample evidence that urban population density decreases as you move away from the city center. Indeed, at some distance from the city center there will be a transition from urban land to rural land. To some extent, the way in which the city boundary (and holes in the market created by parks, rivers, lakes, etc.) is modeled can reduce the significance of this problem. For example, dummy points can be placed around the city such that the effective city boundary that results captures the establishments of interest and most of the urban population. As Figure 3 shows, the effective city boundary will be sensitive to both the placement of the dummy points as well as the locations of the peripheral establishments with which they are creating the city boundary. (The dummy point approach to modeling the city boundary is particularly useful if one wishes to analyze changes in market areas over time since this approach allows the total market area of all establishments in the city to increase over time; this comes about as new establishments are opened on the urban periphery, forcing the effective city boundary to shift outward. The fixed fence post approach to modeling the city boundary lacks this flexibility.) The dummy point approach, however, is only a partial cure for the problem caused by non-uniform density, and has no impact on the market area calculations for interior establishments. To obtain a better proxy for an establishment's sales when population is not uniformly distributed, one would wish to estimate an establishment's market area population.

If one wants to find the residential population contained within a given establishment's market area, information additional to that required for the market area calculation is needed; we take the following data to be known:

(a) the oriented vertex set of the market area M under consideration,
(b) the oriented vertex sets of all census tracts, or, if known, of those that overlap M,
(c) the population counts of census tracts.

To get the market area population of M, we shall have to obtain the sum of areas of the actual intersections of M with the census tracts, each multiplied by the corresponding population density. The principal preliminary labor is to calculate the vertex sets of the intersections $M \cap T_k$ (the T_k are the census tracts), with our basic algorithm to find intersections of polygons. In principle, there is no difficulty with this, but the computer has no eye or visual cortex at its disposal to decide a priori whether a certain census tract actually overlaps M or not. Consequently, it must blindly intersect <u>all</u> tracts, which is a lot of work, most of which comes to naught. In our early investigations we used computer-drawn maps to make these decisions with our own eyes and then tell the machine, but if there are 40 maps like Figure 2 (where census tracts have light boundaries and market areas have bold ones), this soon becomes tedious.

A later idea that uses the computer to better advantage is to prescreen the census tracts with the help of the halfplanes that externally represent the convex hull of a polygonal market area: if for any census tract there is at least one such halfplane which contains none of the vertices of the tract, then the tract cannot overlap the market area, and no intersection procedure needs to be initiated.

With the ordered vertex sets of all tract intersections on hand it is a simple matter to calculate their areas and those of the census tracts themselves. The population of the market area M is then the sum of the population counts in each tract T_k, multiplied by the fraction of the area inside M, i.e.,

$$\text{population of market area } M = \Sigma_k \,(\text{size of } M \cap T_k) \cdot (\text{population of } T_k)/(\text{size of } T_k).$$

For certain types of establishments (e.g., supermarkets and shopping centers) the above procedure should yield good approximations to the actual market area populations. However, on occasion one might wish to estimate the number of apartment dwellers and/or office building tenants and/or commercial establishments that lie within a given establishment's market area. For example, to estimate the number of potential customers of a given firm's long-distance services that are produced by bypassing the local network, one would want to find the locations of these potential customers and treat them as density-weighted points (the weights would be the expected number of voice channels demanded at the different locations). Given the configuration of the market, the question for total demand to be expected within a (previously computed Euclidean or Manhattan Voronoi) market area can be answered by determining which of the business locations lie within the firm's market area boundaries. (This assumes that firms charge non-uniform cost-recovering access fees that are an increasing function of Euclidean or Manhattan distance. If uniform cost-recovering access fees are charged, the determination of market areas and their populations becomes a simultaneous problem.) Given the market area M is convex, the computer can make this decision, again based on the external representation: a point of business, say b, lies in the market area M if and only if b lies in all halfplanes that define M, or equivalently if and only if b satisfies all associated inequalities (as shown in (1) in Section 3).

If the market area happens to be not convex, either because of a non-Euclidean metric or because of non-convex topographical features of the city, approximate calculations can be based on the external representation of the convex hull of M (see Von Hohenbalken [6]). Because the convex hull of any set is never smaller than the set itself, demand in a market area will at worst be overestimated, never underestimated.

7. MARKET AREAS WITH DIFFERENTIAL PRICES

To this point in our review, we have maintained the assumption that firms are identical in every respect except the locations of their establishments. To the extent that this is not so, the market areas calculated for some establishments might be overestimated, while others may be underestimated. For example, in our studies of predatory behavior among supermarkets, we calculated market areas on the assumption that firms were identical and charged the same prices. However, some supermarkets will be larger than others, have a wider assortment of products, higher quality service, and/or lower product prices. Consumers may place positive valuations on these store-specific attributes, leading them to patronize a store which is not the nearest one to them. In other words, consumers might be willing to incur higher transportation costs than are necessary in order to consume a set of characteristics that are more highly preferred. Similarly, long-distance carriers are not necessarily identical. They might differ in their access charges, long-distance networks, long-distance toll charges, and quality of service. If consumers are well-informed about these differences, some would no doubt choose to pay a higher access fee to be served by what is perceived to be a higher quality carrier than a lower access fee to a carrier of lesser quality. In terms of the way in which we model market areas, if two neighboring establishments charge different prices but are identical in other respects, the boundary between them will shift towards the higher-priced establishment. If two neighboring establishments charge the same price but one has more highly valued attributes (i.e., is of higher quality) than the other, the boundary between them can be conceptualized as shifting toward the lower quality establishment, thereby enlarging the market area of the higher quality establishment. We can interpret this situation as one firm charging a lower price per constant quality unit. To know with any degree of confidence by how much the boundary between two neighboring establishments would shift due to differences in quality would likely require an analysis akin to those which have appeared in the modal choice and consumer store choice literature. However, it is not algorithmically difficult to recalculate market areas on the assumption that prices (per constant quality unit) differ in a given way. The remainder of this section discusses how this can be accomplished.

The general equation for the boundary between the market areas of two establishments i and j at locations s^i, s^j that charge possibly different prices π_i and π_j is

$$\pi_i + ||s^i - x|| = \pi_j + ||s^j - x||. \tag{2}$$

If distance is measured the Euclidean way, this boundary is one branch of a hyperbola, namely that one that is closer and bent toward the higher-priced establishment (see Figure 7a). If, say, $\pi_j > \pi_i$, this can be seen by writing (2) as

$$||s^i - x|| - ||s^j - x|| = \pi_j - \pi_i,$$

i.e., the locus of points x in the plane that have a positive constant difference of distances from the two "foci" s^i and s^j (this is the usual textbook definition of a hyperbola). Clearly, if $\pi_i = \pi_j$, the hyperbola degenerates to the usual and well-known perpendicular bisector. Figure 7a shows this Euclidean picture for three price pairs.

The above indicates that Euclidean Voronoi diagrams for the unequal prices case are nonlinearly bounded, and thus not tractable by the analytic tools we discussed above. In an earlier paper (Von Hohenbalken and West [10]) we circumvented this difficulty by approximating the hyperbola by its tangent at the point where it cuts the segment between s^i and s^j (see Figure 7a). This procedure leads to triangular patches of intransitivity in the places where three market areas join, but for small price differences and/or dense point sets P these lacunae can be neglected.

If equation (2) of the boundary is looked at with Manhattan eyes, it is seen that a change in relative prices also shifts the boundary, but that it does not destroy its piecewise linearity. This becomes clear by applying the characteristics of the absolute value function to equation (2). Figure 7b shows three price settings, one of them yielding a plateau (see Section 5 for plateaus). A trivial modification of our usual linearity-based Manhattan algorithm will therefore analyze the differential price case perfectly.

The above development again underscores that both metrics have their own advantages with respect to analytical tractability: Euclidean distances induce Voronoi diagrams consisting of convex polygons as long as underlying prices are assumed equal; thus all the advantages of convexity (especially separation and the representability by inequalities) accrue to the equal-price case. The Manhattan distance function allows precise modeling

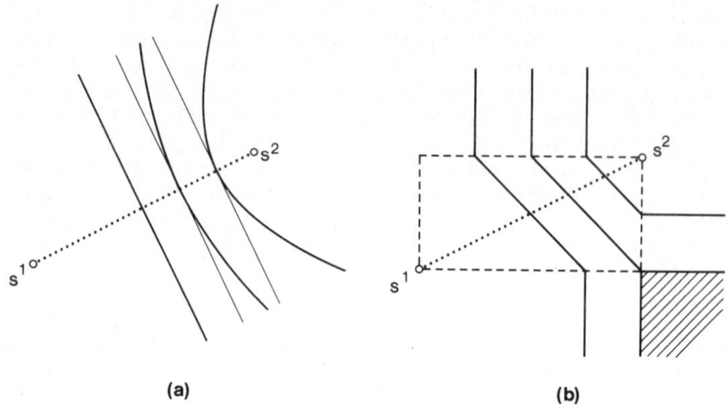

FIGURE 7
Voronoi boundaries when relative prices change.

of rectangular road grids, and its Voronoi diagrams are seen to retain their polygonal character under differential prices. On the debit side there remain nonconvexity, the occurrence of plateaus, and the somewhat slower algorithm.

Finally, an important use of Voronoi diagrams under differential pricing should be mentioned: it allows a comparative static analysis of spatial demand elasticities: by positing say, a 1 percent reduction in its price, a firm can calculate by how many percentage points its market area increases; it just needs to compute and compare the Voronoi diagrams before and after the price change. We have found in our predation studies of supermarkets that such "infringements" will be especially large if a "price cutter" sits close and has "broad fronts" to a competitor.

8. CONCLUSION

The telecommunications industry has undergone and will no doubt continue to undergo significant structural changes due to the breakup of AT&T. What must be of particular concern to the regional telephone companies is the extent to which long-distance carriers will seek to bypass the local network by direct connection with their customers. Large-scale implementation of bypass technologies could lead to the loss of access charges. In this paper, we have suggested that the adoption of bypass technologies might lead to the competition among long-distance firms taking on a distinctly spatial character. This would occur if the delivered price of long-distance telecommunication services becomes a function of the distance between firm and customer. We then reviewed the algorithms which we have developed for studying competition among firms in spatial markets. In particular, we discussed how establishment market areas can be calculated when distances are either Euclidean or Manhattan, and how city limits and other topographical features of a city can be algorithmically modeled. We then considered how market area populations can be obtained and how market areas would be affected by price or quality differences among establishments.

While it was not our intention in this paper to use our techniques to conduct an empirical study of bypass, such an analysis remains a possibility for future research. Our algorithmic tools could be used to calculate intraurban market areas of long-distance firms based on some assumptions regarding their locations and prices. They might also have a role to play in estimating the net benefits to firms from adopting bypass, and the revenues that regional companies might lose on account of bypass.

FOOTNOTES

1 The importance of high volume users has been noted by Meyer et al. [4, p. 186]: "Specifically, over 75 percent of total revenues from business use of long distance are accounted for by less than 8 percent of all customers. Even more dramatically, the 25 largest business customers generate 15 percent of the total interstate business toll revenue, while the 100 largest business customers generate 20 percent of this revenue; interstate revenue, in turn, represents a little over one-quarter of total telephone industry revenues. Similarly, almost 50 percent of long distance residential revenues are from around 10 percent of the users." Meyer et al. also forecast that technological developments "could render local loops even less necessary as hookups to competitive long distance services in the future".

2 The formula represents a discrete implementation of Green's theorem on line integrals; its operation is not dependent on the convexity of the polygon.

3 An algorithm of this sort with any claim to generality must allow for such complications because even $A \cap B$ may consist of many pieces for nonconvex A and/or B.

4 We assume, as usual, that sufficient other points are clustered around the origin (possibly including artificial ones far out), such that its Voronoi set is bounded. For necessary and sufficient densities of points see Von Hohenbalken and West [9].

REFERENCES

[1] Brock, G.W., The Telecommunications Industry: The Dynamics of Market Structure (Harvard University Press, Cambridge, Mass., 1981).

[2] Crandall, R.W. and Owen, B.M., The Marketplace: Economic Implications of Divestiture, in: Shooshan III, H.M., (ed.), Disconnecting Bell: The Impact of the AT&T Divestiture (Pergamon Press, New York, 1984) pp. 47-70.

[3] Jackson, C.L. and Rohlfs, J.H., Access Charging and Bypass Adoption: A Study Prepared for Bell Atlantic, mimeo (1985).

[4] Meyer, J.R., Wilson, R.W., Baughcum, M.A., Burton, E., and Caouette, L., The Economics of Competition in the Telecommunications Industry (Oelgeschlager, Gunn & Hain, Cambridge, Mass., 1980).

[5] Taylor, L.D., Problems and Issues in Modeling Telecommunications Demand, in: Courville, L., de Fontenay, A., and Dobell, R., (eds.),

Economic Analysis of Telecommunications: Theory and Applications (North Holland, Amsterdam, 1983) pp. 181-98.

[6] Von Hohenbalken, B., Least Distance Methods for the Scheme of Polytopes, Mathematical Programming 15 (1978) pp. 1-11.

[7] Von Hohenbalken, B., An Algorithm for Set Operations on Nonconvex Polygons (with an Application to Census Tracts), Discussion Paper No. 8556 (Department of Economics, University of Montreal, Montreal, 1985).

[8] Von Hohenbalken, B. and West, D.S., Block-metric Voronoi Diagrams in R^2, Research Paper No. 83-4 (Department of Economics, University of Alberta, Edmonton, 1983).

[9] Von Hohenbalken, B. and West, D.S., Manhattan Versus Euclid: Market Areas Computed and Compared, Regional Science and Urban Economics 14 (1984) pp. 19-35.

[10] Von Hohenbalken, B. and West, D.S., Predation among Supermarkets: An Algorithmic Locational Analysis, Journal of Urban Economics 15 (1984) pp. 244-57.

[11] Von Hohenbalken, B. and West, D.S., Empirical Tests for Predatory Reputation, Canadian Journal of Economics 19 (1986) pp. 160-78.

[12] West, D.S., Von Hohenbalken, B., and Kroner, K., Tests of Intraurban Central Place Theories, Economic Journal 95 (1985) pp. 101-17.

[13] Wiley, R.E., The End of Monopoly: Regulatory Change and the Promotion of Competition, in: Shooshan III, W.M., (ed.), Disconnecting Bell: The Impact of the AT&T Divestiture (Pergamon Press, New York, 1984) pp. 23-46.

TELECOMMUNICATIONS DEMAND MODELLING
An Integrated View
A. de Fontenay, M.H. Shugard, D.S. Sibley (Editors)

CARRIER BYPASS AND COMPETITIVE STRATEGY

Authors: Craig P. Buxton & Peter Cartwright
Pacific Bell Telephone, San Ramon, California

INTRODUCTION

When individual businesses compete to sell like goods and services to a particular market they are usually lumped collectively into what is termed an industry. Difficulties in the definition arise, however, when attempts are made to articulate the boundaries around the collection created. Authors who have studied the problem of defining industry or business structure and boundaries include: Porter (1986; Baumol, Panzar, and Willig (1982); Sharkey (1982); Brock (1981); Abel (1980); Porter (1980); Caves (1977); Bain (1972); and, Sherer (1970).

In confronting the complexities surrounding definition of business boundaries and industry structure, these and other authors have relied on the economic tools and concepts of industrial organization and competitive strategy. Particular attention has been paid to applying these tools to understand the impact different definitions can have on industry structure, business conduct, and market performance.

Economic theory of competition and industrial organization demonstrates the effects which industry structure has on decisions to diversify across existing product and market boundaries. Perceived barriers to entry in a particular industry may, in fact, be much less formidable than imagined, as changes in technologies, regulation, demand, or competition encourage businesses to translate their production capacity to another industry. The expanding company makes more efficient use of its productive resources by redefining its business boundaries. In the process, it also redefines what is understood to be the relevant industry in which that business operates.

The use of economic concepts in industrial organization helps to demonstrate how leveraging existing or planned capabilities into a new line of business might lead not only to more efficient use of a company's fixed assets, but also help to remove previous barriers to entering an existing or new market not previously entertained.

One of the major problems studied by industrial organization economists relates to drawing the boundary which defines the scope or dimension, of an industry. The major problem appears to be that businesses within a given collection deemed to be an industry can be extremely sensitive to current or even anticipated conditions defined by products not offered, markets not served, or even to predicted movements across industries, among them their own. Nowhere is such a problem better understood and illustrated today than in the economic activities of telecommunications.

Deregulation and the convergence of computer and communications technologies, along with the growth in demand for information processing, has made it more difficult to view telecommunications as just plain old telephone services (POTS). The players in the telecommunications market are not only sensitive to conditions in the telephone industry, they are also quite concerned with the conditions surrounding the computer and information industries as they undergo redefinition. In short, carriers serving either the long distance or local voice and data communication needs are reluctant, if not opposed, to being characterized as just part of the what was conventionally viewed as the telephone industry.

As principal drivers of competition in the telecommunications industry, deregulation and the convergence of communications and computer technologies are necessarily affecting the structure and definition of the participating industries, and the strategic positioning and conduct of firms operating in these industries (Porter, 1985).

Various coalitions such as joint ventures, acquisitions, mergers and other business arrangements are being used by both buyers and suppliers of the technologies used to converge and link information supply and demand. The participating companies are now able to cross horizontally the industry boundaries which once separated, the companies providing communications processing from those providing information processing, computer hardware and software, network connectivity, and even entertainment. Anticipating this evolution, the major long distance carriers - interexchange carriers (IECs) - are pursuing these various business ventures to prepare and position themselves.

The IECs' success in these arrangements, however, depends on their ability to take advantage of changing technologies, deregulation, and perceived needs of different market segments. In many cases, success will translate into a substitution of IEC services for those which are currently made available through the local exchange companies. Such an event is known today as carrier bypass, which if uncovered within the context of industry structure, comeptitive conduct, and market performance, can provide background for economic analysis and policy. Carrier bypass is, therefore, the subject of the analysis that follows.

Before the issue of bypass can be further explored, however, it is first necessary to describe the carrier access, which is what is being "bypassed". As part of the AT&T divestiture agreement, the local exchange telephone companies were required to set up a system of access services to connect the interexchange carrier to the local network. The charges for these connections were called access charges. A collection of access products was established, both for the public switched network services (POTS and Wide Area Toll Service - WATS), and for dedicated private line-type usage. The POTS/WATS access, which is switched at a local telephone company end-office, is called switched access, while the dedicated access products are collectively referred to as special access.

Switched carrier access is a combination of three basic components: the local loop, end-office switching, and inter office transport. The local loop connects the end user to a telephone company central office, local switching refers to the switching function at the central office, and local transport provides the connection between the central office and the interexchange carrier. Figure 1 shows a schematic representation of switched access.

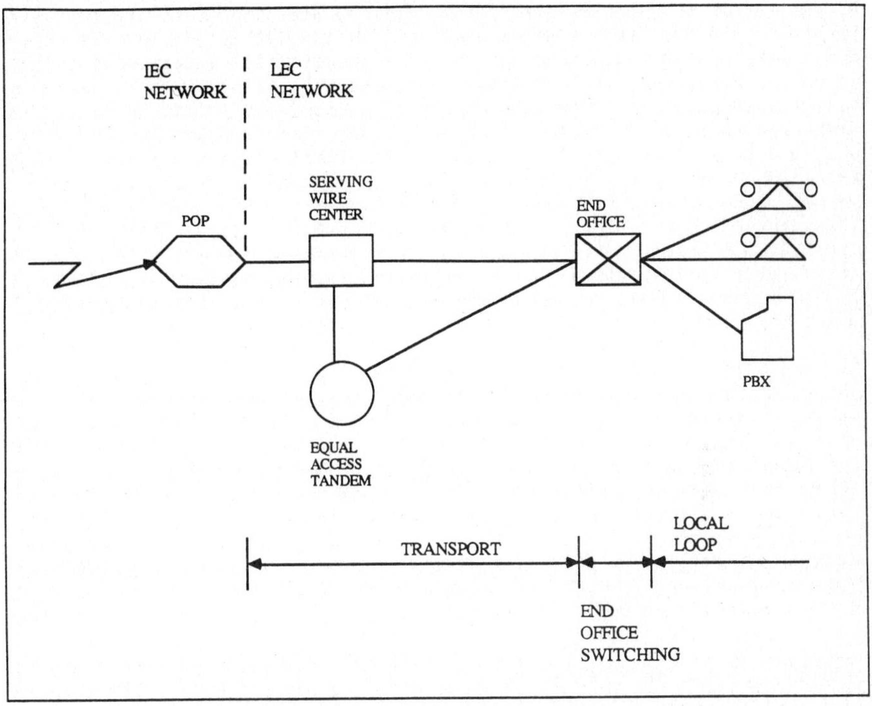

Figure 1. Switched Access Schematic Diagram

Special access is similar to switched access except for the absence of this switching function. A special access circuit is a dedicated circuit from end-user premises to the interexchange carrier. The economics of carrier access and bypass are mixtures of cost-based pricing and welfare economics. The rates for special access are essentially cost-based, with the charge for a circuit being a fixed monthly amount which is a function of the length, capacity, and the features of the circuit.

Rates for switched access, on the other hand are designed with several overriding equity issues in mind. Switched access rates incorporate features to provide "fair" treatment to small interexchange carriers, i.e., "equal price for equal unit of traffic delivered". They also include a surrogate tax (the Carrier Common Line Charge), which supports below-cost rates for basic local exchange services. Switched access is priced on a minute-of-use basis, with the charge-per-minute varying as a function of the transport distance. The inclusion of the Carrier Common Line Charge in switched access rates creates, in some instances, a significant disparity between cost and price.

In competitive markets, large disparities between cost and price in a particular market act as incentives for new entrants to compete with existing suppliers and, in the process, drive prices closer to costs. The carrier access market has been defined as part of the local exchange company franchise, so that the exchange company's prices for access are regulated to achieve desired social goals. In theory, then, the access market is a monopoly market. In fact, however, the access market is open to certain types of competition which come from participants who do not share the same franchise obligations as do the exchange companies, and who can capitalize on the cost price disparity inherent in the access services offered by the exchange companies.

This type of competition has typically been referred to as "carrier bypass". In this paper, the term carrier bypass will be used to refer to competition for any component of the access market, whether it is the entire access connection, or only a single component of access such as switching.

Placing this paper's analysis in the context of industrial organization and, more specifically, competitive analysis, is intended to expand the definition of carrier bypass. Today's definition of carrier bypass has not gone much beyond the context of the pricing anamolies created by the existing tariffs for carrier access. In this analysis, carrier bypass is, instead, considered a part of an IECs larger decision having to do with attempts to create and sustain competitive advantage.

An understanding of an IEC's competitive strategy helps to disclose the fundamental business objectives which underlies its day-to-day tactical activities. Knowledge of this underlying strategy helps provide a consistent, unifying, context for what might otherwise appear to be mutually independent activities on the part of a competitor. Abstracting the important forces from the details helps to identify the interaction among important factors which affect business conduct, competition, and industry structure. It can also entail recognizing strategic effects and motivations which affect decision making any competition in individual product markets.

Porter (1986) suggests that there are essentially two areas in which an IEC can pursue and achieve competitve advantage:

(1) Cost Leadership - The sources of cost advantage are varied and can include the adoption of economies of scale, proprietary technology, and preferential access to key supplies for "production" (e.g., the nationwide 800 service data base); or,

(2) Differentiation - The sources of differentiation stem from an IEC's ability to capture market value by offering customers products and services perceived to be unique to their needs. To differentiate itself from other carriers, an IEC must be capable of satisfying customer demand with the most valued features and functions at a premium price and without jeopardizing its overall relative cost position within the industry.

These two areas provide a backdrop for evaluating an IEC's strengths and weaknesses. Such evaluation is best achieved by rating the impact that an IEC's internal operations and external activities can have on its relative cost and on its ability to differentiate. The case presented in this paper will illustrate how AT&T and other carriers are engaging in carrier bypass as a means to gain competitive advantage through both cost leadership and differentiation.

BACKGROUND

Interexchange carrier access services are a major source of Pacific Bell's revenue, and they provide the Company with important net contribution. From the IEC perspective, access charges constitute the largest single operating expense. It is estimated that access charges constitute 55 to 60 percent of AT&T's current operating expense (Yankee Group, 1985). Because the IECs have targeted California as a prime market for their own revenue and profit generation, both growth and cost reduction are primary operating objectives. Significant population concentration in California also makes the state an attractive market for new product offerings designed to benefit easily aggregated, high-volume voice and data usage.

Reducing the cost of operating in California has become a major IEC objective. Access charges vary somewhat from state to state, but the charges in California are among the highest in the country. Since access costs represent a significant fraction of an IEC's own total operating cost, an IEC will aggressively attempt to minimize these costs as price competition if the long distance market erodes overall profit margins.

While a number of alternatives for reducing total costs may exist for the IECs, the most significant can be classified either costs associated with access to end users or the costs associated with operating the basic backbone network. Not all of the IECs will, however, face the same conditions for reducing these two types of costs.

Throughout most of their history in the industry, the other common carriers (OCCs), e.g., Sprint, MCI, etc., have enjoyed a cost advantage derived primarily from FCC-mandated discounts in access charges. These advantages have been counterbalanced somewhat by the cost of completing significant portions of their traffic on facilities leased from other carriers (primarily AT&T). As equal access has been implemented, the access cost discounts for the OCCs have been reduced significantly and the cost advantage they once enjoyed has been eroding. The primary OCC response to this change has been to expand their backbone networks to obtain better control of network costs. Secondarily, the OCCs have also been attempting to reduce access costs.

AT&T, on the other hand, already has complete interLATA connectivity. Therefore, cost reduction for AT&T translates into selective improvements in technical capability, and efforts in both the marketplace and the arena of public policy to reduce access charges. A primary avenue for reduction of access costs is the substitution of dedicated special access for Pacific's switched access. However, access cost avoidance is not the only factor influencing the IECs to substitute special for switched access; the IECs, particularly AT&T, can use this substitution as an opportunity to effectively capitilize on the capabilities their switches have for providing value-added services to end-users.

An analysis of AT&T's operations and activities in one part of California demonstrates how the increased use of special access is directed at ensuring competitive advantage both through cost leadership and differentiation.

The incorporation of competitive assessment also provides at least a qualitative basis for adjusting results produced from the application of more conventional business tools to a competitive reality where business boundaries are shifting. We will illustrate this process with a case study involving competition in the carrier access market between Pacific Bell and AT&T, and we will attempt to generalize the process used in this case to any similar competitive situation. The process involves three general components:

- o Market Assessment
- o Competitive Assessment
- o Competitive Response

These components are not parts of a programmable, step-by-step procedure. Each component requires feedback from the others, since competitive actions affect the market, and conversely, the market affects competitive actions.

SITUATIONAL BACKGROUND

In October of 1984, AT&T informed Pacific Bell that it intended to establish three new points of presence in the Easy Bay region of the San Francisco Area. These new points of presence (POPS) would serve wire centers which, at that time, were being served by an A&T POP in Oakland. This announcement was one of the first signs that switched carrier access was to become a competitive market and that Pacific Bell needed to develop, as a starting point, a quantitative assessment of the access market.

The area served by the Oakland POP is shown in Figure 2. The proposed POPs were Walnut Creek, which is 14 air miles from the ATT POP in Oakland; Pleasanton, which is 25 air miles from Oakland; and Hayward, which is 15 air miles from Oakland.

MARKET ASSESSMENT

Our objective in assessing the access market was to characterize the set of market conditions that accompanied the AT&T proposal for the East Bay. This characterization would develop the basic data for competitive analysis. In the competitive analysis we would then synthesize a view of the AT&T strategy, and develop an ability to predict the kinds and locations of strategic moves that we could expect AT&T to make in the future. From the perspective of hindsight, the following characteristics seem to have been most important:

- o Switched access pricing
- o Geographical distribution of AT&T access traffic and access costs
- o Configuration of Pacific's access network
- o Local economic conditions
- o Location of major customers
- o AT&T network configuration and network capabilities

The following sections describe the conditions which applied to each of these market variables.

SWITCHED ACCESS PRICING

Switched access is priced per minutes-of-use. Switching (and the charge for the local loop) are priced at a fixed rate per minute. Transport is also priced on a per-minute basis, but the per minute charge is a function of the distance between the carrier's POP and the end user's wire center. At the time we began this study, Pacific Bell's transport charges ranged from \$.0050 per minute to \$.0848 per minute, depending on the transport distance. The distance sensitivity of transport prices created a clear incentive to add points of presence to minimize average transport distances.

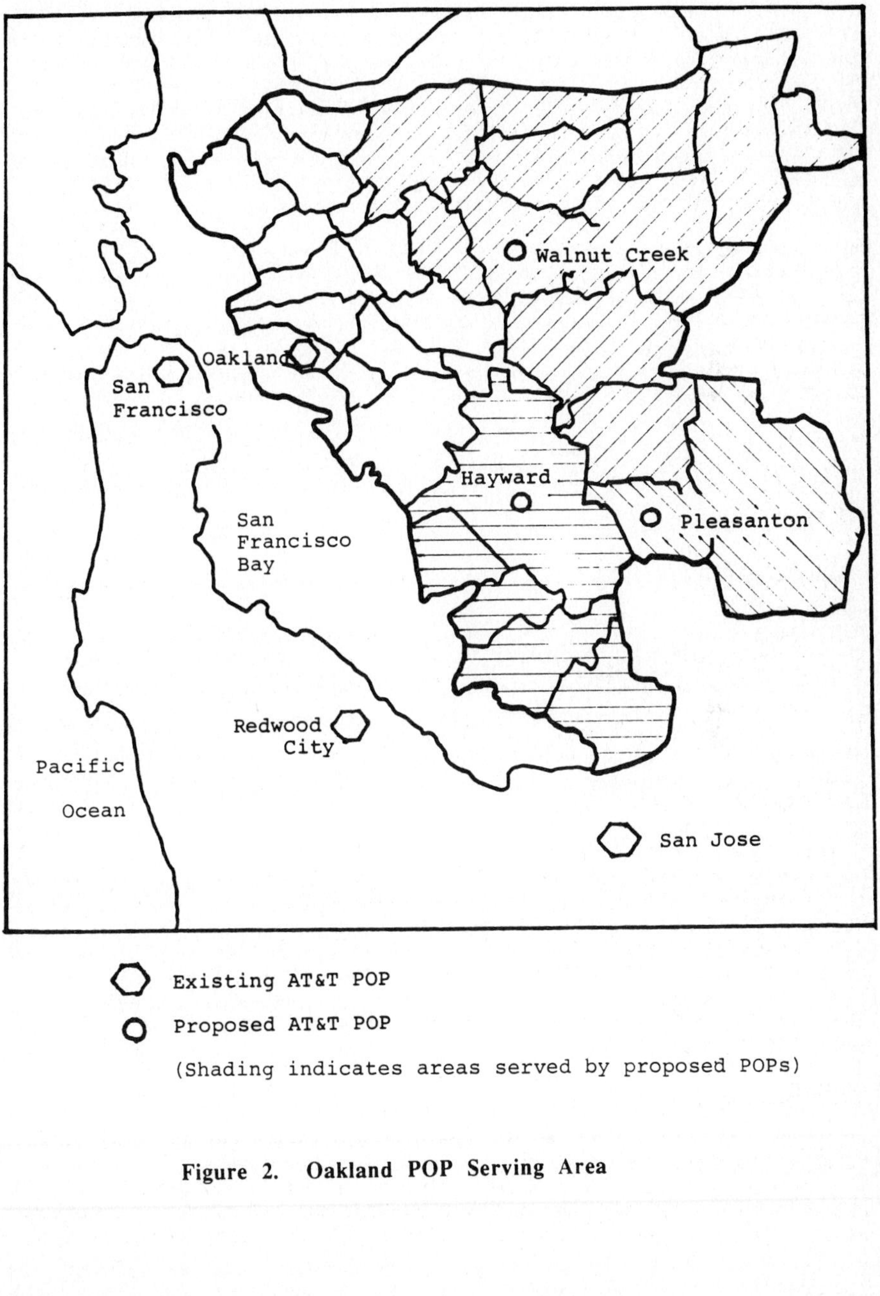

Figure 2. Oakland POP Serving Area

GEOGRAPHICAL DISTRIBUTION OF SWITCHED ACCESS TRAFFIC

Since reducing distance-sensitive local transport charges appeared to be a principal motive for AT&T's requested access reconfiguration, the first step in the analysis was to assess the effect that the proposed POPS would have on AT&T's transport costs. Carrier access usage data were investigated to determine the quantities and the geographical distribution of the Easy Bay switched access traffic. This analysis showed that the areas which would be served by the three new POPs generated 50 perent of the traffic currently being served by the existing Oakland POP. The analysis also showed that the same wire centers generated 70 percent of the Oakland POP's acess transport expense to AT&T. It was estimated that, under rates in effect at the time, the new POP configuration, would reduce AT&T's access transport costs to the affected wire centers by about 50 percent.

The second step of the access cost analysis was to create a model that would predict locations with high potential for new points of presence. If reducing local transport charges is the IECs motivation for adding POPs, the most attractive locations for new POPs will be those where there are aggregations of wire centers which have relatively large access volumes and which are at some distance from existing points of presence. Since the geographical location of wire centers is of key importance, we created a map of what we called "transport revenue at risk". We defined revenue at risk as the existing transport revenue, less what that revenue would be if there were a POP in that wire center, i.e., the zero mileage charge.

For each wire center:

$$\text{Revenue at risk} = \text{Access Minutes} * (R_e - R_o)$$

Where:

R_e = Transport rate, $ per minute, at existing transport distance
R_o = Transport rate, $ per minute, at potential transport distance

This calculation treats each wire center as a potential POP location. In practice, the most economical network configuration would most likely incorporate fewer POPs than one at each wire center. The optimal configuration, strictly from a cost standpoint, must reflect extensive carrier network.

The results of these calculations were plotted on a wire center map to reveal clustering of wire centers with high revenue at risk. The portion of the map which covers the Oakland POP area is shown in Figure 3. The areas with the highest revenue at risk are shown in the darkest shade; those with the least risk are shown in the lightest. The results of this mapping, which show concentrations of transport revenue in the Walnut Creek, Pleasanton, and Hayward areas, correspond very well with AT&T's plans for new POPs.

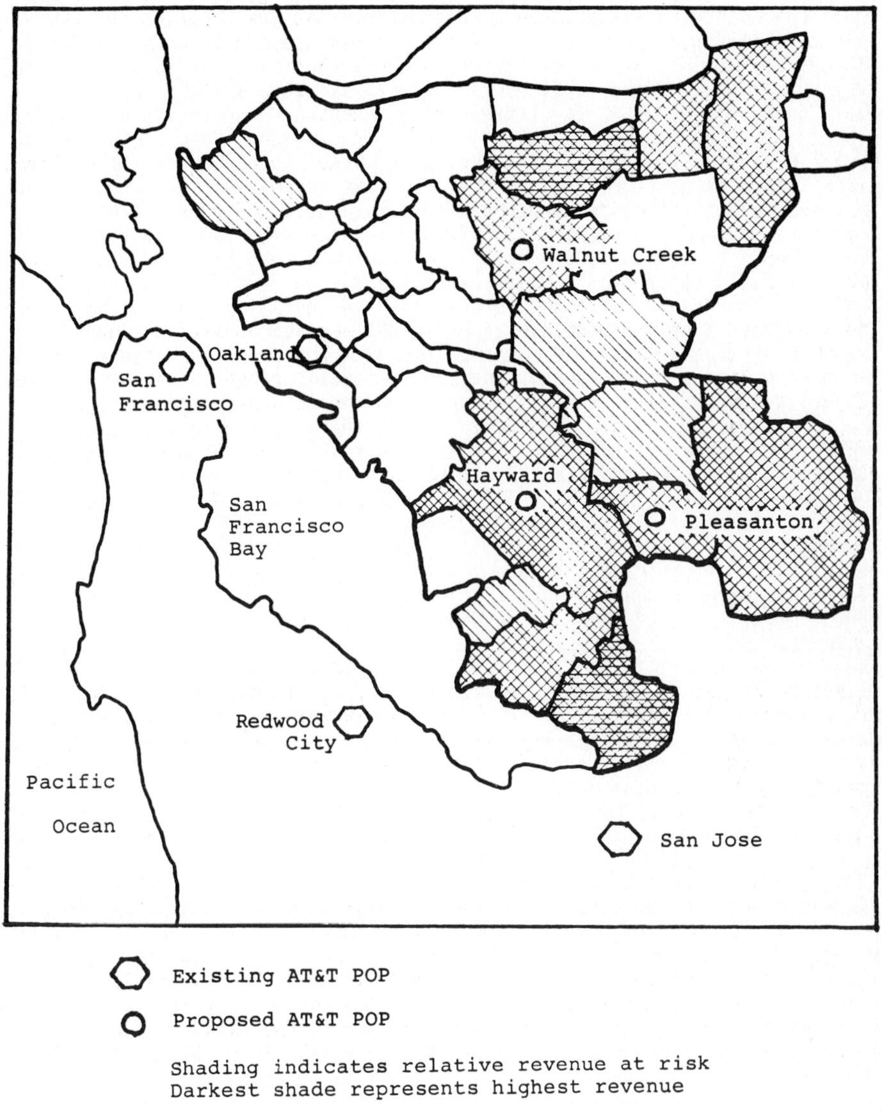

Figure 3. Relative Transport Revenue at Risk

PACIFIC BELL NETWORK CONFIGURATION

Pacific's access network has a kind of hub and spoke configuration in which so-called facility hubs provide points of traffic concentration for access traffic. Access trunks from remote end offices intersect at the facility hubs to become part of a common transport route between the hub and the AT&T POP. A hub is not a point of switching; it is just a location where its convenient to bring together various groups of dedicated lines and carry them into the same facility section.

In the East Bay, Pacific has facility hubs in all three of the sites selected by AT&T for new POPS. Thus it seemed that facility hub locations would be likely targets for additional new POPS. By assuming that further POP additions would be at hub locations, we have been able to conduct hypothetical revenue impact analyses based on re-homing access traffic to nearby facility hubs rather than to more distant POPS.

Assuming that access demand was geographically static and that reducing transport charges was the sole motive for POP proliferation, then an analysis such as the one we have outlined so far would provide adequate data for estimating potential POP locations and formulating a strategic response. One such response would be simply to reduce the distance sensitivity of transport charges overall, but it would lower the benefit of adding POPs to reduce transport length of haul.

Another pricing alternative would be to explore other means of providing transport and pricing the new offering more competitively. Pacific proposed several such alternative arrangements to AT&T for the link between Oakland and Walnut Creek with significant price reductions in comparison with current charges. All of these proposals were declined.

The failure of Pacific Bell's offers of more competitive pricing on the transport routes in question indicated that AT&T probably had objectives that went beyond simple reduction of transport costs. We therefore expanded our analysis to consider the effects of general economic trends in the area, the configuration and capabilities of the AT&T network, and AT&T's competitive position in access.

REGIONAL ECONOMIC CONDITIONS

The areas that would be served by the new points of presence are experiencing rapid economic growth. This growth consists of both industrial and residential/commercial development. The industrial growth is fueled by a vigorous regional economy and by a movement of jobs from San Francisco. Major office developments are creating tens of thousands of new jobs in the area, many of them with national corporations. Thus, this area is being transformed from one characterized by middle and upper income bedroom communities and agriculture, where many workers commuted to Oakland or San Francisco for their work, to one which is itself a center of employment and commerce where telecommunication links to areas outside the region are of increasing importance.

The attractions of the region to an IEC are clear. A growing, mobile, middle and high income residential population is an ideal market for conventional long distance service. More important, however, is the growth in employment by communications-intensive regional, state, and national corporations. New corporate residents in these areas include data processing facilities for Bank of America, headquarters, administrative and technical operations for Chevron Petroleum, and regional offices for Toyota and AT&T.

THE AT&T NETWORK

Carriers face a number of alternative means to extend themselves across existing or new business boundaries. Vertical integration represents one viable means, whether it be backward of forward integration. A carrier considering the alteration or augmentation of its network should consider both backward and forward integration of new facilities into its existing network. The integration problem has both geographical and technological dimensions, and must account for the location of the existing network relative to proposed additions. It must also satisfy the need for technological compatibility of proposed and existing facilities.

Equally important to the technical aspects of the integration problem are the extra financial burdens imposed as the carrier takes on more fixed costs to gain production economies.

In this particular case, the proposed AT&T POPs were all in close proximity to existing AT&T fiber or microwave facilities. Thus it appeared that AT&T's network reconfiguration costs would be minimal. Although fiber and microwave are different transmission media, the underlying protocol for both is similar so that integration of networks with dissimilar transmission media does not present a difficult problem.

The availability of ample in-place transmission capacity had another important effect on AT&T's actions and greatly reduced the investment required to establish the new POPs. Because of the low incremental cost of transmission circuits, AT&T has been able to add the new points of presence without placing switching capability at the new POPs. Each new POP is expected to consist primarily of a digital access cross connect frame, which would accept DS-1 (1.5 Megabits per second) or DS-3 (44.736 Megabits per second) digital circuits from Pacific for efficient transport to the existing switch at Oakland. Each of the new POPs can be thought of as the termination point of an extension chord, which reaches from the existing class-4 switch out toward the customers. The extension chord reduces access transport to Pacific and positions AT&T physically close to many major customers.

Thus, the "extension cord" concept describes a form of vertical integration that reduces access transport costs paid to Pacific. Furthermore, this form of integration provides AT&T with more control over the distributive transport channels that position AT&T physically closer to the major customers previously reached through transport outlets provided by Pacific.

SWITCHED AND SPECIAL ACCESS AS SUBSTITUTE SERVICES

Until recently, the capability of AT&T's network was limited in that it could not originate or terminate switched traffic without using the class-5 switches of the local exchange companies. However, AT&T is developing the capability that will allow it to use special access as a low cost alternative to switched access for certain customers.

This capability also allows the class-4 switches at AT&T POPs to emulate the functions of an exchange company class-5 for the origination and termination of toll services. With this expanded switching capability, AT&T can provide WATS and 800 service without using the local exchange company's switched access service. As a result of this capability, AT&T and the exchange companies end up attempting to provide like goods - the first point of network switching - to the same set of customers and markets and, therefore, find themselves in direct competition with each other over the supply of value added services.

COMPETITIVE ASSESSMENT

Judge Greene's Modified Final Judgment (MFJ) would seem to have cleanly drawn the functional and geographical boundaries between the business of the IECs and the LECs. Our experience, however, has been that such an assumption of a fixed boundary is mostly wishful thinking. Both the changing economics of telecommunications and the heightened competitiveness of AT&T have put the local exchange companies into a position of competing for that part of the business that was made the "exclusive" province of the exchange companies by the MFJ.

Competitive assessment requires a theoretical framework within which to synthesize observed facts to form a picture of the competitor's strategy. The competitive situation which we have described is a competition over definition of the business boundary. The introduction of competition into the carrier access market is producing a change in the structure of the telecommunications industry as it leads to a redefinition of the relevant market in which market share is computed. Thus we have an industry that was previously defined as monopolistic in which several players now compete for a share of the business not all of which existed before divestiture.

Following Michael Porter's framework for competitive analysis, we asked ourselves how the observed events and conditions might fit into an AT&T plan to augment market share through either cost leadership, product/service differentiation, or business focus.

The competition over the business boundaries could be interpreted primarily as an effort by AT&T to reduce its costs. Additional POPs to handle switched access traffic would reduce average transport lengths of haul and thus reduce distance sensitive transport charges. Conversion of certain types of access traffic to special access would also reduce access costs.

Perhaps more importantly, however, AT&T's move to shift the business boundaries can also be interpreted as an effort to provide the primary value-added components of certain services and thereby increase its market share of major customers. If the local exchange company provides the first point of switching, as was envisioned by the MFJ, it acts as the source of many value added service enhancements and maintains its position as the end user's conduit to the long distance network. If the IEC, on the other hand, also acts as the first point of switching through direct connection to the end user from an extensive network of POPs, the IEC greatly enhances its ability to be a full service provider and to be the sole source of telecommunications services. This single-source concept may be very attractive to the business customer faced with ever increasing and more complex telecommunication and information management requirements.

It is also very attractive to AT&T as these same customers are a very significant source of its net income.

The competitive dynamics of the access market are influenced in large part by two recent changes in the economics of telecommunications. Both of these changes have occured since the establishment of the public switched network in its present form, and have created possibilities for reconfiguration of the network.

The public switched network has been designed as a hierarchical arrangement of switches, starting with the end office or class-5 switches, which are linked by interoffice "trunks" in such a way as to minimize total network costs while providing universal connectivity and a high degree of reliability. The first economic change to consider is that switching has become less expensive and easier to maintain.

As a result of these technological changes, it has become economically feasible for medium and large sized business customers to vertically integrate by having their own switches (PBXs) on site. The first point of toll traffic aggregation in such a case occurs at the customer premise rather than at the exchange company's class-5 switch. For such traffic, the class-5 switch may be providing little, if any, added efficiency to the network.

The second economic change to consider is that the cost of interoffice trunking has been dropping. This change is more recent than the one affecting switching, and is largely attributable to the advent of fiber optic transmission. This change means that it is less important than it once was to achieve the maximum efficiency from a trunk. Thus a direct connection between an end user's PBX and an IC, which bypasses the traffic aggregation provided by the local exchange company's class-5 switch, now becomes economically feasible. The attractiveness of such arrangements is enhanced by the "non-economic" price of switched access, which has been set to provide a substantial contribution to recover fixed costs of the local network.

The revised economics of access, coupled with competition for customer presence, are pushing AT&T toward a new network architecture that could be described as "nodal" rather than "hierarchical." This new architecture places a multitude of service nodes close to selected end users, and ties the nodes back to the toll switches in the existing network. The nodes are not true switches, but they allow for efficient operation of the POP extension cord. When an IEC has existing transport facilities in an area, it may be able to use those facilities to establish a service node POP with a relatively modest investment.

CRITERIA FOR CONTEST OVER GEOGRAPHIC BUSINESS BOUNDARIES

To summarize, there are two market conditions that will make an area attractive for POP expansion:

o An aggregation of high IEC volume wirecenters at some distance from existing POPs. Additional POPs may lower access transport costs, depending on the distance-sensitivity of local transport rates.

o The presence of major end -ser customers in areas remote to existing POPs. Geographical proximity to such customers may enhance an IEC's capability for direct connection or service bypass. Such direct linkages may lower access costs and will ensure that value-added services will be provided by the IEC rather than the local exchange company.

In addition, the presence of two network conditions will enhance the likelihood of the carrier expanding its network with new POPs:

- o Excess capability on the toll switch. The segregation of functional responsibility between AT&T and the BOCs at divestiture, as well as OCC competition, may have created significant excess capacity at some AT&T class-4 switches which could be used to emulate class-5 switching functions.

- o Excess or easily augmented transmission capacity. The route distance of an alternative transport route may be of little importance because fiber optic transmission costs are relatively insensitive to distance. In addition, the incremental cost of a circuit may be very low if line-haul facilities already exist. The evolution of technology is toward an ever-increasing number of derived channels on a facility, as through the use of a higher bit rate on an existing fiber optic cable.

COMPETITIVE RESPONSE

Performing competitive analysis is relatively new to the local exchange companies. Before divestiture, most market type analysis was in response to either the desire of the public utility commissions to assess new rates or rate structures, or the desire of engineers to test a new feature or function developed within, and in many cases, for the network. Overall, the concepts and tools employed were those suitable to an industry defined as monopolistic in structure, requiring regulated policies, and judged by priorities for achieving universal service. The motivations and expectations in today's telecommunications environment will, instead, require use of different concepts and tools: new economics and greater competition are changing the features of the market environment (structure), the policies on pricing, product characteristics, and transactions (conduct), and the priorities in resource allocation (performance).

In our case study, we attempted to illustrate the benefits of employing the economic concepts and tools used in industrial organization. In many respects, these concepts and tools gave us a means to broaden our scope of consciousness about what we considered as the essential environmental features, policies, and priorities to study. As a result, we viewed the changes in the ways AT&T and other IECs conduct business with Pacific Bell and other exchange companies, and the advent of increased competition in existing and new, perhaps yet to be conceived, markets - not as obstacles to be removed, but as opportunities to understand and pursue.

The proper response to new entrants in the access market may involve adjustments in price, quality, the product mix, or some combination of these factors. There will always be the possibility that a shift in the business boundaries may take place, regardless of the local exchange company's efforts. Absent significant regulatory and legal changes, this shift will most likely occur in the immediate future among those segments of the market where customers with national telecommunications needs view the IECs as their primary suppliers of integrated, nationwide service.

Nevertheless, in any segment, the potential for contests over business boundaries creates a situation where past experience may be an inadequate basis for predicting a future position for local exchange companies. Changes in such factors as regulation, technology, and business conditions can cause companies to redefine their strategies and upset the existing order. The new order can mean new definitions of what constitutes the revelant market and the appearance of new players in the market. These new definitions can, in turn, imply, searching for new, more relevant information. Departing on the note of what is relevant is best stated by Warren Weaver in his book, The Mathematical Theory Of Communication.

"Information is a measure of one's freedom of choice when one selects a message. The concept of information applies not to the individual messages (as the concept of meaning would), but rather, to the situation as a whole"

BIBLIOGRAPHY

Abell, Derek F., Defining The Business: The Starting Point of Strategic Planning, Englewood Cliffs, New Jersey: Prentice-Hall, 1980.

Bain, Joe S., Essays on Price Theory and Industrial Organization, Boston, Little Brown, 1972

Baumol, William J., John C. Panzer, and Robert Willig, with contributions by Elizabeth E. Bailey, Deitrich Fischer, and Herman C. Quirmbach, Contestable Markets and the Theory of Industry Structure, New York: Harcourt, Brace, Jovanovich, 1982.

Brock, G. W., The Telecommunications Industry: The Dynamics of Market Structure, Cambridge, Harvard University Press, 1981.

Caves, Richard, American Industry: Structure, Conduct, Performance, 4th Ed., Englewood Cliffs, New Jersey: Prentice-Hall, 1977.

Pacific Bell Interstate Acces Tariff, FCC No. 128, Effective June 1, 1986.

Porter, Michael E., Competitive Strategy: Techniques for Analyzing Industries and Competitors, New York: The Free Press, 1980.

Porter, Michael E. Competitive Advantage, New York: The Free Press, 1986.

Scherer, F. M., Industrial Market Structure and Market Performance, Skokie, Ill.: Rand McNally, 1970.

Sharkey, William W., The Theory of Natural Monopoly, Cambridge: The Cambridge University Press, 1982

The Yankee Group, Long Distance Update, Volume 4, Boston: The Yankee Group, 1985.

TELECOMMUNICATIONS DEMAND MODELLING
An Integrated View
A. de Fontenay, M.H. Shugard, D.S. Sibley (Editors)

FORECASTING TECHNOLOGY ADOPTION WITH AN APPLICATION TO TELECOMMUNICATIONS BYPASS *

1. INTRODUCTION [1]

The rate at which technology users innovate by adopting a new technology has important implications for decisions involving the introduction, pricing, and regulation of such technology and for the pace of technical change. Technology introduction is a combination of both production and marketing. Factors that limit the availability of a new technology are supply-side constraints on the rate of technology adoption, while those that affect the acceptance of a new technology are demand-side constraints. Demand considerations have occupied a more central role in the study of new product introduction, although adoption rates are determined by constraints acting from both directions.

This paper examines various demand and supply constraints and how they affect the rate of technology adoption, using examples from the telecommunications industry. These determinants of technology adoption are then applied to bypass adoption.

2. DEMAND-SIDE CONSTRAINTS ON THE RATE OF TECHNOLOGY ADOPTION

When a new technology becomes available, user acceptance determines both the rate of adoption and the eventual market penetration. It is possible to represent the set of all potential users as a distribution, where users are differentiated by one or more characteristics which may or may not be observable. The fraction of potential users that adopt at any date is the fraction with characteristics for which the new technology is more favorable than competing technologies.

If the world were this simple, the adoption of a new technology would proceed in abrupt steps. Immediately after the introduction of the new technology, the number of users would jump and remain at a constant level until some other variable(s) affecting the value of the new or old technology changed. Yet, such abrupt behavior is not the norm. Most experience with adoption models suggests a smoother response to the introduction of a new technology.

2.1. Expected Profit From Innovation

According to basic economic theory, profitable innovations will be adopted, while unprofitable innovations will fail. If so, profitability would only determine whether an innovation has a market. It would have little impact on the speed of adoption. There is some evidence that more profitable innovations are adopted

---

* Dr. Jeffrey H. Rohlfs is senior economist and principal in the firm of Shooshan & Jackson Inc. Dr. Richard J. Gilbert is professor of economics, University of California, Berkeley. The authors acknowledge many useful comments from and discussions with Dr. Charles L. Jackson. Kara T. Boatman provided valuable editorial contributions.

more quickly [2]. However, in practice, many other factors enter into the adoption decision, and expected profitability need not be the most important determinant of the speed of adoption. Furthermore, the incentive effect of profit on technology adoption need not be directly proportional to the expected gain. For example, adoption may not occur unless expected profit exceeds some threshold level, and the level may vary from one individual to another.

Profitability typically will vary from one application to another. A technology may require a critical volume of use before it can be economic. Examples are a new PBX with a large number of circuits, or an accounting system for long distance service whose fixed costs make it economic only in high traffic areas (centralized automatic message accounting had this aspect when it was introduced in the Bell System).

2.2. Risk

When a new technology is first introduced, it represents little more than a claim by its manufacturer as to its promised performance. With no track record, the value of the technology in service is as yet unknown.

Experience generates information about the performance of a new technology. Potential users may differ in their access to new information, and their faith in alternative information sources. These differences have important consequences for the rate of adoption and for the return from alternative marketing efforts.

The continuous accumulation of information about a new technology is a main reason why the use of a new product, if the product fills a market need, will increase in an approximately continuous manner over time. As experience demonstrates and confirms the value of a new technology, other potential users will be induced to experiment with the product. Thus, adoption may exhibit a smooth pattern, rather than occurring in discrete jumps.

2.3. Ease of Innovation

Ease of innovation is a general term that captures the extent to which a new technology requires significant dislocations before it can be put to use. Ideally, a new technology can be plugged in and business goes on as usual -- except cheaper, faster, or better. In some cases, a new technology cannot be adopted unless users substantially alter the way they do business. Examples are traffic service positions and centralized automatic message accounting, discussed below. The need for such alterations makes the innovation less attractive and retards adoption.

Dislocation costs should be included in the expected profit of an innovation. However, these costs are difficult to measure and may not be adequately reflected in cost-benefit calculations. Although imprecise, the costs involved can be very large. Personnel changes may involve severance pay if jobs must be eliminated, and moving expenses if people are transferred. New workers have to be trained, and old workers may have to be retrained. Any personnel dislocations can have adverse impacts on morale and performance.

2.4. The Capital Stock of Potential Users of a New Technology

The term capital stock refers to all assets owned or accessible to the potential user which affect the services available to the user. Significant characteristics are:

a. *The value of services currently available from a potential user's existing capital.* A new technology is an economic choice for a potential user only if the value of the technology, net of its acquisition cost, exceeds the value of the existing technology. To the extent that the existing technology includes capital charges on expenditures that have already been made and cannot be recovered, these costs are "sunk" and should be excluded from a measure of its value. Sunk costs give a competitive advantage to an existing technology, which need compete only on its variable cost, while a new technology must compete on full cost.

b. *The age distribution of the capital stocks owned by potential users.* The age distribution is a determinant of the rate at which potential users' equipment will become obsolete. When an old technology is replaced, capital charges are incurred. This gives the new technology an opportunity to compete with the old on the basis of total costs because capital costs are no longer sunk for the old technology. Therefore, adoption of new technology is particulary appropriate at this time.

The age distribution of the old technology allows for a gradual increase in the rate of new technology adoption as more and more equipment wears out. The speed of adoption is increased if the rate of turnover of the capital stock is faster (that is, if the existing capital wears out more quickly). The speed of adoption also would be increased if a larger portion of the existing capital stock were installed early and therefore would be closer to retirement. Of course, the rate of obsolescence and the other demand-side constraints on the rate of technology adoption depend on the innovator's pricing decisions.

c. *Turnover of the capital stock.* A competitive disadvantage of a new technology is that its purchase involves new capital costs, while continued use of an old technology may be possible without incurring anything other than variable costs. But, whenever events force a turnover in the existing capital stock, continued use of the old technology requires a new purchase and, therefore, new capital expenditures. For example, the relocation of plant facilities is a logical time to upgrade production facilities. In the case of telephone equipment, for which installation costs are large, any move can justify evaluation of new technologies. Thus, the adoption of new telecommunications technologies can be facilitated by customer movement without the need for net growth in economic activity.

3. SUPPLY-SIDE CONSTRAINTS ON THE RATE OF TECHNOLOGY ADOPTION

The supply-side constraint on the rate of technology adoption is the limited capacity to make the new technology available to potential users and to communicate its effectiveness. Supply-side constraints can be affected by several factors, including the following:

3.1. Productive Capacity

The demand for new products is often highly uncertain. Prudent innovators estimate demand conservatively, so that they can limit their losses if demand is weak. If demand for the new product exceeds these conservative expectations, the firm may be unable to fill all its orders. Such circumstances call for rapid expansion, but there are limits to efficient rates of growth.

a. Growth usually requires installation of additional capacity, which can involve many months of delay.

b. Rapid growth may require hiring inexperienced personnel. A firm may face difficulties training new people and must be prepared for mistakes due to lack of experience. Higher expenditures do not necessarily mean greater production.

c. Start-up firms may encounter difficulties financing growth. If a firm's rate of increase in expenditures exceeds its return on capital, growth cannot be financed by profits. Instead, the firm must obtain funding, usually from outside equity and debt sources. Venture capitalists are often conservative and will demand either a clear track record or a large share of potential profits to compensate for the risk.

d. In some cases product introduction can be staged over time or among different market segments. For example, demand for a new durable good may be initially very high, but then decline quickly as the market becomes saturated. An innovator may wish to introduce the product slowly, and choose a pricing schedule that leads users with higher reservation prices to consume earlier. If the product appeals to different market segments, the innovator may develop the product first for the most lucrative segment, with subsequent introduction in other segments if it is successful.

These limitations on productive capacity mean that innovation may be delayed even when demand-side constraints on technology adoption are absent. In such circumstances, it is not appropriate to attribute innovation lags to consumer inertia.

3.2. Producer Reputation

Uncertainty about the performance of a new product is a major obstacle to adoption and an important element is the reputation of the seller. A firm with an established reputation for service and product quality can expect less resistance to the introduction of a new product than a firm without a comparable track record. Given the importance of adoption delay in the profitability of a new product, it is clear why reputation for quality and service is a critical business asset.

A firm concerned with maintaining a reputation for quality and service would take measures to assure user satisfaction with a new product and would attempt to convince potential customers of this commitment. If successful, user uncertainty would play a lesser role in retarding product introduction. An obstacle for the producer is the risk that it will have to incur unexpected expenses in meeting its commitment to customers. This suggests that a producer concerned with reputation maintenance would engage in extensive product testing prior to commercial sales, would follow a gradual introduction program to reduce the risk of "recalls" or major repair expenses in the field, and may include a price premium to cover the costs of guaranteeing a reliable product.

3.3. Industry Concentration and Ease of Entry

The effect of industry concentration on supply-side adoption constraints is uncertain. A firm with a large market share can offer inducements to potential users of a new technology and be relatively certain of capturing most of the resulting benefits.

For example, the firm may engage in promotional pricing, costly advertising, or an expensive warranty program. These activities would facilitate initial acceptance of the product if the product meets consumers' expectations, and as a result all suppliers would benefit. A firm with a small share of the market might be less inclined to offer promotional incentives if the benefits are not specific to the firm's particular brand, but spill over to affect the products sold by competing suppliers. However, innovation may be retarded if potential adopters anticipate that a firm with a large market share will exploit market power after the product has been introduced. The problems that concentration can pose for innovation are much less significant if there are few barriers to entry in the industry.

3.4. Economies of Scale

Substantial economies of scale may limit the ability of new entrants to compete in the introduction of new technologies. In addition economies of scale can put upper limits on the maximum pace of introduction for a technology that is otherwise free of supply constraints. An example is centralized automatic message accounting (CAMA), which provided automated billing for long distance messages. When it was introduced the technology was not economically feasible in areas where long distance traffic volume was light. CAMA systems were installed first in the largest toll centers. Over time, traffic growth and production economies permitted installation of CAMA systems in other toll centers, although the technology was always limited to regions of high traffic density.

3.5. Complementarities and Network Effects

Related to scale economies are "systemic" supply-side constraints resulting from interactions between technologies and user demands. The introduction of one technology may depend critically on the availability of another technology. Unless the introduction of both is controlled by one firm or industry, the supply of one can be a constraint on the demand for the other. The industry may be stuck in a situation where more of both technologies would be desirable, yet neither one is offered [3].

When complementarities and network effects are significant, a firm that is integrated in operations affected by a new technology is better positioned to introduce the technology in commercial applications. The vertical integration of the Bell System prior to divestiture was beneficial for the introduction of some of the technologies discussed in this report. The economics of CAMA, traffic service positions, and automatic intercept systems were highly dependent on system configurations. The integrated Bell system was able to adapt system planning to exploit the benefits of these technologies.

3.6. Promotional Campaigns and Price Competition

Both cost savings and the availability of a new product or service provide incentives to adopt a new technology. In either case, price is a factor (although not necessarily the most important factor) in the adoption decision.

In its simplest terms, the strategic trade-off in introducing a new product is to price low and sacrifice current profits for market growth, or to price high and enjoy temporary profits. In some cases, the problems of new product acceptance can be so severe as to necessitate a period of promotional pricing. Promotional pricing will be more profitable if the benefits from sacrifices made to break down obstacles to user acceptance accrue only to the initiating producer.

4. MEASURING ADOPTION LAGS

It is important to understand the limitations associated with alternative measures of technology adoption. Clearly, the number of customers that have sampled a new product gives no indication of the product's total market penetration. Profitability depends on total market penetration. But total market penetration has to be interpreted with caution if it is used as a measure of adoption lag. The reason is that total market penetration depends on factors such as the rate of turnover of the capital stock, the market growth rate, and anticipated technical change. These factors are typically independent of user inertia with respect to the purchase of a new product. Thus, a small share of the embedded base for a new telephone instrument does not necessarily mean that the product is experiencing customer resistance.

A useful statistic for measuring the influence of user inertia in technology adoption is the market share of new orders because it is in new orders where old and new technologies compete on a roughly equal basis, independent of such factors as the rate of growth of the market or the rate of turnover of the capital stock. Another reason why the share of new orders is a useful statistic is that it serves as a *leading indicator* for the share of total market penetration. If the share of a product in new orders exceeds its share in the total market, its total market share will increase. The opposite is true if the product's share in new orders is less than its share of the total market. Thus, *examining shares of new orders shows the direction of change for total market shares.* (This statement assumes that products depreciate at the same rate.)

Total market shares ultimately follow the pattern established by new orders. While new orders set the pattern, the response of total market shares can be quite slow. The reason is that adoption requires either replacement of the stock of existing equipment or new demand which can be satisfied by the new product. If capital costs are high, or if there are significant installation costs for the new product, replacement demand requires depreciation or turnover of the "embedded base" of existing equipment, and total market penetration may proceed slowly even when a new product is quickly adopted by new buyers.

4. CASE STUDIES

4.1. Bell System Technologies Before Divestiture

a. *The T-1 Carrier.* The T-1 carrier introduced time division multiplexing as an alternative to frequency multiplexing for the transmission of voice signals over medium haul trunks (10 to 100 miles). It had a 24-channel capability on two pairs of wires.

Measured on the basis of new units shipped, the T-1 system was adopted quickly. Introduced in 1962, by 1966 the T-1 system accounted for 50 percent of all channel units. The time lag from 10 percent to 50 percent penetration of the market for new shipments of channel units was only three years. Excluding long haul systems, the 10 percent to 50 percent adoption lag for the T-1 system was only about one year.

Installation of T-1 carrier systems required a substantial capital investment (about $29 million in 1968 US dollars) and produced total net annual savings, including capital charges, of about $2.8 million (US). The rate of return to innovation, above and beyond the cost of capital, was a significant but not remarkable figure of about 10 percent.

The T-1 carrier system illustrates that rapid adoption can occur even with modest expected profit. Several characteristics of the T-1 system were favorable to rapid adoption. At the time the T-1 was introduced, the advantages of digital transmission were well understood. The dislocations necessary to install the system, in terms of personnel and physical resources, were minor. Actual installation of the technology proceeded on schedule, with few "bugs" encountered in the field. [4]

b. *The N-2 and N-3 Carrier Systems.* The N-2 and N-3 carrier systems are advanced versions of the N carrier system used for long haul service. The systems were introduced in 1962. New shipments of N-2 and N-3 channel units reached 10 percent of all shipments of long haul units by 1963, and 50 percent by 1964. By 1966, N-2 and N-3 channel units accounted for more than 75 percent of all shipments of long haul units. This rapid adoption is partly attributable to low risk and ease of innovation. The N-2 and N-3 systems were relatively straightforward extensions of existing N-1 technology. Few bugs were expected or encountered. Like other carrier systems, N-2 and N-3 could be installed without reconfiguring other parts of the network.

c. *Traffic Service Positions.* Traffic service positions (TSP) allow direct dialing by the customer of long distance calls that had previously required operator assistance. Traffic service positions provided considerable cost savings. However, adoption of TSP to achieve these cost savings was far from easy. Operator functions had to be centralized in large toll centers in order to use TSP efficiently. This required considerable network planning and reconfiguration. In addition, adoption of TSP involved substantial dislocations of personnel. Operators had to move from central offices where they formerly worked to toll centers, which were often quite distant, with those unwilling to be moved being placed in new jobs or laid off. Furthermore, the new technology developed some "bugs" in the field and components used on some early TSPS systems had to be replaced [5].

As a result of these difficulties TSP was adopted slowly. In 1970, seven years after the first installation, only 17 percent of main telephones were served by TSP. Indeed, TSP was adopted so slowly that it almost missed the window of opportunity. By the time traffic service positions took off, the new technology, TSPS, was available. Some operator mechanization in the late 1960s may have been postponed in the expectation of benefits from the improvements of TSPS when it became available.

Installation of traffic service positions proceeded more rapidly in the 1970s. By 1974, over 50 percent of main telephones were served by TSP or TSPS. Over 90 percent were served by TSP or TSPS in 1979. The 10-50 percent adoption lag was on the order of four years. These adoption figures represent total penetration, as TSP and TSPS were integrated with existing switching equipment.

d. *Centralized Automatic Message Accounting.* Centralized automatic message accounting (CAMA) is a technology which allows automated billing for step-by-step switches. The savings from CAMA, which are primarily operator wages, are very large. The first CAMA systems were installed in 1953. In 1971, 57 percent of all Bell toll centers were equipped for CAMA. In view of the large cost savings this number seems small. However a number of factors must be considered. Step-by-step switches, for which CAMA is intended, make up only a fraction of all Bell toll centers. Another reason for the small percentage of toll centers equipped for CAMA is scale. CAMA is economical only in applications where the volume of long distance traffic is high.

CAMA illustrates that profitability is not the sole determination of the speed of technology adoption. Many factors have to be considered. The fact that CAMA was a retrofit technology intended for specific systems means that the technology was not appropriate in many applications. Consequently, use in only a small percentage of toll centers does not, itself, indicate slow adoption.

e. *Automatic Number Identification and Long Distance Mechanization.* Automatic number identification permits identification of the calling party in a long distance call without the need for operator intervention. The savings are similar to those for TSP-TSPS and CAMA. In 1960, approximately 32 percent of all telephones equipped for outward direct distance dialing had either automatic number identification or automatic message accounting. By 1962, the percentage exceeded 50 percent, and had reached almost 75 percent by 1965.

While adoption of CAMA proceeded slowly, this was not the case for long distance mechanization in general. In only about three years, the share of main telephones equipped for outward direct distance dialing that had either automatic number identification or automatic message accounting jumped from 10 percent to more than 50 percent.

f. *Automatic Intercept Systems.* Automatic intercept is a computerized system to identify calls made to numbers not in service or requiring referral information and to provide the required information with minimal operator assistance.

Semi-automatic intercept systems, provided by non-Bell suppliers, were first installed in 1965. Fully automatic intercept systems, produced by WECO, were first installed in 1967 [6].

Large centers had to be established in order to use AIS efficiently. This required network planning and reconfiguration. There were also personnel dislocations. These obstacles were similar to those faced by TSP-TSPS. However AIS is a smaller scale technology. AIS also developed relatively few "bugs" in the field [7].

AIS achieved a 10 percent market penetration about five years after first adoption. Two-thirds penetration was achieved by 1977, twelve years after initial introduction.

4.2. Competition in Telecommunications

a. *PBX Markets* [8]. The focus of this case study is not the introduction of a new technology, but the emergence of competition from interconnect companies in telecommunications equipment markets. Interconnect companies have made strong gains in markets for PBXs. By 1983 interconnect companies accounted for 74 percent of all new shipments.

The Federal Communications Commission (FCC) opened up telecommunications equipment markets to competitive entry in 1968 with its Carterfone decision. By 1976, interconnect companies accounted for almost half of PBX line shipments. Prior to 1968 potential competitors were excluded from the U.S. PBX market and consequently had to start from the ground floor when restrictions were lifted. Entry required the establishment of sales and service organizations and the development of new products for the U.S. market. Such tasks can consume a great deal of time and effort. Bearing this in mind, we regard the adoption of interconnect PBXs from 1968 to 1976 as quite rapid.

b. *Electronic Key Telephone Systems* [9]. The market for key telephone systems has undergone considerable change since the commercial introduction of electronic systems in 1979. Installation of new stations has increased at a rate of about 8 percent a year in the 1980s.

The major domestic manufacturers have been slow in moving to the new electronic key telephone systems (EKTS). In part this was a consequence of confusion in the wake of the Modified Final Judgement.

Although the new EKTS are generally more expensive than the electromechanical alternatives, their innovative features (e.g., speed dialing, call forwarding) and ease of installation, along with attractive design, resulted in rapid acceptance. The share of EKTS in new shipments has risen from less than 10 percent in 1979 to over 70 percent in 1983. Most of the EKTS sales are for replacement of existing KTS as users retire their old systems and upgrade to the newer models. Often this accompanies a move or a change in office configurations.

Despite a rapid turnover of existing KTS and rapid acceptance of EKTS, the share of EKTS in the total key telephone system market has increased at a much slower rate. Although EKTS accounted for 70 percent of new shipments in 1983, their share of all KTS stations was only 30 percent in that year. The reason of course is the large embedded base of existing systems, of which AT&T has the largest share.

c. *Business Telephones* [10]. Interconnect companies have achieved steady progress in the market for business telephone instruments. However, interconnect growth has not been nearly so dramatic as in the case of EKTS. The interconnect share of the installed business telephone base increased from less than 1 percent in 1970 to just over 10 percent in 1980 and 12 percent in 1981.

Installation of business telephones is quite easy. Many locations have modular jacks, so the new instrument can simply be plugged in. The risk of adopting interconnect telephones is substantially less than for interconnect PBXs or electronic key systems. Much less can go wrong with simple telephones and defective instruments can be replaced while they are being repaired.

Nevertheless, adoption of business telephones has proceeded at a much slower pace than adoption of interconnect PBXs or electronic key telephone systems. One reason is that the telephones sold by the interconnect companies do not offer substantial improvements in features relative to the established products. Competition in the telephone instrument market is mainly in terms of price. Many cheaper units sacrifice durability and voice quality, while the performance of the new and more expensive units is uncertain. Given that telephone instruments are a small component of business telecommunications costs, the possible economy from new telephones may not be worth the loss of quality for many business customers.

In other words, the slow adoption of business telephones may not be due to customer inertia. Quite possibly, many business customers are aware of interconnect products and would be willing to adopt them to get superior value. However they prefer the high quality they believe they can get from AT&T and other telephone companies to the lower prices they can get from interconnect vendors.

5. APPLICATION TO TELECOMMUNICATIONS BYPASS [11]

The speed of adoption of bypass technology can be analyzed in the context of the preceding discussion. "Bypass" is the local transmission of voice and data communications from customers directly to the inter-LATA carriers without use of local exchange carrier (LEC) facilities. Bypass is a cooperative venture between the customer and the inter-LATA carrier. Both must agree to a long term commitment, since the venture involves investment in durable equipment.

6. DEMAND-SIDE CONSTRAINTS

6.1. Expected Profit From Innovation

Since bypass involves large fixed costs, its profitability depends on the size of the customer and the volume of use. Bypass is generally more profitable for larger customers than for smaller customers. Bypass adoption involves no degradation of quality or inconvenience (e.g., no additional digits must be dialed by the customer). Furthermore, digital transmission is less noisy than analog, and allows high speed data transmission. Vendors can develop portable electronic equipment which demonstrates this improvement in transmission, thereby facilitating the adoption process.

Our estimates of the speed of adoption of bypass services assumes no change in the cost of traditionally regulated services or in the regulation of bypass services. This condition is, however, unlikely. Bypass involves a revenue loss for the utility providing the bypassed service and this loss can be very large if there is excess capacity, so that the marginal cost of the service is low. Hence, providers of services subject to bypass have an incentive to offer competitive rates and/or to lobby for regulations (such as standby charges) that discourage bypass.

6.2. Risk

Since many bypass systems are mature technologies, the risk associated with their adoption is comparable to that of installing a mature carrier system. Regulators may impose costs on customers that choose bypass, in order to protect traditional regulated services (and in particular to protect existing rate structures). Thus, customers may bear a regulatory risk, although the FCC currently opposes such measures. Customers can reduce this risk by returning to the LEC, but the costs of switching back and forth can be considerable.

Bypass customers bear several kinds of risk. First, the bypass equipment may fail to function properly. However, reputable system vendors provide guarantees and contracts to ensure prompt service in the event of a malfunction. Therefore, such technological risk is minimized.

Customers also incur a business risk when they adopt bypass technology -- the risk that the anticipated business gains will fail to materialize. Business risk does not present a problem if the bypass vendor effectively communicates the cost savings that the system will yield. The task of conveying the cost savings of a bypass system is relatively easy to accomplish.

Bypass customers do not bear the risk of lost telephone service, since LEC facilities may be used as carriers of last resort. In fact, the bypass system could be designed so that calls are automatically rerouted if the system fails.

6.3. Ease of Innovation

Installing a bypass system is a simple operation, and can be supervised by the system vendor. The difficulties are comparable to those of installing an electronic key system. Bypass vendors will compete to minimize customer inconvenience, thereby simplifying the adoption process.

6.4. Capital Stock Replaced by Innovation

Bypass systems do not replace equipment owned by either inter-LATA carriers or their customers, as current facilities are primarily leased from the LECs. Therefore, capital stock replacement is not an issue in the case of bypass adoption.

7. SUPPLY-SIDE CONSTRAINTS

Again, our discussion presumes no change in service rates or regulation. Either a competitive response to the threat of bypass or regulatory constraints on bypass can considerably alter supply incentives. Moreover, the possibility of these actions is a significant risk that may deter potential suppliers of bypass technologies.

7.1. Productive Capacity

Bypass systems are stock items that use standard components. Most of the components and many of the subassemblies and subsystems are widely used in the communications and electronics industries. If necessary, production could be rapidly expanded to meet increased demand. Such expansion would require only modest expansion or reallocation of capacity in the electronics industry.

7.2. Producer Reputation

If the bypass vendor guarantees the system, he bears the risk of system failure. However, this risk is minimal and should not retard the adoption of bypass. Furthermore, system failure does not impose drastic costs, since the customer always has access to toll services via local lines.

7.3. Industry Concentration and Ease of Entry

Local Exchange Carriers (LECs) have an incentive to avoid bypass by competitive or regulatory means. If they cannot provide competitive rates for traditional utility service, they have an incentive to offer bypass services, although current regulatory policy prohibits them from doing so. Thus, bypass supply will come from the inter-LATA telecommunications industry, which is likely to be as innovative as regulation permits. AT&T has a dominant market share, so it can benefit fully from innovations. Furthermore, it faces serious competitive threats and the prospect of a rapidly declining market share. Under these circumstances AT&T should be extremely innovative in order to be competitive -- to the extent permitted by regulation. The industry that produces bypass systems is highly competitive and very innovative. Therefore, industry market structure should not impede bypass adoption.

The main impediment to entry is the risk of regulatory and competitive responses to the threat of bypass. The latter are particularly important in light of projections of excess telecommunications service capacity. If changes in either regulations or rate structures makes bypass uneconomical, the risk is borne mainly by the providers of the service.

7.4. Economies of Scale

Bypass systems are characterized by significant economies of scale. As more systems are demanded, more will be produced, and the cost per unit will fall. Furthermore, the costs of certain key technologies used in bypass systems, such as solid state electronics and fiber optics, are also declining due to other factors.

7.5. Complementarities and Network Effects

Microwave-based bypass systems can be installed piecemeal, one customer location at a time. There are no significant complementarities or network effects. However, waveguide bypass systems (e.g., cable I-Nets, fiber optic systems) do possess significant complementarities, and projections of bypass adoption must take these into account.

7.6. Promotional Campaigns and Price Competition

It is assumed that bypass vendors will launch promotional campaigns to induce customers to adopt bypass. Bypass vendors will find active marketing profitable. The vendor must effectively communicate to the potential bypass customer that the system will yield the costs savings that are claimed. If the vendor cannot convince the customer that these gains will materialize, the customer will not purchase the system.

8. CONCLUSION: SPEED OF BYPASS ADOPTION

The benefits of bypass adoption are substantial. Digital data transmission allows significant and measurable gains in productivity. Furthermore, adoption of bypass involves virtually no risk of lost telephone service. Some vendors are reputable firms with established track records. As a result, it seems likely that bypass will be adopted very rapidly. Although declines in access charges make bypass economically attractive in fewer circumstances, adoption will be very rapid for those customers for whom it continues to be profitable. In addition, it is important to note that bypass adoption will occur much more rapidly than the adoption of other common carrier (OCC) services. OCC adoption has occurred slowly, both because OCC service is of lower quality, and because OCC adoption involves dealing with an unfamiliar long distance carrier. Contrary to OCC adoption, bypass adoption does not result in a loss in quality. Furthermore, since AT&T sells bypass systems, bypass adoption does not have to involve an unfamiliar carrier. Therefore, conditional on continuation of existing rate structures and on permissive regulation toward bypass, bypass adoption will continue to rise as the costs of electronics fall and customers gain greater experience with bypass technology.

REFERENCES

[1] Information for sections 1-5 was taken from Gilbert, Richard J. and Rohlfs, Jeffrey H., Forecasting Technology Adoption; Appendix to Access Charging and Bypass Adoption (1985).

[2] Mansfield, Edwin, Industrial Research and Technological Innovation: An Econometric Analysis, (W.W. Norton, New York, 1968) and Oster, Sharon, "The Diffusion of Innovation Among Steel Firms: The Basic Oxygen Furnace," Bell Journal of Economics (1974).

[3] Farrell, Joseph and Solaner, Garth, "Standardization, Compatibility and Innovation," (MIT working paper, 1984) and Rohlfs, Jeffrey H. "A Theory of Interdependent for a Communications Service," Bell Journal of Economics (vol. 5, pp. 16-37, 1974).

[4] Discussion with W. Ruhl and D. Zuck, Bell of Pennsylvania Network Department (June 27, 1984).

[5] Ibid.

[6] Joel, A.E. Jr., et al in Schindler, G.E., ed., A History of Engineering and Science in the Bell System: Switching Technology (1925-1975), (Bell Telephone Laboratories, 1982).

[7] W. Ruhl and D.Zuck, op cit.

[8] Data for this case study are from a market study by Probe Research Inc.

[9] Data for this study are from Northern Business Information, "Key Telephone System Market, 1984 Edition" (1984).

[10] The data for this case study are taken from the North American Telephone Association "1982 Telecommunications Equipment Industry Statistical Review."

[11] Information for sections 6-10 was taken from Jackson, Charles L. and Rohlfs, Jeffrey H., Access Charging and Bypass Adoption, Shooshan & Jackson Inc., Washington, D.C, 1985.

# # #

SECTION III: USER MARKET STRUCTURE

III.2. Pricing and Regulation

TELECOMMUNICATIONS DEMAND MODELLING
An Integrated View
A. de Fontenay, M.H. Shugard, D.S. Sibley (Editors)

CONSUMPTION EXTERNALITIES IN TELECOMMUNICATION SERVICES

Benjamin BENTAL*

Technion-Israel Institute of Technology,
Faculty of Industrial Engineering and Management
Haifa 32000, Israel.

Menahem SPIEGEL**

Department of Economics,
The University of Haifa, Mt. Carmel
Haifa 31999, Israel.

Positive externalities in consumption of telecommunication services are due to the fact that the utility a consumer derives from having access to a communication network increases with the number of other consumers he can communicate with. Under this assumption of interdependence of the utilities of consumers, and the assumptions about the technology, the necessary conditions for socially optimal network size are derived. This optimal size consists of an optimal number of switchboards and optimal number of consumers per switchboard.

---

* Visiting Department of Economics, University of California, San Diego, CA 92003

** Visiting Department of Economics, University of Connecticut, CT 06268. Financial support from the University of Connecticut Research Foundation is gratefully acknowledged.

It is shown that the unconstrained profit maximizing monopolist (fully integrated) will operate the socially optimal network size.

Different market organizations and their effects on the network size are investigated. In particular the divestiture of the fully integrated monopolist into several local independent monopolies is discussed.

1. INTRODUCTION

Telephone services are known to be a product where there exist significant external effects in consumption. These externalities in consumption arise because the utility a consumer derives from being connected to a network is increasing with the number of other consumers he can communicate with. This increase in utility will also result in an increase in the maximum price consumers are willing to pay for joining the network as it gets larger in numbers of subscribers. Given the nature of the externalities, each and every consumer would want to belong to one universal network. If marginal costs of constructing a network were zero, this network would have emerged as the optimal one. A clustering of consumers into one network of optimal size and the outcome of a free entry zero profit competetive environment are mutually inconsistant.

In this paper we develop a model which describes the socially optimal telecommunication network size. This optimal size consists of the optimal number of switchboards and the optimal number of customers per switchboard. Moreover, we investigate how different modes market organizations will affect this optimal network size. The second area we explore is that of an equilibrium after "divestiture." We consider the following two possible market settings. (1) The independent switchboards (firms) are to be connected into a network by a single monopolist. (2) The interswitchboard connection is supplied by a "competitive" industry with the following payment scheme: each connecting firm charge consumers directly for its services and pays the local switchboard a connecting fee.

In a very recent article Katz and Shapiro [1985], solve a model of an oligopolistic competition in a market, where different products, if compatible, are considered to be part of the same network. Our discussion in the last section solves a similar problem under different market conditions. In general, we are concerned with socially optimal network size and with different kinds of regulation the producer might face. In particular we assume that given the infinite number of potential consumers of telecommunication services, the network size is an endogenous variable which is determined by the producer(s) in the industry.

The paper is organized as follows. In Section II we present the model, the utility function of the consumers and the technology and cost of production. The characteristics of a socially optimal network size are considered in Section III. The network size which will be constructed by a fully integrated monopoly and the application of possible regulation are discussed in Section IV. In Section V we describe the nature of the divestiture when each local switchboard is an independent producer but all switchboards are connected to each other. In the following sections we assume that under the divestiture the local switchboards are not interconnected. There is a new industry which will supply the interswitchboard connection. Section VII characterizes the Nash equilibrium solution when connection services are being produced by an independent monopolist, while section VIII presents a Nash equilibrium solution to oligopolistic competition in a market of the connecting industry.

2. THE MODEL

There are two goods in the model. The first is an all encompassing consumption good, and the second consists of communication services. The latter good is a discrete commodity which is either consumed or not. All consumers are alike, and have a utility function with the usual properties defined over consumption of both goods, given by

$$(1) \qquad U = U(x,y)$$

where x - amount of the all encompassing good consumed
y - communication services.

All consumers have an endowment of I of the first good. There is a continuum of consumers.

The "quality" of the communication service, y, depends on the number of other consumers with which any given consumer can communicate. Formally, let z the size of the network. Then

$$(2) \qquad y = y(z)$$

where we assume that $y' > 0$ and $y'' < 0$.

In order to have access to an existing telecommunication network of size z, a consumer may have to give up k units of the first commodity. Accordingly, his utility is $U(I-k,y(z))$. Clearly, the consumer is willing to join the network if

$$(3) \qquad U(I-k,y(z)) \geq U(I,0)$$

where $U(I,0)$ denotes the utility of a consumer who is not connected to the communication network. Let $k = R(z)$ be the reservation price for which (3) holds with equality. It can be shown that $R' > 0$ and that if both the first good and the communication services are normal goods $R'' < 0$.

In our model we assume that telecommunication services are produced by a technology which has the following properties. Each consumer is connected to a local switchboard. The cost of constructing a switchboard depends on the number of consumers (n) connected to the switchboard, $C(n)$, with $C' > 0$ and $C'' > 0$. Switchboards can be connected to each other. The cost of doing so, $F(m)$, is increasing at an increasing rate with the number of switchboards (m) connected. Thus the number of consumers in the network is $z = nm$.

In the next section we solve for the network size which maximizes the total value of consumer surplus.

3. SURPLUS MAXIMIZING

The network size which maximizes the total value of consumer surplus is characterized as follows: A pair (m^*,n^*) where m^* denotes the number of switchboards and n^* the number of consumers connected to each switchboard maximizes total consumer surplus if:

(i) the consumers connected to the network are willing to pay for the costs of constructing it.

(ii) Adding new consumers and switchboards to the network costs more than the amount the existing and the potential new consumers are willing to pay for increasing the network size.

Formally, the pair (m^*,n^*) has to satisfy the following requirements.

$$(4)\qquad \int_0^{m^*n^*} k(t)dt \geq m^*C(n^*) + F(m^*)$$

where k(t) denote price charged to the t-th consumer.

$$(5)\qquad k(t) \leq R(m^*n^*) \qquad \text{for all } t$$

$$(6)\qquad m^*n^*[R(m^*n^*+h) - R(m^*n^*)] \leq C(n^*+h) - C(n^*) \qquad \text{for } h \geq 0$$

$$(7)\qquad C(n^*) - C(n^*-h) \leq$$

$$\leq (m^*n^*-h)\ [R(m^*n^*) - R(m^*n^*-h)] + hR(m^*n^*) \qquad \text{for } h \geq 0$$

$$(8)\qquad in^*R((m^*+i)n^*) + m^*n^*[R((m^*+i)n^* - R(m^*n^*)] \leq$$

$$\leq F(m^*+i) - F(m^*) + iC(n^*) \qquad i=1,2,3,\ldots$$

$$(9)\qquad F(m^*) - F(m^*-i) + iC(n^*) \leq$$

$$\leq in^*R(m^*n^*) + (m^*-i)n^*\ [R(m^*n^*) - R((m^*-i)n^*)] \qquad i=1,2,3,\ldots$$

Condition (4) requires that all costs be covered by consumer connected to the network, while condition (5) sets the cost for each consumer to be less than his reservation price.

Conditions (6) and (7) imply that adding or subtracting consumers to any switchboard is undesireable while conditions (8) and (9) imply that increasing or reducing the number of switchboards is unwarranted.

The way we find the pair (m^*,n^*) is the following. Given any number for switchboards m, find the number of consumers per switchboard n^*. Then, given $n = n^*_m$ iterate on m to find m^*_n until a pair is found such that the choices of m^* given n^* and n^* given m^* are the same. We assume that this point exists. We can characterize the desired switchboard size by the following marginal condition, derived from (6) and (7) by dividing both expressions by h and taking the limit as h approaches zero.

$$(10) \qquad R(m^*n^*) + m^*n^*R'(m^*n^*) = C'(n^*)$$

Expression (10) reveals that marginal cost pricing does not satisfy (10) since $C'(n^*) > R(m^*n^*)$. The fact that the marginal cost of adding the last consumer to a switchboard is bigger than the maximal price a consumer is willing to pay for joining the network implies that existing consumers are subsidizing the last entrant into the network[1]. An analogous conclusion about cross subsidization holds also for the number of switchboards. Clearly, the solution (m^*, n^*) is also Pareto optimal. We show next that a monopoly achieves the same solution.

4. MONOPOLY

A monopolist chooses the number of switchboards (m) and the number of consumers per switchboard (n) so that his profits are maximized. In particular, the monopolist solves

$$(11) \quad \max_{m,n} \{ \pi = mnR(mn) - mC(n) - F(m) \}$$

Since π is the total value of consumer surplus it is clear that the profit maximizing values of m and n are those found in the previous section. The only difference concerns some distributional effects. The monopolist charges a price $R(m^*n^*)$ and leaves no surplus to the consumers, while in the previous section all we required was that the costs of the network be covered by the consumers.

Now suppose the monopolist is regulated. We consider here two types of regulation: quantity and price regulation.

If the regulator sets a minimum network size which is bigger than the optimal one, then the implicit assumption that the profit function is concave implies a reduction of profits. However, the price charged by the monopoly will increase, since the reservation price of the consumers is increasing with the network size.

If the monopolist is price regulated, so that the maximal price he can charge is $p < R(m^*n^*)$, then he is faced with the following profit maximizing problem:

$$(12) \quad \max_{m,n} \{V = pmn - mC(n) - F(m)\}$$

Problem (12) has the following first order condition

$$(13) \qquad p \geq C'(n)$$

Note that condition (13) is independent of the number of switchboards, m. Therefore, we analyze (13) for $m = m^*$.

Define n_a as a solution to

$$(14) \qquad p = R(m^*, n_a)$$

Clearly, $V = \pi$ (defined in eq(11)) for $n \leq n_a$ and $V < \pi$ for $n_a > n$. Consequently, the price regulated monopoly chooses a switchboard size n such that $n_a \leq n < n^*$.
In addition, the number of switchboards cannot grow since the cost of connecting additional switchboards cannot be covered by the reduced revenue each switchboard generates.
Therefore, the price regulated monopolist serves less consumers.

5. DIVESTITURE

In this section we consider the case where the m-swithcboard fully integrated monopolist is forced to divest so that each switchboard becomes an independent profit maximizing producer. The switchboards remain connected to each other at no cost to them or to their customers. We assume that each producer plays as a Nash player in deciding how many consumers he will serve.

The problem of the j-th switchboard firm is to choose his

optimal switchboard size (n); assuming that the choices of all other switchboard firms is unaffected by his decision. Specifically,

$$(15) \qquad \max_{n_j} \{ Z = R(n_j + \sum_{\substack{i=1 \\ i \neq j}}^{m} n_i) n_j - C(n_j) \}$$

Concentrating on the symmetric solution, we obtain that the optimal switchboard size will be determined by:

$$(16) \qquad R(mn) + nR'(mn) - C'(n) = 0$$

Comparing (16) to (10) reveals that the Nash player fails to take into account the externality he creates for all other firms connected to him. As a result, the switchboard size chosen by the Nash player is smaller than the one chosen by the integrated monopolist. The price charged under this kind of divestiture will be lower than that charged by the integrated monopolist.[2]

The results of the above analysis will hold as long as profits of each Nash player remain non-negative after divestiture. Clearly, if the profits due to divestiture decrease too much then the equilibrium discussed above does not obtain. The alternative equilibrium depends on the exact specification of the game the switchboard firms will play.

6. INTERSWITCHBOARD COMMUNICATION

In this section we assume that initially switchboards are not connected to each other. The interswitchboard connection is established by other independent profit maximizing firm(s). Under this arrangement we have to be very careful about describing the way in which the different participants share in the proceeds arising from the externalities. We concentrate on one particular such sharing scheme which assumes that each producer can charge its customers a price which will not exceed the value of his services to the consumer. We assume that the local switchboard firm after being connected to other switchboards is not allowed to charge its customers a price beyond that charged without the interswitchboard connection. On the other hand, the switchboard firm is allowed to collect a connection fee from the producers of interswitchboard connections.[3] In particular, that payment may depend on the switchboard size (n). Accordingly, every switchboard maximizes a profit function of the form

$$\max_{n} \{ \pi = nR(n) - C(n) + G(n) \} \tag{17}$$

where G(n) denotes the connection fee.

The firms providing the interswitchboard connection, in addition to paying the aforementioned connection fee, carry the cost of the interswitchboard connections. Their revenue is being collected directly from their customers.

We consider two market schemes. First, the interswitchboard connections are provided by a single firm. Second, the interswitchboard connection industry consists of an exogenously given number of firms.

7. CONNECTING MONOPOLY

The single firm providing switchboard connections has to determine how many switchboards (b) to connect. To do so it solves the following profit maximizing problem;

$$\max_{b} \{ s = bn [R(bn) - R(n)] - F(b) - bG(n) \} \tag{18}$$

The individual switchboard size (n) is the result of the maximization problem (17). The first order condition for (17) is

$$R(n) + nR'(n) - C'(n) + G'(n) = 0 \tag{19}$$

Let us denote the solution of (17) by $\hat{n}$. Clearly, if $m^* > b$ and $G'(n) = 0$, then $\hat{n} < n^*$. In general, $\hat{n} \neq n^*$. Using (10) and (19) we get that $\hat{n} = n^*$ only if

$$G'(n^*) = R(m^*n^*) - R(n^*) + m^*n^*R'(m^*n^*) - nR'(n^*)$$

The choice of b (the solution of equation (18)) is discrete and in principle has to follow the method described in section 3. For convenience we treat b as a contionuous variable. Accordingly, the first order condition for (18) is

$$\hat{n}[R(b\hat{n}) - R(\hat{n})] + b\hat{n}^2R'(b\hat{n}) - F'(b) - G(\hat{n}) = 0 \tag{20}$$

In general $\hat{b}$ which solve (20), given $\hat{n}$, is different from m^*.

A comparison of (11) and (18) reveals two differences between the profit functions of the integrated monopoly and the switchboard connection monopoly. First, the integrated monopoly carries the correct switchboard cost $C(n)$, while the connection monopoly faces a connection charge of $G(n)$. The second, and more important, problem is the fact that the connecting monopoly cannot benefit from the total value of the externality existing in the network.

Note that the first problem cannot be removed by setting $G(n) = C(n)$, since this policy would create an incentive for the switchboard firm to expand indefinitely, since its cost is paid by the connecting monopoly. This is not a feasible solution.

The question which arises is whether the connecting monopoly connects more of less switchboards than the integrated monopoly. This question may be broken down into two components. First, we may try to compare the solutions to (11) and (18) for the optimal number of switchboards assuming that $\hat{n} = n^*$. Next, we may assess the effect of the individually chosen switchboard size (n) on the number of switchboards connected by the connecting monopoly, $\hat{b}$. The first problem can be addressed by noting that at $\hat{n} = n^*$ and $\hat{b} = m^*$ the left hand side of equation (20) simplifies to a negative expression given by

$$(21) \qquad -n^*R(n^*) - G(n^*) + C(n^*) < 0$$

Accordingly, if $\hat{n} = n^*$, $\hat{b} < m^*$. The connecting monopoly, which does not take into account fully the externality he creates, connects too few switchboards.

The second effect can be assessed by derivating (20) with respect to n. This derivative has an ambiguous sign. It contains $R(bn)$ as one of its positive elements and an expression equivalent to $-C'(n)$ as a negative element. If the positive outweighs the negative elements, so that the externality effect captured by the magnitude of $R(bn)$ is sufficiently big, then $\hat{n} < n^*$ implies $\hat{b} < m^*$. Again, if the externality, which is not fully captured by the connecting monopoly, is large, his decision is to connect "too-few" switchboards. On the other hand, if the marginal costs of an individual switchboard are dominant, then the connecting monopoly, that does not carry any of these costs, tends to connect too many switchboards.

8. COMPETING CONNECTING FIRMS

In this section we assume that there are t independent firms competing in the contestable market for interswitchboard connection. Each firm offers consumers belonging to any particular switchboard its connection services with other switchboards. These services allow the subscribing consumers to communicate with any consumer belonging to all switchboards connected by this connecting firm. The firms compete among themselves at any given switchboard by offering consumers a combination of the number of accessible switchboards and a subscription price for their services. Every connecting firm bears the cost of constructing its connection network $F(b)$ and a connecting fee to every switchboard it serves, $G(n)$.

The profit of the j-th connecting firm is

$$\pi_j = \sum_{i=1}^{b_j} [n_{ij}\, p_{ij} - G(n_i)] - F(b_j) \tag{22}$$

when n_{ij} is the number of subscribers to the j-th connecting firm at switchboard i, p_{ij} is the subscription price paid by

each subscriber at switchboard i who subscribes to the j-th firm and b_j is the number of switchboards connected by the j-th firm.

Clearly, p_{ij} is restricted by the number of switchboards (b_j) it connects and the action taken by firm j's competitors. In particular,

$$\text{(23)} \quad \begin{aligned} & U(I - p_{ij} - R(n),\ y(b_j\ n)) \geq \\ & \geq \max_{h \neq j} U(I - p_{ih} - R(n),\ y(b_h\ n)) \geq \\ & \geq U\ (I - R(n),\ y(n)) \qquad \text{for all } i = 1,..,b_j \end{aligned}$$

Since the cost to the connecting firm of adding subscribers belonging to any given switchboard is zero, each connecting firm has an incentive to add (connect) as many customers as possible from that switchboard. Therefore, every firm is willing to reduce its own price given that of its competitors in order to capture more consumers. As a result, in equilibrium profits of all connecting firms are reduced to zero.

We concentrate here only on symmetric equilibria. Accordingly, we look at $p_{ij} = p$, $n_{ij} = n/t$ for all i and j and $b_j = b$ for all j.

Treating b as a continuous variable, the first order condition of maximizing (22) at a symmetric solution is

$$\text{(24)} \quad pn/t - F'(b) - G(n) = 0$$

From (24) we obtain

$$b = H(p,t) \tag{25}$$

where $H_1 > 0$ and $H_2 < 0$.

The equilibrium profit is found by substituting (25) into (22). The requirement that equilibrium profits be zero yield

$$H(p,t)[pn/t - G(n)] - F(H(p,t)) = 0 \tag{26}$$

A change in the number of firms t has an immediate effect on p given the number of switchboards connected by each firm. Since profits are zero in equilibrium, these firms have only fixed costs. However, the number of customers served by each firm decreases, and therefore the price charged by each firm has to increase. Of course, this fact by itself implies that the number of connecting firms cannot be too large, for otherwise the second inequality in (23) is violated. Customers prefer remaining without the service of the connecting firms if the price of that service is too high.

If the number of connecting firms is not at its limit, the increase in price may induce every individual firm to increase the number of switchboards connected (from (25)). However, the direct effect of increasing the number of competitors is to reduce the number of switchboards connected. The end effect on that number, therefore, remains ambiguous. On the other hand we obtain from (26) that the price charged to customers rises unambiguously. This somewhat surprising conclusion implies that even an increase in total number of customers who obtain service due to a possible increase in the number of switchboards connected is insufficient to increase each individual connecting firm revenue sufficiently to cover its costs at the existing prices. Given the nature of our model, equilibrium in the market for interswitchboard connection implies that prices charged in this market will increase together with the number of firms operating in the market. Although there is an upper limit to the number of competitors a given market can support, it is preferable to have entry into this market regulated to keep costs and prices lower.

ACKNOWLEDGEMENTS

We would like to thank David Haddock, Dan Landau, Benjamin Shitowitz, and David Sibley for helpful commments. The paper has benefited from seminar presentations at the Economics Department Colloquium at the University of Connecticut and at European Economic Association meetings.

FOOTNOTES

1. A similar result of the non-optimality of marginal cost pricing in the case of externalities was presented by Rohlfs 1979.

2. This result holds also if the number of independent switchboards m approaches infinity provided that $\lim_{m \to \infty} mnR'(mn) = a$ where $0 < a < \infty$.

3. This connection fee paid by the inter-switchboard connecting firm to the switchboard firms is determined exogenously (by the FCC).

REFERENCES

1. Katz, Michael L. and Shapiro, Carl. "Network Externalities, Competition, and Compatibility" The American Economic Review . Vol 75 #3 (June, 1985): 424-40.

2. Rohlfs, Jeffrey. "Economically Efficient Bell-System Pricing" Bell Laboratories, Economic Discussion Paper #138. Jan, 1979.

TELECOMMUNICATIONS DEMAND MODELLING
An Integrated View
A. de Fontenay, M.H. Shugard, D.S. Sibley (Editors)

A THEORETICAL FRAMEWORK FOR DESIGNING ACCESS CHARGES

Michael A. Einhorn
Department of Economics
Rutgers University
Newark, New Jersey

1. INTRODUCTION

One of the most important economic problems now facing the telecommunications industry involves the proper design of prices to charge callers for using local exchange carrier (LEC) switched access facilities to reach their interexchange carriers' points of presence. Local companies fear that high usage charges on switched access lines will induce large long-distance callers to utilize alternative circuits to reach their long-distance carriers; these bypass circuits could include microwave systems, coaxial cables, fiber optic lines, and LEC-provided special access facilities. If large long-distance customers bypass the local company's switched access facilities, enough revenue must be generated from the other telephone services to cover the company's nontraffic sensitive (NTS) costs (of approximately $25 per line) which, at present, are not adequately covered by monthly flat-rate fees. Should large customers bypass, remaining customers will face an even more burdensome revenue requirement and some small customers may be driven off the system altogether; future rounds of bypass and attrition may then follow. (For an examination of whether the bypass-price hike-bypass spiral ever reaches an equilibrium, see Brock (1984). For a discussion of the costs of possible bypass technologies, see Bell Communications Research (1984), Jackson and Rohlfs (1985).)

This paper provides a theoretical framework for a second-best economically efficient solution to the above problem; I shall assume that some solutions that may be economically preferable (e.g., higher monthly subscriber line charges) are nonetheless politically unacceptable. Since only multiline business customers would realistically consider replacing switched access circuits with bypass alternatives, this paper will explicitly consider only these customers. Multiline customers will have the option of choosing bypass circuits which have, at sufficiently high levels of usage, lower true costs than LEC switched access lines. Single-line customers are presumed to pay subscriber line and usage charges that are prescribed on the basis of social considerations rather than economic theory; consequently, an NTS burden exists which revenue from multiline customers must in some way meet.

Long-distance access prices for multiline customers will be chosen to maximize overall social surplus subject to the constraint that the local company's recovered revenues are sufficient to cover both its incurred variable cost and its necessary NTS cost subsidy. Therefore, although bypass at a very high level of usage may be economic, uneconomic bypass may result as well if prices for switched access usage are above their associated marginal cost.

The conclusions of the paper are as follows. For each customer-owned switched access line, the local company should offer a choice between (at least) two calling plans. The first plan is suitable for lightly used lines (called small lines); after paying a predetermined subscriber line charge, small line users pay a per minute access fee that must exceed its associated marginal cost. In a second plan more suitable for heavily used lines (called large lines), customers would pay a higher subscriber line charge (above its associated marginal cost); however, these lines would have per minute access fees that would be below their associated marginal costs.

This paper is organized as follows. Assuming that bypass is an economically more efficient technology for large circuits at some level of usage, Section 2 presents a simple model that determines an economically efficient schedule of access charges; politically unrealistic restrictions are imposed upon small user line charges. Section 3 modifies the above results to incorporate more realistic small user considerations. Section 4 concludes with some theoretical and practical extensions of the paper.

2. A Simple Model

This section will develop a simple model of efficient pricing of switched access lines that are used by multiline customers; single-line customer access and usage are assumed to be priced in a manner that requires a nontraffic-sensitive cost subsidy from multiline customers. We shall demonstrate that if one calling plan were offered to all multiline customers, usage should be priced below marginal cost.

Assume that each multiline customer has the choice of reaching a long-distance carrier with switched access lines and/or bypass alternatives that are provided either by the local company or other firms. Bypass systems, although available to anyone in principle, entail substantial initial costs and consequently would appeal only to large long-distance callers. Assume that market competition and the threat of entry keep the prices of all bypass alternatives at their associated marginal costs.

In this section, we shall assume that regulators may set both flat-rate end-user access fees (subscriber line charges) and per minute access charges at constrained efficient levels without regard to the effect upon small multiline customers. Also assume that both the total number of customers and the total number of access circuits is fixed; however, the proportion of these circuits that is switched access is variable and depends on the respective prices of switched access and bypass service.

The local company is neither purely competitive nor profit-maximizing; it is a regulated monopoly subject to a binding constraint on its level of net earnings or profits. Total revenue from customers must cover variable and fixed costs. In particular, prices to multiline business customers must be designed in order to recover usage-sensitive costs, customer-related costs, and the necessary NTS cost subsidy. Because of the NTS cost recovery problem, telephone services to multiline customers cannot be priced at first-best marginal cost prices; distortions are necessary. (The marginal cost for additional switched access usage is positive due to the incremental capacity expansion costs that results under switched access usage. We assume that "traffic-sensitive" costs have been reallocated between marginal and fixed components in a manner that is more consistent with economic theory. Under current separations procedures, per minute or per call "traffic-sensitive" costs do not accurately measure per minute or per call marginal cost.)

The local company's per line and per call usage fees for switched access lines are A and P. The corresponding marginal costs are Z and C. The numbers of switched access lines and calls are respectively N and Q.

The costs and prices of bypass technologies will differ from those of switched access. Because of competition and the threat of entry, the price of each bypass service is assumed to be equal to its marginal cost; this includes, for simplicity, the prices of special access lines that are installed by the local company. For a prototype bypass vendor, the end-user access charge and its usage fee are, respectively, A* and P*; the respective costs are Z* and C*. For most bypass systems, P* = C* = 0; therefore, A* must exceed A (or else everyone would bypass). In addition each long-distance carrier charges a price for usage of its own long-distance facilities; without loss of generality, we assume that these prices are zero.

A user will value access circuits differently depending upon probable long-distance usage requirements on each. In considering whether to install a bypass circuit or a switched access line, a customer must compare the consumer surplus that it would enjoy under each alternative. Each circuit can be indexed with an i. If switched access, the relevant consumer surplus from circuit i is:

$$cs_i = \int_o^{q_i} (V_i(x) - P)dx - A \tag{2.1}$$

where:

$V_i(x) = value\ of\ xth\ unit\ usage,\ circuit\ i$
$q_i = usage\ of\ circuit\ i\ under\ switched\ access;\ q_i = q_i(P)$

For a bypass circuit i, a relevant consumer surplus is:

$$cs_{i*} = \int_o^{q_{i*}} (V_i(x) - P^*)dx - A^* \tag{2.2}$$

where:

$q_{i*} = usage\ of\ circuit\ i\ under\ bypass;\ q_{i*} = q_{i*}(P^*)$

Customers will sometimes be completely indifferent between bypass and switched access. For these customers, $cs_i = cs_{i*}$. We employ the frequently made assumption that demand curves (for usage on different circuits) do not cross one another (e.g., see Faulhaber and Panzar (1977)). Given this assumption, we may define a unique maximum usage level q_m on a switched access line above which bypass takes place. For this marginal switched access line, $cs_m = cs_{m*}$. That is,

$$\int_o^{q_m} (V_m(x) - P)dx - A = \int_o^{q_{m*}} (V_m(x) - P^*)dx - A^* \tag{2.3}$$

Without loss of generality, the index i can be designed so that i = 0 for the smallest circuit and i = 1 for the largest; there is no need to ascribe any cardinal properties to this intensity index. i has a continuous distribution function F(i) and a density function f(i) = dF(i)/di. i_m will represent the intensity level of a marginal switched access line; for this circuit, $cs_m = cs_{m*}$ (i.e., a customer owning circuit i_m is indifferent between bypass and switched acccss for this circuit).

Aggregate consumer surplus CS_T is the sum of consumer surpluses for switched access lines CS and bypass circuits CS*:

$$CS_T = CS + CS^* \tag{2.4}$$

where:

$$CS = \int_o^{i_m} cs_i\, dF(i)$$

$$CS^* = \int_{i_m}^{1} cs_{i*}\, dF(i)$$

Local company revenues from its switched access lines comprise end-user charges AN and usage charges PQ. The associated costs would be ZN and CQ. Revenue needed from these customers for NTS cost recovery is K. A necessary constraint is that:

$$K \leq (A - Z)N + (P - C)Q \tag{2.5}$$

We assume that the demand for access lines is inelastic at the minimal user intensity. We shall relax this assumption in Section 4. With this assumption, bypass can only occur at the higher usage intensities.

It is then possible to express N and Q:

$$N = \int_o^{i_m} dF(i) = F(i_m) \tag{2.6a}$$

$$Q = \int_o^{i_m} q_i(P) dF(i) = Q(P, i_m) \tag{2.6b}$$

Without a binding K constraint, the socially optimal prices for long-distance access and usage would be Z and C. If the level of K under first-best prices were inadequate, second-best pricing rules would prescribe price distortions to maximize overall social surplus while simultaneously meeting the binding constraint in eq. 2.5 (e.g., see Baumol and Bradford (1970), Ng and Weisser (1974), Rohlfs (1979)). Accordingly, prices A and P can be chosen to maximize aggregate social surplus subject to a binding K constraint; aggregate social surplus would include CS, CS*, producer surplus to the local company PS, and producer surplus of the bypass vendors PS*. (These respective producer surpluses are $PS = (A - Z)N + (P - C)Q$ and $PS^* = (A^* - Z^*)N^*$.) The relevant Lagrangian is then:

$$L = CS + CS^* + PS + PS^* + g[K - (A - Z)N - (P - C)Q] \tag{2.7}$$

Eq. 2.7 can be differentiated with respect to prices P and A to derive suitable Kuhn-Tucker conditions and efficient prices; from here, the resulting levels of Q and N can be determined. Equivalently, eq. 2.7 can be represented as a function of i_m and usage by inframarginal switched access lines Q_M (i.e, switched access lines with $i < i_m$). Since N is a continuous function of i_m (see eq. 2.6a), it can be used instead of i_m as a decision variable. Since N and Q_M are both functions of P and A, the optimal quantities can then be translated to optimal prices. We shall follow the latter route.

The relevant Kuhn-Tucker conditions for an N, Q_M optimum are:

$$\partial L/\partial N = cs_m - cs_{m^*} + ps - ps_{m^*} - g[(A - Z) + (P - C)(\partial Q/\partial N) \tag{2.8a}$$
$$+ (\partial A/\partial N)N + (\partial P/\partial N)Q] \leq 0;\ N(\partial L/\partial N) = 0;\ N \geq 0$$

$$\partial L/\partial Q_M = (P - C) - g([P - C + (\partial A/\partial Q_M)N + (\partial P/\partial Q_M)Q] \leq 0; \tag{2.8b}$$
$$Q_M(\partial L/\partial Q_M) = 0; Q_M \geq 0$$

$$\partial L/\partial g = K - (A - Z)N - (P - C)Q \leq 0;\ g(\partial L/\partial g) = 0 \tag{2.8c}$$

where:

ps_m = producer surplus of circuit i_m on switched access line $= A - Z + (P - C)q_m$
ps_{m^*} = producer surplus of circuit i_m on bypass alternative $= A^* - Z^*$

To simplify eq. 2.8a, note three things. First, it must be true that for the marginal circuit i_m, $cs_m = cs_{m*}$. Second, it follows that $\partial Q/\partial N = q_m$, where q_m represents usage by circuit i_m on a switched access line. Third, since bypass prices are assumed to be at marginal cost, $A^* = Z^*$; therefore, $ps_{m*} = 0$.

With some straightforward but long algebra, eqs. 2.8a and 2.8b can be solved for 1/g:

$$1/g = [A - Z + (P - C)q_m + Pqu]/[A - Z + (P - C)q_m] \quad (2.9)$$
$$= (P - C + Pv)/P - C)$$

where:

$$q = Q/N$$
$$u = (e_{u1} + re_{u2})$$
$$v = (e_{v1} + re_{v2})(Q/Q_M)$$
$$e_{u1} = (\partial P/\partial N)(N/P)$$
$$e_{u2} = (\partial A/\partial N)(N/A)$$
$$e_{v1} = (\partial P/\partial Q_M)(Q_M/P)$$
$$e_{v2} = (\partial A/\partial Q_M)(Q_M/A)$$
$$r = AN/PQ$$

Simplifying eq. 2.9:

$$A - Z + (P - C)q_m = (P - C)qx \quad (2.10)$$

where:

$x = u/v$

Assuming that the K constraint (eq. 2.5) holds with equality:

$$(A - Z) = K/N - (P - C)q \quad (2.11)$$

Substituting eq. 2.11 into eq. 2.10 and rearranging terms obtains:

$$(P - C)Q/K = 1/(x + 1 - q_m/q) \quad (2.12)$$

Solving for $(A - Z)N/K$:

$$(A - Z)N/K = (x - q_m/q)/(x + 1 - q_m/q) \quad (2.13)$$

Equations 2.12 and 2.13 define the optimal shares of $(P - C)Q$ and $(A - Z)N$ in K. Each share depends only upon x and q_m/q. In addition, x and q_m/q are related to one another. A mathematical appendix demonstrates that x can be written:

$$x = (dQ_M/dP)(P/Q)/[(\hat{N}_a/N)P(q - q_m)] \quad (2.14)$$

where:

$\hat{N}_A = \partial N / \partial A$

The numerator in eq. 2.14 is negative; therefore, x is positive (zero, negative) when $q_m < (=, >) q$. For this paper's concerns, the interesting case is $q_m > q$; i.e., the most heavily used circuits are marginal. Since x is negative, the denominators in eqs. 2.12 and 2.13 are negative; therefore, P < C and A > Z.

This is an important conclusion. If regulators are worried about bypass technologies uneconomically attracting the largest long-distance users away from switched access lines, the appropriate constrained social welfare maximizing prices are given by eqs. 2.12 and 2.13. Note that the usage-sensitive price P should be below its associated marginal cost C. In this way, the local company to some extent can compete with the bypass technologies. Furthermore, $A - Z$ is positive; i.e., the customer charge A must be higher than its associated marginal cost Z. Depending on the size of A, this might be particularly harsh on small multiline customers. The following section considers some remedies regarding these small multiline customers.

Throughout this section, we have assumed that competition and/or the threat of entry drove bypass access price A* to its associated marginal cost Z*; bypass includes special access lines that local companies offer. In reality, prices for special access lines and other bypass alternatives may continue to exceed somewhat their associated cost; as a result, $ps_{m^*} = A^* - Z^* > 0$. The resulting first-order conditions would then become considerably more complicated. However, if $A^* - Z^*$ is below a certain amount, it still will be appropriate to price switched access usage below its associated marginal cost; the size of this margin can be derived but we shall not do so in this paper.

Alternatively, regulators may determine that the producer surplus of bypass vendors PS* is not a legitimate part of social welfare; consequently, PS* would drop out of eq. 2.7 and ps_{m^*} would drop out of eq. 2.8a. Under these circumstances, the basic results of this section would be entirely unaffected by the possibility that A* > Z*.

3. A Two-Schedule Model

This section considers the possibility that the optimal flat-rate price A_1 that was obtained in Section 2 is deemed inappropriately high for lightly used lines; consequently, lightly used lines are constrained to pay no more than A_0 as a flat-rate charge. Given this restraint on the customer charge and the K constraint in eq. 2.5, usage price P could then be designed to ensure an adequate level of NTS cost recovery:

$$P_{AC} = (K - (A_0 - Z)N)/Q \tag{3.1}$$

A_0 and P_{AC} would be the relevant access and usage prices for each switched access circuit. This is a simple modification of average cost pricing.

However, overall social surplus can be improved if each customer were offered a choice between two calling plans for each line. In one schedule, a user can pay a flat-rate charge of A_0 and a per call charge of P_0. In the second, a user can pay a flat-rate fee of A_1 and a per call charge of P_1; prices P_0, P_1, and A_1 all can be chosen in order to maximize social welfare subject to the constraint that fixed costs K are covered. The number of small lines is N_0 and calls on these lines are Q_0. The number of large lines is N_1 and their calls are Q_1.

The total level of customer surplus from each of the two tariffs j ($j = 0,1$) and bypass circuits can be expressed:

$$CS_j = \int_{i_{Lj}}^{i_{Uj}} cs_{ij}\, dF(i) \tag{3.2}$$

$$CS^* = \int_{i_{U1}}^{1} cs_{i*}\, dF(i)$$

where:

i_{Lj} = *minimal usage intensity* for *tariff j*

i_{Uj} = *maximal usage intensity* for *tariff j*

$cs_{ij} = \int_0^{q_{ij}} (V_i(x) - P_j)dx - A_j$

q_{ij} = *usage by line i under tariff j*

Express Q_j and N_j for each of the two switched access tariffs j:

$$Q_j = \int_{i_{Lj}}^{i_{Uj}} q_{ij}(P_j)dF(i) \tag{3.3a}$$

$$N_j = \int_{i_{Lj}}^{i_{Uj}} dF(i) = F(i_{Uj} - F(i_{Lj}) \tag{3.3b}$$

We continue to assume that the demand for access lines is inelastic at the minimal intensity; i.e., $i_{L0} = 0$.

Prices P_0, P_1, and A_1 must be set to maximize the sum of consumer surplus (e.g., CS_0, CS_1, and CS_2) and producer surplus (PS_1 and PS_2) subject to the constraint that net revenue is sufficient to cover NTS costs K. This constraint can be expressed:

$$K \leq (P_0 - C)Q_0 + (P_1 - C)Q_1 + (A_0 - Z)N_0 + (A_1 - Z)N_1 \tag{3.4}$$

Given the assumption that A_0 is exogenously fixed by regulators, the revenue constraint can be reexpressed:

$$K' \leq (P_0 - C)Q_0 + (P_1 - C)Q_1 + (A_1 - Z)N_1 \tag{3.5}$$

where:

$K' = K - (A_0 - Z)N_0$

The relevant Lagrangian can be expressed:

$$L = CS_0 + CS_1 + CS_2 + PS_1 + PS_2 + g[K' - (P_0 - C)Q_0 - (P_1 - C)\,Q_1 - (A_1 - Z)N_1] \tag{3.6}$$

As in Section 2, optimizing conditions can be obtained by differentiating eq. 3.6 with respect to Q_{MO}, Q_{M1}, and i_{U1}; N_1 can be used instead of i_{U1}. The relevant Kuhn-Tucker conditions are:

$$\partial L/\partial Q_{M0} = P_0 - C - g[P_0 - C + (\partial P_0/\partial Q_{M0})Q_0] \le 0;\ Q_{M0}(\partial L/\partial Q_{M0}) = 0 \tag{3.7a}$$

$$\partial L/\partial N_1 = A_1 - Z + (P_1 - C)q_{m1} - g[A_1 - Z + (P_1 - C)q_{m1} + \tag{3.7b}$$
$$+ (\partial A_1/\partial N_1)N_1 + (\partial P_1/\partial N_1)Q_1] \le 0;\ N_1(\partial L/\partial N_1) = 0$$

$$\partial L/\partial Q_{M1} = P_1 - C - g[P_1 - C + (\partial A_1/\partial Q_{M1})N_1 + (\partial P_1/\partial Q_{M1})Q_1] \le 0; \tag{3.7c}$$
$$Q_{M1}(\partial L/\partial Q_{M1}) = 0$$

Solving for 1/g in each of the three equations in eq. 3.7:

$$1/g = [A_1 - Z + P_1 - C)q_{m1} + P_1 q_1 u]/[A_1 - Z + (P_1 - C)q_{m1}] \tag{3.8}$$
$$= (P_1 - C + P_1 v)/(P_1 - C)$$
$$= (P_1 - C + P_0 w)/(P_0 - C)$$

where:

$$w = e_{w1} + s e_{w2}$$
$$e_{w1} = (\partial P_0/\partial Q_{M0})Q_0/P_0$$
$$e_{w2} = (\partial A_0/\partial Q_{M0})Q_0/A_0$$
$$s = A_0 N_0/P_0 Q_0$$
$$q_{m1} = \partial Q_1/\partial N_1$$
$$q_1 = Q_1/N_1$$

The remaining terms in eqs. 3.7 are defined after eq. 2.9. Since A_0 is fixed, $\partial A_0/\partial Q_{M0} = 0$; therefore, $e_{w2} = 0$. The term e_{w1} is the inverse of the own-price elasticity of Q_{M0} and therefore must be negative. Therefore, w is negative.

Simplifying eq. 3.8:

$$A_1 - Z + (P_1 - C)q_{m1} = (P_1 - C)q_1 x = (P_0 - C)q_1 y P_1/P_0 \tag{3.9}$$

where:

$$x = u/v$$
$$y = u/w$$

The mathematical appendix proves that u is negative; v (x) is positive (negative) when $q_{m1} > q_1$, zero when $q_{m1} = q_1$, and negative (positive) when $q_{m1} < q_1$. As noted, w is always negative. Assuming that the K' constraint holds with equality:

$$A_1 - Z = K'/N_1 - (P_1 - C)Q_1/N_1 - (P_0 - C)Q_0/N_1 \tag{3.10}$$

Since $(P_1 - C)/(P_1 v) = (P_0 - C)/(P_0 w)$ (see eq. 3.9), rewrite eq. 3.10:

$$A_1 - Z = K'/N_1 - (P_1 - C)q_1 [1 + wP_0Q_0/vP_1Q_1] \tag{3.11}$$

Substituting eq. 3.11 into 3.9 and solve for $(P_1 - C)Q_1/K'$:

$$(P_1 - C)Q_1/K' = 1/D \tag{3.12}$$

where:

$$D = wP_0Q_0/(vP_1Q_1) + x + 1 - q_{m1}/q_1$$

Solving for $(P_0 - C)Q_0/K'$ and $(A_1 - Z)N_1/K'$:

$$(P_0 - C)Q_0/K' = wP_0Q_0/(vP_1Q_1D) \tag{3.13a}$$

$$(A_1 - Z)N_1/K' = (x - q_{m1}/q_1)/D \tag{3.13b}$$

Since $q_{m1} > q_1$, $v > 0$ and $x < 0$; since $w < 0, D$ must be negative.

Given these signs, three things are immediately evident from eqs. 3.12, 3.13a, and 3.13b. First, since D is negative, $P_1 < C$; this follows from eq. 3.12. Second, since both the numerator and the denominator in eq. 3.13a are negative, $P_0 > C$. Third, since both x and D are negative, $A_1 > Z$; this follows from eq. 3.13b. Since $P_1 < P_0$, $A_1 > A_0$ must hold; otherwise, everyone would choose calling plan A_1, P_1.

Accordingly, the usage price P_0 in the small line calling plan should be above its associated marginal cost C; the subscriber line charge for the small-line plan is A_0, which regulators exogenously set. For a higher subscriber line charge A_1, customers should be able to obtain a large-line calling plan with a usage price P_1 that is below its marginal cost C. In this way, the local company can compete to some extent with bypass technologies that offer large long-distance callers other ways of reaching their long-distance carriers.

4. Conclusion: Theoretical and Practical Extensions

This section concludes the paper with some additional points regarding Section 3 and its major conclusion.

Of course, the demand for switched access lines is price-elastic at the minimal usage intensity i_{L0}. This would affect the resulting level of P_0 but $P_1 < C$ still results. Therefore, one switched access tariff would still have usage prices below marginal cost.

Regarding large lines, note that the most heavily used switched access line i_m is the marginal one; consequently, it has usage q_{m1}. For this line, revenue less costs equals $A_1 - Z + (P_1 - C)q_{m1} = (P_1 - C)q_1x$; the equality follows from eq. 3.9. As the appendix demonstrates, x is negative if $q_m > q_1$; therefore, $(P_1 - C)q_1x$ is always positive. Therefore, the net revenue of the most heavily used switched access line is positive. Since $P_1 < C$, the resulting net revenue for usage $q_{i1} < q_{m1}$ is $A_1 - Z + (P_1 - C)q_{i1} > (P_1 - C)q_1x > 0$. Therefore, net revenue from all large switched access lines must be positive as well. Consequently, no large switched access line is subsidized "in the whole" (although usage is) and each makes a positive contribution toward NTS cost recovery.

The price P_0 that small lines pay for usage is not necessarily below or equal to the average cost price P that would prevail in eq. 3.1. Furthermore, depending upon A_0 and P_0, some small lines could be better off under average cost pricing. If P_0 were deemed to be unsuitably high, it could be constrained not to exceed some upper bound, perhaps P_0 in eq. 3.1. However, a flat rate schedule P_{AC} should not be implemented for all customers; it is still possible to make each customer better off by offering an optional large-line calling plan with $P_1 < C$. The small line tariff must incorporate the constraint that usage price $P_0 \leq P_{AC}$ as well as the prescribed flat-rate fee A_0. (see Willig (1978), Brown and Sibley (1986)).

In the large-line plan, there is some level of long-distance usage $q_x > q_{m1}$ where $A_1 - Z + (P_1 - C)q_x = 0$; however, a switched access line that uses q_x or more is expected to bypass and theoretically presents no danger of subsidy. Nevertheless, $q_i \geq q_x > q_{m1}$ could result on a switched access line, for example, during a temporary period of abnormally high phone usage; if marginal usage price P_1 were kept below marginal cost C for usage above q_x, these lines would lose money. Given this contingency, it seems reasonable to ensure against subsidization during any surge in usage, even if temporary. This is most easily achieved simply by adding to the large-line calling plan a tail block with a usage price of C; the breakpoint to the tail block would occur at usage level q_x where $(A_1 - Z) + (P_1 - C)q_x = 0$

There is no need to stop at two calling plans; local companies can always secure Pareto-improvements by introducing more (see Willig, Brown and Sibley). Regardless of the design of the final schedule, usage price for the largest switched access lines should be below the associated marginal cost.

APPENDIX

This appendix derives eq. 2.14 in Section 2 and the statements that follow it.

The number of switched access lines N depends upon both prices A and P. Usage by inframarginal switched access lines Q_M can depend upon both A and P as well. Therefore:

$$Q_M = Q(P, A) \tag{A.1}$$

$$N = N(P, A) \tag{A.2}$$

Taking derivatives:

$$\begin{bmatrix} dQ_M \\ dN \end{bmatrix} = \begin{bmatrix} \hat{Q}_p & \hat{Q}_A \\ \hat{N}_p & \hat{N}_A \end{bmatrix} \begin{bmatrix} dP \\ dA \end{bmatrix} \tag{A.3}$$

Designate the 2×2 matrix of derivatives on the right hand side as M; the hats represent derivatives. We can solve for $(dP \ dA)'$:

$$\begin{bmatrix} dP \\ dA \end{bmatrix} = \begin{bmatrix} \hat{Q}_p & \hat{Q}_A \\ \hat{N}_p & \hat{N}_A \end{bmatrix}^{-1} \begin{bmatrix} dQ_M \\ dN \end{bmatrix} \tag{A.4}$$

$\hat{N}_A$ is the derivative of long distance switched access lines N with respect to the price A. From eq. 2.6a, this can be written:

$$\hat{N}_A = f(i_m)(\partial i_m | \partial A) < 0 \tag{A.5}$$

$\hat{N}_P$ is the derivative of N with respect to the price of service P. This can be written:

$$\hat{N}_P = f(i_m)(\partial i_m | \partial A) q_m = \hat{N}_A q_m \tag{A.6}$$

$\hat{Q}_A$ is the price elasticity of usage by inframarginal switched access lines with respect to A. Assuming no income effects on inframarginal usage,

$$\hat{Q}_A = 0 \tag{A.7}$$

$\hat{Q}_P$ is the derivative of Q_M with respect to its own-price P; this derivative must be negative.

Substituting eqs. A.5, A.6, and A.7 into eq. A.4 and using Cramer's Rule obtains:

$$\partial P / \partial Q_M = \hat{N}_A / \det M < 0 \tag{A.8}$$

$$\partial P / \partial N = 0$$

$$\partial A / \partial Q_M = -\hat{N}_p / \det M = -\hat{N}_A \, q_m / \det M > 0$$

$$\partial A / \partial N = \hat{Q}_P / \det M < 0$$

where:

$$\det M = \hat{N}_A \, \hat{Q}_p > 0$$

Beginning with eq. 2.7, the text derives eqs. 2.12 and 2.13, which determine the shares of Q and N in meeting costs K:

$$(P - C)Q/K = 1/(x + 1 - q_m / q) \tag{2.12}$$

$$(A - Z)N/K = (x - q_m / q)/(x + 1 - q_m / q) \tag{2.13}$$

where:

$$x = u/v$$

$$u = e_{u1} + r e_{u2}$$

$$v = (e_{v1} + r e_{v2})(Q/Q_M)$$

$$e_{u1} = (\partial P / \partial N)(N/P) = 0$$

$$e_{u2} = (\partial A / \partial N)(N/A)$$

$$e_{v1} = (\partial P / \partial Q_M)\,(Q_M / P)$$

$$e_{v2} = (\partial A / \partial Q_{M)}\,(Q_M / A)$$

$$r = AN/(PQ)$$

depend, eventually, upon P, Q_M, A, N, and the four partial derivatives $\partial P/\partial N$, $\partial A/\partial N$, $\partial A/\partial Q_M$, and $\partial P/\partial Q_M$ that are defined in eq. A.8. We substitute appropriately and use some straightforward but long algebra to derive:

$$x = (\partial Q_M / \partial P)(P/Q)/\left[(\hat{N}_A / N)P\,(q - q_m)\right] \tag{A.9}$$

This is equation 2.14 in the text. The numerator $(\partial Q_m / \partial P)(P / Q)$ is negative. In the denominator, $\hat{N}_A$ is negative (see eq. A.5). When the largest lines bypass, usage of the marginal line exceeds usage of the average; i.e., $q_m > q$. Therefore, $X < 0$.

BIBLIOGRAPHY

Baumol, W.J., and D. Bradford, (1970), "Optimal Departures from Marginal Cost Pricing", *American Economic Review,* June, 1970, pp. 265-83.

Bell Communications Research, (1984), *The Impact of End-User Charges on Bypass and Universal Telephone Service,* Livingston, New Jersey, September, 1984.

Brauetigam, R.R., (1979), "Optimal Pricing with Intermodal Competition", *American Economic Review,* March, 1979, pp. 38-49.

Brock, G.W., (1984), *Bypass of the Local Exchange: A Quantitative Assessment,* Office of Plans and Policy, Federal Communications Commission, Washington, D.C., September, 1984.

Brown, S.J., and D.S. Sibley (1986), *The Theory of Public Utility Pricing,* Cambridge University Press, Cambridge, England.

Faulhaber, G. R., and J.C. Panzar, (1977), "Optimal Two-Part Tariffs with Self-Selection", Bell Laboratories Economic Discussion Paper #74, January, 1977.

Goldman, M.B., H.E. Leland, and D.S. Sibley, (1984), "Optimal Nonuniform Prices," *Review of Economic Studies,* April, 1984.

Jackson, C.L. and J.H. Rohlfs, (1985), *Access Charging and Bypass Adoption,* Shooshan and Jackson, Inc., Washington, D.C., 1985.

Ng, Y.K., and M. Weisser, (1974), "Optimal Pricing with a Budget Constraint -- The Case of a Two-Part Tariff", *Review of Economic Studies,* July, 1974, pp. 337-45.

Oi, W., (1971), "A Disneyland Dilemma: Two-Part Tariffs for a Mickey Mouse Monopoly," *Quarterly Journal of Economics,* February, 1971, pp. 77-86.

Ramsey, F., (1927), "A Contribution to the Theory of Taxation," *Economic Journal*, 1927, pp. 47-61.

Rohlfs, J., (1970), "Economically Efficient Bell System Prices," Bell Laboratories Economic Discussion Paper # 131, June, 1979.

Willig, R.D., (1978), "Pareto-Superior Nonlinear Outlay Schedules," *Bell Journal of Economics,* Spring, 1978, pp. 56-69.

TELECOMMUNICATIONS DEMAND MODELLING
An Integrated View
A. de Fontenay, M.H. Shugard, D.S. Sibley (Editors)

PERFORMANCE INDICATORS FOR REGULATED INDUSTRIES

by W.E. Diewert
Dept. of Economics
Univ. of British Columbia

1. INTRODUCTION

How should a natural monopoly (such as telephone, electricity, water, natural gas and some transportation services) be regulated in the interests of society? This question has been discussed by economists for well over 100 years, but there is still no agreement on how to answer it. Fifty years ago, economists thought that forcing the monopolist to sell each product at a price equal to marginal cost would suffice to bring about an "ideal" allocation of resources.[1] The deficits that this policy would lead to for enterprises with large fixed costs were to be supported by the general taxation powers of the state.

The price equals marginal cost solution to the control of a monopoly was successfully attacked on a number of grounds: (i) Fleming [1945; 336] and Wilson [1945; 456] pointed out that there was no natural index available to evaluate the performance of managers, or put another way: how can the regulator know what each marginal cost is? (ii) Wilson [1945; 457] also raised a problem that was later stressed by Domar [1974; 4]: how can the regulator motivate the manager of the regulated industry to take the "right" course of action? (iii) Wilson [1945; 458-459] also noted the fact that any deficits generated by the regulated industry were to be covered out of general revenue and this would create a tremendous incentive for an empire building manager to expand unduly. In the end, political bargaining would determine the allocation of resources due to the imprecision of the price equals marginal cost rule in an intertemporal context. (iv) Coase [1945; 113] [1946; 176] pointed out that the Pigou-Hotelling-Lerner solution to the monopoly problem would redistribute income to consumers of products in which fixed costs form a high proportion of total costs. (v) Finally, Hotelling [1939; 155] and Coase [1946; 179] both noted that if taxes on fixed factors could not cover the government's revenue needs, then covering the deficits of regulated enterprises will have adverse efficiency effects in the rest of the economy. Thus marginal cost pricing, by itself, does not seem to be a solution to the problem of regulating monopolies. Scott [1952] [1978; 154] noted that the level of profits serves as an adequate indicator for the performance of a firm that is subject to constant or diminishing returns to scale (or, in more modern terminology, we would say that the technology set of the firm is convex). However, profit is a totally inadequate performance indicator for a firm that is subject to increasing returns to scale (or more generally has a nonconvex technology set). Thus Scott [1978; 155] suggested using the change in profits (where inputs and outputs are valued at constant prices) as an appropriate indicator of firm performance for a firm that has the potential to engage in monopolistic behaviour. This criterion has the advantage that it depends only on observable price and quantity data (and not on difficult to observe things like "marginal cost" or "the elasticity of demand"). Similar performance indicators for regulated firms have been developed by Vogelsang and Finsinger [1979], Finsinger and Vogelsang [1981] [1982] [1985], Tam [1981] [1985], Gravelle [1985] and Sappington and Sibley [1985].

Setting price equal to marginal cost is a necessary condition for solving a somewhat vaguely specified social maximization problem. The new literature on performance indicators makes further progress by specifying alternative objective functions or bonus functions that a manager of a regulated enterprise is supposed to maximize (subject to various constraints specified by the social planner). In most cases, an ideal objective function is not empirically observable; in such cases, it is necessary to form approximations to the ideal functions that are observable. The specification of an approximation for a bonus function is called an incentive scheme.

The literature on performance indicators and incentive schemes seems very promising but there are at least two problems with it: (i) the welfare measures are either vaguely specified or are based on the Dupuit [1844] and Marshall [1920] consumer and producer surplus concepts (i.e., areas under market demand and supply curves) and (ii) the schemes are partial equilibrium in nature. Thus the primary purpose of the present paper is the development of performance indicators for firms that are subject to increasing returns to scale (or a nonconvex technology) in a rigorous general equilibrium setting.

We conclude this section by outlining the contents of subsequent sections.

In section 2, we introduce our notation and assumptions that describe a general equilibrium model of a region that has a public utilities sector and trades with the rest of the world, taking world prices as fixed.

In section 3, we show how a Pareto optimal allocation of resource may be characterized. We also define various marginal cost and willingness to pay functions that will plan a role in our subsequent analysis.

In section 4, we consider various general equilibrium bonus functions or performance indicators for the manager of the regulated sector. At first sight, it may seem that we have handed the manager the impossible task of optimizing performance over the entire economy. However, this pessimistic view neglects the fact that since we will assume that the nonutilities part of the economy behaves competitively, market prices summarize the required information about the allocation of resources in these markets in an efficient manner, a point stressed by Arrow [1964] and Arrow and Hurwicz [1977]. Thus in sections 5 and 6, we develop first and second order approximations to our theoretical bonus functions. These approximations turn out to depend only on observable price and quantity data.

Since most firms produce many outputs and utilize many inputs, our general equilibrium model contains a rather large number of variables. For readers who are unaccustomed to dealing with a sea of symbols, we present a geometric treatment of our analysis (for a special case of very low dimensionality) in section 7.

Section 8 concludes with some comments on various performance indicators that have been suggested by other authors.

2. A REGIONAL GENERAL EQUILIBRIUM MODEL WITH A REGULATED SECTOR

Consider an economy that occupies a definite geographical area. We assume that there are three classes of goods in this economy: (i) I interregionally traded goods, (ii) M utility services (e.g., telephone, water, gas, electricity, sewage, cablevision, postal, bus, rail and air services, etc.) and (iii) N other domestic goods (types of labour, capital, natural resource, transportation and retailing services).

There are four classes of economic agent in our regional economy: (i) K competitive firms, (ii) H households, (iii) one consolidated utilities sector which supplies the M utility services mentioned in the previous paragraph in a possibly monopolistic manner, and finally, (iv) a consolidated government sector which supplies certain public goods such as defense, law administration, protection services and possibly education and medical services. The government sector also taxes the other economic agents and may also provide subsidies and grants. The competitive producers take prices as given and maximize profits. We assume that these producers have convex technology sets. Each household takes interregional and local prices (and commodity taxes) as given and maximizes a utility function subject to a budget constraint.

There are three sets of feasibility constraints in the regional economy which correspond to the three classes of goods.

The first such constraint is:

$$(1) \qquad w\cdot[\Sigma_{k=0}^{K} x^k - \Sigma_{h=0}^{H} a^h + \Sigma_{h=0}^{H} \bar{a}^h] \geq 0$$

where $w \equiv [w_1,\ldots,w_I] \gg 0_I$ is a strictly positive vector of "world" prices that the regional economy faces; $x^k \equiv [x_1^k,\ldots,x_I^k]$ is sector k's net output vector of interregionally traded goods for $k = 0, 1,\ldots,K$ where sector 0 is the public utilities sector and sectors 1 to K are the competitive firms (if $x_i^k < 0$, then sector k is utilizing good i as an input); $a^h \equiv [a_1^h,\ldots,a_I^h]$ is the net consumption vector of household h for interregionally traded goods (household supplies are indexed with a negative sign) for $h = 1,\ldots,H$ while a^0 is the government's net requirements vector for traded goods (government supplies are indexed with a negative sign);

$\bar{a}^h \equiv [\bar{a}_1^h,\ldots,\bar{a}_I^h] \geq 0$ is household h's nonnegative endowment vector of traded goods for $h = 1,\ldots,H$, and $\bar{a}^0$ is the government's endowment vector. Constraint (1) says that exports minus imports in the regional economy evaluated at world prices should be nonnegative.[2]

Our second set of constraints is

$$(2) \qquad \Sigma_{k=0}^{K} y^k - \Sigma_{h=0}^{H} b^h + \Sigma_{h=0}^{H} \bar{b}^h \geq 0_M$$

where $y^0 \equiv [y_1^0,\ldots,y_M^0]$ is a nonnegative vector of public utility services produced by the regulated sector; $y^k \equiv [y_1^k,\ldots,y_M^k] \leq 0_M$ is sector k's demand for public utility services vector for $k = 1,\ldots,K$ (demands are indexed by negative signs); $b^h \equiv [b_1^h,\ldots,b_M^h] \geq 0_M$ is household h's demand for public utility services for $h = 1,\ldots,H$ (household demands are indexed by positive signs); $b^0 \geq 0_M$ is the government sector's demand for public utility services, and $\bar{b}^h \equiv [\bar{b}_1^h,\ldots,\bar{b}_M^h] \geq 0_M$ is household h's endowment vector of public utility services for $h = 1,\ldots,H$ while $\bar{b}^0 \geq 0_M$ is the government's endowment vector (if utility service m cannot be stored, then $\bar{b}_m^h = 0$ for all h). The M constraints in (2) just tell us that the supplies of public utility services should be equal to or greater than the corresponding demands by firms, households and the government.

The third set of feasibility constraints is:

$$(3) \qquad \Sigma_{k=0}^{K} z^k - \Sigma_{h=0}^{H} c^h - \Sigma_{h=0}^{H} \bar{c}^h \geq 0_N$$

where $z^k \equiv [z_1^k,\ldots,z_N^k]$ is sector k's net supply vector for local goods (if sector k is utilizing good n as an input, then $z_n^k < 0$) for $k = 0,1,\ldots,K$; $c^h \equiv [c_1^h,\ldots,c_N^h]$ is household h's (or the government's if $h = 0$) net demand vector for local goods and services (if household h supplies a type of labour service that corresponds to good n, then $c_n^h < 0$), and $\bar{c}_1^h \equiv [\bar{c}_1^h,\ldots,\bar{c}_N^h] \geq 0$ is household h's (or the government's if $h = 0$) nonnegative endowment vector of local goods. If local good n is a type of labour service, then we assume $\bar{c}_n^h = 0$ for all h; i.e., labour services cannot be stored. The N constraints in (3) say that the supply of each local good should not be less than the corresponding demand.

We assume that the preferences of household h can be represented by a quasiconcave, continuous, nondecreasing utility function f^h defined over a closed, convex set Ω^h which is a subset of R^{I+M+N} for $h = 1,\dots,H$.

The technology sets of the competitive sectors are closed convex subsets S^k of R^{I+M+N} for $k = 1,\dots,K$. The technology set for the public utility sector is a closed subset S^0 of R^{I+M+N}. We also assume that each technology set satisfies a free disposal property. We do not assume that S^0 is necessarily convex, since this would rule out the increasing returns to scale or decreasing cost phenomena that characterize the provision of public utility services in real life economies.

3. PARETO OPTIMALITY AND EFFICIENCY

Consider the following constrained maximization problem:

$$(4)\quad \max_{a^h,b^h,c^h,x^k,y^k,z^k} \{f^1(a^1,b^1,c^1) : (1),(2),(3);\ (x^k,y^k,z^k) \in S^k \text{ for } k = 0,1,\dots,K;\ f^h(a^h,b^h,c^h) \geq u_h^* \text{ for } h = 2,\dots,H;\ b^h \geq 0_M \text{ and } (a^h,b^h,c^h) \in \Omega^h \text{ for } h = 1,2,\dots,H;\ y^0 \geq 0_M \text{ and } y^k \leq 0_M,\ k = 1,\dots,K\}.$$

In (4), the government net demand vectors a^0, b^0 and c^0 are held fixed. The above constrained maximization problem may be described as follows. We attempt to maximize the welfare of household 1 subject to the following constraints: (i) the demand equal to or less than supply constraints (1), (2) and (3) in the previous section must be satisfied; (ii) each household (except of course household 1) must attain at least a prespecified level of welfare u_h^* for $h = 2,\dots,H$, and (iii) each sector must choose a technologically feasible set of inputs and outputs.

Under appropriate regularity conditions, a finite maximum for the objective function in (4) will exist; call this maximum u_1^*. The production vectors (x^{k*},y^{k*},z^{k*}) for $k = 0,1,\ldots,K$ and the net consumption vectors (a^{h*},b^{h*},c^{h*}) for $h = 1,\ldots,H$ that solve (4) represent a Pareto optimal allocation of resources for the regional economy; i.e., no household's welfare can be increased without decreasing the utility or welfare level of one or more other households. In an ideal world, we would like to restrict attention to these Pareto optimal equilibria.

We shall find it convenient to characterize efficient equilibria in a somewhat different way. Consider the following constrained maximization problem:

$$(5)\quad \max_{a^h,b^h,c^h,x^k,y^k,z^k} \{\Sigma_{k=0}^{K} w\cdot x^k - \Sigma_{h=0}^{H} w\cdot a^h + \Sigma_{h=0}^{H} w\cdot \bar{a}^h : (2),(3);$$

$$(x^k,y^k,z^k) \in S^k \text{ for } k = 0,1,\ldots,K;\ f^h(a^h,b^h,c^h) \geq u_h^*,\ b^h \geq 0_M \text{ and}$$

$$(a^h,b^h,c^h) \in \Omega^h \text{ for } h = 1,\ldots,H;\ y^0 \geq 0_M \text{ and } y^k \leq 0_M \text{ for } k=1,\ldots,M\}.$$

The u_h^* which appear in (5) are the same as the u_h^* in (4). In both problems, we hold government net demand (a^0,b^0, c^0) fixed. In problem (5), we no longer have the foreign exchange constraint (1). However in (5), we are now attempting to maximize the amount of foreign exchange that the regional economy can produce subject to the aggregate demand less than supply constraints (2) and (3), the technology constraints for each sector and subject to household h attaining the welfare level u^*_h for $h = 1,\ldots,H$. Note that (5) has an extra utility constraint, $f^1(a^1,b^1,c^1) \geq u^*_1$, that was not present in (4).

If $f^1(a,b,c)$ is increasing in the components of the vector a, then it can be verified that the solution sets to (4) and (5) coincide.

To make further progress, we make two sets of definitions.

For $k = 0,1,\ldots,K$ and $w \gg 0_I$, define the sector k <u>restricted profit function</u> π^k by

$$(6)\quad \pi^k(w,y^k,z^k) \equiv \max_x \{w\cdot x : (x,y^k,z^k) \in S^k\} .$$

If there is no x such that $(x,y^k,z^k) \in S^k$, define $\pi^k(w,y^k,z^k) = -\infty$. Our free disposability assumptions on the sets S^k imply that $\pi^k(w,y^k,z^k)$ is non-increasing in the components of y^k and z^k; see Diewert [1973; 293] for a proof.

For h = 1,...,H, define the household h <u>restricted expenditure function</u> e^h by

$$(7)\quad e^h(u_h,w,b^h,c^h) \equiv \min_a \{w\cdot a : f^h(a,b^h,c^h) \geq u_h,\ (a,b^h,c^h) \in \Omega^h\} .$$

If there is no vector a such that the constraints in (7) are satisfied, then define $e^h(u,w,b^h,c^h) = +\infty$. Our assumption that the utility function f^h is nondecreasing in its arguments implies that $e^h(u,w,b^h,c^h)$ is nonincreasing in the components of b^h and c^h; see Diewert [1986; 172] for a proof.

Now let us fix y^k,z^k for k = 0,1,...,K and fix b^h,c^h for h = 1,...,H and maximize (5) with respect to x^k, k = 0,1,...,K, and a^h, h = 1,...,H. Using definitions (6) and (7), it can be seen that we may rewrite (5) as follows:

$$(8)\quad B(w,u^*) = \max_{y^k,z^k,b^h,c^h} \{\Sigma_{k=0}^K \pi^k(w,y^k,z^k) - \Sigma_{h=1}^H e^h(u_h^*,w,b^h,c^h)$$

$$- w\cdot a^0 + \Sigma_{h=0}^H w\cdot a^{-h} : (2),(3);\ b^h \geq 0_M,\ h = 1,...,H;\ y^0 \geq 0_M \text{ and}$$

$$y^k \leq 0_M \text{ for } k = 1,...,K\} .$$

Problems (5) and (8) may be regarded as central planning problems for the regional economy. In both problems, the planner attempts to maximize the amount of foreign exchange that the economy can produce subject to the constraints of technology and subject to a prespecified welfare level constraint for each household.

In order to characterize an optimal solution for (8), we make the somewhat restrictive assumption that the π^k and the e^h are once differentiable with respect to their arguments when evaluated at a solution to (8).[3] Let (y^{k*},z^{k*}), $k = 0,1,\ldots,K$ and (b^{h*},c^{h*}), $h = 1,\ldots,H$ solve (8). Then assuming that an appropriate constraint qualification condition is satisfied, the Kuhn-Tucker [1951] Theorem gives us the existence of multiplier vectors p* and r* such that the following Kuhn-Tucker conditions are satisfied:

$$(9)\quad \nabla_y\pi^0(w,y^{0*},z^{0*}) + p^* \leq 0_M\ ;\ y^{0*} \geq 0_M\ ;\ y^{0*}\cdot[\nabla_y\pi^0(w,y^{0*},z^{0*}) + p^*] = 0;$$

$$(10)\quad \nabla_y\pi^k(w,y^{k*},z^{k*}) + p^* \geq 0_M\ ;\ y^{k*} \leq 0_M\ ;\ y^{k*}\cdot[\nabla_y\pi^k(w,y^{k*},z^{k*}) + p^*] = 0,\quad k = 1,\ldots,K;$$

$$(11)\ -\nabla_b e^h(u_h^*,w,b^{h*},c^{h*})-p^*\leq 0_M;b^{h*} \geq 0_M;\ b^{h*}\cdot[\nabla_b e^h(u_h^*,w,b^{h*},c^{h*})+p^*] = 0,\quad h = 1,\ldots,H;$$

$$(12)\quad p^*\geq 0_M;\ \Sigma_{k=0}^K y^{k*}-b^0-\Sigma_{h=1}^H b^{h*}+\Sigma_{h=0}^H \bar{b}^h\geq 0_M;\ p^*\cdot[\Sigma_{k=0}^K y^{k*}-b^0-\Sigma_{h=1}^H b^{h*}+\Sigma_{h=0}^H \bar{b}^h]=0;$$

$$(13)\quad \nabla_z\pi^k(w,y^{k*},z^{k*}) + r^* = 0_N\ ,\ k = 0,1,\ldots,K\ ;$$

$$(14)\ -\nabla_c e^h(u_h^*,w,b^{h*},c^{h*}) - r^* = 0_N\ ,\ h = 1,\ldots,H\ ;$$

$$(15)\quad r^*\geq 0_N;\Sigma_{k=0}^K z^{k*}-c^0-\Sigma_{h=1}^H c^{h*}+\Sigma_{h=0}^H \bar{c}^h\geq 0_N;r^*\cdot[\Sigma_{k=0}^K z^{k*}-c^0-\Sigma_{h=1}^H c^{h*}+\Sigma_{h=0}^H \bar{c}^h]=0;$$

where $\nabla_y\pi^k(w,y^{k*},z^{k*}) \equiv [\partial\pi^k/\partial y_1^k,\ldots,\partial\pi^k/\partial y_M^k]$ is the vector of first order partial derivatives of π^k with respect to the components of y, etc.

If the solution vectors $y^{0*},b^{1*},\ldots,b^{H*}$ are strictly positive and the solution vectors $y^{1*},\ldots,y^{K*}$ are strictly negative, then conditions (9), (10) and (11) reduce to the following more familiar Lagrange multiplier conditions:

$$(16)\quad p^* = -\nabla_y\pi^k(w,y^{k*},z^{k*}) = -\nabla_b e^h(u_h^*,w,b^{h*},c^{h*});\ k = 0,1,\ldots,K;\ h=1,\ldots,H.$$

When $k = 0$, $-\nabla_y\pi^0(w,y^0,z^0) \equiv p^0(w,y^0,z^0) \equiv [p_1^0(w,y^0,z^0),\ldots,p_M^0(w,y^0,z^0)]\geq 0_M$

is a vector of quasi marginal cost functions for producing marginal units of the various types of utility services. Thus the first set of conditions in (16), $p^* = p^0(w,y^{0*},z^{0*})$, is the counterpart to the familiar price equals marginal cost rule that occurred in the early Pigou-Hotelling-Lerner literature. We call these functions quasi marginal cost functions because they are not equal to the usual marginal cost functions that occur in the literature. The true (net) cost function for sector 0 equals $-\tilde{\pi}^0$, where $\tilde{\pi}^0(w,y^0,r) \equiv \max_z\{\pi^0(w,y^0,z) + r\cdot z\}$, and the system of true marginal cost functions for sector 0 is defined by $\tilde{p}(w,y^0,r) \equiv -\nabla_r\tilde{\pi}^0(w,y^0,r)$. This distinction between quasi and true is not a semantic one, since Arrow and Hurwicz [1977; 81-82] showed that an efficient allocation of resources will not in general be achieved by marginal cost pricing. This very important point was elaborated on by Guesnerie [1975; 13-14] and Arnott and Harris [1977]. Of course, the distinction between quasi and true vanishes if N=0 so that there are no domestic goods. For k>0, $-\nabla_y\pi^k(w,y^k,z^k) \equiv p^k(w,y^k,z^k)$ is sector k's vector of producer willingness to pay functions. Alternatively, $p^k(w,y^k,z^k)$ can be regarded as a system of (inverse) demand functions for public utility services for k = 1,...,K.[4] For h = 1,...,H, $-\nabla_b e^h(u_h^*,w,b^h,c^h) \equiv P^h(u_h^*,w,b^h,c^h)$ is a vector of household h consumer willingness to pay functions for marginal units of the various types of utility services. Under our hypotheses on the utility functions f^h, it can be shown that[5] $P^h(u_h^*,w,b^h,c^h)$ is a system of inverse conditional Hicksian [1946; 331] demand for public utility services functions (conditional on c^h) for household h.

Equations (13) and (14) may be rewritten as follows:

$$(17)\quad r^* = -\nabla_z\pi^k(w,y,^{k*},z^{k*}) = -\nabla_c e^h(u_h^*,w,b^{h*},c^{h*});\ k = 0,1,\ldots,K;\ h=1,\ldots,H$$

where $r^* \equiv [r_1^*,\ldots,r_N^*]$. Equations (17) may be interpreted as follows. For $k = 0,1,\ldots,K$, $-\nabla_z\pi^k(w,y^k,z^k) \equiv r^k(w,y^k,z^k)$ is a vector of net quasi marginal cost functions for sector k for producing extra units of the various domestic goods (if $z_n^k < 0$ so that the nth domestic good is being used as an input by sector k, then $r_n^k(w,y^k,z^k) \equiv -\partial\pi^k(w,y^k,z^k)/\partial y_n^k \geq 0$ is the sector k willingness to pay for an extra unit of this input). For $h = 1,\ldots,H$, $-\nabla_c e^h(u_h^*,w,b^h,c^h) \equiv R^h(u_h^*,w,b^h,c^h)$ is a vector of household h <u>conditional net willingness to pay functions</u> for marginal units of the various domestic goods (if $c_n^h < 0$ so that household h is supplying the nth domestic good as a type of labour service, then $R_n^h(u_h^*,b^{h*},c^{h*}) \equiv -\partial e^h(u_h^*,w,b^{h*},c^{h*})/\partial c_n^h \geq 0$ is the wage rate that will just induce household h to supply an extra unit of this service without causing a change in household h's welfare level). Thus in an efficient equilibrium in our regional economy, for each good n, each sector k will equate the net quasi marginal cost of producing an extra unit of domestic good n to r_n^* and each household h will equate its net conditional willingness to pay for an extra unit of good n to r_n^* also. These conditions (17) usually do not appear in the regulation of monopolies literature, since most of the analysis has taken place in a partial equilibrium environment[6] which neglected factor markets and interregional trade.

If the reader prefers to work with a model of a closed economy, this can be accomplished as follows: set $I = 1$ and $w = 1$ and interpret the single good in category one as a domestic numeraire good that is either produced or utilized by every producer and household. Under this interpretation, the profit functions π^k become ordinary production functions and the restricted expenditure functions $a^h = e^h(u_h,b^h,c^h)$ are inverse functions for the utility functions $u_h = f^h(a^h,b^h,c^h)$. The reader should now go back and reinterpret the constrained maximization problems (5) and (8).

We conclude this section by providing interpretations for the vectors of partial derivatives of the restricted expenditure and profit functions with respect to the components of the vector of interregional prices w.

If $e^h(u_h^*,w,b^{h*},c^{h*})$ is differentiable with respect to the components of w, and if a^{h*} denotes the a^h solution to (5) or (7), then:

(18) $a^{h*} = \nabla_w e^h(u_h^*,w,b^{h*},c^{h*})$. (Shephard's Lemma).

The derivative property (18) is a modification of a property due originally to Hotelling [1932; 594] and Shephard [1953; 11]; see Diewert [1986; 171]. In a similar manner, it can be shown[7] that if $\pi^k(w,y^{k*},z^{k*})$ is differentiable with respect to the components of w, then the solution vector x^{k*} of interregionally traded goods to the maximization problems defined by (5) or (6) is equal to the vector of partial derivatives of π^k with respect to the components of w; i.e.,

(19) $x^{k*} = \nabla_w \pi^k(w,y^{k*},z^{k*})$. (Hotelling's Lemma).

4. ALTERNATIVE OBJECTIVE FUNCTIONS FOR THE REGULATED SECTOR

Consider an observed period 0 equilibrium for our regional economy. Suppose that the firm net output vectors for the three classes of goods (interregional, public utility, and other domestic) are x^{k0},y^{k0},z^{k0} for $k = 0,1,\ldots,K$. Suppose that the observed household net demand vectors for the three classes of good are a^{h0},b^{h0},c^{h0} for $h = 1,\ldots,H$. Suppose further that the period 0 world price vector is $w^0 \gg 0_I$ and that each producer and consumer is competitively optimizing with respect to the interregionally traded goods, conditional on their allocation of domestic goods and public utility services. Thus if we denote the welfare level of household h by u_h^0, we have

(20) $u_h^0 = f^h(a^{h0},b^{h0},c^{h0})$; $w^0 \cdot a^{h0} = e^h(u_h^0,w^0,b^{h0},c^{h0})$; $h = 1,\ldots,H$ and

(21) $w^0 \cdot x^{k0} = \pi^k(w^0,y^{k0},z^{k0})$; $k = 1,\ldots,K$.

Let us define $\alpha = a^0 - \Sigma_{h=0}^H \bar{a}^h$, $\beta \equiv b^0 - \Sigma_{h=0}^H \bar{b}^h$ and $\gamma \equiv c^0 - \Sigma_{h=0}^H \bar{c}^h$. These are net government demand vectors less the corresponding endowment vectors. We shall hold α, β and γ fixed throughout the analysis which follows in subsequent sections.

The period 0 allocation of public utility services and other domestic goods is summarized by the following equations (we assume that there are no free goods):

$$(22)\quad \Sigma_{k=0}^{K} y^{k0} - \Sigma_{h=1}^{H} b^{h0} - \beta = 0_M \;; \qquad \Sigma_{k=0}^{K} z^{k0} - \Sigma_{h=1}^{H} c^{h0} - \gamma = 0_N \;.$$

We do not assume that resources pertaining to the regulated and domestic good markets are optimally allocated in period 0. The amount of "foreign" exchange that the regional economy has produced in period 0 is (we no longer demand that interregional trade be exactly balanced):

$$(23)\quad f^0 \equiv \Sigma_{k=0}^{K} \pi^k(w^0, y^{k0}, z^{k0}) - \Sigma_{h=1}^{H} e^h(u_h^0, w^0, b^{h0}, c^{h0}) - w^0 \cdot \alpha \;.$$

The maximal amount of foreign exchange the economy could produce and still give each household the period 0 level of utility is given by the solution to the following constrained maximization problem which is analogous to (8):

$$(24)\quad f^* \equiv \max_{y^k, z^k, b^h, c^h} \{\Sigma_{k=0}^{K} \pi^k(w^0, y^k, z^k) - \Sigma_{h=1}^{H} e^h(u_h^0, w^0, b^h, c^h)$$

$$- w^0 \cdot \alpha : \Sigma_{k=0}^{k} y^k - \Sigma_{h=1}^{H} b^h - \beta \geq 0_M \;; \Sigma_{k=0}^{K} z^k - \Sigma_{h=1}^{H} c^h - \gamma \geq 0_N \;;$$

$$y^0 \geq 0_M \;; y^k \leq 0_M,\ k = 1,\ldots,K \;; b^h \geq 0_M,\ h = 1,\ldots,H\} \;.$$

A natural measure of the waste of resources inherent in the observed period 0 equilibrium is

$$(25)\quad W_A \equiv f^* - f^0 \geq 0 \;.$$

The inequality in (25) follows from the feasibility of the observed equilibrium for the problem (24). The waste measure defined by (25) is a variant of a measure introduced by Allais [1943] [1977].

In principle different regulatory schemes could be judged by the amount of waste that they engendered. However, we shall use the Allais measure of waste for a different purpose: namely, as a bonus or performance evaluation function that can be used to evaluate the performance of the manager of a regulated industry.

Consider the following candidate for an objective function:

$$(26)\quad G^0 \equiv \Sigma_{k=0}^{K}\pi^k(w^0,y^{k1},z^{k1}) - \Sigma_{h=1}^{H}e^h(u_h^0,w^0,b^{h1},c^{h1}) - w^0\cdot\alpha$$

$$- [\Sigma_{k=0}^{K}\pi^k(w^0,y^{k0},z^{k0}) - \Sigma_{h=1}^{H}e^h(u_h^0,w^0,b^{h0},c^{h0}) - w^0\cdot\alpha] .$$

We regard the period 0 equilibrium as being fixed. The manager of the regulated sector is supposed to choose price and quantity vectors for period 1 that maximize G^0. The period 0 equilibrium is characterized by the interregional price vector w^0, the welfare vector $u^0 \equiv [u_1^0,\ldots,u_H^0]$ and the resource allocation equations (22). We do not assume that the period 0 equilibrium is efficient, so we will sometimes refer to it as the observed distorted equilibrium. From (23), we see that the terms in square brackets in (26) add up to f^0, the amount of "foreign" exchange the regional economy produced in period 0. We assume that the world price vector w^0 and the period 0 real income vector u^0 will carry over to period 1 and we ask the manager of the regulated sector to maximize (26) subject to the following restrictions:

$$(27)\quad \Sigma_{k=0}^{K}y^{k1} - \Sigma_{h=1}^{H}b^{h1} - \beta \geq 0_M ;$$

$$(28)\quad \Sigma_{k=0}^{K}z^{k1} - \Sigma_{h=1}^{H}c^{h1} - \gamma \geq 0_N ;$$

$$(29)\quad y^{01} \geq 0_M ;\ y^{k1} \leq 0_M,\ k = 1,\ldots,K ;\ b^{h1} \geq 0_M,\ h = 1,\ldots,H.$$

It is reasonable to assume that the manager of the regulated sector can control the allocation of resources in the M regulated public utility service markets. If the manager can also control the allocation of goods in the N domestic markets, then the control variables for the manager are y^{k1}, z^{k1}, $k = 0,1,\ldots,K$ and b^{h1}, c^{h1}, $h = 1,\ldots,H$ and maximizing (26) with respect to these variables, subject to the constraints (27) - (29), is equivalent to maximizing (24) and so the optimized objective function for (26) will be

$$(30)\quad G^0 = f^* - f^0 = W_A \geq 0\ ,$$

which is the amount of extra foreign exchange the economy could produce, keeping each household at its period 0 level of welfare. Thus we are giving the manager of the regulated sector the task of eliminating waste in the economy for period 1. Note that we are agreeing with Schmalensee [1979; 8] that the dominant objective of a regulatory scheme should be the improvement of economic efficiency.

Is it reasonable to suppose that the manager can control the allocation of goods in domestic markets? The answer is yes if sectors 1 to K and each household behave competitively and the government eliminates all taxes and subsidies on domestic goods in the period 1 economy. After the manager chooses his net output vector z^{01} for domestic goods, the market will optimally allocate domestic resources across the competitive firms and households, where the optimality is conditional on the choices of the manager of the regulated sector. Thus in the case where there are no distortions on domestic markets in period 1, it is perfectly reasonable to ask the manager to maximize G^0 with respect to all quantity vectors subject to the restrictions (27)-(29).

However, it may not be reasonable to suppose that the vectors u^0 and w^0 will persist in period 1. Suppose that the period 1 distribution of real income ends up being given by the vector $u^1 \equiv [u_1^1,\ldots,u_H^1]$ and the vector of world prices turns out to be $w^1 \equiv [w_1^1,\ldots,w_I^1] \gg 0_I$. Then another possible objective function for the manager of the regulated sector is

$$(31)\quad G^1 \equiv \Sigma_{k=0}^K \pi^k(w^1,y^{k1},z^{k1}) - \Sigma_{h=1}^H e^h(u_h^1,w^1,b^{h1},c^{h1}) - w^1\cdot\alpha$$
$$- [\Sigma_{k=0}^K \pi^k(w^1,y^{k0},z^{k0}) - \Sigma_{h=1}^H e^h(u_h^1,w^1,b^{h0},c^{h0}) - w^1\cdot\alpha]\ .$$

The first 3 sets of terms on the right hand side of (31) represent the amount of foreign exchange actually earned by the economy in period 1. The 3 sets of terms in square brackets represent a hypothetical amount of foreign exchange that would be earned in the economy if the distribution of real income remained fixed at the actual period one distribution u^1, and if the economy faced the world price vector w^1, but the allocation of regulated and domestic goods remained at their period 0 levels. The manager is supposed to maximize(31) subject to the constraints (27) - (29).

Note that $G^1 = G^0$ if $w^1 = w^0$ and $u^1 = u^0$; thus under static conditions, the two objective functions coincide.

Note also that the manager would probably prefer G^0 as an objective function, since in this case, he or she does not have to forecast what world prices w^1 and the distribution of income u^1 will be in period 1. On the other hand, the government would probably prefer G^1 as a performance indicator for the manager, since in this case, the forecast risks are imposed on the manager. Thus we could view the maximization of $(1/2)(G^0 + G^1)$ as a risk sharing arrangement between the government and the manager.

Unfortunately, each of the two indicators defined above contains terms which cannot be evaluated using just observable price and quantity data for the two periods. Thus in the following two sections, we shall impose differentiability assumptions on our functions and attempt to find approximations to our theoretical indicators that can be evaluated using only market data.

5. FIRST ORDER APPROXIMATIONS

Recall G^0 defined by (30). We need to form approximations for the differences $\pi^k(w^0,y^{k1},z^{k1})-\pi^k(w^0,y^{k0},z^{k0})$ and $e^h(u_h^0,w^0,b^{h1},c^{h1})-e^h(u^0,w^0,b^{h0},c^{h0})$. We shall assume that the price and quantity data pertaining to periods 0 and 1 are observable and we further assume that the functions $\pi^k(w^t,y^{kt},z^{kt})$, k=0,1,...,K, and $e^h(u_h^t,w^t,b^{ht},c^{ht})$, h=1,...,H are differentiable with respect to all arguments for t = 0,1. In particular, we assume that the following vectors of derivatives exist and are equal to corresponding nonnegative price vectors p^{kt} and r^{kt} that sector k faces in period t and P^{ht} and R^{ht} that household h faces in period t for public utility services and domestic goods respectively:

(32) $p^{kt} \equiv -\nabla_y \pi^k(w^t,y^{kt},z^{kt})$, $r^{kt} \equiv -\nabla_z \pi^k(w^t,y^{kt},z^{kt})$; t=0,1; k=0,1,...,K;

(33) $P^{ht} \equiv -\nabla_b e^h(u_h,w^t,b^{ht},c^{ht})$, $R^{ht} \equiv -\nabla_c e^h(u_h,w^t,b^{ht},c^{ht})$; t=0,1; h=1,...,H.

A first order Taylor series approximation to $\pi^k(w^0,y^{k1},z^{k1})$ is:

(34) $$\pi^k(w^0,y^{k1},z^{k1}) \simeq \pi^k(w^0,y^{k0},z^{k0}) + \nabla_y \pi^{k0} \cdot (y^{k1} - y^{k0}) + \nabla_z \pi^{k0} \cdot (z^{k1} - z^{k0})$$

$$= w^0 \cdot x^{k0} - p^{k0} \cdot (y^{k1} - y^{k0}) - r^{k0} \cdot (z^{k1} - z^{k0})$$

where π^{k0} denotes $\pi^k(w^0,y^{k0},z^{k0})$ and we have used (21) and (32). We assume that p^{k0} and r^{k0} are price vectors (excluding any fixed charges) that sector k faces for regulated goods and domestic goods respectively in period 0 for k = 1,...,K.

An alternative first order approximation is:

(35) $$\pi^k(w^0,y^{k1},z^{k1}) \simeq \pi^k(w^1,y^{k1},z^{k1}) + \nabla_w \pi^{k1} \cdot (w^0 - w^1)$$

$$= w^1 \cdot x^{k1} + x^{k1} \cdot (w^0 - w^1)$$

$$= w^0 \cdot x^{k1}$$

where we have used the counterpart to (21) for period 1 and Hotelling's Lemma (19) applied to period 1 data.

A first order approximation to $e^h(u_h^0,b^{h1},c^{h1})$ is:

(36) $$e^h(u_h^0,w^0,b^{j1},c^{h1}) \simeq e^h(u_h^0,w^0,b^{h0},c^{h0}) + \nabla_b e^{h0} \cdot (b^{h1}-b^{h0}) + \nabla_c e^{h0} \cdot (c^{h1}-c^{h0})$$

$$= w^0 \cdot a^{h0} - P^{h0} \cdot (b^{h1}-b^{h0}) - R^{h0} \cdot (c^{h1}-c^{h0})$$

where $\nabla_b e^{h0}$ denotes $\nabla_b e^h(u_h^0,w^0,b^{h0},c^{h0})$ and we have used (20) and (33). Recall that P^{h0} and R^{h0} are price vectors (excluding any fixed charges) that household h faces for regulated goods and domestic goods respectively during period 0.

Substitute (36), (35) for k = 0 and (34) for k = 1,...,K into the definition of G^0, (30), and we obtain the following first order approximation to G^0:

$$(37)\quad G^0 \simeq w^0\cdot(x^{01} - x^{00}) - \Sigma_{k=1}^{K} p^{k0}\cdot(y^{k1} - y^{k0}) + \Sigma_{h=1}^{H} P^{h0}\cdot(b^{h1} - b^{h0})$$

$$- \Sigma_{k=1}^{K} r^{k0}\cdot(z^{k1} - z^{k0}) + \Sigma_{h=1}^{H} R^{h0}\cdot(c^{h1} - c^{h0}) .$$

In order to "simplify" (37), we assume that there are no domestic tax distortions so that all producers and consumers face the same domestic goods price vector r^t in period t. Using also (22), we find that (37) reduces to the following first order approximation for G^0:

$$(38)\quad G^0 \simeq w^0\cdot(x^0 - x^{00}) + r^0\cdot(z^{01} - z^{00}) - \Sigma_{k=1}^{K} p^{k0}\cdot(y^{k1} - y^{k0}) + \Sigma_{h=1}^{H} P^{h0}\cdot(b^{h1} - b^{h0}) .$$

Let us analyze the terms on the right hand side of (38). $w^0\cdot x^{00} + r^0\cdot z^{00}$ is the regulated sector's period 0 net revenue from supplying world and domestic goods (other than the regulated public utility services) minus input cost. If the regulated sector does not produce any outputs other than the M regulated services, then $w^0\cdot x^{00} + r^0\cdot z^{00}$ is <u>minus</u> period 0 cost for sector 0. In general, these terms will represent minus the net cost of producing the regulated services vector y^{00} in period 0. The terms $w^0\cdot x^{01} + r^0\cdot z^{01}$ represent minus the net cost of producing the period 1 regulated services vector y^{01} but valued at period 0 prices. Thus $w^0\cdot(x^{01} - x^{00}) + r^0\cdot(z^{01} - z^{00})$ represents minus the change in cost for the regulated sector, where all inputs are valued at period 0 prices. The term $-p^{k0}\cdot(y^{k1} - y^{k0})$ represents the net increase in sector k's utilization of public utility services valued at period 0 prices (remember, $y^{kt} \leq 0_M$ for k = 1,...,K by our sign conventions for the treatment of inputs and outputs). The term $P^{h0}\cdot(b^{h1} - b^{h0})$ represents the increase (decrease if negative) in household h's consumption of public utility services valued at period 0 prices (our sign conventions in this case imply $b^{ht} \geq 0_M$).

In an entirely analogous manner, we can derive the following first order approximation to the performance indicator G^1 defined by (31):

$$(39)\quad G^1 \simeq w^1 \cdot (x^0 - x^{00}) + r^1 \cdot (z^{01} - z^{00}) - \Sigma_{k=1}^{K} p^{k1} \cdot (y^{k1} - y^{k0}) + \Sigma_{h=1}^{H} p^{h1} \cdot (b^{h1} - b^{h0})$$

Note that the right hand side of (39) is analogous to (38) except that period 1 prices and tax rates have replaced the corresponding period 0 prices and tax rates. The first two inner products on the right hand side of (39) represent the regulated firm's <u>cost reduction</u> going from period 0 to 1 evaluated at period 1 prices, the next K terms represent the net increase in the utilization of public utility services by the competitive firms in the economy evaluated at period 1 prices (a measure of <u>producer benefit</u>), and the last H terms represent the net increase in the consumption of public utility services by households evaluated at period 1 prices (a measure of <u>consumer benefit</u>).

Recall that if the manager of the regulated sector solves the problem of maximizing G^1 subject to the various resource constraints, then the period 1 economy will be efficient. If we replace the unobservable performance indicator G^1 by the approximation on the right hand side of (39), then the manager will be motivated to choose his outputs, inputs and prices for regulated goods in such a way so that the allocation of resources will be approximately efficient in period 1. The manager is allowed to set arbitrary prices for the public utility services, but he is obligated to supply whatever quantities are demanded at those prices.

If instead of using the right hand side of (39) as a <u>performance indicator</u>, we wish to use it as a <u>bonus function</u> for the manager of the public utilities sector, then we need only use a suitable monotonic transformation of the right hand side, so that the required payments to the manager will not be enormous.

The performance indicators defined by (58) and (59) can be evaluated in principle using only observable price, quantity and tax data for the two periods.

6. SECOND ORDER APPROXIMATIONS

In order to provide second order approximations to our theoretical performance indicators defined in section 4, we shall proceed in a somewhat indirect manner. We shall obtain expressions for an average of G^0 and G^1 that are exactly correct if the profit functions π^k and the expenditure functions e^h have certain functional forms. These functional forms will be flexible; i.e., they can approximate arbitrary twice continuously differentiable profit or expenditure functions to the second order.[8] The reader who is familiar with the index number literature will realize that our procedure is analogous to the construction of a superlative index number formula.[9] The main difference is that instead of approximating ratios of functions, we now approximate differences of functions by empirically observable formulae.

Consider the following functional form for a restricted profit function π where we combine our old vectors y and z into the M+N dimensional vector s; i.e., $s^T \equiv [y^T, z^T]$ where s^T indicates the transpose of the column vector s:

$$(40) \quad \pi(w,s) \equiv b \cdot w + (1/2)(w \cdot \theta)^{-1} w \cdot Aw + w \cdot Bs + (1/2)(w \cdot \theta) s \cdot Cs$$

where $\theta > 0_I$ and b are I dimensional vectors of parameters, A is an I by I symmetric positive semidefinite matrix, B is an I by M+N matrix of parameters and C is a symmetric M+N by M+N matrix. Note that we are assuming that the vector of parameters θ is nonnegative with at least one component positive. Using a result in Diewert and Wales [1985; 20], it can be shown that the positive semidefiniteness of A implies that the $\pi(w,s)$ defined by (40) is a convex function of w, as it must be in order to be a valid profit function.

Proposition 1: π defined by (40) is a flexible functional form for any preassigned choice of $\theta > 0_I$; i.e., it can provide a second order approximation to an arbitrary twice continuously differentiable restricted profit function at an arbitrary point w^*,s^* where $w^* \gg 0_I$. Moreover the parameters of the matrix A can be further restricted to satisfy the following I linear restrictions without destroying the flexibility property:

$$(41)\quad Aw^* = 0_I \ .$$

The functional form defined by (40) is a very convenient one for econometric purposes, since if the restrictions (41) are satisfied, it has the flexibility property with a minimal number of parameters. Moreover, linear regression techniques may be used to estimate the unknown parameters. Finally, the convexity in prices property that a profit function must satisfy may be imposed globally without destroying the flexibility of the functional form by using a technique explained in Diewert and Wales [1987].

The following two propositions require us to evaluate profit functions at the deflated prices $\tilde{w}^t$ defined by:

$$(42)\quad \tilde{w}^0 \equiv w^0/w^0\cdot\theta \ ; \ \tilde{w}^1 \equiv w^1/w^1\cdot\theta \ ; \ \theta > 0_I \ .$$

The vector $\theta \equiv [\theta_1,\ldots,\theta_I]$ which appears in (42) can be an arbitrary nonnegative (but nonzero) vector. Once θ is chosen, we use it to define π by (40).

<u>Proposition 2</u>: Let the restricted profit function π be defined by (40) and define the deflated world price vectors $\tilde{w}^t$ by (42). Then the following identities hold for all vectors s^0 and s^1:

$$(43)\quad \pi(\tilde{w}^1,s^1) - \pi(\tilde{w}^1,s^0) + \pi(\tilde{w}^0,s^1) - \pi(\tilde{w}^0,s^0)$$

$$= [\nabla_s\pi(\tilde{w}^0,s^0) + \nabla_s\pi(\tilde{w}^1,s^1)]\cdot[s^1 - s^0] \quad \text{and}$$

$$(44)\quad = [\tilde{w}^0 + \tilde{w}^1]\cdot[\nabla_w\pi(\tilde{w}^1,s^1) - \nabla_w\pi(\tilde{w}^0,s^0)] \ .$$

The proof is by a lengthy series of straightforward calculations, remembering that $\tilde{w}^t \cdot \theta = 1$ for $t = 0,1$.

Consider the following functional form for a restricted expenditure function e where we now define the vector s to be our old household vectors b and c; i.e., $s^T \equiv [b^T, c^T]$:

$$(45) \quad e(u,w,s) \equiv b \cdot w + (1/2)(w \cdot \theta)^{-1} w \cdot Aw + w \cdot Bs + (1/2)(w \cdot \theta) s \cdot Cs$$
$$+ w \cdot au + (w \cdot \theta) s \cdot cu + (1/2)(w \cdot \theta) c_0 u^2$$

where $\theta > 0_I$, a and b are I dimensional vectors of parameters, c is an M+N vector of parameters, A is an I by I negative semidefinite symmetric matrix, B is an I by M+N matrix of parameters, C is an M+N by M+N positive semidefinite symmetric matrix and c_0 is a scalar parameter. Recall[10] that restricted expenditure functions, e(u,w,s), have to be concave in their price arguments w and convex in their quantity arguments s, under our assumption that the household's utility function is quasiconcave. The curvature conditions that we have imposed on the matrices A and C will ensure that e defined by (45) satisfies the appropriate theoretical curvature restrictions globally.

The following proposition is a counterpart to Proposition 1.

Proposition 3: e defined by (45) is a flexible functional form for any preassigned choice of $\theta > 0_I$. Moreover, the parameters of the matrix A can be restricted to satisfy the I restrictions (41) without destroying the flexibility result.

The following proposition is a counterpart to (43) in Proposition 2 and may be proven in the same manner.

Proposition 4: Pick $\theta > 0_I$, define the restricted expenditure function e by (45) and define the deflated interregional price vectors $\tilde{w}^t$ by (42). Then the following identity holds for all s^0 and s^1:

$$(46)\ e(u^1,\tilde{w}^1,s^1) - e(u^1,\tilde{w}^1,s^0) + e(u^0,\tilde{w}^0,s^1) - e(u^0,\tilde{w}^0,s^0)$$

$$= [\nabla_s e(u^1,\tilde{w}^1,s^1) + \nabla_s e(u^0,\tilde{w}^0,s^0)]\cdot[s^1 - s^0] .$$

Recall definitions (32) and (33). Define the corresponding deflated price and distortion vectors by:

$$(47)\ \tilde{p}^{kt} \equiv p^{kt}/w^t\cdot\theta\ ,\ \tilde{r}^{kt} \equiv r^{kt}/w^t\cdot\theta\ ,\ \tilde{r}^t \equiv r^{0t}/w^t\cdot\theta;\ t=0,1;\ k=0,1,\ldots,K;$$

$$(48)\ \tilde{P}^{ht} \equiv P^{ht}/w^t\cdot\theta\ ,\ \tilde{R}^{ht} \equiv R^{ht}/w^t\cdot\theta\ ; \qquad t = 0,1,\ h = 1,\ldots,H.$$

<u>Proposition 5</u>: Let $\theta > 0_I$ be fixed. Let each restricted profit function π^k be in the class of functional forms defined by (40) and let each restricted expenditure function e^h be in the class of functional forms defined by (45). Assume that the economy's domestic resource constraints (3) hold with equality in each period. Define the theoretical performance indicators $\tilde{G}^0$ by (26) (except use the deflated price vector $\tilde{w}^0$ in place of w^0) and $\tilde{G}^1$ by (31) (except use $\tilde{w}^1$ in place of w^1). Then the following identity holds <u>exactly</u>:

$$(49)\ \tilde{G}^0 + \tilde{G}^1 = (\tilde{w}^0+\tilde{w}^1)\cdot(x^{01}-x^{00}) + (\tilde{r}^0+\tilde{r}^1)\cdot(z^{01}-z^{00}) - \Sigma_{k=1}^K(\tilde{p}^{k0}+\tilde{p}^{k1})\cdot(y^{k1}-y^{k0})$$

$$+ \Sigma_{h=1}^H(\tilde{P}^{h0}+\tilde{P}^{h1})\cdot(b^{h1}-b^{h0}) \equiv 2G^* .$$

The right hand side of (49) can in principle be calculated knowing price and quantity data for the economy for the two periods. Note that G* is the average of the first order approximation to G^0, (38), and the first order approximation to G^1, (39), except that the prices in (38) and (39) are replaced by the corresponding deflated prices. Thus there is no need to belabour the interpretation of G*: the same old cost reduction, producer benefit and consumer benefit terms show up, expressed at average deflated prices rather than at individual period prices. If $u^0 = u^1$ (so there is no change in the distribution of real income) and w^1

is proportional to w^0, then it is easy to show that $\tilde{G}^0 = \tilde{G}^1$. Thus under these hypotheses, G* is exact for the common theoretical performance indicator $\tilde{G} = \tilde{G}^0 = \tilde{G}^1$.

In general, note that G* is defined in terms of observable price and quantity data. Moreover, G* is exactly equal to the theoretical performance indicator $(1/2)(\tilde{G}^0 + \tilde{G}^1)$ if the industry profit functions have normalized quadratic forms (39) and the consumer restricted expenditure functions have normalized quadratic forms (45). Since these quadratic functional forms can provide second order approximations to arbitrary profit and expenditure functions? G* has a second order approximation property and thus may be called a superlative performance indicator in analogy with the index number literature.

In order to help the reader interpret the rather formidable formula (49), we present the geometry of a very special case of our analysis in the following section.

7. THE GEOMETRY OF A SPECIAL CASE

Let I be arbitrary, M = 1, N = 0, K = 0 and H = 1. Hence there is an arbitrary number of interregionally traded goods, one public utility good, no domestic goods, no competitive firms and only one household. Furthermore, we assume that the welfare of the single household remains unchanged and that world prices remain unchanged, so that $u \equiv u^0 = u^1$ and $w \equiv w^0 = w^1$. Under these conditions, our theoretical performance indicators defined in section 4 collapse to the same thing, G say. There are only two functions in the regional economy, $\pi^0(w,y^0)$, the manager's restricted profit function, and $e^1(u,w,b)$, the single household's restricted expenditure function. The adding up constraint for public utility services implies that $y^0 = b$, where b is household one's demand for public utility services. Since b is the only managerial choice variable, let us simplify the notation and define $\pi(b) \equiv \pi^0(w,b)$ and $e(b) \equiv e^1(u,w,b)$. In period 0, the allocation of the single public utility good is b^0 and in period one, the allocation is b^1. The manager's problem is to maximize

$$(50)\quad G \equiv \pi(b^1) - e(b^1) - [\pi(b^0) - e(b^0)]$$

with respect to $b^1 \geq 0$. If the observed $b^1 > 0$ solves this maximization problem and the functions π and e are differentiable at the solution, then the following first order condition will be satisfied:

$$(51)\quad -\pi'(b^1) = -e'(b^1) \equiv P^1 .$$

The left hand side of (51) represents the regulated firm's marginal cost of producing an extra unit of the public utility service while the right hand side represents the household's willingness to pay for an extra unit and represents a point on the consumer's Hicksian [1946; 331] (real income constant) demand curve. P^1 denotes the price the household faces in period 1 while $P^0 \equiv -e'(b^0)$ denotes the price the household faced in period 0.

Since we are assuming that $w^1 = w^0$, it is harmless to assume that $w^0 \cdot \theta = w^1 \cdot \theta = 1$. Under these conditions, the approximation to G defined by one half of (49) is:

$$(52)\quad (1/2)(w^0+w^1)\cdot(x^{01}-x^{00}) + (1/2)(P^0+P^1)\cdot(b^1-b^0)$$

$$= \pi(b^1) - \pi(b^0) + (1/2)(P^0+P^1)\cdot(b^1-b^0)$$

where the equality in (52) follows from $w^0 \cdot x^{00} = \pi(b^0) = w^1 \cdot x^{00}$ and $w^1 \cdot x^{01} = \pi(b^1) = w^0 \cdot x^{01}$, using also $w^0 = w^1$.

The algebra above is illustrated in Figure 1.

Figure 1.

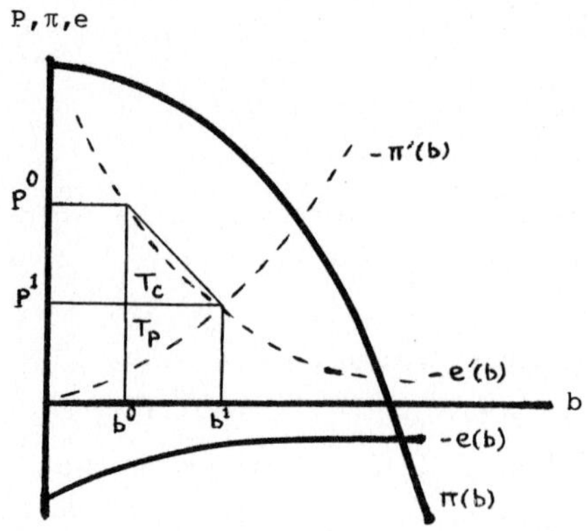

The top solid curve in Figure 1 is $\pi(b)$ while the bottom solid curve is $-e(b)$. The verticle sum of these two curves is maximized at b^1 where the marginal cost curve, $-\pi'(b)$, intersects the Hicksian demand curve, $-e'(b)$.

Note that the area under $-\pi'(b)$ from b^0 to b^1 equals $-[\pi(b^1) - \pi(b^0)]$ while the area under $-e'(b)$ from b^0 to b^1 equals $-[e(b^1) - e(b^0)]$. Using these relations, it can be seen that the theoretical performance indicator (or measure of efficiency gain) G is equal to the triangular regions T_c plus T_p where the upper boundary of T_c is the dashed line. On the other hand, it can be seen that the approximate measure of gain defined by (52) is equal to T_c^* plus T_p but now the upper boundary of the triangular region T_c^* is the straight line instead of the dashed line. Of course, if the expenditure function e is in the class of functional forms defined by (45), then the triangular regions would coincide. The first order approximation to G^0 (see (38)) can be shown to be $T_p + (P^0 - P^1)(b^1 - b^0)$ while the first order approximation to G^1 (see (39)) is T_p.

We leave to the reader the task of adapting the geometry to situations where the manager chooses b^1 to the right of the intersection of the marginal cost and Hicksian demand curves and to situations where the technology set is not convex and thus the marginal cost curve can be downward sloping. The main impression we want to leave with the reader is that the quadratic approximations defined in section 6 are very much better than the linear approximations defined in section 5.

8. ALTERNATIVE PERFORMANCE INDICATORS

Early contributors to the incentive scheme and performance indicator approach to the control of monopoly include Scott [1952], Domar [1974], Vogelsang and Finsinger [1979] and Tam [1981]. In our notation, the three most interesting performance indicators were defined by:

$$(53) \quad I_S \equiv w^1 \cdot (x^1 - x^0) + p^1 \cdot (y^1 - y^0) + r^1 \cdot (z^1 - z^0) \; ;$$

$$(54) \quad I_{FV} \equiv p^1 \cdot (y^1 - y^0) - [C(y^1, w^1, r^1) - C(y^0, w^0, r^0)] \; ;$$

$$(55) \quad I_T \equiv (1/2)(p^1 + p^0) \cdot (y^1 - y^0) - [C(y^1, w^1, r^1) - C(y^0, w^0, r^0)]$$

where x^t, y^t, z^t denote the net output vectors for internationally traded goods, public utility services and domestic goods respectively for the monopoly sector in period t; w^t, p^t, r^t denote the corresponding period t price vectors that the public utility sector faces for the three classes of goods, and $C(y^t, w^t, r^t)$ denotes the net cost of producing the vector of public utility services y^t in period t. I_S defined by (53) is our mathematical interpretation of Scott's [1978; 155] verbal description of his performance indicator. Note that our performance indicator G^1 defined by (39) reduced to it if $H + K = 1$ or if all users of the public utility services are charged the same price for the same service. I_{FV} defined by (54) is a performance indicator due to Finsinger and Vogelsang [1981; 400] [1982; 285] [1985; 165]. If $w^0 = w^1$ and $r^0 = r^1$ and $C(y^t, w^t, r^t) = [w^t \cdot x^t + r^t \cdot z^t]$ for t = 0,1 so that the Arrow-Hurwicz problem does not apply, then it can be shown that $I_S = I_{FV}$. The performance indicator I_T defined by (55) is due to Tam [1985; 284]. Sappington and Sibley [1985] have defined a closely related indicator. Note that I_T is related to our performance indicator G* defined in (49); in fact the indicators are identical if: (i) we use normalized prices in (55),

(ii) H + K = 1 or all users of the same public utility service are charged the same price and (iii) $w^0 = w^1$ and $r^0 = r^1$.

The above performance indicators may be subjected to a number of criticisms. (i) Scott's indicator has no formal economic or mathematical justification and the other two indicators were justified using a suspect Marshallian [1920; 487] consumer surplus concept. (ii) Scott's indicator has only a first order approximation property while the other indicators have formal approximation properties only under additional assumptions. In contrast, our performance indicator defined by the right hand side of (49) had a second order approximation property. (iii) I_{FV} and I_T are not inflation proof; i.e., they are not invariant to scale changes in the level of prices in any period. (iv) The last two indicators are based on a partial equilibrium analysis and moreover, they are subject to the Arrow-Hurwicz objection if the monopolist's production possibilities set has certain types of nonconvexities.

In spite of the above criticisms, it is clear that our analysis owes a great deal to this pioneering incentive scheme literature.[11]

9. CONCLUSION

Much of the literature on evaluating the efficiency of a regulated public utility monopoly and the literature which proposed optimal regulatory schemes made use of the consumer surplus concept. It is well known that the consumer surplus concept suffers from a number of theoretical deficiencies, particularly when it is applied in a many consumer context. Thus there appears to be a significant theoretical deficiency in the literature.

In our paper, we proposed a new method for regulating a public utility monopoly, one that is solidly based on microeconomic theory. Our proposed method involved giving the public utility manager the "right" objective function to maximize. This social objective function is a very broad one; it is the amount of extra money or foreign exchange that the regional economy that is served by the public utility could earn while keeping each household at a specified level of welfare. This theoretical objective function is a variant of one originally proposed by Allais [1943].

A difficulty with our proposed regulatory scheme is that the theoretically desirable social objective function is unobservable. However, we provide first and second order approximations to the theoretical objective function which depend only on observable price and quantity data. Thus we end up with a regulatory scheme that is "practical" in the sense that no elasticity information is required in order to implement it and it should achieve an efficient allocation of resources (to the second order). Thus the deadweight loss generated by currently used regulatory methods should be reduced by our proposed method.

In the course of our derivations, we defined some new classes of flexible functional forms for restricted profit and expenditure functions (see Propositions 1 and 3 above) that should prove to be useful in applications.

Finally, it seems appropriate to conclude by quoting Zajac [1978; 47] on the significance of deadweight loss due to inadequate regulatory schemes:

> "A deadweight loss to an economist is like an unexploited energy source to an engineer. In both cases lack of tools or the presence of constraints may thwart the realization of potential benefits. But the larger the potential benefits, the more justified is an effort to obtain more flexible or additional policy instruments to convert potential benefits into realized benefits. Of course, the obstacles may be so great that even herculean efforts are insufficient to bring about the conversion -- a situation of great frustration to both the economist and the engineer."

FOOTNOTES

1. See Walras [1980; 83], Pigou [1920; 278], Hotelling [1938; 256], Lerner [1944; 182], Meade [1945] and Fleming [1945].
2. If the region were indebted to the rest of the world, then the 0 in (1) could be replaced by a parameter b, the amount of world currency required to service the debt. Notation: $w \cdot x \equiv \Sigma_{i=1}^{I} w_i x_i$.
3. This means that each producer and consumer either produces or utilizes an interregionally traded good at the optimal solution.
4. See Diewert [1986; 25].
5. See Diewert [1986; 175].

6. Notable exceptions include Guesnerie [1975], Brown and Heal [1979] [1980] and Drèze [1980].
7. See Hotelling [1932; 594], Gorman [1968] and Diewert [1986; 142].
8. See Diewert [1974; 113] for the concept of a flexible functional form.
9. See Diewert [1976] and Denny and Fuss [1983a] [1983b].
10. See Diewert [1986; 170-176].
11. The material in this chapter is treated at greater length in Diewert [1985].

References

Allais, M. (1943), A la recherch d'une discipline économique, Tome I, Paris: Imprimerie Nationale.

Allais, M. (1977), "Theories of General Economic Equilibrium and Maximum Efficiency," pp. 129-201 in Equilibrium and Disequilibrium in Economic Theory, E. Schwoediauer (ed.), Dordrecht: D. Reidel.

Arnott, R. and R. Harris (1977), "Increasing Returns and the Inefficiency of Cost Minimization," Discussion Paper #278, Department of Economics, Queen's University, Kingston, Canada.

Arrow, K.J. (1964), "Control in Large Organizations," Management Science 10, 397-408.

Arrow, K.J. and L. Hurwicz (eds.), (1977), Studies in Resource Allocation Processes, New York: Cambridge University Press.

Brown, D.J. and G. Heal (1979), "Equity, Efficiency and Increasing Returns," The Review of Economic Studies 46, 571-585.

Brown, D.J. and G. Heal (1980), "Two-Part Tariffs, Marginal Cost Pricing and Increasing Returns in a General Equilibrium Model," Journal of Public Economics 13, 25-49.

Coase, R.H. (1945), "Price and Output Policy of State Enterprise: A Comment," Economic Journal 55, 112-113.

Coase, R.H. (1946), "The Marginal Cost Controversy," Economica 13, 169-182.

Denny, M. and M. Fuss (1983a), "The Use of Discrete Variables in Superlative Index Number Comparisons," International Economic Review 24, 419-421.

Denny, M. and M. Fuss (1983b), "A General Approach to Intertemporal and Inter-spatial Productivity Comparisons," Journal of Econometrics 23, 315-330.

Diewert, W.E. (1974), "Applications of Duality Theory," pp. 106-171 in Frontiers of Quantitative Economics, Vol. II, M.D. Intriligator and D.A. Kenrick (eds.), Amsterdam: North-Holland.

Diewert, W.E. (1976), "Exact and Superlative Index Numbers," Journal of Econometrics 4, 115-145.

Diewert, W.E. (1985), "Efficiency Improving Incentive Schemes for Regulated Industries", Discussion Paper 85-34, Department of Economics, University of British Columbia, Vancouver, Canada, November.

Diewert, W.E. (1986), The Measurement of the Economic Benefits of Infrastructure Services, Berlin: Springer-Verlag.

Diewert, W.E. and T.J. Wales (1987), "Flexible Functional Forms and Global Curvature Conditions," Econometrica 55, 43-68.

Domar, E.D. (1974), "On the Optimal Compensation of a Socialist Manager," The Quarterly Journal of Economics 88, 1-18.

Drèze, J.H. (1980), "Public Goods with Exclusion," Journal of Public Economics 13, 5-24.

Dupuit, J. (1944), "De la mesure de l'utilité des travaux publics," Annales des Ponts et Chaussées 8, translated as "On the Measurement of the Utility of Public Works," International Economic Papers 2 (1952), 83-110.

Finsinger, J. and I. Vogelsang (1981), "Alternative Institutional Frameworks for Price Incentive Mechanisms," Kyklos 34, 388-404.

Finsinger, J. and I. Vogelsang (1982), "Performance Indices for Public Enterprises," pp. 281-296 in Public Enterprise in Less-Developed Countries, Leroy P. Jones (ed.), New York: Cambridge University Press.

Finsinger, J. and I. Vogelsang (1985), "Strategic Management Behavior Under Reward Structures in a Planned Economy," Quarterly Journal of Economics 100, 263-269.

Fleming, J.M. (1944), "Price and Output Policy of State Enterprise: Comment," Economic Journal 54, 328-337.

Gorman, W.M. (1968), "Measuring the Quantities of Fixed Factors," pp. 141-172 in Value, Capital and Growth: Papers in Honour of Sir John Hicks, J.N. Wolfe (ed.), Chicago: Aldine Publishing Co.

Gravelle, H.S.E. (1985), "Reward Structures in a Planned Economy: Some Difficulties," Quarterly Journal of Economics 100, 271-278.

Guesnerie, R. (1975), "Pareto Optimality in Non-Convex Economies," Econometrica 43, 1-29.

Hicks, J.R. (1946), Value and Capital, Oxford: Clarendon Press.

Hotelling, H. (1932), "Edgeworth's Taxation Paradox and the Nature of Demand and Supply Functions," Journal of Political Economy 40, 577-616.

Hotelling, H. (1938), "The General Welfare in Relation to Problems of Taxation and of Railway and Utility Rates," Econometrica 6, 242-269.

Kuhn, H.W. and A.W. Tucker (1951), "Nonlinear Programming," pp. 481-492 in The Proceedings of the Second Berkeley Symposium on Mathematical Statistics and Probability, J. Neyman (ed.), Berkeley: University of California Press.

Lerner, A.P. (1944), The Economics of Control, New York: Macmillan.

Marshall, A. (1920), Principles of Economics, London: Macmillan.

Meade, J.E. (1944), "Price and Output Policy of State Enterprise," Economic Journal 54, 321-328.

Pigou, A.C. (1920), The Economics of Welfare, London: Macmillan.

Sappington, D.E.M. and D.S. Sibley (1985), "Regulatory Incentive Schemes Using Historic Cost Data," unpublished paper.

Schmalensee, R. (1979), The Control of Natural Monopolies, Toronto: D.C. Heath.

Scott, M.F.G. (1952), "Criteria of Efficiency in Nationalized Industries," B. Litt. Thesis, Oxford.

Scott, M.F.G. (1978), "A Profits Test for Public Monopolies," pp. 152-168 in Policy and Politics, D. Butler and A.H. Halsey (eds.), London: Macmillan.

Shephard, R.W. (1953), Cost and Production Functions, Princeton, N.J.: Princeton University Press.

Tam, M.S. (1981), "Reward Structures in a Planned Economy: The Problem of Incentives and Efficient Allocation of Resources," Quarterly Journal of Economics 96, 111-128.

Tam, M.S. (1985), "Reward Structures in a Planned Economy: Some Further Thoughts," Quarterly Journal of Economics 100, 279-289.

Vogelsang, I. and J. Finsinger (1979), "A Regulatory Adjustment Process for Optimal Pricing by Multiproduct Monopoly Firms," The Bell Journal of Economics 10, 157-171.

Walras, L. (1980), "The State and the Railways," Journal of Public Economics 13, 81-100.

Wilson, T. (1945), "Price and Outlay Policy of State Enterprise," Economic Journal 55, 454-461.

Zajac, E.E. (1978), Fairness or Efficiency, Cambridge, Massachusetts: Ballinger Publishing.